COLLECTIONS OF
WORLD RESTAURANTS

世界餐厅集成

《世界餐厅集成》编写组 ◎ 编
常文心 代伟楠 ◎ 译

辽宁科学技术出版社
·沈阳·

图书在版编目（CIP）数据

世界餐厅集成 / 《世界餐厅集成》编写组编；常文心，代伟楠译 . —沈阳：辽宁科学技术出版社，2017.6
ISBN 978-7-5591-0134-1

Ⅰ . ①世… Ⅱ . ①世… ②常… ③代… Ⅲ . ①饭店－室内装饰设计－世界－图集 Ⅳ . ① TU247.4-64

中国版本图书馆 CIP 数据核字 (2017) 第 072472 号

出版发行：辽宁科学技术出版社
　　　　　（地址：沈阳市和平区十一纬路 25 号 邮编： 110003）
印 刷 者：辽宁新华印务有限公司
经 销 者：各地新华书店
幅面尺寸：240mm×330mm
印　　张：64
插　　页：4
字　　数：100 千字
出版时间：2017 年 6 月第 1 版
印刷时间：2017 年 6 月第 1 次印刷
责任编辑：杜丙旭　于峰飞
封面设计：周　洁
版式设计：周　洁
责任校对：姚喜荣

书　　号：ISBN 978-7-5591-0134-1
定　　价：498.00 元

联系电话：024-23280367
邮购热线：024-23284502
http://www.lnkj.com.cn

地域交织的多彩美食馆

近年来，随着人们经济收入增加，工作压力也随之增加，因此更重视休闲与美食。由于市场需求强烈，各种类型的餐馆，不论是一般消费的快餐店或是中等消费的家庭餐厅，抑或是高级食府，如雨后春笋般在世界各地迅速开张营业，掀起世界餐饮业的一股股浪潮。

不同的地理及风土人情促成了不同的美食文化，丰富多彩的餐馆承载了迥异万千的美食。在国际化潮流势不可挡的今天，不同地域的餐馆互相影响，与各个地区的日常生活、文化背景相结合，因地制宜、相得益彰的创造出你中有我、我中有你的具有活力的世界各地的餐馆。来源于另一个国家的餐厅风格不但可以在另一个国家成为换换口味的有趣享受，并且还能促进该国餐厅主旋律的变革。餐厅的故事从这里开始，它的篇章会一篇接一篇的精彩上演。

以专业化的视角放眼全球，深度剖析餐厅设计经典案例。本书精选全球6大洲50多个国家餐厅设计的精品之作500个，每个案例配有详实的设计简介、灵感来源等项目说明及实景图片3~5张，平面图1张，为读者准确详实的了解项目提供高品质的保证。类别为：餐厅、酒吧、咖啡店、甜品店、快餐厅、团体餐厅、连锁餐厅。以其不同的角度展现了设计师对其当地饮食文化在餐厅设计中的创新性应用。

本书突出时效性、全球性、地域性、专业性，有助于全球读者寻找设计灵感，了解新材料、改造旧项目传承餐厅设计文化。

Colourful Gourmet Restaurants All Over the World

Nowadays, with the increase of people's income and work pressure, people pay more attention to leisure and gourmet food. Therefore, out of the market's need, all kinds of restaurants, from ordinary fast-food restaurants to middle-level family restaurants and high-end boutique restaurants, come to open one after another, creating an upsurge in world's catering industry.

Different geographies and customs generate different gourmet cultures, while different restaurants offer various gourmet food. In these days of rapid internationalisation, restaurants from different regions influence each other and merge with local customs and culture backgrounds, creating a vigorous catering industry all over the world. A foreign restaurant style may not only become the local amusing enjoyment, but also promote the revolution in the theme of restaurants. The restaurant's story begins here and will evolve more interesting gradually.

This book offers readers a visual feast with a collection of the world's most classic restaurant projects. It selects 500 projects from more than 50 countries of 6 continents all over the world. Each project is illustrated with 3 to 5 photos, a plan and a brief introduction. It is catergorised in 7 parts, including Restaurant, Bar, Café, Desert Shop, Fastfood, Canteen and Chain Restaurant, revealing designers' innovative applications in restaurant design in different views.

This book is featured with its timeliness, globalisation, regionalisation, and professionalisation to help readers from all over the world to find inspiration, approach new materials and refurbish the old restaurants to inherit their cultural heritage.

Location of the selected projects of *Atlas of World Restaurants*
《世界餐厅地图》项目分布

1. Canada 加拿大
2. USA 美国
3. Mexico 墨西哥
4. Peru 秘鲁
5. Brazil 巴西
6. Chile 智利
7. Argentina 阿根廷
8. Sweden 瑞典

9. Denmark 丹麦
10. UK 英国
11. Ireland 爱尔兰
12. The Netherlands 荷兰
13. Belgium 比利时
14. Luxembourg 卢森堡
15. France 法国
16. Spain 西班牙

17. Portugal 葡萄牙
18. Germany 德国
19. Poland 波兰
20. Czech Republic 捷克
21. Austria 奥地利
22. Switzerland 瑞士
23. Italy 意大利
24. Hungary 匈牙利

Contents /目录

Photo: David Whittaker

Ron Wong, Olga Evstifeeva, Jason

Cedar Restaurant

The central column is an architectural representation of a cedar trunk with lit panels above typifying the canopy of the tree. Branches float against a stone wall that leads to the private dining area, where the dappled green of forest leaves creates an enclosure of calm renewal. The yellow slotted wall denotes the setting sun with the fiery red sky above. Floating tree trunks celebrate the tree in its natural form juxtaposed to the refined wood used throughout the restaurant. The fireplace is a welcome invitation for the Creator. The design challenge for this project was to respectfully bridge the gap between contemporary design and customs of native legacy.

雪松餐厅

餐厅中央的方柱代表雪松树干，嵌板象征着树冠。从石墙上悬浮出来的树枝延伸到私人就餐区，有斑纹的绿色的树叶营造出宁静的氛围。黄色的墙壁象征着下落的夕阳和火红的天空。悬浮树枝的造型十分逼真，餐厅内的设计将木材的精细发挥得淋漓尽致。壁炉的设计是设计师的又一个亮点。设计这一项目面临的严峻挑战就是要中和当代设计特征与当地风俗习惯之间的差异。

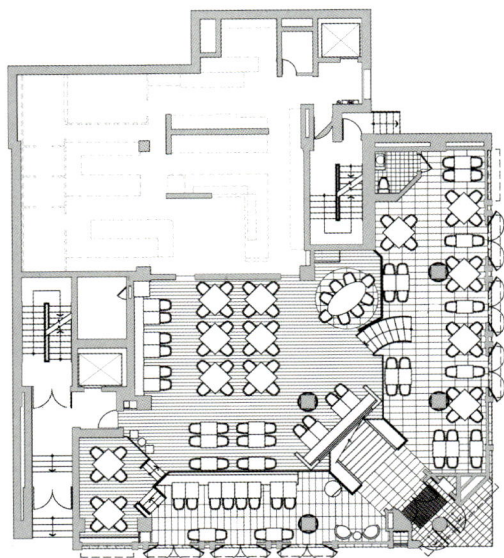

Restaurant

Canada

Toronto

2006

Photo: David Whittaker

II BY IV

Luce

Despite a challenging site with massive architectural elements, vast window areas, and soaring height, this restaurant is characterised by intimacy, human scale and an exquisitely simple luminosity. At a distance, this interior presents as smoothly executed large brush strokes, but closer inspection reveals multiple levels of subtle texture, much like the menu itself. This dramatic, gallery-like marketing centre reflects the distinctive community surrounding the stylish development. The constant display of local artists' work, the striking original wall sculpture that eventually will be reused in the building lobby and even the framed marketing drawings salute the creative spirit and artistic lifestyle.

露丝餐厅

餐厅云集了多种设计元素，有巨大的窗户和超高的举架。餐厅规模适中，简洁的灯光设计营造出温馨的氛围。大气的室内设计中有着微妙的层次感，和餐厅的菜单有着异曲同工之妙。画廊般的营销中心设计独特，展示着当地艺术家的作品。最为突出的是雕塑墙壁，在大堂和营销中心都能看到它的身影，显示了创新精神和唯美的生活方式。

Restaurant

Canada

Toronto

2006

Photo: Bruce Mau

Bruce Mau

Embryo

The client wanted something very organic that would capture the interest and imagination of any one that steps into it. Functionally the space works like this. The overall concept of the design was to represent 'pre-life' or life at a cellular scale. Although most of the elements were carved in hard plastic, the designer tried to make a 'soft' space by distorting all the elements that would have given one a sense of orientation regarding scale or the common shape of objects. Both the client and the designer are most happy with the fact that for once they were able to stick with the concept and finally realised it.

茧餐厅

客户希望打造一个有机建筑，激发人的兴趣和想象力。空间的功能达到了这一目的。设计主要的理念表现了"细胞时期的生命"，大部分的设计元素都以硬塑料诠释，并呈扭曲状，营造柔软的感觉。设计师非常满意的就是他们坚持和发展了设计理念，最终将理念转化为现实，打造出一个成功的设计餐厅。

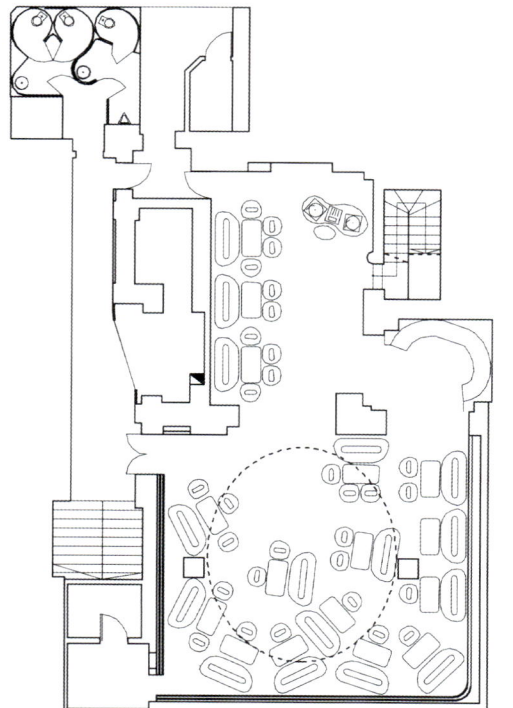

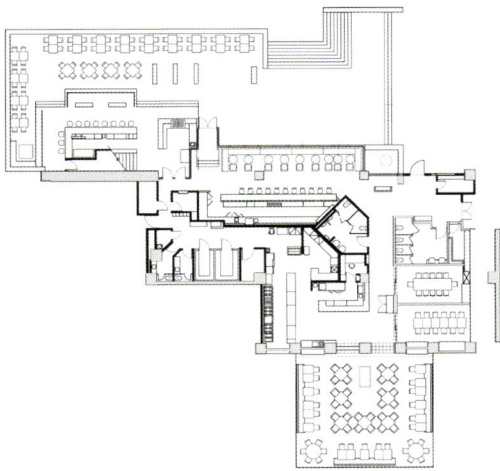

Vertical

This much sought-after Toronto hospitality space is located at the base of Canada's tallest commercial building. Vertical features a spacious and open dining room that presents a subtle contemporary mood with warm woods and floor-to-ceiling windows. The design team has created an interior with a restrained sophisticated palette. It is slick and modern, yet comfortable. Based loosely on the muted tones of men's suits, the interior's colour palette presents a tone-on-tone foundation with highlights of rich colour and plush seating to soften the clean surfaces. The large natural sandstone bar becomes a sophisticated centrepiece with the addition of a highly polished epoxy resin finish.

垂直餐厅

多伦多这家最受欢迎的餐厅位于加拿大最高的商贸大厦一楼。宽敞开放的就餐空间，运用木质家具和落地窗营造出现代的氛围。餐厅室内的设计，色彩搭配和谐，充满现代感，又不失优雅舒适之感。就像男士服装的搭配一样，采用同一色系的色彩，突出其丰富性和舒适度。表面使用环氧树脂材料抛光的巨大天然砂岩吧台是本餐厅项目彰显奢侈华丽的设计亮点。

Restaurant

Canada

Toronto

2007

Photo: Mackay|Wong Strategic Design

Mackay|Wong Strategic Design

Restaurant
Canada
Quebec
2007
Photo: Jasmin Frechette
Jasmin Frechette, Chantal McDuff

Zone AD – Restaurant

The restaurant is within a 30,000-square-foot business and commercial complex that are located on the ground floor. The grill house type restaurant is composed of three sections, the main dining area, a bar area and the services (kitchen, Water Closet, etc). The two main spaces both have access to an exterior terrace area looking onto the seaway. The interior such as the floor, wall and ceiling, is designed with environmental friendly and non-toxic materials. The lighting for the entire project had to respect the client's specifications and also had to be energy effective. The furniture incorporated within the restaurant were all custom designed to be an integral part of the space.

AD餐厅

餐厅位于一幢三层商贸大楼的一楼。这家烧烤餐厅包括三个部分，主要用餐区、酒吧区和服务区（厨房、洗手间等）。用餐区、酒吧区都直接通向户外阳台，客人可以观赏美丽壮阔的海滨景色。餐厅室内的地板、墙壁和棚顶都使用无毒害的环保材料。餐厅所有照明系统都按照客户的要求使用节能灯照明。所有家具都是特别定制的，用来搭配室内的设计，成为空间主体部分。

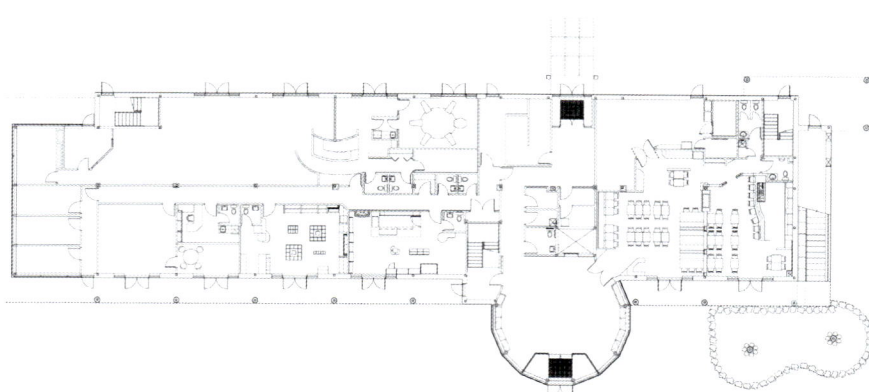

Photo: Lemaymichaud Architecture Design

Restaurant Le Local

The challenge was met: on arrival, guests now encounter a contemporary, sheltered wooden concept terrace created in the former parking lot and furnished by Jardin de ville, which draws them towards the restaurant's entrance. The restaurant space opens to a bar whose design includes bright skylights and chic lounge decor set before an open-concept windowed facade. The industrial and grandiose aspect of the environment is made apparent by the accentuated grey ceilings, original metal beams, and polished concrete floor, with stone and brick walls contributing to the overall ambiance of the space.

Le Local餐厅酒吧

来到餐厅，客人首先经过由原停车场改造的现代木制带棚阳台，接着看到周围装饰的小花园，然后不由自主来到了餐厅的入口。餐厅直通酒吧。酒吧内明亮的天窗、开放式的带窗幕墙，幕墙前面是别致的装饰。独创的金属横梁、抛光的混凝土地面、石砖墙再加上灰色的天花板更加营造了壮丽辉煌的工业化环境。原建筑的窗户经改造后激发出一种亲密、舒适的现代化氛围。

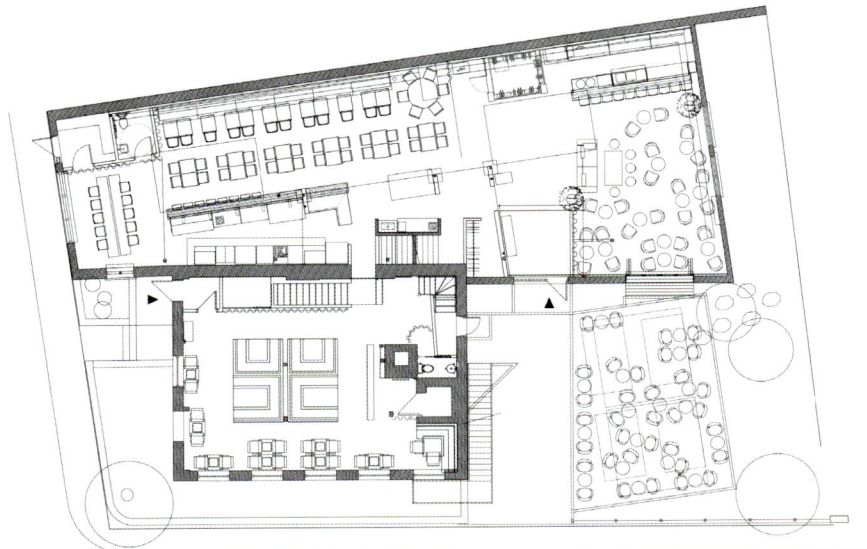

Lemaymichaud Architecture Design

Restaurant

Canada

Montreal

2010

Photo: Jean Malek

Stéphane Proulx, Jean de Lessard

Restaurant Kazumi

Origami is the ancient and celebrated Japanese art of paper folding: by folding one or several sheets of paper, one creates a shape, a decoration. This art originated in ceremonies in which folded pieces of paper were used to ornament jugs of sake. It is in this spirit of art and celebration that the Jean de Lessard design firm approached the interior design of the Kazumi Japanese restaurant. The designer's intuition was perfect: the angular shape, pierced with sushi-pink translucent glass, on its own creates the ambience and tone of the restaurant. The colourful opening also gives a glimpse of the intense work being done in the kitchen. A colourful, fresh locale – just like Chef Tri's cuisine! As soon as they enter, customers are enveloped in a corridor of clear glass set into brightly striped walls.

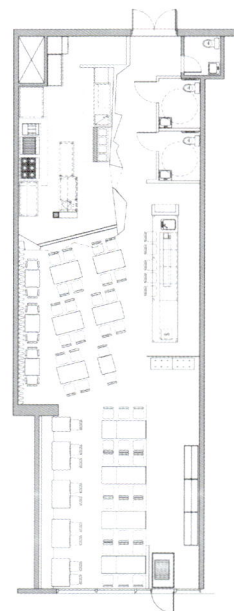

和美日式餐厅

Origami 是一种古老而著名的日式折纸艺术。它指的是用一张或几张纸折出物体的形状或用作装饰。这种折纸艺术起源于日本米酒庆典，在庆典上人们用折纸来装饰酒壶。Jean de Lessard 设计事务所正式本着这种艺术和庆典的设计精神来为伟霞餐厅进行室内设计。设计师的设计灵感堪称完美：各种角度、形状各异的折纸，穿插于粉色寿司吧半透明的玻璃间，营造出日式餐厅特有的氛围与基调。五颜六色的开放式厨房设计使顾客能够看到员工们紧张工作的情形。

Restaurant

Canada

Toronto

2008

Photo: Tom Arban

Kuwabara Payne McKenna Blumberg Architects

Nota Bene

Nota Bene is the synthesis of entrepreneurial and design thinking to create a highly successful dining model where social, cultural, ceremonial and business needs converge. Nota Bene is located in a contemporary office tower. The business model focused on three essential objectives: inviting and accessible environment, high quality food and wine at accessible price points, and impeccable, friendly service. A well-designed interior was the first platform for realising the other two objectives. If a design detail did not enhance functional performance or support the business model it was modified or rejected. For example, proposed stained walnut for the floors were substituted for Jatoba Brazilian cherry wood, a more resilient and cost-effective product. Human scale and comfort was of the utmost importance.

注意餐厅

注意餐厅是经营和设计思维的综合体，是一个集社会、文化、礼仪和商业因素于一身的成功餐厅模型。餐厅专注于三方面：吸引人而便于进入的环境、高品质而价格合理的餐饮以及完美亲切的服务。精心设计的室内环境是实现其他两个目标的基础和平台。不能提高餐厅功能性能或支持商业模型的设计细节将会被改进或替换。例如，彩色胡桃木地板被更划算的巴西樱桃木所替代。人性尺度和舒适度是项目设计中最重要的元素。

Restaurant

Canada

Montreal

2009

Photo: MCramer

Nda Architecture

Le Cartet Restaurant

The colours and materials used in the composition of the project were kept simple: wood, black steel and white surfaces. These three materials are all characteristics of the original space. Combined, they define the project, making a strong statement in fluidity, flow and repetition. First, the space was painted entirely in white: brick walls, columns, and ceiling, too. The only exception is the original varnished wood floor, left as is. Stripped down, the space is restructured by two elements, wood and steel. The metal element is made up of two long strips of black steel that envelop all the services of the restaurant - café, bar, kitchen and washrooms - for its entire length. This black line links the black steel vestibules, located at each end of the restaurant. The furnishings and architectural elements were chosen expressly to play up these smooth, flat surfaces.

乐卡特餐厅

项目选用简单的材料与色调：木头、黑钢和白色墙面。这三种材料的结合，界定了自然、流畅而又相互呼应的空间氛围，构成了创意空间的全部特色。首先，包括砖墙、柱子和天花板在内的整个空间都涂成了白色。唯一例外的就是原来空间内的涂漆木地板。剥去表层，整个重建空间主要包括两个元素：木头和钢材。咖啡厅、吧台、厨房和洗手间包层材料的金属元素主要是两片长条黑钢片。黑钢条将餐厅两端的黑钢材料装饰的前厅连接起来。

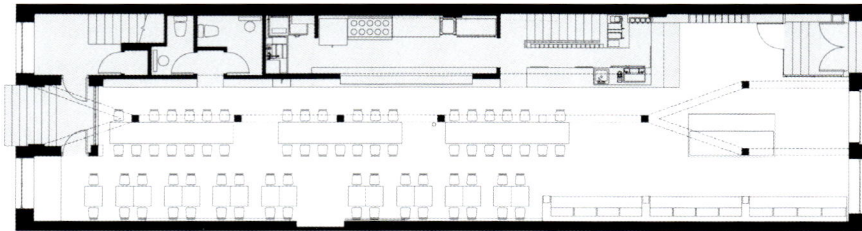

Continental Bistro

In 2007, the beloved Montreal Continental Bistro was destroyed in a fire. Designer Zébulon Perron was commissioned to reinvent the restaurant within a new and very different space in the trendy Plateau neighbourhood. The challenge was to recapture the essence of the old restaurant all the while avoiding a nostalgic caricature of it. The original venue was animated by the very particular spirit of the 1930's American style: Streamline. Thus, it was important to express movement, speed and a worldly ethos. This was partly achieved with custom designed lighting fixtures that are reminiscent of an airplane. Pursuing the same idea, a large scale topographic mural was also imagined by the designer. Another important design criterion was the perennity of the new décor. Therefore, special attention was paid to the selection of materials.

大陆酒店酒吧

2007年，大陆酒店毁于一场大火。设计师受委托在时尚的高原街区重建餐厅。项目面临的挑战是在重现旧餐厅的本质的同时又不能让人过多的想起它。原来的设计带有明显的20世纪30年代美国风格：流线型。飞机造型的定制灯具达成了这种效果。另一个设计标准是装饰的持久性，因此材料的选择至关重要。只有能够经过时间考验的高质量材料才能被应用在项目中。

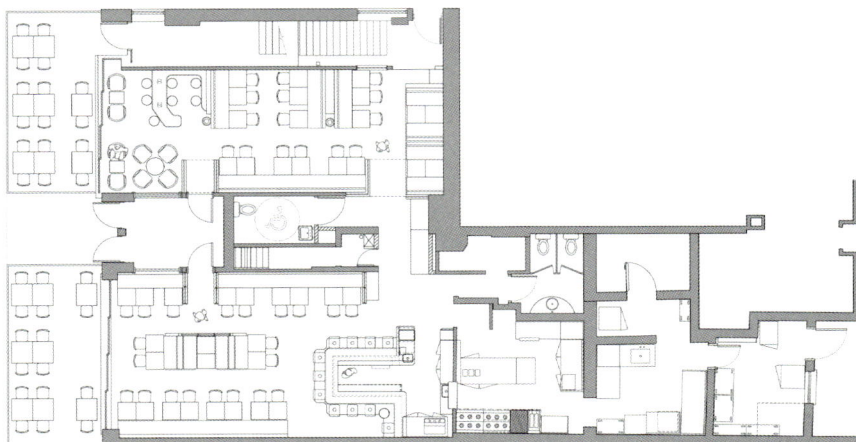

Bar

Canada

Montreal

2008

Photo: Zébulon Perron

Zébulon Perron

Restaurant

Canada

Montreal

2010

Photo: Jessica Lee Gagné

Zébulon Perron

Hachoir Restaurant

Located in the trendy Plateau Mont-Royal neighbourhood in Montreal, Hachoir is a popular and casual restaurant that specialises in tartars, gourmet burgers and private importation wines. The décor is chic and modern yet cosy and unpretentious. Durable materials such as solid white oak, brass, raw steel, glass and small mosaic tiles leaves one with a sense of quality. The look is raw yet sophisticated with some eye-catching elements. The custom designed brass light fixtures hover over the dark coloured dining area bathing the patrons in a warm glow. The lively bitro-esque ambiance is accentuated by the smells and sounds of the open kitchen where a big, bright yellow surgical lamp hangs over the pass.

绞肉机餐厅

绞肉机餐厅是一家深受欢迎的休闲餐厅，专营美味的汉堡和私人进口的美酒。餐厅的装饰时尚而舒适，现代而不铺张。白橡木、黄铜、生铁、玻璃和小马赛克地砖等耐用材料的应用让项目充满了质感。整个空间原始而精妙，不乏吸引眼球的元素。特别定制的黄铜灯具悬垂在深色的餐饮区上方，投下温暖的光晕。开放式厨房里悬挂着一盏巨大的手术灯，厨房散发出的香味增添了生动的酒馆气氛。

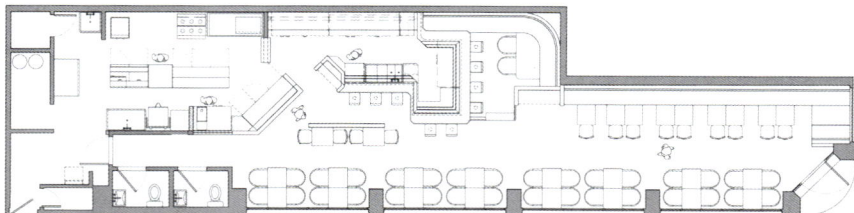

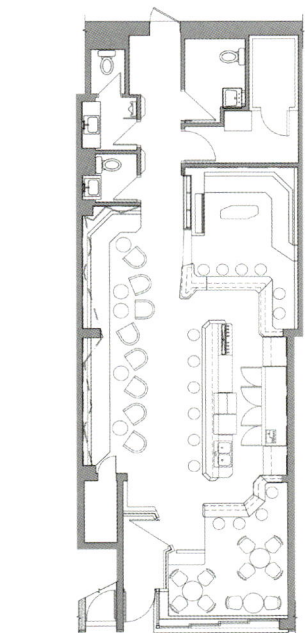

Plan B Bar

It was with the intention of creating a timeless space that the design for Plan B was elaborated. The mandate was to create a refined and convivial neighbourhood bar located in the trendy Plateau Mont-Royal district in Montreal, Canada. With that in mind several sub-spaces were created to encourage contact between patrons and to break the monotony of this tiny local. A large mirror installed on the back wall creates the illusion of depth, an effect made even more convincing with the presence of a half moose trophy installed against that very same mirror thus creating visual continuity. Also a long zig zag shaped mirror runs against the side wall fragmenting the space and offering multiple views to the onlooker.

B计划酒吧

B计划酒吧试图打造一个永恒的经典空间，设计师被要求在蒙特利尔的皇家高地区打造一个精致而欢愉的本地酒吧。设计师在酒吧内打造了一些附属空间，以便有客人之间的相互交流，也打破了空间的单一性。后墙上的镜子营造出深度的幻觉，半个麋鹿战利品更像是装在镜子上，从而打造了视觉的连续性。一面锯齿形的长镜子沿着侧墙将空间分割成小块，为旁观者提供了多重视角。

Bar

Canada

Montreal

2006

Photo: Zébulon Perron

Zébulon Perron

Lounge

Canada

Orillia

2008

Photo: Mgm Mira

Humà Design

Revolution Lounge

The Revolution Lounge itself measures 715 square metres (7,700 Square feet) and is divided into several smaller lounges. At the centre, the architect Stephanie Cardinal used the three support pillars to create central point of attraction. Inspired by the song Lucy in the Sky with Diamonds, she surrounded the pillars with triangular panels, some of which are made of gleaming steel, while others are white screens. With this covering, she has transformed each pillar into the point of a diamond frozen as it shatters, dragging the ceiling with it in its fall. Under the pillars, straight-lined and pink Sacco (1969 design) banquettes alternate with white, interactive tables (2005 design).

革新餐厅

革新餐厅占地715平方米，分为几个小休息厅。餐厅中间，建筑师Stephanie Cardinal 设计了三个支柱作为设计的亮点。受歌曲"天空中佩戴钻石的露西"的启发，设计师在柱子周围镶嵌了一些三角形的嵌板，有些嵌板由钢片制成，其他则是一些白色隔板。设计师用这种包装材料，使支撑柱如同闪耀的钻石一般，瞬间降落，拖拽着天花板，使其支离破碎。柱子下面，粉色直线条Sacco 长沙发与白色曲线桌子相互搭配。

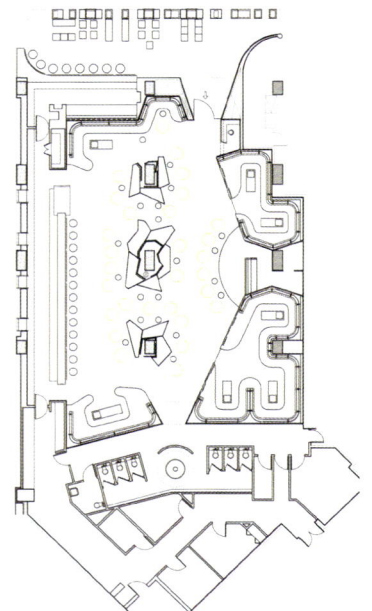

Rosalie Restaurant

Located in downtown Montreal this popular Italian restaurant was renovated by Zébulon Perron in 2009. Some elements of the old décor where salvaged as others where modified or added for maximum impact. For instance, the furniture, colour and lighting schemes where completely redone. A large custom designed yellow suspension lamp was created to fill the double storey space at the front of the restaurant. To fashion a dialogue another custom designed 'T' shaped lamp of the same colour was installed on a large communal table. A new open kitchen and pizza oven bar was installed near the front window so the passersby may observe the chefs in action and to easily serve the large and very popular front terrace in the summer. The designer's intention was to craft a modern bistro with a classic sensibility.

罗莎莉餐厅

这家意大利餐厅位于蒙特利尔的商业区，由设计师泽伯伦•派隆于2009年进行了翻新工程。设计师保留了一些旧装饰，也改造、添加了一些新装饰。家具、色彩和灯光设计进行了完全改造。一盏特别定制的黄色吊灯填满了餐厅前部的两层空间，另一盏黄色的T形灯设在长餐桌上方。一个新的开放式厨房和比萨吧被设在前窗，因此路人可以看到厨师们的动作，也更方便厨房在夏天为露天平台提供服务。设计师的目的是打造一个具有经典感的现代餐厅。

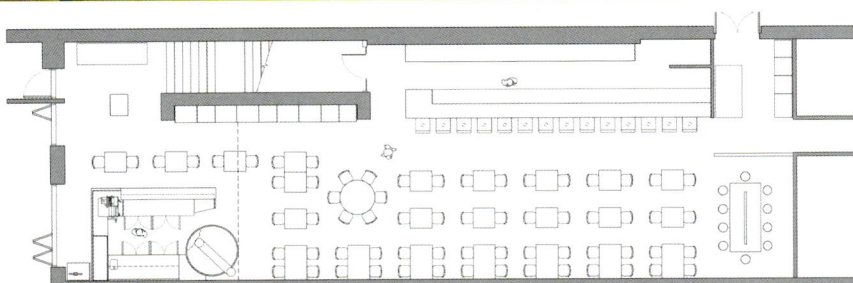

Restaurant

Canada

Montreal

2009

Photo: Zébulon Perron

Zébulon Perron

Restaurant & Bar

Canada

Montreal

2008

Photo: Zébulon Perron

Zébulon Perron

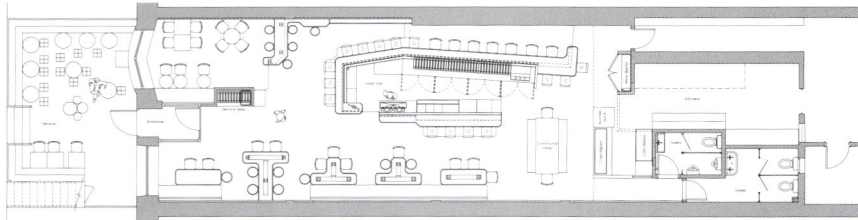

Buvette Chez Simone

In accordance with the owner's intents, Zébulon Perron aimed at making Buvette Chez Simone an unpretentious and convivial gathering place, therefore avoiding the prevalent minimalist and sometimes impersonal contemporary style. The inspiration was the European 'buvettes'; unassuming and occasionally makeshift places where people gather for a drink. At the centre of the space an imposing bar, around which patrons can congregate, sets the tone; the atmosphere of the establishment is meant to be genial but not stiff. The space was deliberately planned to encourage contact between people by offering them a variety of scenarios. A raised section is furnished with small tables for more intimate encounters while on the lower levels are found an assortment of 'T' shaped tables, standing bars and large communal tables to accommodate bigger groups.

蒙特利尔西蒙娜酒吧

设计师将西蒙娜酒吧打造成一个含蓄而又欢乐的聚会场所。设计灵感来源于欧洲的"buvettes"一词，意思是不显眼的人们聚会用的临时地点。酒吧中心位置是一个壮观的吧台，顾客可以坐在四周饮酒聊天，营造了亲切友好的酒吧氛围。空间设计通过设置一系列情境有意地增进了人与人之间的交流。酒吧内突起的部分摆放着供密友用的小餐桌，而低处则摆放了一些T形桌、独立吧台和供多人聚会用的聚会桌。

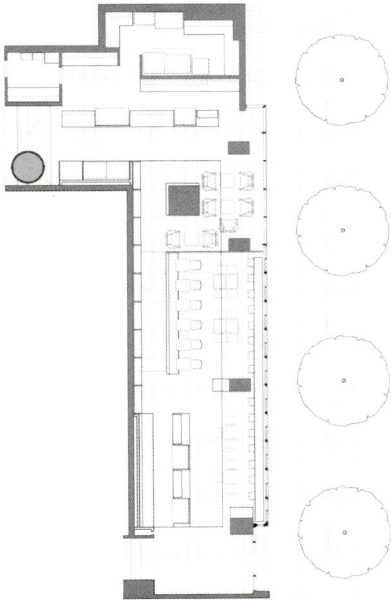

Giovane Café + Bakery + Deli

Giovane is a place to spend time. A café, bakery, and deli located on the street level of the new Fairmont Pacific Rim Hotel in downtown Vancouver, British Columbia, this luxury café opens both to a downtown street and directly into the hotel lobby. Serving both the city and the visiting hotel guests, the concept combines a host of services and amenities including custom coffee, sandwiches and desserts, and a collection of books, linens, children's toys and unique gift items. Inspired by the cafés of Europe where hours can be spent sipping lattes, losing oneself in a good book, and enjoying simple elegances, the split-level café projects a youthful vibe that encourages guests to leave their day at the door. Fitting really when you consider that the name Giovane means 'young' in Italian.

基奥云尼咖啡厅+面包店+熟食店

基奥云尼餐饮店是一个适合消遣的地方。基奥云尼咖啡厅+面包店+熟食店坐落于不列颠哥伦比亚省温哥华市中心太平洋边际费尔蒙酒店所在的大街上。这家奢华的餐厅面朝市中心大街，后门直通费尔蒙酒店的休息大厅。以城市居民及过往行人为消费群体，咖啡厅为顾客提供定制咖啡、三明治、甜点等一系列餐厅服务及包括书籍、亚麻制品、儿童玩具等别致礼品在内的一系列配套服务设施。

Restaurant

Canada

Vancouver

2010

Photo: Michael Boland Seng Tsoi

Mcfarlane | Green | Biggar Architecture+Design

23

Restanrant & Bar

Canada

Montreal

2008

Photo: Rockwell Group

Rockwell Group

Aloft

A design concept set to raise the bar in affordable select-service hospitality, offering airy, bright loft-like guest rooms, enhanced technology services, landscaped outdoor spaces for socialising day and night, and an energetic lounge scene. Envisioning a series of urban oasis on the American roadside, Rockwell has emphasised sophistication, community, functionality, and comfort. In the Aloft restaurant & bar, the urban living experience continues into the primary spaces – a relaxed and social lounge space and light-filled guest rooms. The self-service check-in, lobby pool table, sunken living room area with fireplace, backyard patio visible from picture windows throughout the hotel and ultra-comfortable beds all contribute to the feeling of being 'at home' at Aloft.

Aloft酒店餐厅

设计理念为打造一流的服务、通透明亮的阁楼式客房、健全的技术设施、24小时的室外集会场所以及活力充沛的酒吧。餐厅酒吧：城市的生活体验在私人空间内同样感受得到——无论是在酒吧还是客房内。酒吧内的自助桌球案、带有壁炉的客厅、透过落地窗清晰可见的院子、舒适的大床，到处洋溢着家的温馨。

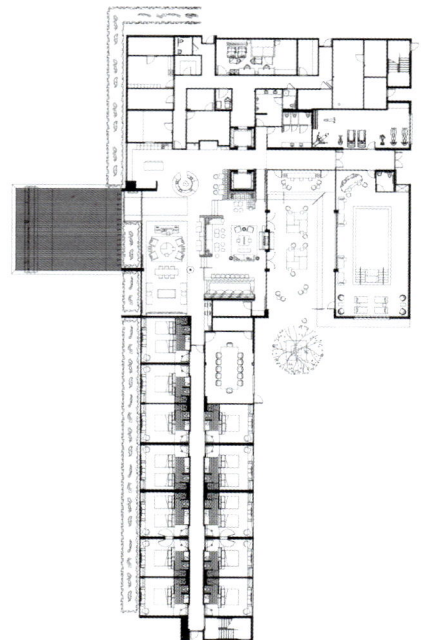

RH Restaurant & Bar in Andaz West Holywwod

A new 'Glass Pavilion', framed in skeletal steel, extends RH, the hotel's bar and restaurant out onto the Sunset. The space was transparented and became a shared experience between the public and private realm. The pavilion's language is suggestive of the modern design and architecture found in the Hollywood Hills, and work captured in the photographs of Julius Shulman. The pavilion features a custom tumbled mosaic patterned floor inspired by the work of Erwin Hauer and Roberto Burle Marx and an art installation on its exterior by the LA-born artist Jacob Hashimoto.

RH餐厅酒吧

酒店的餐厅酒吧名为"RH"，采用钢框架结构，仿佛一座"玻璃展亭"延伸到日落大道上来。餐厅的空间干净通透，成为公共与私人活动的共享空间。这座"展亭"的"语言"暗示了好莱坞的现代设计与建筑，包括摄影大师朱利斯·舒尔曼的作品。特别定制的马赛克图案的地板，灵感来自欧文·豪尔和罗伯托·布尔勒·马克斯的作品，室外的一个艺术造型出自雅各布·桥本之手，这位艺术家出生于洛杉矶。

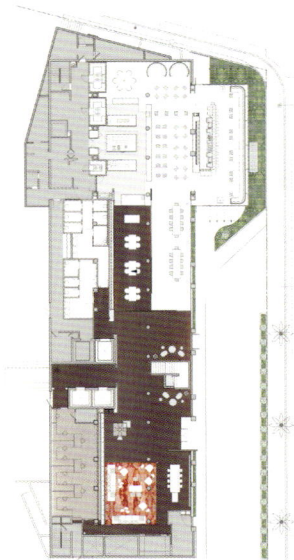

Restaurant & Bar

USA

Los Angeles

2007

Photo: Avro|ko

Janson Goldstein Llp

Café

USA

New Jersey

2007

Photo: James D'Addio, Hastings-on-Hudson

Ikon.5 Architects

Cyber Café

The Café was designed to maximise its campus exposure and to create a new front door to the library. The Café interior is defined by two zones: the very open fully glazed front, with a sloping ceiling which opens up to the outdoors; the moodier interior space, where the Café is inserted into the existing library building, which contains the servery, built-in seating and tables, and features a colour-changing back-lit translucent wall. The open nature of the café allows the feature wall to be clearly seen from the campus quad and beyond.

网络咖啡厅

为了适应世界不断发展的高科技和信息交流，并提升人气，蒙特克莱尔州立大学的图书馆增加了咖啡厅。咖啡厅分为两个部分：正面是包括倾斜的天花板在内的全景玻璃幕墙；图书馆内部的部分则略显严肃，内有备餐室、内置式座椅、餐桌和半透明的背光式背景墙，配有不断变化的灯光。咖啡厅采用开放式设计，在校园的任何位置都能看到里面的情况。

Lounge

USA

Miami

2006

Photo: Light Architecture

Light Architecture

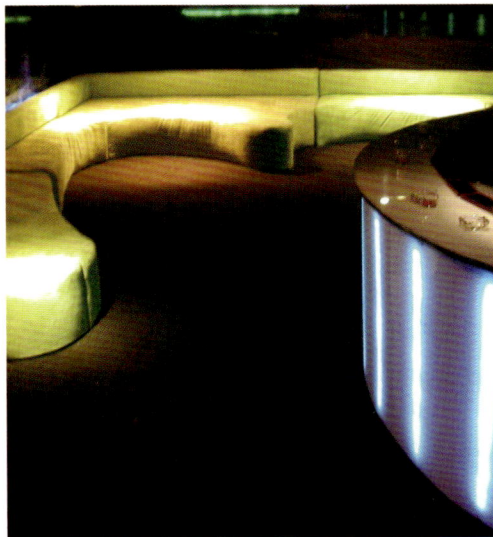

Mynt Lounge

In the Mynt Lounge the four walls of the box are de-materialised through wall projections that contribute to immerse the user in an almost submarine environment. The long wall for projections runs parallel to the bar counter, where the upholstery buttons are replaced by luminous signals, which can be synchronised as wished. Two DJ stations are placed in the opposite far ends of the hall; in the background two large projection walls can be seen. A secluded space, refurbished with long sinuous sofas, serves as lounge and chill-out ambient. A fine mint-scented mist is diffused in the room, recalling the club's name. This is one of designer's satisfying projects.

薄荷休闲酒吧

薄荷酒吧的四面墙使用发光图案墙壁，营造出深邃幽暗的海底情境。发光墙壁与吧台平行，墙上发光的信号灯取代了装饰按钮，实现了理想中功能与美观的和谐统一的效果。两个DJ台分设在大厅的两端，其后是两面巨大的发光背景墙壁。一个僻静的角落，蜿蜒摆放着长长的转角沙发，用于客人休憩。酒吧混杂着酒精和薄荷的香气，使人回想起酒吧名字的由来。

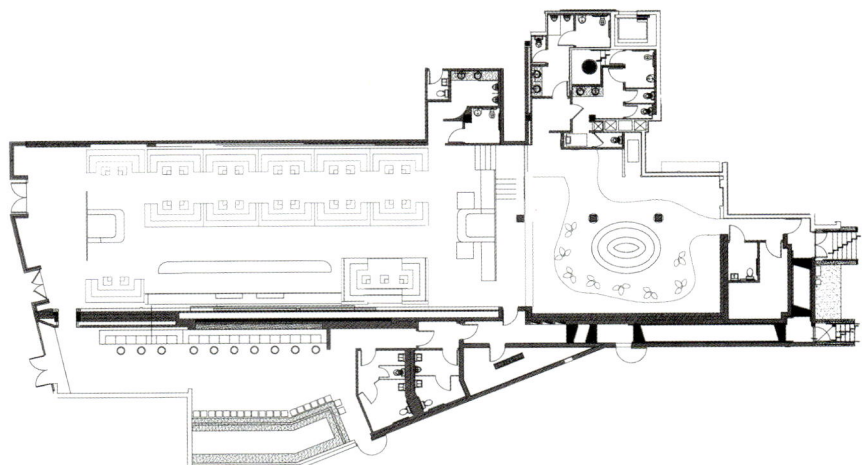

Restaurant

USA

Miami

2007

Photo: Eric Laignel

Zeff Design

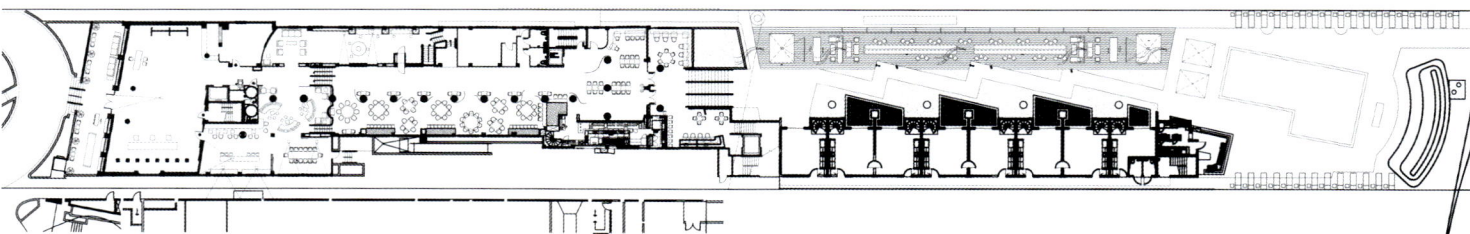

Social Miami

Social Miami is a modern version of the classical European salon scene. Zeff has designed an elegant and inspiring world composed of a gallery dining room, sumptuous gaming room, video art garden, and a pool lounge surrounded by silvery palms. Outdoor Social Miami has a garden, private living rooms, outdoor café, pool and VIP lounge. The main garden is surrounded by three glass walls that showcase interactive video technology creating an intense energy. This main lounge is flanked by two private living room sections. Guests can sunbathe during the day at an authentic sand beach or at the pool on a large wooden deck.

迈阿密社交俱乐部餐厅

迈阿密社交俱乐部是古典欧洲沙龙场景的现代版本。泽弗设计事务所设计出一种非常优雅又令人振奋的世界，包括画廊餐室、华丽的休闲室、视频艺术花园和由银棕榈环绕的游泳池。户外有一处花园、私人起居室、户外咖啡馆和VIP休闲所。主要花园由三面玻璃墙围绕，交互视频创造出剧烈的能量。主要的休闲所与两处私人休息室相连接。

Bar

USA

New York

2007

Photo: Arthur Casas

Arthur Casas

World Bar

The building is the Trump World Tower, located on First Avenue, across from the United Nations. This establishment was named World Bar exactly because of that proximity. Needless to say, the building has a number of details in gold, and the fact that the bar is located on the ground floor left designer with no other choice but to explore the golden colour: it is present in the fabrics and in the tone of the wood, Canadian oak, creating a visual wholeness. The designer loves being there at the end of the day with a dry martini while thinking that few places in the world are worthy, as New York is, of feeling golden.

世界酒吧

这个建筑是川普世界大楼，在第一大道上，联合国总部对面。这里被命名为世界酒吧也正是因为这样的地理位置。不用说，建筑的若干细部结构由黄金打造，目前这种色调体现在酒吧内的纺织品装饰和以加拿大橡木为主调的木材构造中，形成了视觉上的统一。设计师喜欢在傍晚时分来到这个酒吧享受一杯马提尼干红。世界上只有少数的地方配得起这种感觉，因为纽约就应该是金色的。

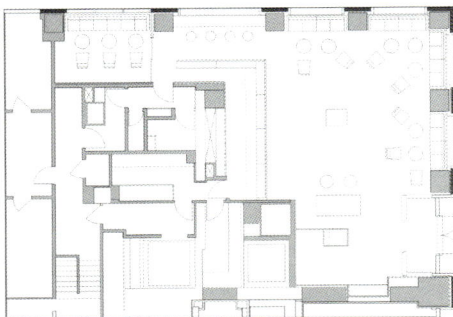

Bar

USA

Atlanta

2007

Photo: Mark O'Tyson, O'Tyson Photo

TVS Interiors

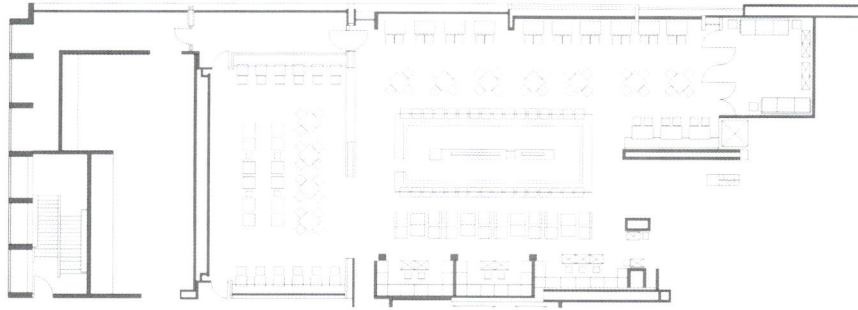

High Velocity

The typical sports bar is redefined by this design with its sleek lines and contemporary portrayal of the fast-moving world of sports. Horizontal lines depicted in the zebra wood veneer and mirrored ceilings around the bar reflect a fast-paced atmosphere and the speed of success. Twenty-one 40-inch plasma screens are located throughout the space with a backdrop of back-lit super graphic images running the perimeter of the seating areas. Beer becomes a theme with the facility's pilsner-coloured palette and a beer chandelier featuring bottles from a local Atlanta brewery. The large viewing area is laid out with oversized leather recliners and warm wood floors.

高速酒吧

典型的运动吧是采用快节奏的圆形流线设计和当代的描绘设计。此项目的地面铺有斑马木地板，玻璃质的天花板给人一种高速、快节奏的感觉。作为背景，40英寸的等离子电视围绕在座位的周围，给人一种超级的视觉感受。屋内的设计以啤酒为主题，空间的主色调采用啤酒色，吊灯的形状也和亚特兰大啤酒厂的酒瓶有异曲同工之处。一面大墙上铺有48麦芽啤酒的广告，引人入胜。

Restaurant

USA

New York

2007

Photo: Avro | Ko

Avro | Ko

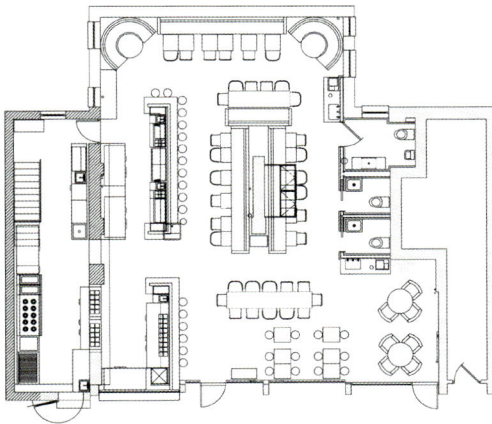

The European Union

A gastro-pub located in New York's East Village, the E.U. was an exploration into the simple beauty of utility. The E.U. differed from the other restaurant projects in that its design was not derived from a visual concept, but a functional concept. The original gastro-pub was born of necessity, a place where young British chefs with limited resources could have their own restaurants. Its aesthetics was based on creating something useful and charming out of ostensibly undesirable and useless parts. Thus, the final concept for the E.U. generated from a blending of the functional, operational aspects of a gastro-pub with the design language of the traditional service culture of the English manor kitchen.

E.U. 餐厅

作为一个位于纽约东部山村的酒馆，E.U. 餐厅是对简洁实用性的大胆尝试，完全颠覆了传统的英国厨房结构。E.U. 餐厅这个项目区别于其他餐厅项目的一个主要原因就是这个项目的设计理念不在于给人以视觉冲击，而是更多地转到实用性上来。E.U. 餐厅的设计理念集合了酒馆的实用性与实际经营，以及英国庄园式厨房传统的服务文化等元素。

Per Se

The highly anticipated Per Se is a complex design portrait of Thomas Keller, reflecting this world-class chef's unique blend of hospitality and personal style. At once elegant, yet accessible, this sophisticated Manhattan enclave, and on the third floor of the new Time Warner Centre, is an exercise project in contrasts, featuring a wealth of natural materials — wood, stone, metal and glass, enriched by custom fabrics and finishes in a soothing palette of chocolate, caramel and cream tones. The space is serene, but not spare; tailored yet luxurious. Classic modern design elements help set the stage for this rare culinary experience.

珀赛餐厅

公众高度期盼的"珀赛"是托马斯·凯勒复杂的设计画像，它反映了世界级厨师的款待方式与个人风格的独特混合。起初的考究风格和后来的平易近人风格在位于新时代华纳中心四楼的餐厅中对比应用。餐厅使用了大量自然的材料，如木头、石头、金属和玻璃，用特制的织物加以丰富，完工后就像温柔的巧克力、焦糖和冰淇淋拼盘。

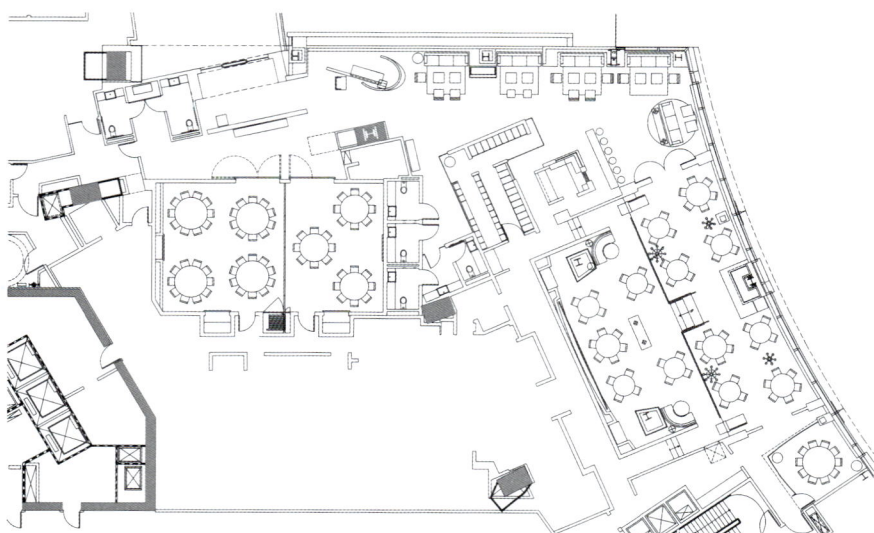

Sapa

Inspired by the quaint Vietnamese French-colonial hill town of Sapa, this sprawling 8,000-square-foot French-Vietnamese restaurant in Chelsea is an exercise in gracious harmony and duality. The lofty white space, originally a church administration office, embraces both French and Vietnamese elements, co-existing in both design and cuisine to create a new dining experience. Sapa is not about the fusion of disparate elements, but the integration of independent elements: the refined, neoclassical French aesthetics with the ornate patterns of traditional Vietnamese craft, the elegant and rustic, textured and plain, East and West. The expansive space is arranged in different areas to avoid being overwhelming.

萨帕餐厅

萨帕餐厅这个项目的设计灵感来源于法国殖民地时期越南萨帕山镇的建筑风格。在英国切尔西，越南法国殖民风格的餐厅很受大众欢迎。萨帕餐厅的设计不是为了把不同元素融合在一起，而是各种独立元素的综合：精致的、新古典主义的法国美学与越南传统工艺华丽图案，高雅与乡土气息，纹理突出与朴实无华，东方与西方在这里汇集到了一起。

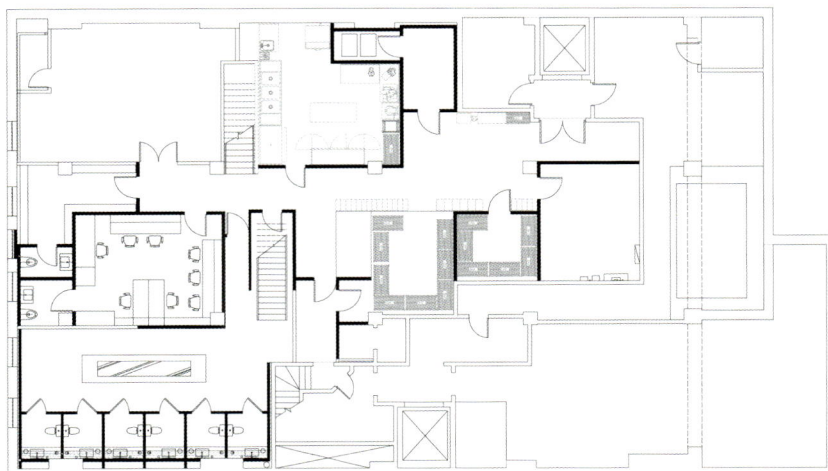

Restaurant

USA

New York

2006

Photo: Avro | Ko

Avro | Ko

Restaurant

USA

Las Vegas

2007

Photo: Avro|Ko

Avro|Ko

Social House

Social House is a departure from the standard Las Vegas restaurant in that it rests on a strong conceptual framework. Inspired by the microcosmic character of an old, walled Asian metropolis, completely self-sufficient like a city-within-a-city, the sprawling Social House space hints at the different elements of the industrial market place. The concept lends itself perfectly to the awkwardly long, winding space, a challenge that designers addressed by organising various bars and areas along the length of the space, each representing an important aspect of society. The idea of a traditional herbal pharmacy informs the entryway.

大众餐厅

大众餐厅没有遵循美国拉斯维加斯标准餐厅的设计风格，其设计构思新颖独特。餐厅的设计灵感来源于亚洲古都的建筑风格，非常类似于一个城中之城，内部空间采用了多种不同的商场设计元素。为了解决餐厅内部过于长而宽敞的问题，设计师在两边设计了各种酒吧和休闲区域，而每一个区域都暗示着社会的某一个方面。

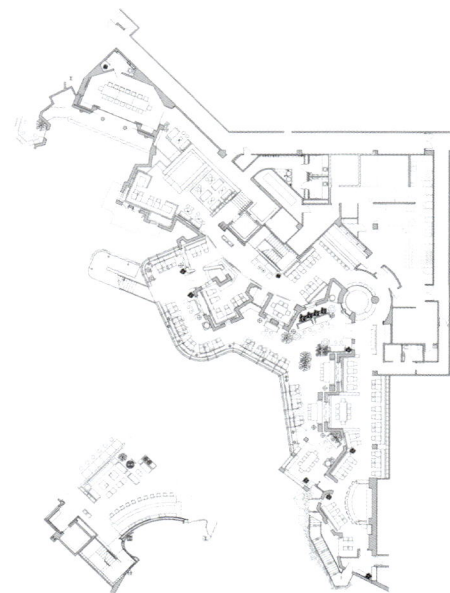

Restaurant

USA

New York

2007

Photog: Avro | Ko

Avro | Ko

The Stanton Social

An ode to the Lower East Side of the turn-of-the-century, The Stanton Social restaurant and lounge embraces its past while looking towards the future with its modern design and shared plates menu. The space was treated in two separate, but cohesive, parts. The main dining room, located on the ground floor, imparts a more masculine feel, influenced by the construction of men's suits and visualised through the rich, dark leather banquettes, herringbone patterns, and leather belts used to fasten the banquette pillows. Herringbone resurfaces at the back wall of the restaurant, where its geometric pattern was transformed into a wine storage wall.

斯坦顿餐厅

整个餐厅内部空间包括独立却又连接着的两个部分。主餐厅位于一楼,通过饱满暗色调的皮革座椅、鱼骨形图案和固定座椅靠垫的皮带体现着一种阳刚之气,这种设计受到了男士服装和形象设计的影响。餐厅后面鱼骨纹路墙上的图案暗示墙的后面就是藏酒室。这种图案在二楼也被应用到,以提供感官的连续性和一楼与二楼之间的连接。

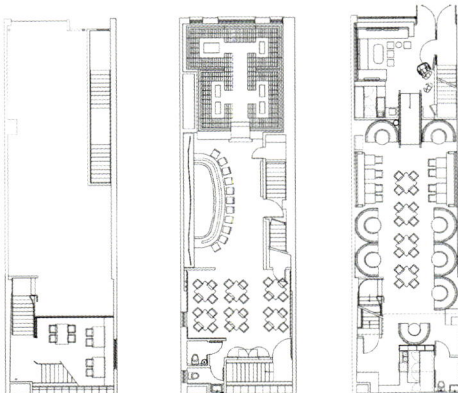

Restaurant

USA

Los Angeles

2006

Photo: M. Laignel

Dutton Architects

Opaline

Opaline is located on a corner of one of the most active restaurant locations in Los Angeles. The design attempts to complement the menu; to be simple, honest, comfortable, and well executed. The floors are polished concrete. The ceiling consists of stained poplar slats, which hide the mechanical ducts and electrical conduits, and allow for ambient sound to escape and be trapped. The ceiling reads then as a continuous fabric, with no light cans or mechanical registers. Cork walls help acoustically soften the space. At the centre of the dining room, wrapping an existing column is a light sculpture of fibreglass. Dividing the two rooms is a stained stud wall, with an etched glass and wood slat screen.

奥帕兰餐厅

奥帕兰餐厅位于洛杉矶最活跃的餐厅聚集地。室内设计简洁、朴实、舒适，并有良好的可操作性。地面由水泥抛光制成。天花板上装饰着杨树板，里面藏着机械和电气的管道，同时起到隔音的作用。天花板没有安装任何灯具或机械设备。软木墙壁使空间内的声音变得柔和。餐厅中央的柱子上装饰着刻花的玻璃钢。木条和玻璃隔断将两个房间分开。

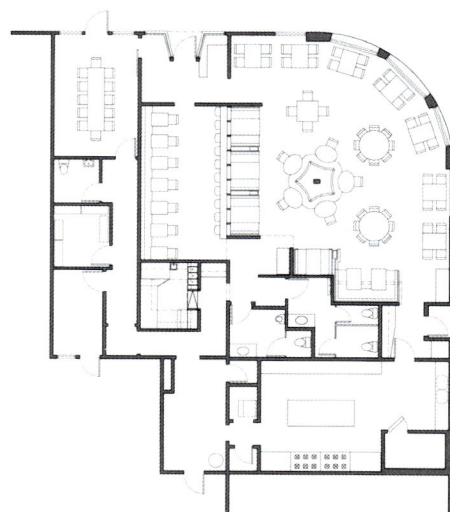

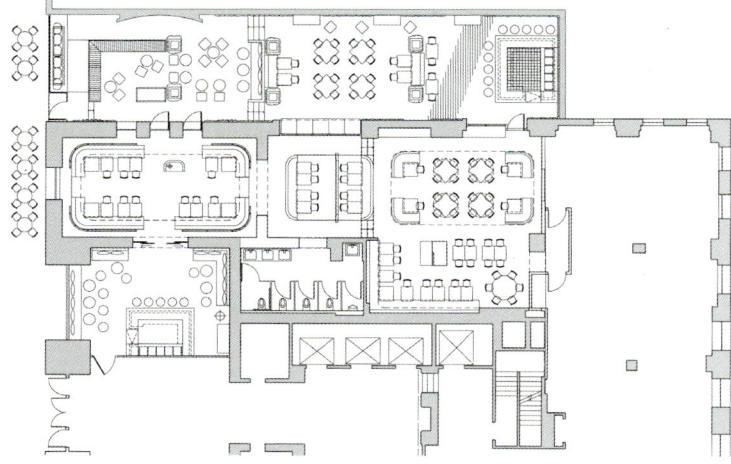

Washington Square

The restaurant space occupies half of the ground floor and includes a private garden, which was used as the access path to the restaurant entry. The exterior garden is neutral in colour, with modern simple furnishings and a white bar; the walls of the outdoor garden have a full range of colours from green ivy to custom ceramic coloured tiles. The first interior space after the garden is the first dining room. Guests upon entering will be taken to their table in this room, the blue dining room or the main dining room. Still set within the classic walnut panelled gallery space is the 'Blue Room'.

华盛顿广场

餐厅占据了一楼一半的面积，私人花园内设有通往餐厅入口的通道。花园采用中性的色彩和现代简洁的家具，酒吧则以白色为主。绿色的常春藤和瓷砖是花园墙面的主要装饰。花园后是一间餐厅。客人从入口进来以后，会被带到这间餐厅内坐下。这里又称作蓝色餐厅或主餐厅。经典的胡桃木板将这里装饰得如画廊般精美。

Restaurant

USA

Philadelphia

2005

Photo: Eric Laignel

Rockwell Group

Café

USA

San Francisco

2006

Photo: CCS Architecture

CCS Architecture

Lettus

This 50-seat counter-service café opened in San Francisco's Marina district. The contemporary design is both casual and upscale, using sustainable materials that are raw and refined. A central element is the juice and salad bar made from hickory, tile, glass and brushed zinc; it is an expression of the fresh food being served. The main dining area is surrounded with a reclaimed hickory slat ceiling and walls with warm linear lighting. One entire wall of the space is composed of an adjustable shelving wall to hold the product of the food being served, which also uses the food as an art piece.

莱特斯咖啡厅

坐落于旧金山马瑞纳区的咖啡厅，可同时容纳50人。设计采用天然精致的环保材料，虽属休闲风格却极富档次。咖啡厅以果汁-沙拉吧为中心，由山胡桃木、瓷砖、玻璃和磨光锌板装饰，突出新鲜食物。就餐区上方是打磨过的胡桃木制天花板，墙壁周边配有温暖的线形灯光。咖啡厅内一面主墙上嵌入活动板，可以托放食物，如同展示艺术品。

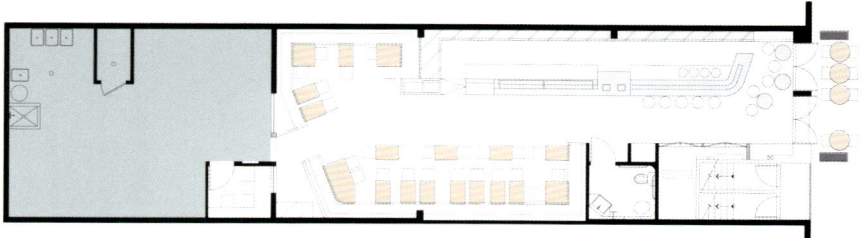

Restaurant

USA

Providence

2006

Photo: Dileonardo International

Dileonardo International

L'Epicureo Restaurant

A style suited to the name L'Epicureo would be created being timeless in design. The design was created to tantalise the imagination by exposing a series of chandeliers that accent the linear layout along with old world master reproductions from renaissance and pre-renaissance period helping to reinforce the owners' desire for this European setting. The lounge area has contemporary stained glass panels separating the lounge from the restaurant seating. Even the table settings were carefully selected. The chandelier and artwork support the name and work well in the arts district in a casual formality, which one might find in any southern European city.

欧莱伊壁鸠鲁餐厅

以永恒为主题的设计风格正适合欧莱伊壁鸠鲁餐厅。一系列的枝形吊灯装饰强调了室内的线性布局，配合使用文艺复兴及前文艺复兴时期的古式仿真品，从而满足了客户欧式风格的要求。设计中采用现代风格的彩色玻璃镶板将休息区与就餐区隔开。桌子的设置精挑细选，而吊灯和艺术品的使用则突出了餐厅的主题，并为艺术区营造出一种简单随意的氛围，极具南欧风格。

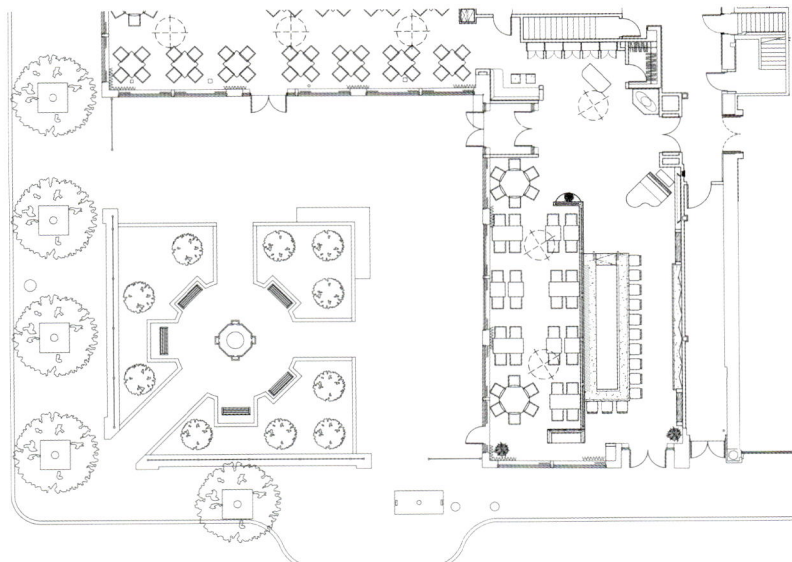

Bar

USA

Atlanta

2007

Photo: Mark O'Tyson

TVS Associates

Pulse Bar

The design goal was to activate and enhance the atrium space functionally and architecturally. The centrally located bar and lounge, this translucent sculpture with its soaring sail form was supported by four steel columns anchored in the floor. The bar itself is cladded in custom-cut resin panels backlit with LED lights, contemporary and clean with white marble panels and a white backpainted curved glass top. Surrounding the bar, a walnut wood floor gives contrast to modern white finishes. Casual and intimate seating arrangements encircle the facility. The towering sail-like structure includes a large resin panel as a projection screen. Clean lines and unique forms create a new, activated place.

脉搏酒吧

设计旨在提升核心空间的功能性。位于中心的酒吧兼休息室，是个扬帆样式的半透明雕塑结构，其帆体以固定于地面的四个钢柱为支撑。酒吧选用树脂面板作为装饰材料，并且采用LED灯照明。帆体运用白色大理石和弧形玻璃作顶。酒吧四周的胡桃木地板与新潮白色材质形成强烈对比，随意而舒适的座椅环绕在周围。此高耸的帆状结构还配有一个大型树脂板作为投影屏幕。

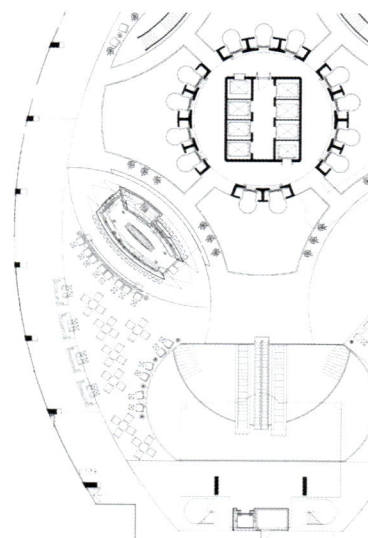

Café

USA

New York

2006

Photo: Z-A

Brett Snyder, Cheng + Snyder in collaboration with Guy Zucker, Z-A

15 Piece Coffee Bar

The café is organised into three 'strands'- a front service bar, a back storage counter, and a standing bar. Each strand is composed of a series of interlocking wood 'blocks'. The size of the blocks is constrained by the dimensions of a single plywood sheet. 15 blocks precisely accommodate the café's 15 functional requirements. Portions are carved out of each block to create spaces of exchange between the server and the customer. These subtractions generate interlocking connections between the blocks, allowing them to be easily stacked and assembled on site. Each strand was built on its own surface.

15咖啡

咖啡馆分成三部分：前面的服务台、后面的储藏柜和一个吧台。每一部分都是由一系列连锁的木块组成，其大小由单块胶合板的尺寸决定。15块恰好顺应了咖啡馆的15种功能要求，并在每个木块镂空的部分留出了服务者与消费者交流的空间。这些除去的部分形成了块之间的相互连接，容易现场堆积和组合。

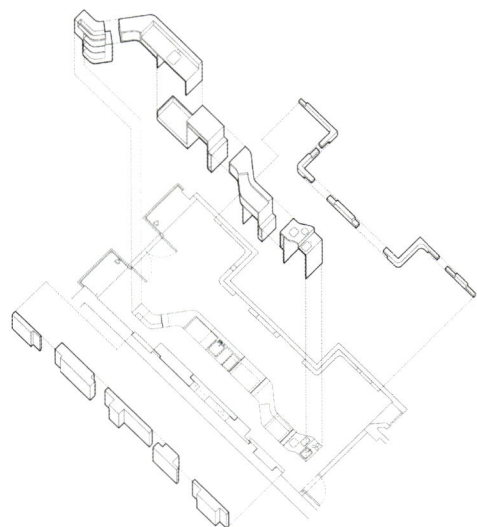

Restaurant

USA

Portland

2007

Photo: Cesar Rubio, San Francisco, CA
Angie Silvy, San Francisco, CA

Cass Smith

Bay 13

The sliding door panels are reminiscent of the previous door system that allowed the historic portion of the Crane Building to operate as a warehouse. Two ceramic tile walls with large porthole windows break the rhythm of the storefront to give the elevation proportion and scale and help to obscure the new seismic braces that have been added to strengthen the buildings' structure, and to emphasise the seafood focus of the restaurant. Materials such as metal siding and roofs were selected to respect the historic nature of the building. A muted palette of yellows, beiges, and greys are incorporated into the exterior so as to correlate well with the existing brick.

海湾13号

滑动门板令人不禁想起建筑最初的门系统；瓷砖墙壁上安装的高大舷窗打破了店面原有的整齐，突出了空间的比例，并将新增的防震结构隐藏起来，同时彰显了餐厅内的特色海鲜食物。设计师精选的金属外墙材料强调了原有建筑的历史特色，外观饰以黄色、米色及灰白色与现存的砖瓦结构相融合。

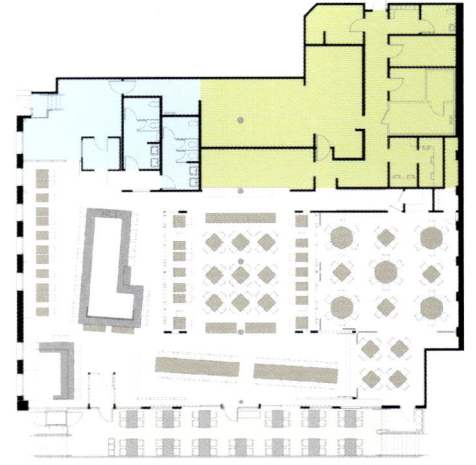

Restaurant

USA

San Francisco

2007

Photo: CCS Architecture

CCS Architecture

Terzo

Inspired by the wine and tapas bars of Europe and the Mediterranean, CCS Architecture crafted a chic, 96-seat space with a palette of natural materials. Terzo's design has a warm, modern sensibility befitting both the food and the San Francisco neighbourhood setting. Understated and casual, the rooms nonetheless give rise to a specific, thriving atmosphere. The front portion of the restaurant resembles a European tapas bar with a modern façade. A custom glass and stainless steel display case invites guests to sit and enjoy small bites in the Mediterranean fashion. Wall cabinets of wine and the Edison lamps continue the warm look of the dining room.

Terzo餐厅

设计师受到欧洲及地中海地区的酒吧和小餐馆风格的影响，使用天然材料，打造了这个仅有96个座位的时尚空间。餐厅内洋溢着温暖、现代的气息，既适合旧金山的城市氛围，又突出了这里的食物。朴素、随意但却不乏精致，餐厅的前半部分如同欧式小酒吧，一个定制的玻璃不锈钢陈列台吸引着客人坐下来品尝美味。装满酒品的壁柜和螺纹式灯泡更加突出了餐厅的温暖气息。

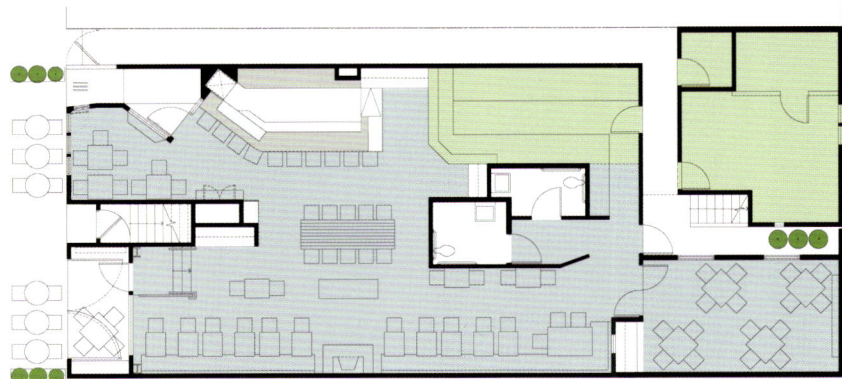

Restaurant

USA

New York

2007

Photo: Studio Gaia, Inc

Studio Gaia

All Day Buffet Restaurant

All Day Buffet Restaurant is conceived as a sequence of contextually different spaces which addresses visitors with experiences known from nature. The restanrant is in vivid colours with walls decorated in stylised graphic impressions from nature (eyesight). Here is the perfect gathering place for the lighting, music and delicious food, reflecting the harmony of the restaurant. It also gives the feeling of coming home before going home.

全天自助餐厅

餐厅别具一格的空间，为客户提供全方位的感官体验。以欢快的色彩为基调，从视觉上将客人的热情点燃。这里是完美的聚会场所，将灯光、音乐及美味融为一体，体现了餐厅的和谐，同时又会让人很兴奋，也可以带来家的温暖。

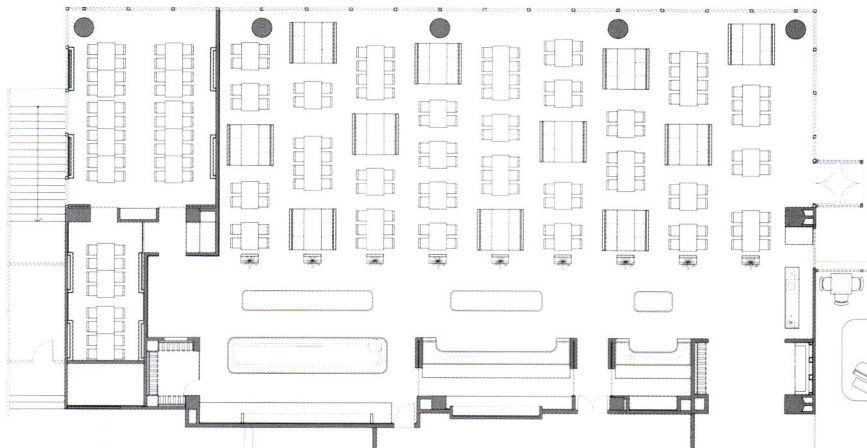

The SW Steakhouse

The SW steakhouse is located inside of Wynn Las Vegas Resort and Country Club (which is located in Las Vegas, Nevada. The resort covers 870 square metres and has been hailed as the pre-eminent luxury destination in Las Vegas since 2005). This high-end steakhouse is both classy and romantic with soft lighting, sleek marble floors and views of the hotel lagoon. Diners enjoy a quiet setting in comfortable leather booths and at tables with high-backed chairs. Chef Davis Walzog offers a menu that is a unique and contemporary twist on a classical American steakhouse: Steaks are hand-selected from USDA wet-aged prime beef, cooked in an open flame and topped with SW's signature sauces.

SW牛排餐厅

餐厅位于 "Wynn" 度假酒店和乡间俱乐部内部。柔和的灯光、光润的大理石地板以及远处的湖泊美景,为餐厅打造出一种时尚而又浪漫的氛围。顾客落座于舒适的座椅,享受着静谧的氛围。厨师长Davis Walzog为美式牛排提供了一种独一无二的新吃法:农业部牛排的首选牛肉,经由手工精选,明火烹调,覆以SW招牌调味料。

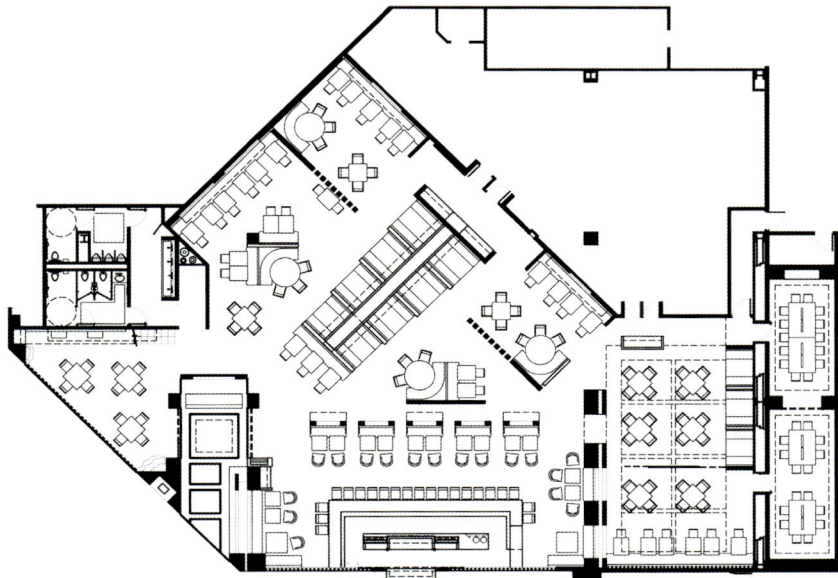

Restaurant

USA

Las Vegas

2006

Photo: Vicente Wolf

Vicente Wolf

Photo: Benny Chan/Fotoworks

John Friedman Alice Kimm Architects

The Brig

Presented with a run-down, 52-year-old landmark bar with an exterior mural that serves as a gateway to Venice's Abbot Kinney Blvd., the designer's goals for this project were threefold. First, intensify the entire building's ability to act as a gateway to Venice; second, maintain the original, ad hoc quality of the bar's interior; and third, create a public space with a sensuous mood that welcomes the open, diverse spirit of Venice. This project is a combination of the raw and refined, the old and new, the highly deliberate and ad hoc. It maintains the edginess of the original bar. That it does so while adding a new sensuality and fluidity makes it a highly complex and sophisticated environment.

禁闭酒吧

这家酒吧有52年的历史，已经破败不堪了，有一幅壁画权当通往Abbot Kinney大道的通道，设计师的目标有三个。第一，强化整幢大楼的功能，使它真正成为通往威尼斯的必经之地；第二，保持酒吧内部的原有风格；第三，创造一个充满美感的公共空间，彰显威尼斯开放另类的风情。设计师保持了酒吧原有轮廓的同时，又增添了性感和流动性，创造了一个极其复杂又不落俗套的氛围。

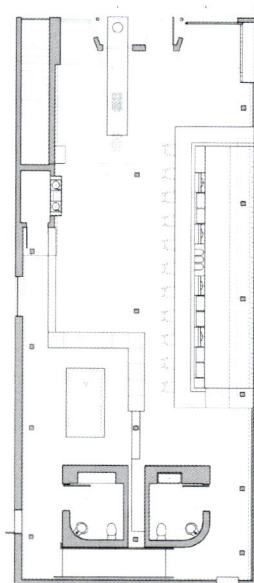

Thor

Thor, the restaurant and lounge in the Hotel on Rivington, captures the spirit of the lower East Side with a diverse menu and lively bar scene lasting well into the night. Its name derived from the acronym for the Hotel on Rivington, Thor offers a market-fresh menu built on the foundation of American bistro classics. Named 'Best New Hotel Restaurant' by Time Out New York, Thor received a glowing review from The New York Times. The modern space, by Dutch designer Marcel Wanders, features an airy dining room with a 21-foot glass ceiling that offers view of the neighbourhood's historic tenement buildings. Most of the action is centred in Thor's bar and lounge, a hub for locals, hotel guests and an essential stop for anyone visiting the lower East Side.

托尔餐厅

托尔餐厅位于莱温顿酒店内，这里提供种类繁多的菜肴，餐厅内还有一个一直活跃至深夜的酒吧。餐厅中的菜式新鲜无比，秉承了美式小酒馆的精致风格。托尔餐厅被纽约时代杂志评选为"最佳酒店餐厅"。餐厅的时尚空间由荷兰著名设计师马塞尔万德斯打造，餐室通透明亮，玻璃天花板高达21英尺，通过这里可以欣赏到附近著名的建筑景观。

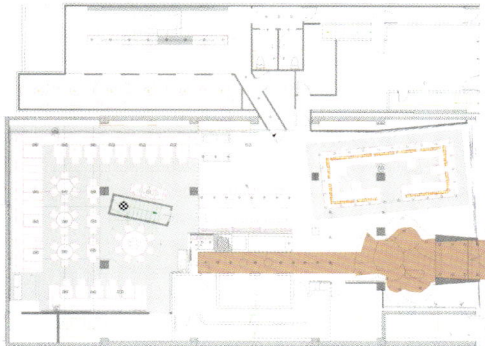

Restaurant

USA

New York

2005

Photo: Inga Powilleit

Marcel Wanders

Restaurant

USA

Los Angeles

2006

Photo: Tom Bonner

Belzberg Architects

Patina Restaurant

The interior of the restaurant was to reflect the same sense of spectacle perceived in the concert hall. The fluidity of the theatre curtain served as a formal inspiration - echoed throughout the restaurant walls and ceiling. Likewise, the concert hall presented an acoustical challenge; every surface had to be acoustically isolated from the rest of the building. Therefore, while the metaphorical curtains were to serve a formal purpose, they were also to function as sound-deadening devices, both in the ceiling and in the walls. The ceiling also plays on the grand occasion of walking through a theatre. These undulating ribs act as a light sieve, which give the appearance of a glowing ceiling above.

帕蒂那餐厅

餐厅内有着与音乐厅相同的氛围。剧场式的窗帘与餐厅的墙壁和天花板十分相衬。音乐厅的声学效果经过特殊处理，每个空间都必须隔音。因此，窗帘、天花板和墙壁除了它们本身的职责外，还起到隔音的作用。天花板给人以宏伟壮观的感觉，一条条起伏的突起起到滤光器的作用，形成闪闪发光的效果。

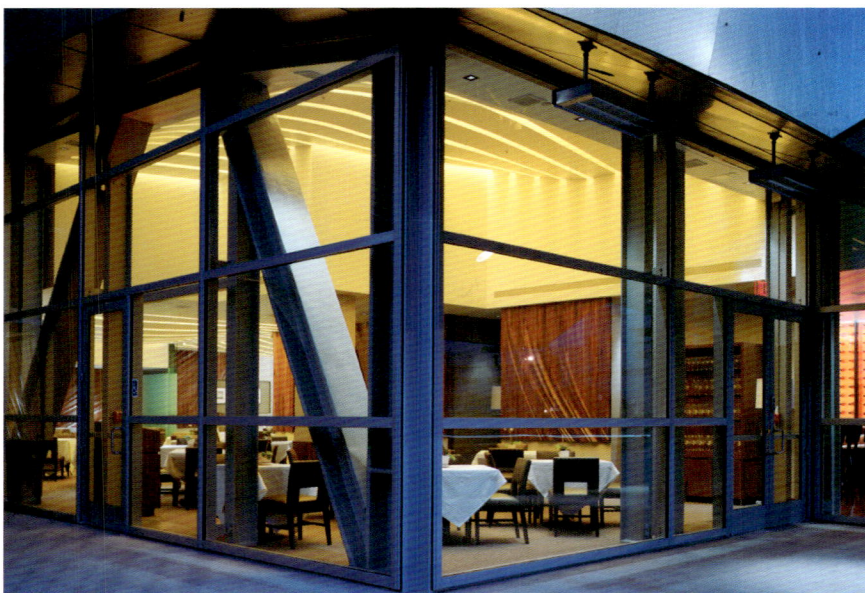

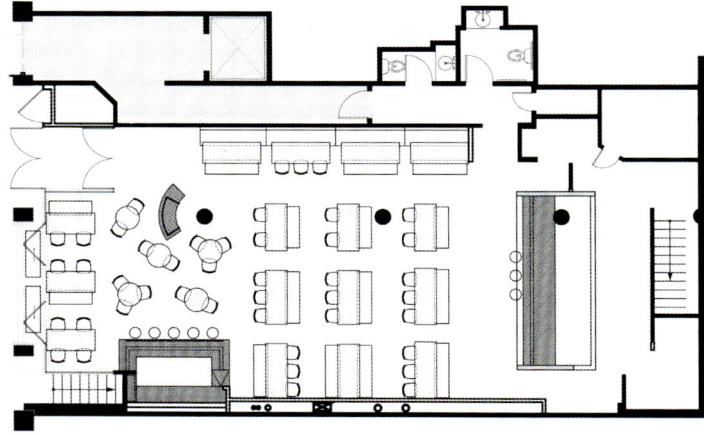

Stand

The designer thought that the idea was to craft a sophisticated design that has longevity and would be comfortable, nothing that he did not want to create a stark and minimalist look. Long wooden tables are actually chunks of butcher block with saw cuts on the top. Stained black, but kept a little raw and somewhat unfinished. It helps create a feeling of simplicity while also emphasising the food. A monochromatic colour scheme plays throughout, which according to Waisbrod, helps the furnishings slip quietly into the background. "It is sophisticated, back-to-basic design," the designer says. "It is very friendly…reflecting the neighbourhood attitude."

经典餐厅

长木桌实际取材于屠夫街市的大块木板，褪色的黑色加之裸露着实木质地，有点像半成品。这样有助于营造质朴的感觉以强调食物的特色。单一色调贯穿始终，使室内陈设与背景融为一体，完全符合威斯宝的设计理念。设计师认为："该项目的设计经久不衰，回归本质，并且极富有亲和力，种种这些都反映了该地区的生活态度。"

Restaurant

USA

New York

2006

Photo: David Joseph

Ilan Waisbrod \ Studio Gaia

Restaurant

USA

Mississippi

2005

Photo: Doug Snower

Jordan Mozer and Associates, Ltd

Annamae

The entry to Anna Mae is through an internally illuminated cubist arch of fused art glass framed in raw steel, up a pathway of limestone and bronze tile to a Pomelle-Sapelli host podium opposite a collection of Asian ceramics, bells and sculptures. The entry is bright; but the Dining Room is dark and cosy and populated by an array of animated elements rendered in natural materials. The designers employed a small army of artists to make almost every element in Anna Mae from scratch. The chairs and booths are animated. They are fabricated in hand carved maple, cast magnesium-aluminium alloy and upholstered seats and backs. The floor is bamboo, stained the colour of chocolate.

安娜·梅餐厅

餐厅入口有一个灯光拱门，由玻璃和钢架搭建而成，有一条石灰石小径通往接待台，对面是亚洲的陶瓷制品、钟和雕塑。餐厅里灯光十分昏暗，给人舒适惬意的感觉，这里充满活力的装饰元素全部由天然材料制成。散台和包间都设计得富于生气。使用的材料为手工雕刻的枫木板条和镁铝合金框架。地面铺着巧克力色的竹地板。

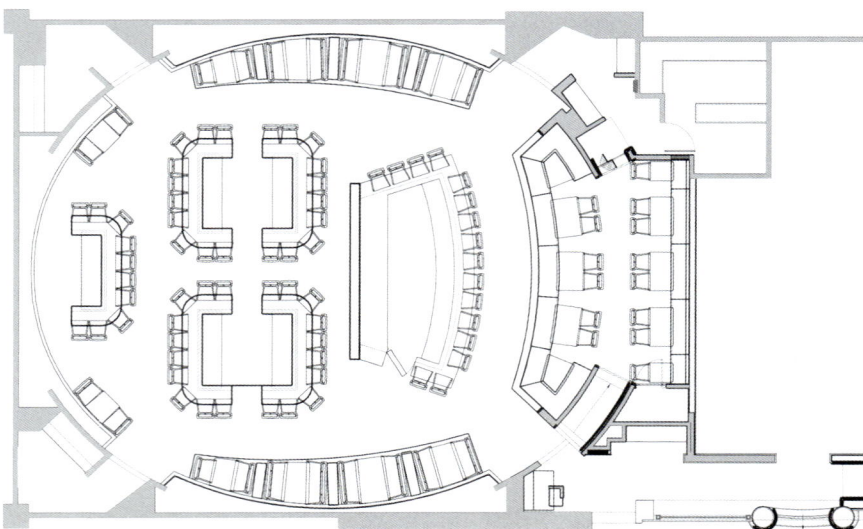

Café Gray

The restaurant combines traditional and contemporary elements and ideas that are reflected in its inventive cuisine and design. With the classic brasserie providing the point of departure, Café Gray is a cleverly crafted, updated version, using familiar materials such as dark wood, leather, mirror and stone in creative ways. This warm, elegant space features curvy, serpentine leather banquettes. Accents in ochre, lavender and deep violet add to the restaurant's rich palette while materials such as chrome and stainless steel were used to detail the banquettes, bar and tables. Beautifully beveled mirror tiles and lace-etched glass panels maximise the light, views and reflection within the space.

格雷咖啡馆

咖啡馆的设计结合了古典与现代的特色，并体现了独特性。格雷咖啡馆以古典餐馆风格为基调，以精湛的工艺和创新的形式，重新演绎了黑色木材、皮革、镜子和石材等常见材料的应用方式。弧线造型和蛇纹皮革营造出高雅的氛围。咖啡馆内色彩丰富，以土黄色和深浅不一的紫色为主。镜面瓷砖和玻璃板令空间内的灯光、视野和反射度大大增强。

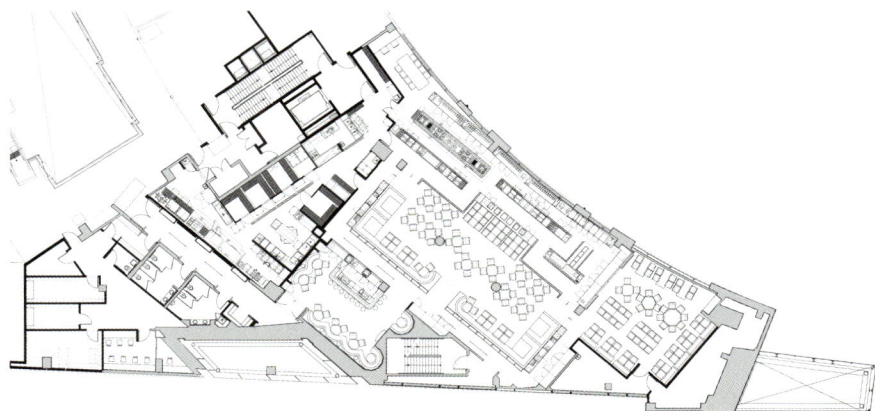

Café

USA

New York

2006

Photo: Rockwell Group

Rockwell Group

Restaurant

USA

Chicago

2008

Photo: Studio Arthur Casas

Arthur Casas

C-House

The designers did not want to create a very formal environment; just to the opposite, they wanted to transport the costumes of this restaurant to a seaside environment for a moment, without leaving aside the architecture elements/decoration of a cosmopolitan city such as Chicago. The designers' style is contemporary but they try to use material more familiar to the human being, material that makes us feel more comfortable and even protected. As for the colour scheme in restaurants, they prefer always to use yellowish hues / earth colours to stimulate hunger. The colours are a mix of reds, browns, beige oranges that reminds rustic old hand-made fabrics.

C餐之屋

设计师不想创造出一种非常正式的环境；恰恰相反，他们想要把这家餐厅搬到海边的环境中，使其具有海滨的外表，同时又不忘建筑元素和装饰元素，不负其所处之地——芝加哥这样的国际都市。设计师的风格很现代，但同时又用了很人性化的材料，使我们感到舒适并且受到保护。至于餐厅中的色彩设计，设计师一贯偏爱黄色系暖色，这样能促进食欲。颜色有红、棕、浅橘黄，这种颜色混合使人想起具有乡村特色的手工制作的材料。

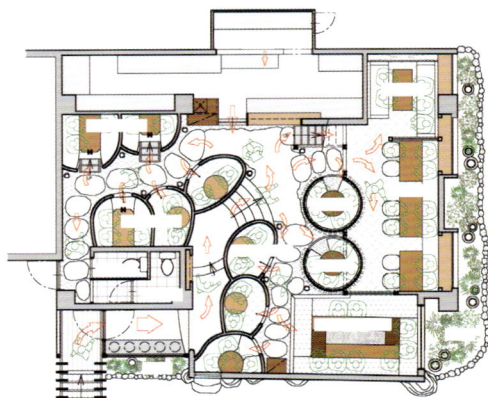

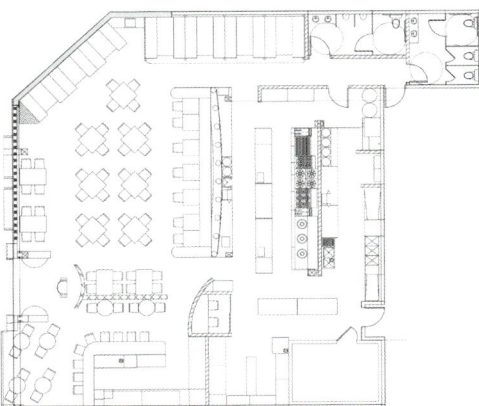

Restaurant

USA

Olympia

2006

Photo: Mahdi Montgomery

Catch Design Studio

Racha

The golden tree represents the king and gold highlights everywhere expand this concept. The focal point of the restaurant is a series of pillars, accented in gold, which support the giant golden bells. Gold leaves hang down to represent the children of the king, the Thai people. The raindrop textured glass booths, framed in the traditional Thai style, give privacy and the feeling of a Thai house. The high ceiling is punctuated with five gigantic, glowing, three-tier, golden lamps symbolising the royal canopy at the Thai palace. The façade is flanked by two stylised elephants, the national symbol of Thailand, carved from sandstone by Thai artists.

帝王岛

金色，彰显皇家风范。这里，用黄金树来代表帝王，低垂的树叶象征着帝王的子民。餐厅中由金色梁柱支撑的编钟是整个空间的焦点所在，大气磅礴，别具一格。采自泰国传统风格的雨滴质感玻璃亭，营造了泰国住宅式的私密空间。天花板上悬挂着的五个三层金色吊灯象征着泰国皇家华盖，富丽堂皇。大厅入口设置的两尊大象雕塑，由泰国艺术家利用砂岩精心雕刻而成，作为整个泰国民族的象征。

Restaurant

USA

New York

2009

Photo: ADO

Antonio Di Oronzo (Bluarch Architecture + Interiors)

Juliet

The conceptual framework behind the design of Juliet is based on the symbols and the tales of "One Thousand and One Nights" told by the legendary Persian queen Scheherazade. This restaurant is a shimmering space of gold cladding materials and lacquered furnishings. A "flying carpet" of gold, mirrored tiles is laid over the entire main room and folds over the walls and the bar. The space vibrates with the mosaic mirror, and the gloss black laser-cut ribs lining the walls describe a warping, organic profile... just as Scheherazade's fluid tales the ribs offer a shifting narrative.

朱丽叶餐厅

餐厅的设计理念源于著名神话故事《一千零一夜》，金色的覆层材料以及喷漆装饰元素打造了一个金光闪闪的世界。主餐厅内，金色瓷砖作为主要元素装饰在墙面和吧台上。马赛克风格的镜子以及墙壁上抛光黑色激光切割图案更是别具特色，定制的餐桌上漆着深深的蓝色，活泼而极尽奢华。

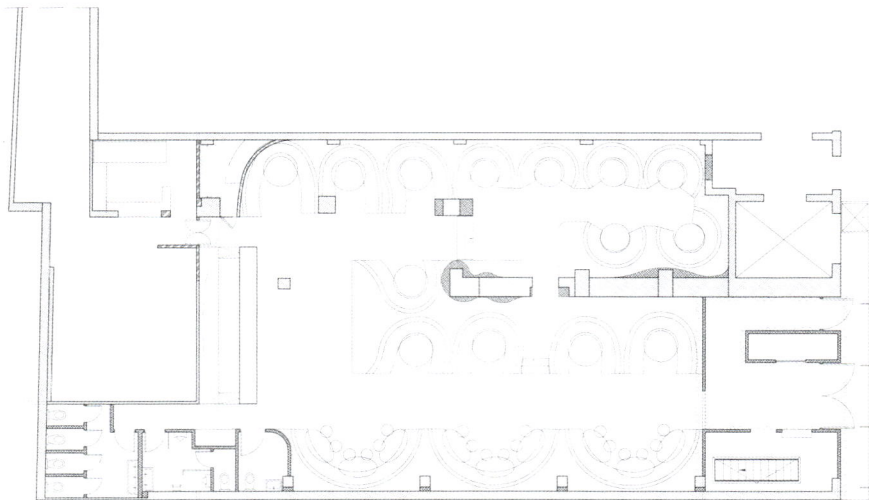

Hannah's Family Bistro

The restaurant was separated into various experiences. The marble Sushi Bar with its water feature is the focal point upon entry into Hannah's. The lounge is anchored by a circular reflective glass-topped bar with embedded metallic sheers. The dining room is divided in half by booths that are shrouded with sheer draperies hanging from wooden trusses above; gently-moving wicker ceiling fans and brown Venetian plaster are all features designed to create a sense of tropical home for the diners. The rear dining room is flanked by large scale Asian influenced artwork and noodle artifacts.

汉娜之家餐厅

餐厅被分为不同的区域，每一部分带来不同的体验。寿司吧紧邻入口设置，构成了空间的焦点；大厅内摆放着半圆形的玻璃面吧台，格外引人注目；就餐区一分为二，布帘从木梁上悬垂下来用作隔断；藤条天花板及褐色灰泥材质营造了热带家居氛围，让食客们备感亲切。

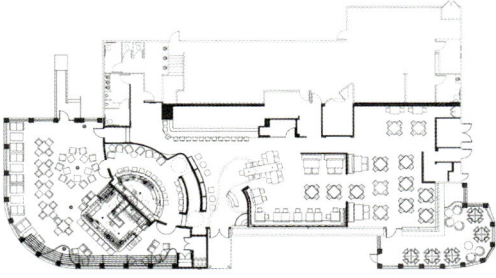

Restaurant

USA

Nevada

2006

Photo: Darius Kuzmickas

Darius Kuzmickas

Restaurant

USA

Culver City

2008

Photo: Alen Lin

Preen Inc.

Akasha

The objective of Akasha is to provide a great place to eat that mainlines healthy and ecologically responsible options for the diner, such that, the demand for such choices grows and imparts a culture of health to Los Angeles. The designers used all local and natural materials to create a new sustainable identity that is unique to this project. Although not formally submitted, the project was designed to the LEED Silver Standard. The designers maximised the industrial aspects of the space, the steel girders, wood ceiling joists and a historically fashioned board-form concrete wall, and added in comfortable, familiar materials that can improve with wear, like limestone, copper, hand-forged steel, reclaimed wood and subway tile. Six-foot mobile chandeliers use low energy to light the space.

阿卡莎餐厅

阿卡莎餐厅旨在提供一个完美的餐饮场所，以保证顾客饮食健康，提供多种环保进餐选择为目标。在这种目标的指引下，阿卡莎向洛杉矶传播着一种健康的饮食文化。为实现该项目的可持续发展，设计全部采用当地天然原材料。尽管未正式申请，该项目设计完全满足"能源与环境设计认证银奖标准"。设计师最大限度地利用了空间的工业价值——钢制梁木、木制搁栅还有古典怀旧的板型石灰墙，再加上像石灰岩、手工铸造的钢材、再生木和地铁瓦等经久耐用的原材料。

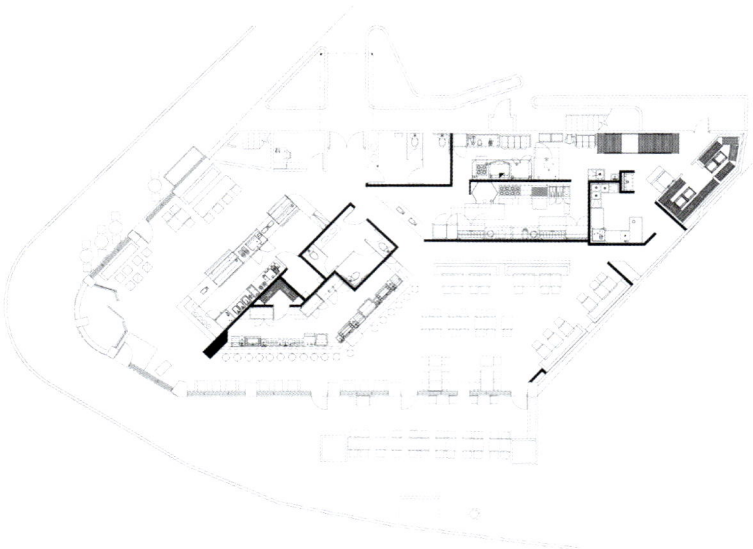

Hatfield's

Hatfield's is located in the former location of Citrus, the 1980's restaurant which launched Chef Michel Richard's career and was designed by Architect, Bernard Zimmerman. The building maintains two distinct street facing facades, articulated as superficial, punctuated planes to the building mass behind them. For Hatfield's, the designers sought to add to this 1980's expression with a layer of landscaping, specifically, using growing vines as an additional layer of depth to the planar facade. As with all living things, they will need some time to develop.

哈特菲尔德餐厅

哈特菲尔德餐厅的前身是由米歇尔·理查德厨师发起、建筑师伯纳德·齐默曼在20世纪80年代设计的餐厅——橘子餐厅（Citrus）。建筑的两个立面都面向大街，两个立面作为表层中断面与后面的建筑群相连。设计师为了使这个80年代的建筑呈现出别样的风景，让葡萄藤爬满了建筑的两个正立面。这些有生命力的装饰，随着时间的推移会愈加茂盛。

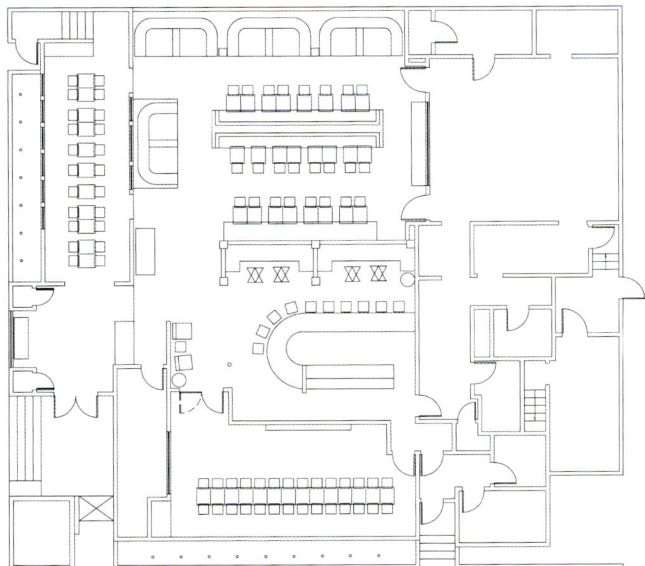

Restaurant

USA

Los Angeles

2006

Photo: Diane Cu and Todd Porter

Preen Inc

Restaurant

USA

Los Angeles

2008

Photo: Jessica Boone

Tag Front

Takami Restaurant

Beyond the host/hostess stand is the restaurant bar. With a wood-slat ceiling and a glass wine room, it features Japanese-fabric panels on the bar front with end-grain mesquite top. Behind the bar is a back-lit aluminium wall with random circular holes and a patina finish. Lounge seating is provided by a jakara root table with monkey stools. The main dining area features custom furniture with a sushi and robata bar. The sushi and robata bars were separated by a back-lit aluminium volume with a random circular hole pattern and a patina finish. Capping both ends of the bar is an interior lit natural-reed laminated glass.

塔卡米餐厅

绕过接待处就是餐厅吧台。室内木制天花板、玻璃酒柜、屋顶材料为梅斯基特年轮木材，并选用日式嵌板作为正面吧台的主要要材料。吧台后面是带背光的铝墙，墙面经过抛光，上面带有直径大小不等的圆孔。休息大厅配有雅加达产的桌子和带有猴子图案的座椅。主餐饮区以寿司吧台和罗巴塔吧台等一些特制家具为特色。两个吧台用带背光的铝墙隔开。吧台尾部罩着天然芦苇材料制成的发光玻璃。

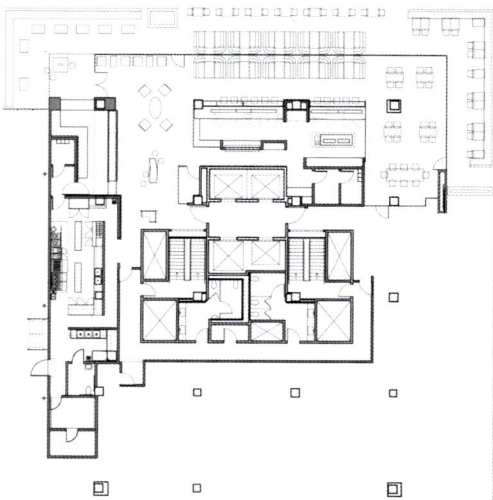

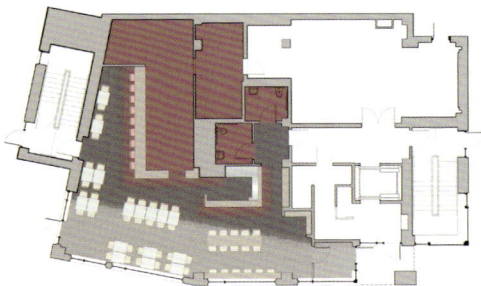

Myers + Chang

Conceived as a 'funky, indie Asian diner' by its creators, a pair of renowned Boston restaurateurs, Myers+ Chang occupies the ground floor of a new mixed-use building, located on a prominent street corner in Boston's South End neighbourhood. The owners envisioned a lively and eclectic neighbourhood eatery, serving a personal interpretation of Chinese, Thai, and Vietnamese specialties in a casual and fun environment. The restaurant space was conceptually divided in two. The front relates directly to the street with its glass facade, high ceiling, and black swing-arm light fixtures while the rear is more intimate with porcelain-tiled walls, food bar, and open kitchen. A random pattern of mosaic floor tiles that fade from light to dark and tones of metallic paint further distinguish these two areas.

"迈尔斯+常"餐厅

"迈尔斯+常"餐厅位于波斯顿南恩德地区主大街街角处一栋综合大楼的一层，由两个著名的波斯顿餐厅老板设计。为把餐厅设计成时尚、个性的亚洲餐厅，设计者构想了一个充满生气的折衷主义街区餐厅，以一种轻松愉快的用餐环境，个性地诠释了中式、泰式和越南的餐厅文化。餐厅概念地分为前后两部分，前面：玻璃面墙、高天花板、摇臂灯，与街道直接相连；后面：瓷瓦墙、食品台和开放式的厨房，烘托出更加亲密的氛围。

Photo: Bruce Martin; Michael Stavaridis

Hacin + Associates, Inc.

Restaurant

USA

San Francisco

2006

Photo: Cesar Rubio

Mr. Important Design

Mercury

The futuristic design is nothing short of stunning. Sensuality & indulgence are the theme. Curved stainless steel doors glide open as though they are on ice. When they close they complete the circular foyer, covered with burnished silver material that appears to glow from within. Everything sparkles. The opening into the dining room frames three chandeliers made of chrome wires hung with elongated teardrops of hand blown glass, shimmering like a thousand diamonds. On either side are columns covered with strips of shiny aluminium woven up like a basket. Up a flight of chrome-banister stairs in back, flanked by frosted glass panels, is a room blanketed with a mirror inlaid with bubbles. The chairs are upholstered in white leather, a dramatic contrast to the main dining room that is painted black.

水星餐厅

餐厅极为现代的设计，魅力无穷。愉悦与放任是该项目的主题。滑开弧形不锈钢门，犹如门在冰上的感觉。关上门，里面是一个圆形门厅，门内侧材料为抛光银质材料，仿佛光从门内发出。整个餐厅都熠熠生辉。餐厅上空装饰着铬线制成的三个吊灯，上面挂着手工吹制的玻璃珠子。珠子闪闪发光，仿佛千颗钻石一般。餐厅的两边是覆盖着铝制篮子条纹的圆柱。

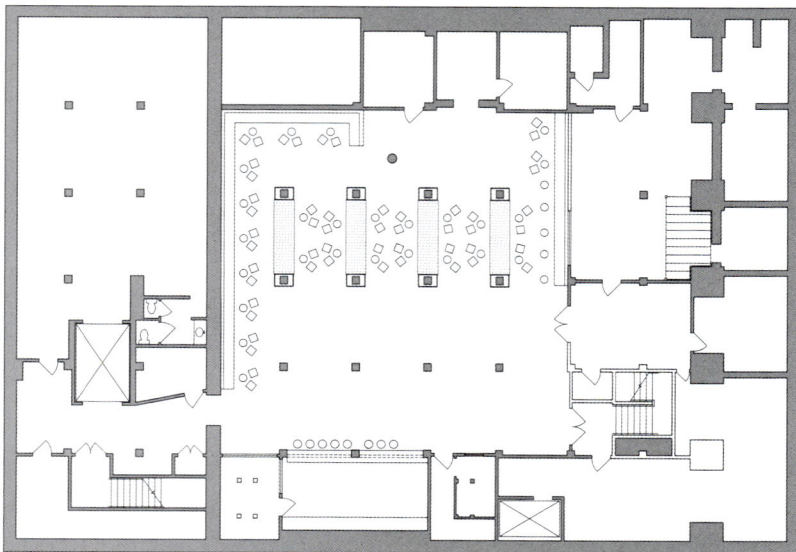

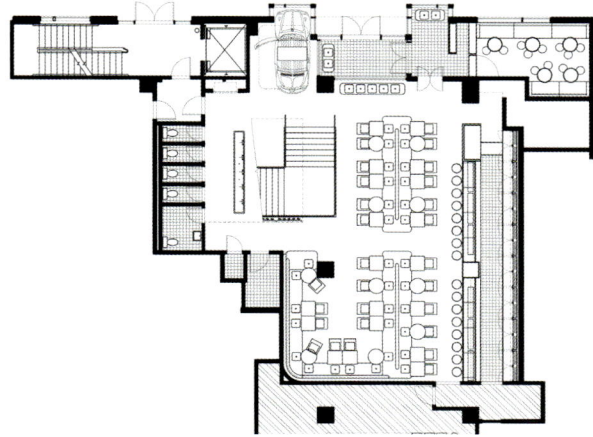

Distrito

Entering the dining space through swinging saloon doors, guests are greeted by flamingo pink walls and a bright yellow resin bar top with scorpions cast inside. Neon signs and marquee-style signs above the bar announce a few of the drinks and menu items. A working jukebox provides the soundtrack for the restaurant's first-floor lounge, and hot pink, sparkling gel bar stools offer ample seating. Distrito's prime table is a booth crafted from a green Volkswagen bug, the common taxicab in Mexico City, seating for up to four guests underneath pink acrylic-rod light fixture. Another prominent design feature is a wall constructed from more than 600 masks of the lucha libre, or free-fight professional wrestlers.

联邦区餐厅

顾客从摇摆的酒吧门进入餐厅后，首先看到的是粉红色的墙壁和嫩黄的树脂吧台台面，里面还印有蝎子图案。酒吧上面的霓虹灯广告牌和跑马灯式的广告牌公布着部分酒品与菜单。正在工作的自动点唱机为一楼大厅播放音乐，耀眼的桃粉色塑胶凳子为客人提供了充足的座位。餐厅最主要的桌子是用墨西哥最普通的出租车——大众甲壳虫出租车制造工艺制成的展示桌，可供四个人同时坐在粉色的亚克力杆灯下。另一个显著的设计就是用600个墨西哥自由摔跤用的摔跤面具制成的墙。

Restaurant

USA

Philadelphia

2008

Photo: Fanny Alliè

Crème Design

Restaurant

USA

Florida

2008

Photo: Morrisonseifertmurphy

Morrisonseifertmurphy

Restaurant in Hyatt Key West Resort & Spa

This newly-renovated Key West resort is taking the Key West experience to a new level. The hotel is conveniently located off the activity of Duval Street and provides the guest with welcoming contrast and escape from the typical Key West experience. The doors and the windows connect the interior of the restaurant with the outdoor, where the beautiful scenery can be directly seen when dining inside. The wooden floor and the tables and chairs in the restaurant seem quite simple, but would stand close observation. The white bar counter is well integrated into the wall, without extra ornament, reminiscent of the outside scenery where the sea meets the sky.

基韦斯特海厄特度假村&水疗中心餐厅

这家新近翻修的基韦斯特度假村将人们在基韦斯特的体验提升到一个新高度。这家酒店的位置很好，远离杜瓦尔大街上的喧闹。餐厅有开敞的门窗与户外联系，在餐厅中就可以看到门外优美的自然景色，餐厅中的木质地板和木质桌椅看似简单，却也非常考究，没有过多的装饰，白色的吧台和墙壁连成一片，如同户外海天相连的美景。

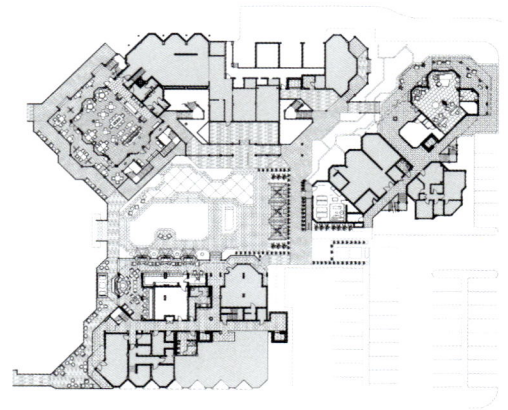

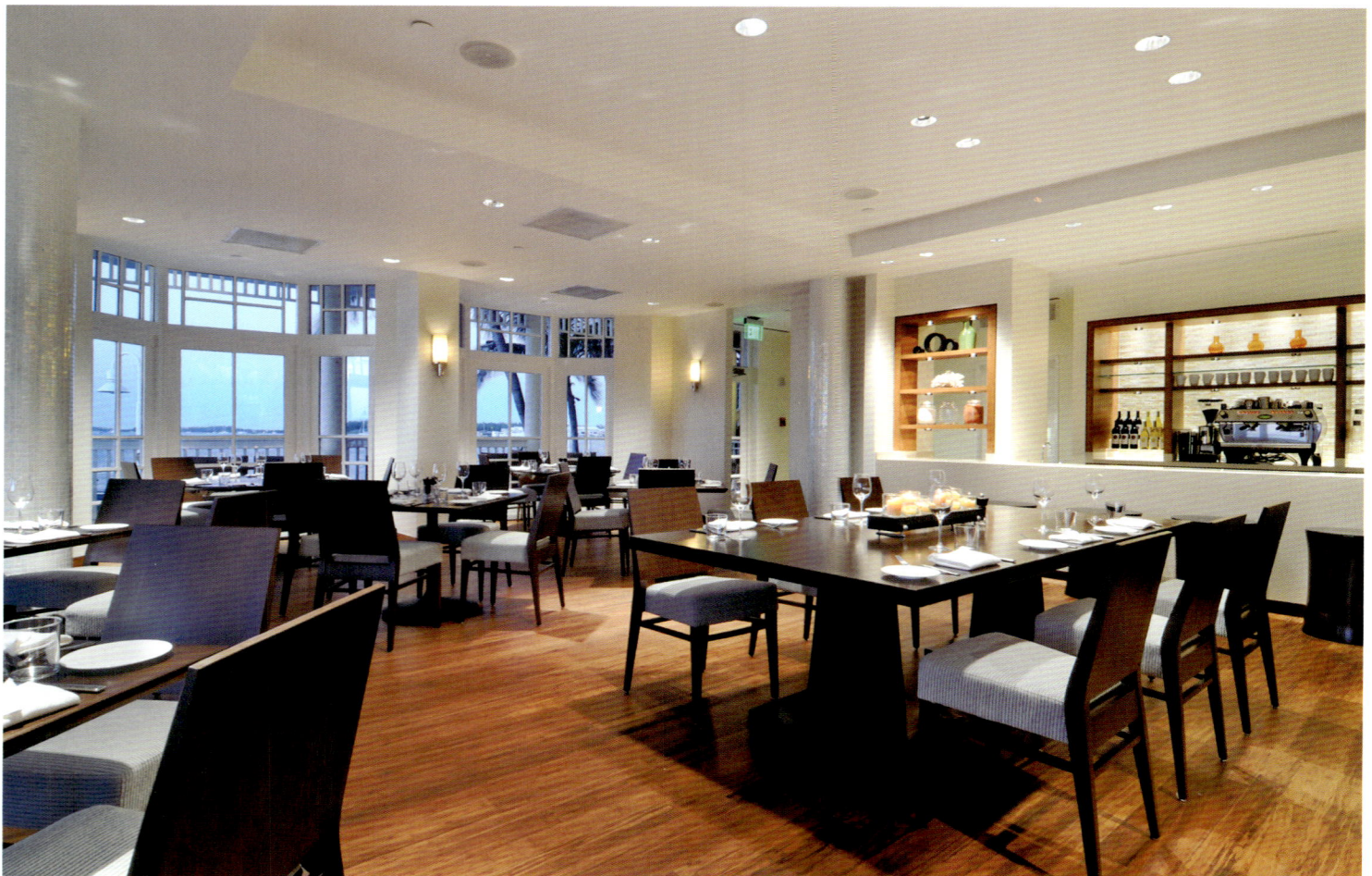

Restaurant

USA

Las Vegas

2006

Photo: GRAFT

GRAFT

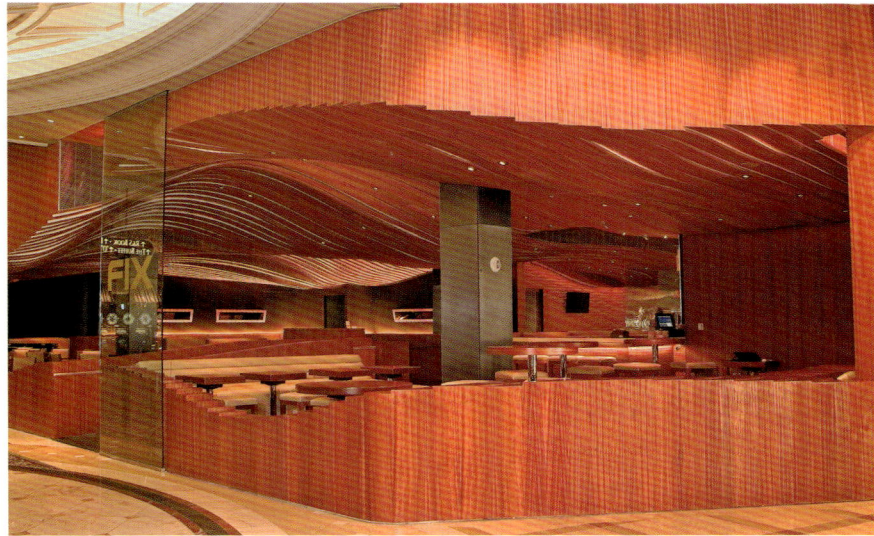

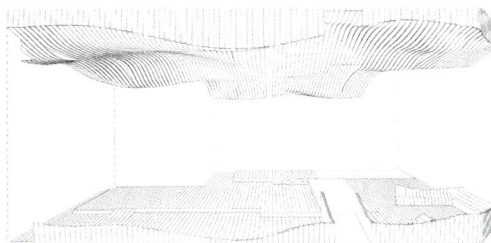

FIX Restaurant

GRAFT took it as a pleasure to be asked to develop a restaurant utilising a modern vernacular within the context of traditional European nostalgia: the Bellagio Hotel in Las Vegas. Forced to maximise revenue through seating, there was little space left to develop design within the traditional confines of the room. So the designers turned their attention to the ceiling. As a complex and mysteriously glowing topography, the ceiling plane acts as a facade for the restaurant and guides visitors into the space. The precision-milled ceiling assembly is comprised of Padouk wood slats that create a lattice of waves, alternately converging into smooth surfaces and diverging to create openness and translucency. Its undulating rhythm creates both generous and intimate spatial conditions while invisibly incorporating 'eye shaped' diffusers for air, sprinklers, lights and speakers. Not a mere ceiling, but an intelligent and beautiful interface.

FIX 餐厅

天花板作为餐厅门面，以其复杂而奇幻的外形将顾客吸引到餐厅内部。精良考究的天花板采用非洲紫檀木条拼合而成。波浪的斜纹格子渐渐过渡为光滑表面,两种图案相互交替营造出开放、半透明的空间氛围。这种波浪般的节奏既烘托出了大气且亲密的空间感,同时又遮盖了眼睛形状的空气扩散器与灭火器、灯和扬声器等装置。它不仅是一个单纯的天花板,更是一个智慧而又精美的界面。

Restaurant

USA

Novato

2009

Photo: Rien van Rijthoven

Stanley Saitowitz/Natoma Architects Inc.

Toast

Toast Restaurant serves comfort food for breakfast, lunch and dinner. This is the second branch at the new Hamilton place shopping mall in Navato, north of San Francisco. The mall is one-storey Mediterranean kitsch in a parking lot. The experience of entering Toast is like walking inside a loaf of bread… like swimming in sparkling Champaign… The yeast that creates this fizzy interior world is particle board, perforated with random shaped holes, which covers all surfaces. Walls, ceilings, furniture are all wrapped in this bubbling, bread-like material. The low canopied entry transitions to a 40' volumetric space with a hanging column suspended over the bar. This column has storage for glasses raised in the air. The bar mirrors this cubic cupola below, with seating on three sides. Behind the bar the ceiling drops to the main dining area.

吐司餐厅

吐司餐厅为顾客供应一日三餐的贴心饮食。此餐厅是三藩市以北的Navato市汉米尔顿购物中心的第二家分店。此购物中心位于停车场内一层的地中海式风格建筑。走进吐司餐厅仿佛进了一条面包里，又好像在泡沫香槟里游泳一般。实现整个空间泡沫效果的酵母原料就是刨花板，板上打满了直径不等、形状各异的小孔。墙面、天花板和家具都笼罩在这种面包一样的泡沫中。

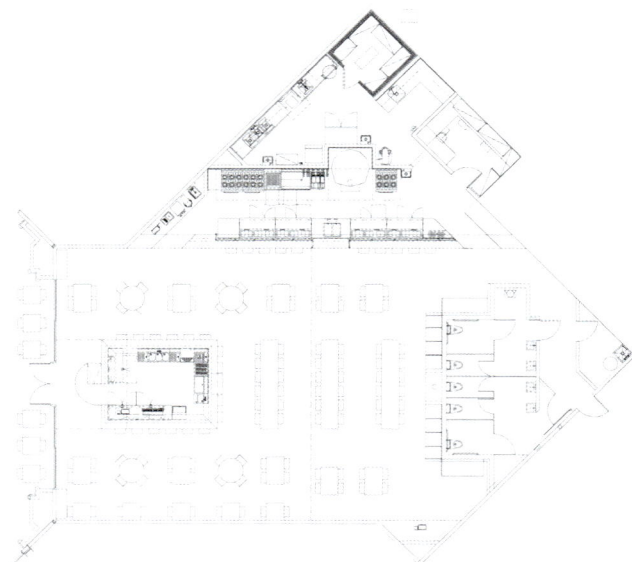

Conduit

Conduit Restaurant emerged from the found circumstances. The ground floor commercial space in a new residential building had a low ceiling and a tangled maze of plumbing, sprinkler and electrical conduits serving the residences above. To cover these pipes would have further reduced the space. Instead, even more conduits were layered over the existing to counteract and remediate the situation. At the entry is a long fireplace. Behind, table seating fills the room. A series of conduit screens in galvanised or copper colour divide the tables. On the right is an open bar made of stacked bars of conduit. Glass shelves support the bottles.

管道餐厅

管道餐厅位于一栋现有的建筑。这栋新居民楼的一层商业区，天棚很矮，各种管道、喷水器和电子管道交缠在一起为楼上的居民提供水电。如果将这些管道覆盖上，就会大大缩小室内空间。取而代之，设计师在现有的管道上装饰了更多管道，使其相互交错、互相综合。餐厅入口处是一个长壁炉。一系列的镀锌和黄铜色管道屏将餐桌分为两部分。右面是由分段管道围成的开放式吧台。

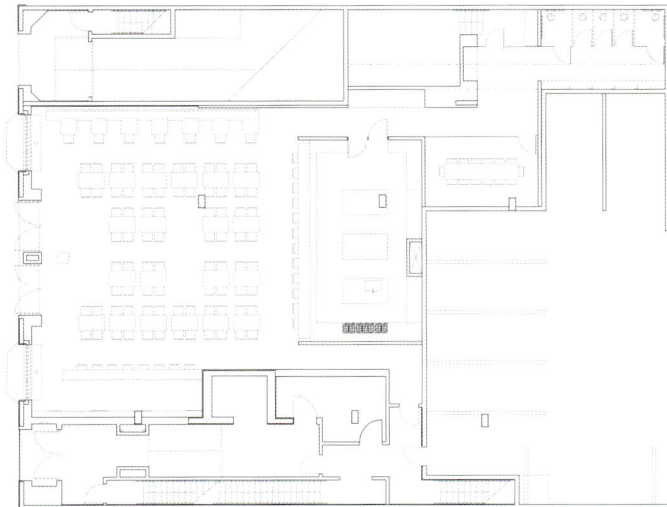

Restaurant

USA

San Francisco

2007

Photo: Rien van Rijthoven

Stanley Saitowitz/ Natoma Architects Inc.

Café

USA

Atlantic

2008

Photo: Darius Kuzmickas

WESTAR Architects

Reflections Café

The Café is a 24 hour dining environment that is made up of seven seating zones that divide the 383-seat dining room into quaint environments that create unique dining experiences for the repeat clientele. Curavalicious banquets sit within cosy rooms decorated with flowers and pearl shell pendants. Slate clad walls with sensual Pullman booths create private pockets that overlook the room. Curved booths are awash with light from arc lights while two types of chairs are placed below fifty dancing glass pendants. The backlit undulating ceiling was inspired by the waves of the adjacent ocean. At times the sensuality of the curved wood beam ceiling almost touches and cradles the guests while at other times it raises and invigorates with its unreachable height.

映像餐厅

此餐厅24小时营业，餐厅内部七大用餐区将383个座位分布在各个古雅精致的用餐环境中，为顾客打造了别致的用餐体验。流线型宴会桌摆放在花朵与珍珠垂饰装饰的温暖室内。配有愉悦感官的铂尔曼展台石板墙为顾客营造了可环视整个餐厅的私人空间。曲线型展台与弧光灯光线保持水平，两种不同样式的餐桌整齐地摆放在55个舞动的玻璃垂饰物下面。背光式波浪型屋顶的设计灵感来源于临近海洋波浪。

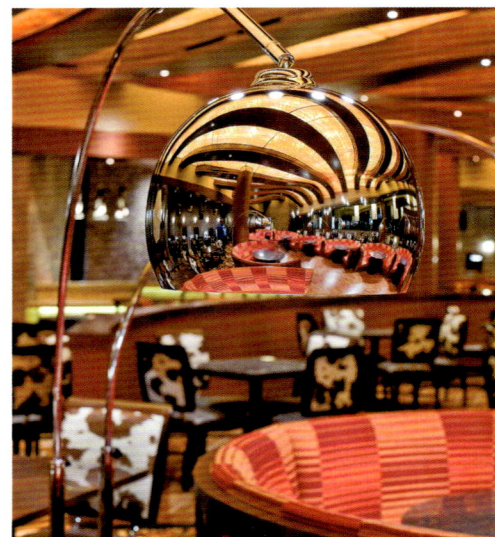

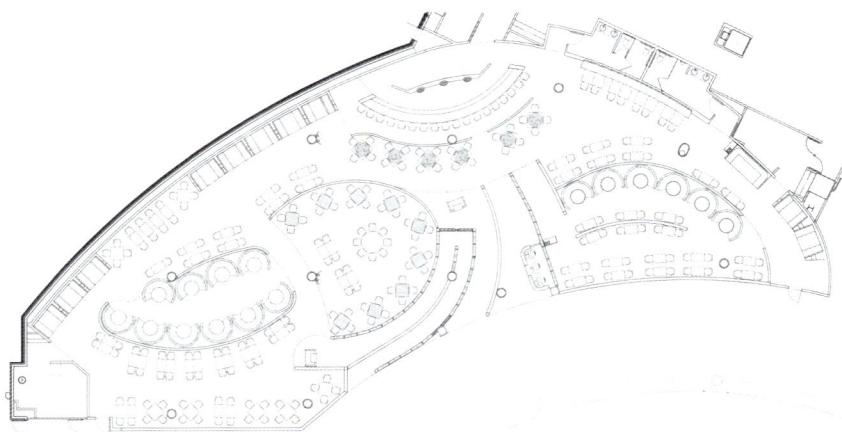

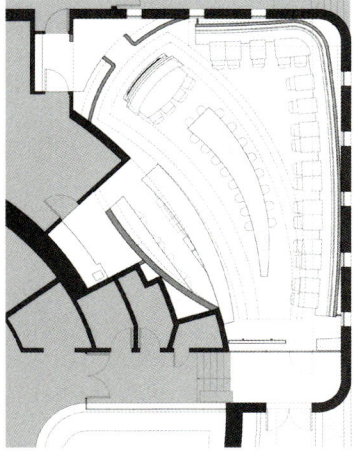

The Wright, Guggenheim Museum

Named after the Fifth Avenue Museum's architect, Frank Lloyd Wright, the 58-seat restaurant was designed by Andre Kikoski Architect and features a modern American menu by David Bouley proégé Rodolfo Contreras featuring local seasonal dishes like seared diver scallops in sea urchin sauce and slow roasted suckling pig with quince and violet mustard."It was both an incredible honour and an exhilarating challenge to work within Wright's iconic building,"said Kikoski. "Every time we visit, we see a new subtlety in it that deepens our appreciation of its sophistication. We sought to create a work that is both contemporary and complementary."

古根海姆博物馆莱特餐厅

以第五大道博物馆设计师弗兰克•劳埃德•莱特的名字命名的莱特餐厅有58个座位，由建筑师Andre Kikoski 设计。餐厅以现代美国菜为特色，由名厨大卫•博雷（David Bouley）的接班人鲁道夫•孔特雷拉斯(Rodolfo Contreras)主理，包括本地时令菜肴，如扇贝蘸海胆酱、木瓜紫罗兰芥末烤乳猪。 Kikoski认为，"在莱特的标志性建筑里面工作既是一个令人难以置信的荣誉，也是一种令人振奋的挑战。在每次访问期间，我们都能看到它新的微妙之处，这进一步加深了我们对成熟干练的理解。我们寻求建立一个工作，既有时代性又有补充性"。

Restaurant

USA

New York

2009

Photo: Peter Aaron

Andre Kikoski Architect

Photo: CCS Architecture

CCS Architecture

Lake Chalet Enlivens Lake Merritt Neighbourhood

CCS Architecture transformed the Mediterranean Revival style building into Lake Chalet Seafood Bar and Grill, a day and nighttime destination that has brought a surge of new energy to Lake Merritt. With 325 indoor and 100 outdoor seats, the restaurant visually and physically embraces the waterside setting, with its walking and jogging path, gondola rides and 'necklace of lights' dating to 1925. New architectural elements, materials and colours pay homage to the lake and the long history of boating at the site. On warm evenings, Lake Chalet's dock patio is packed with patrons enjoying cocktails with views of the lake and setting sun reflecting off the East Bay hills.

为梅里特湖地区注入生机的沙莱湖餐厅

CCS建筑事务所将地中海复古风格建筑改造成沙莱湖海鲜烧烤酒吧。沙莱湖酒吧24小时营业，为梅里特湖街区注入了生机与活力。餐厅拥有室内座椅325张，室外座椅100张，着实将整个湖边美色囊入怀中。顾客可以漫步湖边小路，享受湖上"贡多拉之旅"，欣赏始建于1925年的"项链灯"。新的建筑元素、材料与颜色的结合，表达了对梅里特湖以及当地的悠久的划船历史的无限敬意。

Photo: Kris Tamburello; New York, NY
Melissa Werner; San Francisco, CA

CCS Architecture

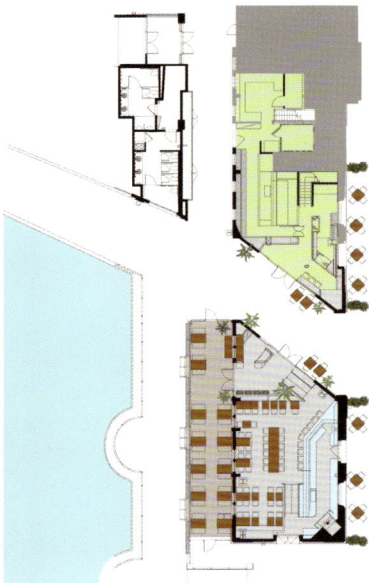

The Plant: Café Organic at Pier 3

The Plant: Café Organic occupies two historic, waterfront buildings at Pier 3, straddling what was once a railroad passage, which CCS Architecture has modified to create a full-service, 112-seat restaurant and a separate, counter-service café. The Plant is one of the 'greenest' restaurants in San Francisco—and one of the few in the country with a rooftop solar PV system for on-site, electrical energy production. CCS inserted light, delicate interiors within the existing pier warehouses, using reclaimed wood, recycled-content tiles and an eclectic mix of zinc, cold-rolled steel, and stainless steel to finish out the spaces.

咖啡树餐厅——3号码头有机餐厅

CCS建筑事务所设计的咖啡树餐厅占据3号码头两大历史性海边建筑，横跨原铁路通道，主要由两部分组成：一部分是拥有112张座椅的全方位服务餐厅，另一部分是一个独立的柜台式服务的咖啡厅。咖啡树餐厅是旧金山市最环保的餐厅之一，也是美国为数不多的屋顶带有供现场发电装置的太阳能光伏系统餐厅之一。CCS建筑事务所使用再生木、可重复利用瓷砖，并综合利用锌、冷轧钢材和不锈钢等材料装饰空间，将原有的码头货栈改成明亮而精致的室内空间。

Restaurant & Bar

USA

California

2007

Photo: Dim Balsem

UXUS

Ella Dining Room & Bar

Ella Dining Room & Bar serves 'Modern American Bistro' cuisine. The owners wanted the restaurant to become 'Sacramento's living room', an urban oasis where lawmakers and other diners can go and unwind after a long day's work. The design objective for Ella's is: create a brand that embodies the principles of 'Rustic Luxury', and that celebrates an elegant, relaxed contemporary lifestyle. 'Rustic luxury' is a synonym for purity, the essential beauty and goodness contained in simple things. It is about the pleasure and sensuality of real materials, and about the inherent comfort of a natural, effortless style. 'Rustic luxury' is not a simplistic reduction. It is the magical crystallisation of two apparent opposites, simplicity and complexity.

艾拉餐厅&酒吧

艾拉餐厅&酒吧提供"现代美式酒吧"的烹饪美食。店主想让酒吧成为所谓的"萨克拉门托的起居室"——一个供国会议员和其他进餐者一天辛苦工作后放松消遣的都市绿洲。艾拉的设计目标是创造一个能体现"乡村奢华"理念的酒吧品牌,并以此来颂扬优雅、轻松的当代生活方式。"乡村奢华"是纯真的同义词,是一切简单事物所特有的美丽与善良。"乡村奢华"并非只是简单的回归自然,是简单与复杂神奇的结晶。

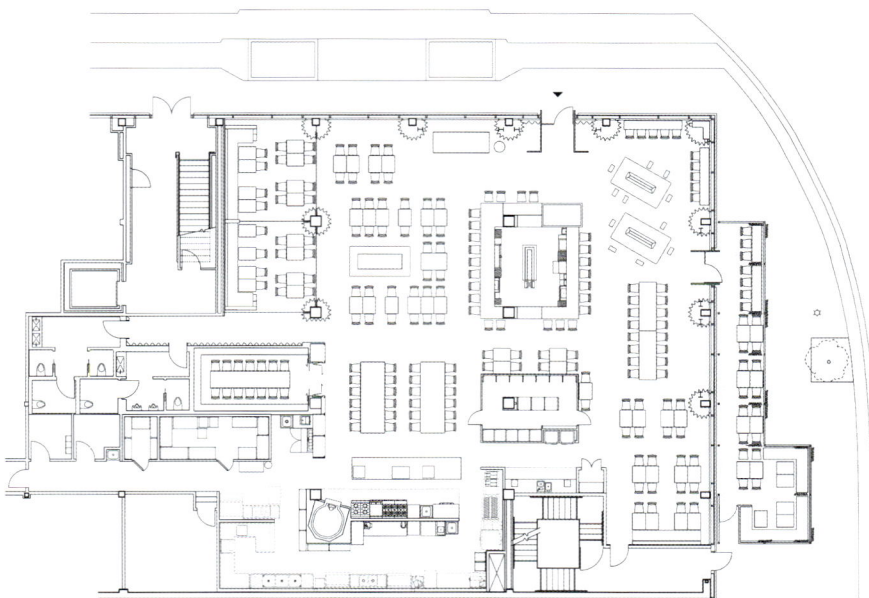

Lucky Fish

A contemporary design was achieved using an interesting palette of materials and textures. Traditional materials were used with a nod to Japanese style; however, they were used in non-traditional ways. The entry area is marked by floor-to-ceiling stone panels that lead up to a hanging sculptural fixture. The sculptural fixture fills the entire ceiling of the entry and patio and features custom-designed wood planks. The wood planks were created with detailing and joinery that hint to an ancient culture, but were designed and used in a more contemporary manner. The interior conveyor belt running nearly the depth of the space is flanked by a custom-designed counter that uses resin-covered bleached river stone. The counter has the texture and depth of a bed of stone, but is completely smooth.

幸运鱼餐厅

运用有趣的材料与图纸调和技术完成了一项极富现代感的设计。设计以另类的方式将传统材料与日式风格装饰完美结合。餐厅入口处的石头嵌板从地面一直铺到天花板上悬挂的雕塑工艺品。入口与露台的天花板上都饰有雕刻工艺品，并配有特制的木支架。木制支架上的点缀物和细木工制品象征着对古代文化的颂扬，但却以现代制造工艺设计。遍布整个空间的传送带旁是经漂白的河石材料制成的特制柜台，石头表面有树脂包层。柜台表面虽为石头质地却极为光滑。

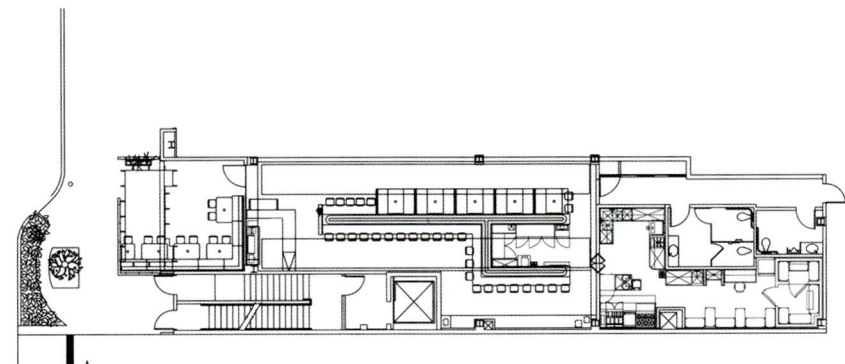

Restaurant

USA

San Francisco

2008

Photo: Eric Axene

Tag Front

Restaurant

USA

California

2007

Photo: Randall Michaelson

Akar Studios

Solana Beach

A former bicycle shop situated in an unassuming commercial strip along the sunny Pacific Coast Highway, the space required a major transformation to emerge in its present incarnation – a dark, seductive environment billed as 'one of the sexiest dining rooms in San Diego'. This exploration of contrasts was a fundamental part of the design process. The dining room is sumptuously appointed in rich zebrawood veneer with tall, opulent booths, a deep chocolate brown ceiling, dark ebony wood floors and muted accents in mocha and taupe. Custom designed fabric lamps are scattered throughout the dining room, their pleats bunched together and wrapped up with an informal, almost organic elegance that plays against the polished refinement of the rest of the space.

索拉纳海滩餐厅

餐厅原是一家自行车店，坐落于阳光普照的太平洋海岸公路沿线一条不起眼的商业街上。因此，整个空间需要彻底改造才能实现到现代餐厅的完美过渡。餐厅色调昏暗、情调诱人，被称为"圣地亚哥最性感的餐厅之一"。这种改造前后的对比设计是整个设计过程的基础。餐厅用彰显奢华的斑马纹木饰面装饰，装配豪华展台、深巧克力色的天花板、黑檀木地板、摩卡兼深灰色的静音板等。特制的布灯挂满了整个餐厅，灯上的褶皱花纹规整有机地缠绕在一起，显得格外优雅。

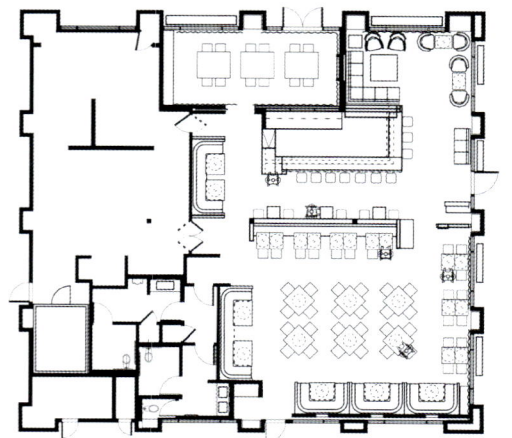

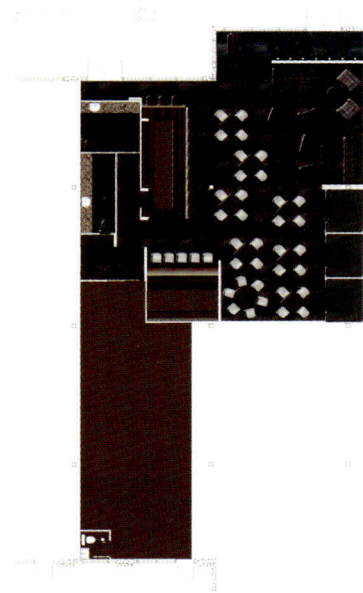

WA Restaurant

This 3200-square-foot restaurant in Orlando Florida has proven unlike any other concept that this area has been exposed to over the course of its development. Completed in September of 2009 the fusion of Japanese and French cuisine was complimented by a serene and open design integrating all aspects of WA restaurant. There is little separation between the dining area, sake bar, lounge and sushi bar allowing for a seamless flow of aesthetic and energy. Similar to the established design style of the Asian regions, the space is expressive without being ostentatious. The project exemplifies the viability of sustainable design within a corporate setting through the use of environmentally conscience materials.

WA 餐厅

这家位于弗罗里达•奥兰多的餐厅面积达3200平方英尺。奥兰多多年的发展史证明，以往任何一家餐厅都不能与这家餐厅的设计理念相媲美。该餐厅实现了日式与法式菜肴的完美结合，遍布餐厅的庄重与开放式设计更是锦上添花。用餐区、清吧、休息厅与寿司吧间分界不是很明显，因此自然流畅地展示出了设计美感与内涵。和亚洲地区特有的餐厅设计风格一样，空间氛围自然展露却不浮夸。

Restaurant

USA

Florida

2009

Photo: Thom LaRussa

WDG Interior Architecture

Restaurant

USA

Tigard, Oregon

2007

Photo: Steve Cridland

Skylab Architect

Sinju Restanrant

Sinju restaurant is a series of spaces connected by a long, narrow passage. From the exterior entrance, a red path morphs into a long red hallway, then up the wall, becoming fireplace and chimney, and continues into a glowing red ceiling band. In the same way Nori is cut into thin strips to band together sushi, this red band joins the interior spaces together. Like a moth to a flame, the red pathway guides the visitor from outside-in pointing to the heart of the lounge facing the lacquer-red-tiled fireplace. The 2700-square-foot restaurant takes cues from spatial values of traditional Japanese architecture to maximise density with contrasted open spaces. Seating 123, the restaurant appears more like an outdoor courtyard than an interior space. The backlit ceiling is a web of convex and concave pyramids that feel more like cloud cover than a rigid ceiling.

珍珠餐厅

一条细长的走廊连接着餐厅的各个部分。餐厅入口，一条红色小路渐渐变为长长的红色门厅，然后沿着墙壁变成壁炉和烟囱，最后演变成闪耀的红色的天花板。就像把紫菜切成条做寿司一样，这条红带子将整个室内空间连接在一起。犹如飞蛾扑火一般，红色小路将顾客从餐厅入口带到休息大厅的正中央，对面是红漆壁炉。2700平方英尺的餐厅设计来源于传统的日式建筑的空间价值理念——用开放空间的相互对比最大化地利用空间。餐厅拥有123个座位，给人感觉不是室内餐厅，更像户外小院。

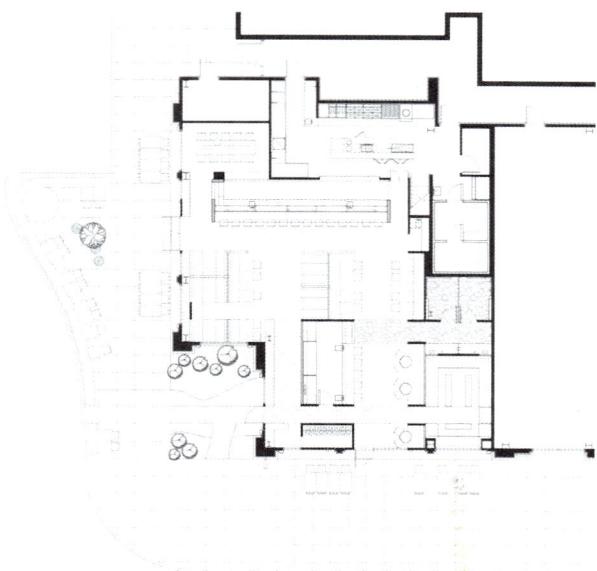

Restaurant

USA

Santa Monica

2007

Photo: Keith Kaplan

Franklin Studios, Inc.

Tengu

Located across the street from the Santa Monica Pier, this stretch of Ocean Avenue is one of the most popular tourist destinations in the Los Angeles area. That being said, one of the strongest forces shaping the design of the restaurant was the need to connect the restaurant with the stream of pedestrians walking by on the sidewalk. To this end, the central core of the space, which houses the kitchen and bathrooms, was thought of as a giant lantern, with a dark wood lattice floating in front of an up-lit wall. The graphic nature of this lattice pattern was bold enough to be easily read from passing cars along the boulevard. Another gesture to connect the restaurant with the street was to open up both the sushi bar and the liquor bar to the patio.

天狗餐厅

坐落于圣塔尼卡港的街对面，天狗餐厅作为海洋大街（Ocean Avenue）的延伸，是洛杉矶地区最重要的旅游目的地。据说，以过往行人为消费主体是餐厅设计最强大的动力之一。为了实现这个目标，设计师将涵盖厨房与浴室的餐厅中心区设计成一个大灯笼。将餐厅与街道连接在一起的另外一个设计就是将寿司吧与酒水吧向露台开放。

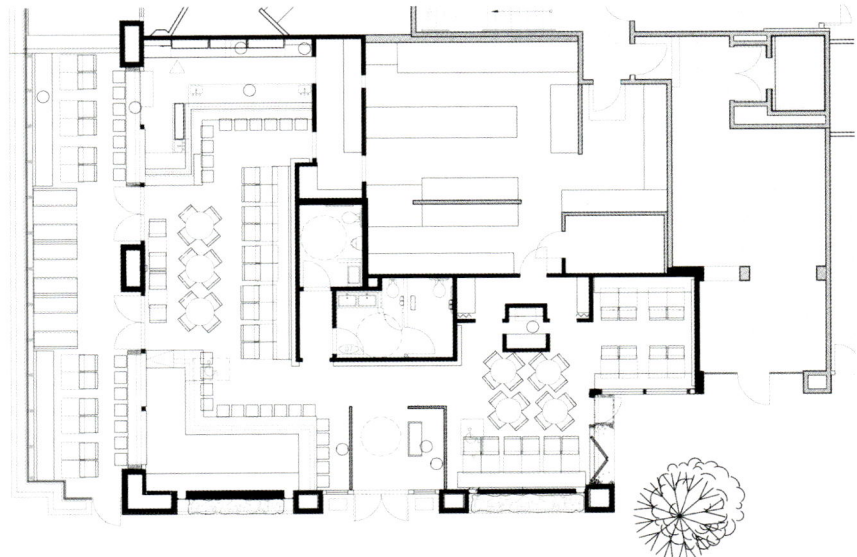

Restaurant & Lounge

USA

Santa Monica

2007

Photo: Kelly Gleason

Franklin Studios, Inc.

Abode Restaurant and Lounge

Created by Kelly Gleason and Anastasia Israel, Abode is a small, sophisticated gem tucked away in an intimate courtyard off the Ocean Avenue. The quiet, private nature of the space, combined with the stately fourteen foot high ceilings led Franklin Studios to the idea of a refined oasis, or garden, hidden away from the tourist traffic along Santa Monica's main drag. The austere, abstract geometries of high-sixties design encase the dining room and lounge-bar. The concept of garden is brought out abstractly by regularising organic forms into ornate patterns. The colour pallet is rich and earthy, and the stone slabs provide a natural source of wildly varied texture.

宅院餐厅

安静而隐蔽的空间加上庄严地14英尺高的天花板，使富兰克林设计事务所想到了圣莫尼卡主商业区外的都市绿洲、花园等远离城市喧嚣的美丽景色。60英尺高简朴而抽象的室内餐厅包括餐厅与休息吧两部分。通过将材料的有机外形设计成华美图案，设计师演绎了其抽象的花园设计理念。绿色的草垫散发着自然浑厚的泥土气息，石板也有着极为特别的天然质地。

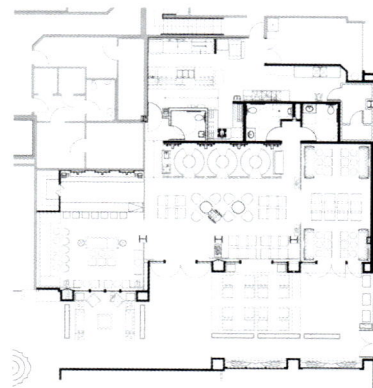

Restaurant

USA

San Francisco

2009

Photo: CCS Architecure

CCS Architecure

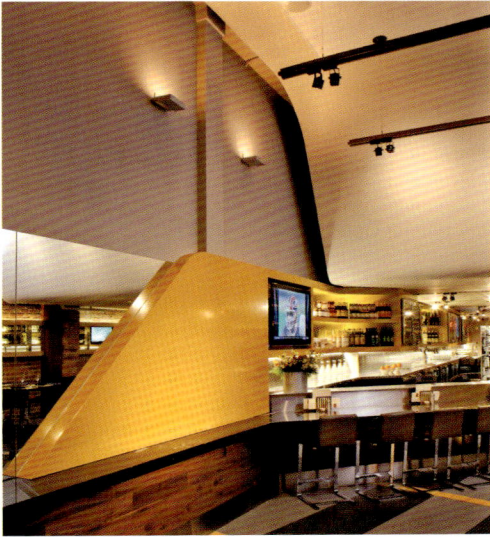

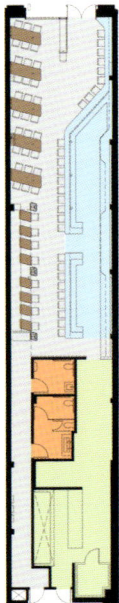

Barbacco Eno Trattoria, San Francisco

Barbacco, just opened in San Francisco's Financial District, is the 'sexy little sister' of Perbacco, the acclaimed 2006 Italian restaurant on California Street. Owners Umberto Gibin and chef Staffan Terje saw the need for a more casual option to serve the neighbourhood, and when a space next door became available, they engaged CCS Architecture to develop a trattoria that would convert to an enoteca or wine bar at night. The result is Barbacco Eno Trattoria. Barbacco, with 66 seats, occupies a long, narrow room within the 1912 Hind Building. Inspired by traditional wine bars in Milan and Rome, the space is sleek, urbane and welcoming. A dramatic shaped ceiling, evoking the lines of a sports car, links the higher ceiling at the front of the restaurant with the lower ceiling towards the rear.

旧金山巴巴客伊诺餐厅

位于旧金山金融区刚刚开业的巴巴客餐厅是2006年加利福尼亚大街备受欢迎Perbacco意式餐厅的姊妹餐厅。餐厅老板Umberto Gibin 和主厨 Staffan Terje 一直希望拥有一个适合街区消费人群的更加休闲的餐厅，因此旁边有空地后，他们便邀请CCS建筑事务所为其设计一个可在午夜变成酒吧的餐厅，巴巴客伊诺餐厅由此而生。巴巴客餐厅拥有66张座椅，位于1912年建成的欣德大厦的一间细长的房间内。受米兰与罗马传统酒吧文化的启发，餐厅氛围井然有序、温文尔雅而又给人一种宾至如归的感觉。

Restaurant & Lounge

USA

Los Angeles

2007

Photo: Eric Axene

Tag Front

Blue Velvet Restaurant and Lounge

Tag Front was commissioned to design Blue Velvet, a contemporary 7,000-square-foot restaurant and lounge in an up-and-coming area of downtown. Located at the foot of a hotel that was recently converted living spaces, it was to be designed to easily mesh with the existing building, yet maintain a separate identity. Blue Velvet consists of three main areas – the main dining area, lounge, private dining and the cloud room. The exterior of the space was designed to create a defined space for the restaurant, but still feel part of the overall building. Horizontal members made from apitong wood make up the awning, creating a separation between the restaurant and the building. The wall leading up the entry is made up of bead-blasted anodised aluminium panels of varying blue hues.

蓝丝绒餐厅与休息厅

泰格•弗朗特负责设计位于繁华市中心的蓝丝绒餐厅。该餐厅是一家面积达7000平方英尺的现代餐厅。坐落于刚刚翻修的酒店底层，餐厅设计既要与整个酒店风格保持一致又要拥有它独立的个性。蓝丝绒餐厅主要包括三大区域：主用餐区、休息大厅、私人用餐区和彩云阁。外部空间的设计彰显餐厅的精致独特，而又与整栋建筑浑然一体。由菲律宾阿匹栋木材制成的天棚将餐厅与整个酒店隔离开来。餐厅入口墙材料为不同蓝色调的喷丸式阳极氧化铝质嵌板。餐厅的玻璃墙外就是游泳池，透过玻璃墙可以清楚地看到游泳池的全景。

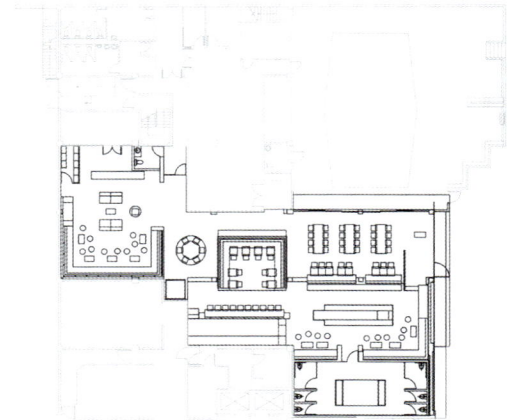

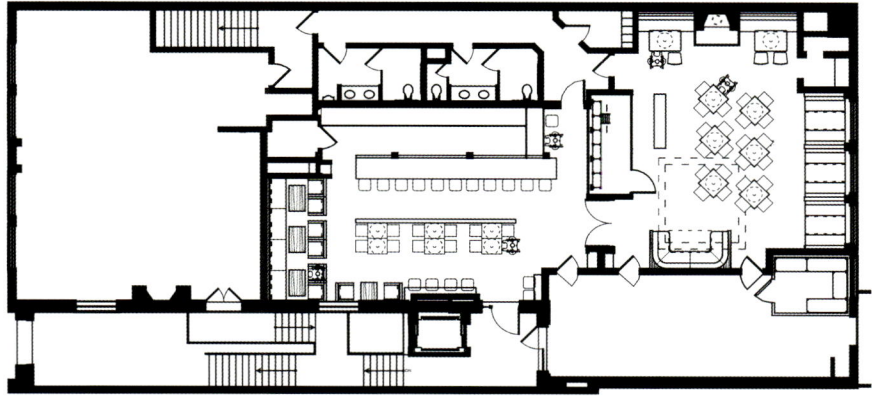

Vertical Wine Bistro, Pasadena

Occupying the second floor of a historic building in Old Town Pasadena, the space, previously a cigar lounge owned by actor George Hamilton, underwent a radical transformation to become this exclusive Wine Bistro and restaurant serving high-street and sophisticated food. The focal point of the entire venue is the long bar that specialises in exclusive wines served by the glass. The bar is accented with raw steel and frosted glass wine racks that hover above the front counter. The back of the bar wall houses walnut wood boxes for displaying the large variety of wines available on the premises. The main dining space is separated from the wine bar by a large glass walled wine storage area that is climatically controlled to 55 degrees at all times.

垂直酒馆

垂直酒馆位于帕萨迪纳老城一栋历史性建筑的二层，前身是演员乔治·汉米尔顿开的雪茄吧。经彻底整修后，变成了现在的个性酒吧与餐厅。餐厅提供从街区小吃到高级食品等一系列餐饮服务。整个项目的核心是长长的酒吧，专门销售用玻璃杯装的特制酒。酒吧最突出的设计就是盘旋在前台上方的粗钢与磨砂玻璃材料制成的酒架。酒吧的后墙上挂着核桃木盒子，上面摆放着酒吧里的各种酒。主餐区与酒吧中间是一个大玻璃墙围成的酒品存储区，里面的温度永远控制在55摄氏度左右。

Restaurant & Bar

USA

Pasadena

2007

Photo: Randall Michaelson

AkarStudios: Sat Garg, design principal, Keiko Okada and Anabel Oreiro, designers, Matt Lutz, Graphic designer

Restaurant & Bar

USA

San Diego

2009

Photo: Ramona D' Viola

AkarStudios: Sat Garg, design principal, Matt Lutz, Chris Jones designers

Crescent Heights, San Diego

A ground-level space with a contemporary design, the venue comprises of a restaurant, lounge, bar and an outdoor patio. The 170-cover location has been designed as a series of distinct yet interacting spaces to accommodate different sized groups and private events. One of the defining design aspects of the space happens to be the ceiling itself, comprising of a series of undulating ceiling planes that covers the entire dining and bar space. The interior wall planes have been rendered in hues of brown and taupe colours, offset by the polished white marble-tabletops and chocolate brown leather-upholstered chairs. Whereas walnut wood flooring defines the main seating area, polished concrete in the bar and circulation spaces and lush carpeting highlight the lounge space.

圣地亚哥新月餐厅

拥有当代设计理念的开放式餐厅—圣地亚哥新月餐厅由三部分组成：餐厅区、休息区、酒吧和室外露台。餐厅面积达170平方米，内部各个区域既独具特色而又相互呼应，既能满足个人用餐也能接待其他各种不同用餐群体。餐厅最为个性的设计当属覆盖整个餐厅和酒吧的曲线天花板。内部墙体主色调为棕色和灰褐色，与抛光白色大理石桌面和巧克力色毛皮座椅铺垫形成鲜明对比。主用餐区选用核桃木地板，酒吧和走廊铺设磨光混凝土地板，而休息区则以青色地毯为特色。

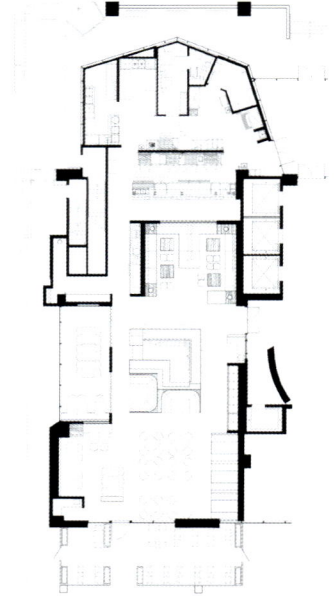

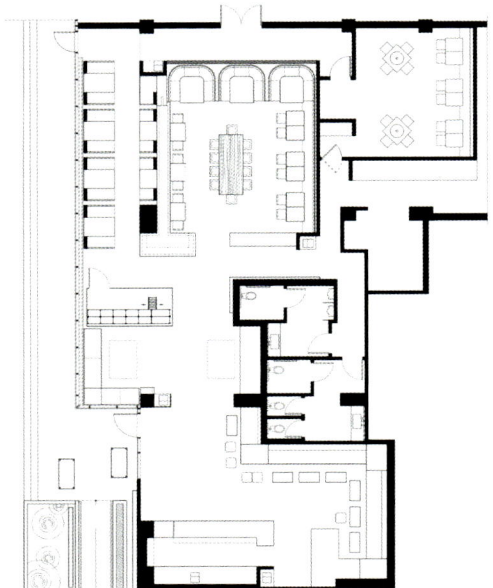

Restaurant & Bar

USA

Beverly Hills

2008

Photo: Randall Michaelson and Derek Rath

AkarStudios: Sat Garg, design principal, Matt Lutz, Chris Jones, designers, Matt Lutz, graphic designer

Tanzore, Beverly Hills

The new Indian venue is located in a large 1970's modern building on the Restaurant Row of Beverly Hills. The elongated space contains a series of interconnected areas integrating a bar, lounge, restaurant and a private room that carry their own distinct identities. From the very outset, colour and texture play an important part in the overall appearance of the interior spaces. The walls and ceilings have been swathed in bold and colourful Indian hues to create a befitting backdrop for the collection of Indian artifacts that have been selectively positioned within the venue. Offsetting the bar and lounge areas are a series of dining spaces that were separated by an intervening water channel. Lighting has been integrated as a critical element of the overall design, both as a visual object and also as a provider of luminance, warmth and mood.

比佛利山TANZORE餐厅

这家新印度餐厅位于比佛利山餐馆街上一栋20世纪70年代的现代大楼里。细长的空间包含几个相互连接的区域——酒吧、休息厅、餐厅和一间包间,每一部分都有各自独特的用途。颜色与材料质地对发挥整个室内设计的外观效应发挥了重要作用。墙面与天花板都用重彩色调的材料包裹,与餐厅内各种印度工艺品的布置背景相映衬。灯光设计在整个设计中发挥着至关重要的作用,它不仅仅提供照明,而且为餐厅增添了浓浓暖意与屡屡情怀。

Restaurant

USA

Hollywood

2009

Photo: Farhad Samari and Jessica Boone

Tag Front

BOA

In the fourth iteration of BOA Steakhouse, Tag Front expanded on the themes evident in the design of the original restaurant located nearby. The design team used elements prominent in the original design as accents to maintain a continuity of design vision; however, the newest space has its own feel and provides the ambiance necessary as the flagship location. Located on the street level of a recently renovated high-rise, the challenges facing Tag Front included creating a unique restaurant that maintained its own identity while complementing the existing building. The interleaving of unique textures, colours and themes – coupled with a stimulating use of space – resonates throughout the restaurant. The main areas of the space include a dining room, a private dining room, an extensive patio dining area, a bar and lounge area.

BOA牛排餐厅

Tag Front 建筑事务所在BOA牛排餐厅第四家连锁店的设计中沿袭了附近原餐厅的设计主题。设计团队继续使用原设计的突出元素来发挥其设计理念，而新空间又有自己的特色，极具旗舰餐厅的氛围。与新改建的高层建筑坐落于同一大街，Tag Front 需要设计一个既彰显身份又与周围建筑搭配的别致餐厅。特殊材料、颜色和主题的搭配使用与室内空间的创新利用贯穿了餐厅设计的始终。餐厅的主要区域包括一个主餐厅、一个私人包间、一个广阔的就餐露台、一个吧台和休息区。

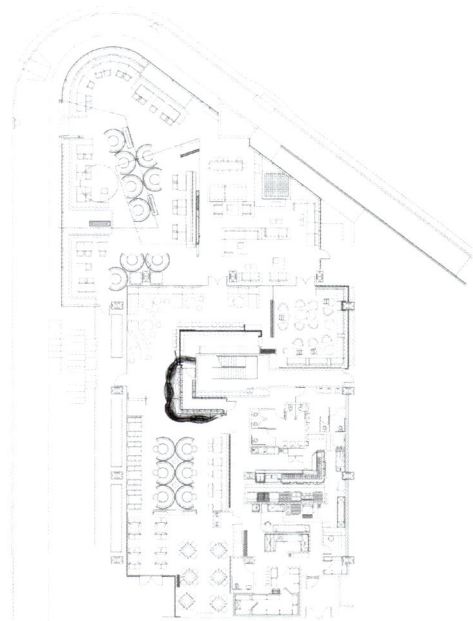

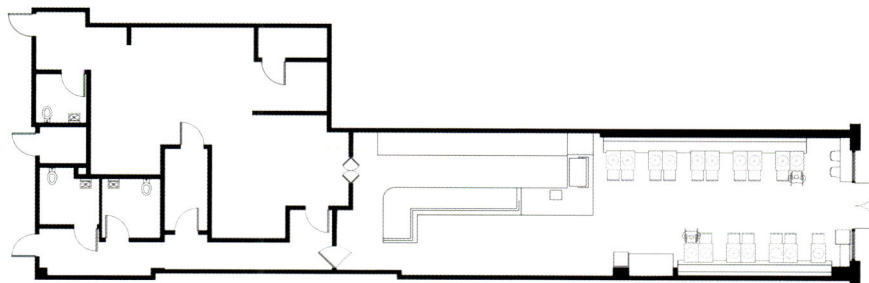

Fresh Cutt, Los Angeles

Featuring a dramatically lit open kitchen and serving counter, the quick-service Mediterranean concept is a result of a major transformation of a space that previously housed a salad-bar concept. The former venue was completely altered to create a sense of freshness that would appeal to an aspirational audience. With an array of contemporary eco-friendly materials and finishes, the design has been kept clean and simple, with the main counter becoming the focal point. AkarStudios created a branded identity and graphics programme for the launching of this new concept. Made up of two interconnecting spaces with a high ceiling, the first being an eating area that has a parallel row of banquet seating along the walls facing an outdoor patio. The second space is the exhibition kitchen—a long cook line whose focal point is the front preparation counter displaying an array of fresh and locally-sourced natural ingredients.

崭新餐厅

以熠熠生辉的开放式厨房与服务台为特色，此地中海快餐厅的前身为一家沙拉店，设计师将原空间进行了彻底改造，形成了全新的餐厅氛围，来迎合那些有消费观念的顾客群体。现代环保材料的使用使得整个设计显得干净而简朴，以此突出设计核心——主柜台。Akar设计事务所为这个新餐厅提供品牌与图形设计。餐厅的整体布局是高高的天花板下两个互相连接的区域。第一部分为餐饮区，正对着室外露台的墙边平行摆放着一排宴会桌。另一部分是展示厨房，厨房的核心设计是前方的备餐台，上面放着当地的天然原材料。

Fast Casual Dining

USA

Los Angeles

2009

Photo: Randall Michelson

AkarStudios: Sat Garg, Design Principal,
Sean W. Morris, Graphic Designer

Photo: Derek Rath

AkarStudios: Sat Garg, design principal, Matt Lutz, Chris Jones designers, Sean W. Morris, graphic designer

Seoul Bros., Pasadena

Designed in clean, simple lines, the homogenous space has modern finishes running throughout. The wood covered ceiling and the dark brown flooring combine to create a sophisticated dining space for this fast-food venue. The bold imagery of the wall murals, created by AkarStudios for this space, adds an exotic flair to this urban location. The restaurant is set on establishing a more modern-day environment. The quick-service venue uses modern materials to impart a sense of welcoming, relaxing and accessible environment. Large communal tables, sofa seating and modern wall graphics are some of the design elements of this space. Food production is the focal point – an open exhibition kitchen that is reminiscent of an Asian street food market allows customers to watch food being freshly prepared in a performance setting, which can make the food look more delicious and give the customers a better appetite.

首尔BROS餐厅

采用干净、简约的流线型设计理念，整个室内空间布满了现代装饰。木制天花板与深棕色地板交相辉映、为快餐店创造了一个精致的餐饮空间。由AKAR工作室设计的壁饰的夸张意向为这家城市餐厅注入了异国情调。本餐厅旨在营造一个更加现代的用餐环境。这家快餐店用现代装饰材料，为顾客营造一个轻松、自在的用餐环境，给顾客一种宾至如归的感觉。大餐桌、沙发座椅及现代的墙面装饰都是本餐厅的特色元素。而厨房设计更是整体设计的点睛之笔——开放的厨房展示空间，使人想到亚洲的食品市场，顾客可以看到厨师准备新鲜食物的过程，增加顾客的食欲。

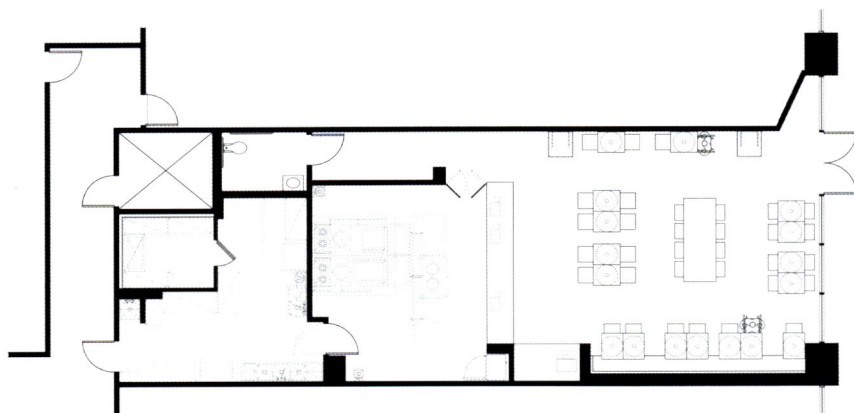

Restaurant

USA

Beverly Hills

2007

Photo: Fotoworks / Benny Chan

Clive Wilkinson Architects

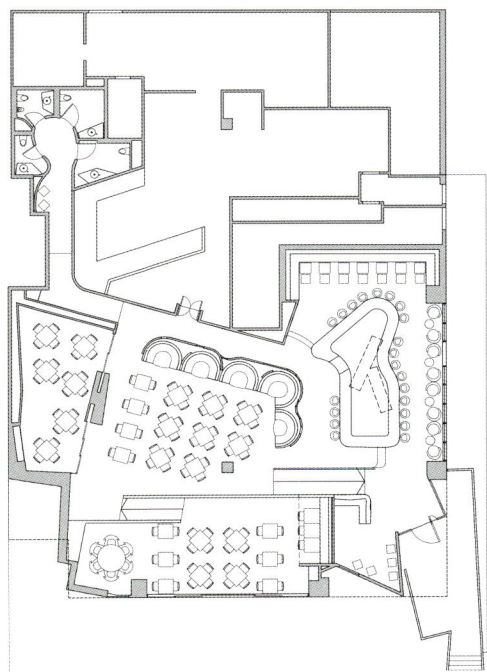

Paperfish Restaurant

Clive Wilkinson Architects was commissioned to design their new fresh seafood restaurant. Designers proposed the name 'Paperfish' to underscore the restaurant's function and to align with the concept of treating the plain white architecture as thin leaves of curving drywall to evoke fluidity, water flowing and pooling, and to give the sense of being immersed below the sea. The word 'paper' also evoked fish cooked 'en papilotte', and this has become the restaurant's signature dish. The complete brand identity was developed simultaneously with the architectural design. Designers researched plates and uniforms, glassware and upholstery, door handles and light fixtures, logos and fonts all at the same time. The restaurant experience needed to be effortless and smoothly integrated, with an architecture that could form a subtle and calming backdrop to the dining event. The environment evokes the ocean in movement, but the effect is almost subliminal as the colour blue is absent. The shades of red to orange relate to colour changes on the body of a carp fish.

纸鱼餐厅

威尔金森建筑师事务所负责设计全新的纸鱼餐厅。设计师建议使用"纸鱼"这一名称，以凸显餐厅的性质。设计师还将白色的修饰材料制成弯曲的墙和轻薄的叶子，激发流动感，使人们想到水、池塘，仿佛置身于大海深处。名字中的纸代表将鱼用纸包裹烹调，这也是餐厅的一道招牌菜。餐厅的品牌设计与建筑设计同时完成。设计师同时设计了餐具、制服、玻璃器皿、坐垫、门把手与灯具等。餐厅设计需要与整个建筑设计实现自然流畅的结合来形成微妙而又安静的用餐环境。

Restaurant & Bar

USA

California

2009

Photo: Fanny Alliè

Crème Design

Lucy's

Sam Hazen, formerly the Executive Chef of two of the most popular eateries in the United States, Tao Las Vegas and Tao New York, has teamed up with La Dolce Vita Hospitality (LDV) to form Sam HazeManagement (SHM) and under this new umbrella, launch Lucy's Cantina Royale, a Baja style restaurant this fall. Adjacent to the Local West bar, also owned by LDV, this introduction will complete the social renewal of Penn Plaza. SHM currently oversees the front and back of house operations for Lugo Caffe, a trendy Italian brasserie, and Local West a burger bar in Midtown in New York City. It is opening Lucy's Cantina Royale, a Baja style cantina this fall.

露西餐厅

美国两大最受欢迎小型餐厅拉斯维加斯道餐厅与纽约道餐厅前任执行总监山姆•哈森与La Dolce Vita Hospitality(LDV)合作组成山姆•哈森策划公司，两大公司强强联手，金秋季节启动了包姚式餐厅项目——露西餐厅。 毗邻LDV旗下的另一家酒吧Local West Bar，露西餐厅的建成将会是宾夕法尼亚广场社交场所的革新。山姆•哈森公司目前正监管两家酒吧项目的建筑正背面建设。这两家酒吧分别是时尚的意大利啤酒店卢戈咖啡厅与位于纽约市中心地方西式汉堡店。

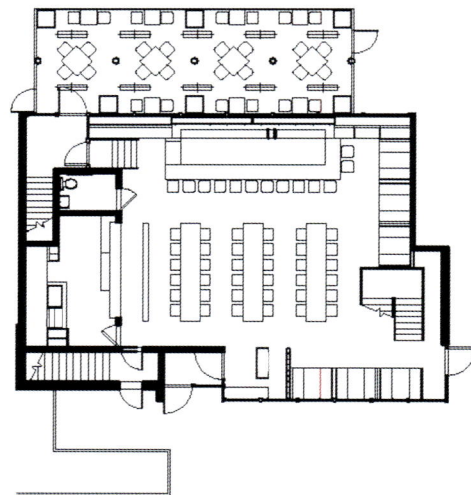

Restaurant

USA

Philadelphia

2010

Photo: Fanny Alliè

Crème Design

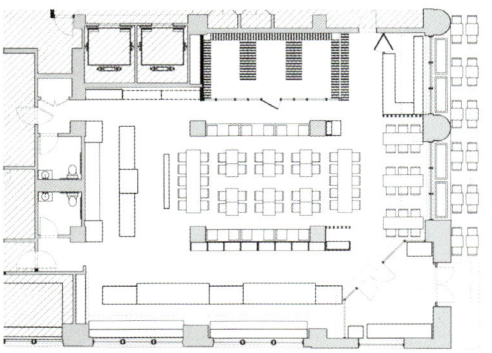

Garces Trading Company

Garces Trading Company, the newest venture by Iron Chef Jose Garces, is a gourmet café with sit-in dining and also a market with an extensive repertoire of products. The market is filled with Garces' fresh breads, 7 different coffee blends, Spanish, French and Italian cheeses, with house-made jams and jellies to compliment, house-cured and imported Charcuterie, and oils and vinegars from Spain, Italy and Greece. The site will also house a flower stand, bakery, and Philly's first wine shop leased from PLCB to a chef, to complete the GTC experience. Merchant ships and European markets are the inspiration points for the Garces Trading Co. design by Crème. The high, black tin ceiling and walls of white subway tile create a comfortable and airy atmosphere.

伽瑟斯贸易公司餐厅

伽瑟斯贸易公司是铁人厨师约瑟·伽瑟斯创建的全新餐厅。该餐厅由两部分组成：室内美食餐厅与商品齐全的市场。市场内销售伽瑟斯新鲜面包、7种不同的咖啡、西班牙、法国与意大利奶酪、自制果胶果酱，自制熟食与进口熟食，还有从西班牙、意大利、希腊进口的不同商品等。市场内还有一个花架、一个面包店和费城第一家酒馆，使顾客的伽瑟斯之行更加充实。欧洲商船与市场是伽瑟斯贸易公司的设计灵感来源。高高的黑锡天花板与白色瓷砖墙营造出舒适而又通风的室内环境。

Restaurant

USA

Chicago

2008

Photo: Frank Oudeman

D-Ash Design

Mercat a La Planxa

For Mercat, the designers drew from modern Spanish aesthetics to create a vibrant ambience reminiscent of Barcelona's bustling markets. The entrance to the restaurant is through a small bar on Michigan Avenue. The guest enters this intimate space, clad in walnut, to evoke the feeling of a classic Spanish bar. The back of the bar is surrounded by a curved screen composed of multi-coloured hex blocks that is also the back of the large circular stair that leads to the second floor dining room. The dining room is a grand two-story space. The designers created a layout which radiates from the space so that all three levels of the main restaurant allow guests to have prime views through the giant windows that overlook Grant Park.

Mercat a La Planxa 餐厅

Mercat 餐厅的设计师运用西班牙现代美学营造出巴塞罗那小市场式怀旧风格而又生机勃勃的餐厅氛围。顾客从密歇根大街一个小酒吧进入到Mercat餐厅。进入这个核桃木装饰的亲密空间，就会使人想起古典西班牙酒吧。一楼酒吧背面与通往二楼餐厅的大圆型楼梯的背面都用曲线型屏风装饰。屏风由五彩斑斓的六角木组成。二楼餐厅又分为两层。设计师放射性的布局设计使整个三层主餐厅的顾客都能透过玻璃观赏到格兰特公园的全景。

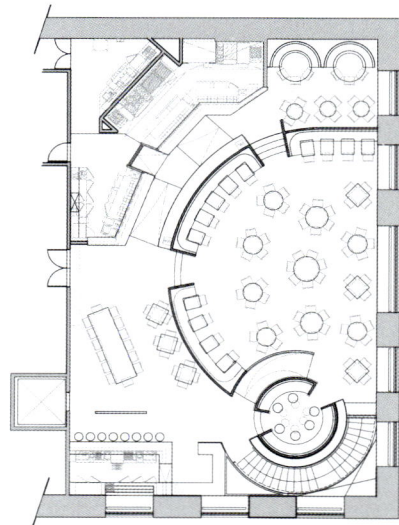

Restaurant

USA

Seattle

2006

Photo: Sally Painter, Thomas M. Barwick

Corso Staicoff, Inc.

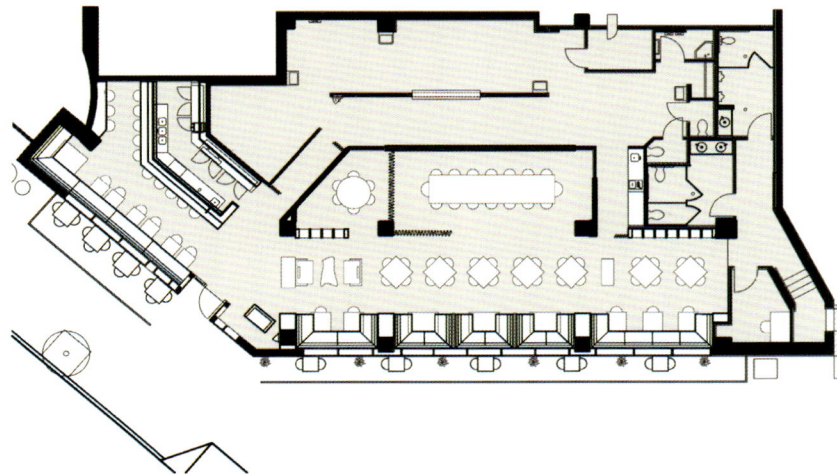

Barolo Restaurant

The client requested a restaurant that blends old world charm and modernism in an extremely narrow 2300-square-foot space with very high ceilings. Strategically placing expansive mirrors provided the illusion of a much larger space. Lighting design played a major role – ornate crystal chandeliers light the communal table while minimal spots illuminate the massive wine cases, creating a heightened sense of drama. Backlit onyx bar counters and plinths provide a soft, subtle glow. The mixture of diaphanous white sheers, drippy candles, baroque mirrors, green velvet booths and dark concrete floors created a feeling of opulence and style.

巴罗洛餐厅

客户希望餐厅的设计能够将古典的美和现代主义结合在这顶棚高挑、非常狭窄的214平方米的地方。将一个大的公共餐桌作为餐厅的中心；降低了顶棚的高度，以创建出空间的层次感；策略地布置了几面大镜子，如同空间被扩大了一般。在这里，灯光的设计起了很大的作用：华丽的水晶枝型灯照亮了公共餐桌，而小的灯光则为大面积的葡萄酒柜提供照明，抬升了一种戏剧化的氛围。玛瑙红吧台和柱基处则采用了平缓、精妙的背光照明。透明的白纱帘、流泪的蜡烛、巴洛克风格的镜子、绿色天鹅绒的隔间、黑色的混凝土地面，这些给客人一种庄严辉煌而又时尚的感觉。

Restaurant

USA

Lincoln

2008

Photo: Julian Taylor Design Associates

Julian Taylor Design Associates

Sakura

Entry to the venue comes from street level. Immediately guests are greeted by a comfortable but sophisticated fixed seating area with an oriental theme, ideal for daytime eating and socialising. The décor is comprised of bamboo screens and oriental cherry blossom wallpaper with shimmer screen flanking the small stairs which lead down to the main bar area. With a combination of high and low fixed seating customers can relax over a casual dinning experience while sampling some of the best cocktails on the high street. The open staircase leads directly to level two where a more nighttime operation can be found for those seeking a premium night-time club experience.

樱花餐厅

餐厅的入口与街道保持水平。进入餐厅，舒适而又做工精良的东方主题座椅给顾客宾至如归的感觉。樱花餐厅是白天用餐与社交的理想场所。餐厅用竹子屏风与东方樱花壁纸装饰。通往楼下主餐区的小楼梯的两侧还有明亮耀眼的屏风。高矮不等的座椅使顾客在品尝商业街最佳品质鸡尾酒的同时可以轻松愉快地用餐。开放式的楼梯径直通向二楼餐厅，那是专门为那些喜欢优质午夜俱乐部生活的顾客而准备的夜晚用餐场所。

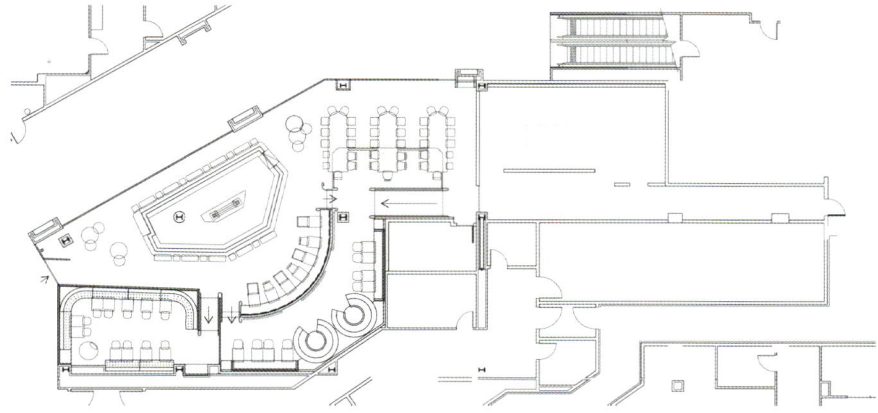

Johnny Smalls Restaurant and Bar

Named after the marijuana/tobacco cigarette, Smalls makes you feel like you've smoked a few with wild 'Dali-esque' décor: psychedelic murals on the walls of men, mushrooms, and monkeys, wooden chairs that look like hands, and stained glass lighting helping you read a pan-earth not-so-big-plates menu boasting Spanish tapas, Asian dim sum, Italian antipasti, Mediterranean mezze, Indian thali, Mexican antojitos, and classic American starters… People could enjoy various cuisines all over the world.

约翰·斯莫尔斯餐吧

约翰·斯莫尔斯餐吧的名字来源于一种香烟，其内部的达利式装修也让人云里雾里：墙壁上绘有人像、蘑菇、猴子，木椅子像手掌一样。在彩色玻璃灯光下阅读着诱人的菜单：西班牙小吃、亚洲点心、意大利小菜、地中海冷盘、印度菜、墨西哥菜和经典美国开胃菜……在这里，可以尽享世界各地的美食。

Restaurant & Bar

USA

Las Vegas,

2010

Photo: Jeff Dow

Mr. Important Design

Bar

USA

Boston

2008

Photo: Projects Design Associates

Projects Design Associates

Sel de la Terre

Sel de la Terre is a French Bistro in the new Mandarin Oriental Hotel in Boston MA. While each of the three Sel de la Terre locations has an interior design specifically tailored to their space, similar design elements at every location provide the signature Sel de la Terre look and feel. To enhance the Provencal bistro atmosphere the designers include authentic, rustic elements and artwork that add warmth to each restaurant space. The designers paired these elements with modern accents and architecture that reflect our New England home.

盐世界酒吧

盐世界酒吧是波斯顿新文华东方酒店的一个法式酒吧。盐世界系列的三个酒吧中，每一个都有适合其地理位置的个性设计，然而，每一个地点的酒吧都有相似的设计元素，这些元素都能体现如酒吧名字盐世界的外观和感觉。为增加普罗旺斯式的酒吧氛围，设计中融合了真正的乡村元素和艺术作品，为每个餐厅增加了丝丝暖意。设计师在这些古典设计之中又加入了反映新英格兰住宅风格的现代建筑元素。

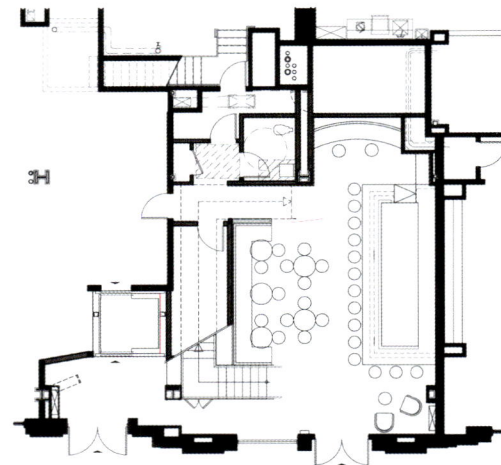

Restaurant

USA

California

2010

Photo: Sharon Risedorph

Arcsine Architecture & Bellusci Design

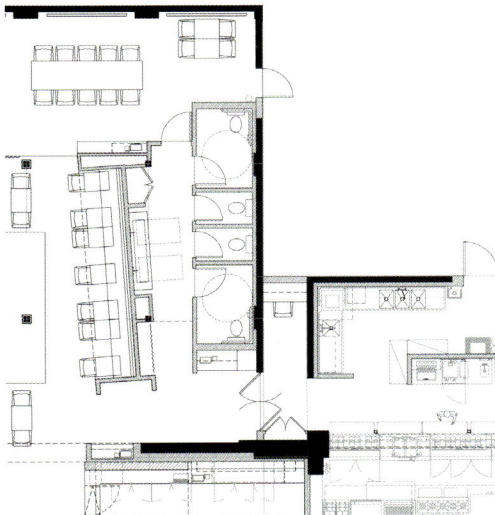

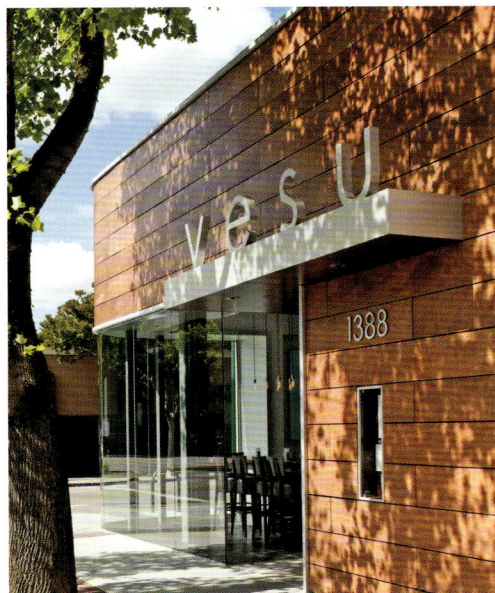

Vesu Restaurant

Arcsine Architecture and Bellusci Design transformed a piano store with little curb appeal into a vibrant and engaging restaurant, fitting the bustling downtown Walnut Creek. Arcsine wrapped the building with a contemporary, wood-panel façade, complimented by a curving wall of frameless glass that entices pedestrians to look in and feel as if they are already inside. An Arcsine-designed canopy of FSC Certified Eucalyptus leads patrons from the entrance to the lounge and extends all the way to the back bar. Bellusci Design continued the contemporary motif, selecting furniture and interior finishes in a warm, monochromatic spectrum with plenty of texture.

Vesu 餐厅

Arcsine Architecture 建筑事务所和Bellusci Design 设计公司共同将庄重风格的钢琴店改造成了与核桃溪市繁华景象相匹配的活力餐厅。反正弦外观的立面铺上现代风格的木板，曲线的墙壁加上无框架的窗玻璃，使人不由自主地将视线投向餐厅内部，仿佛已经置身于餐厅内部一般。入口上方反正弦结构的顶棚森林管理委员会（FSC）认证的桉木。Bellusci Design 设计公司延续了其现代设计风格，家具和室内设计都采用纹理丰富、色彩温暖的单色色调设计。

Lounge

USA

San Francisco

2009

Photo: Kinney Chan

Kinney Chan

Infusion Lounge

Infusion Lounge will redefine upscale nightlife not just for San Francisco, but for the industry itself. Found at the epicentre of the City's international icon, Union Square, Infusion Lounge opens as an sophisticated destination spot. The Asian-inspired sub-rosa ultra lounge executed by Hong Kong's hottest designer, Kinney Chan, will offer a full menu of Cal-Asian fusion small plates paired with exquisite ambiance and exceptional service. Six thousand square feet encompasses a private VIP room, fully equipped dance floor and an impressive main lounge. Infusion is a true ultra lounge catering to both dancing hipsters and sophisticated young professionals looking to relax in style.

Infusion 酒吧

Infusion酒吧将对旧金山乃至整个餐饮业的高层都市夜生活理念进行重新定义。Infusion 酒吧位于三藩市高级商业区内国际著名的联合广场中心。这个充满亚洲风的极致酒吧由著名香港设计师陈德坚先生设计，在这里，顾客会品尝到全套的亚加风味小盘式菜肴，享受到别致的服务，感受精致的餐厅氛围。6000平方英尺的酒吧内有一间私人贵宾室、装饰精良的舞池，还有一间与众不同的酒吧。Infusion 酒吧是一个真正的极致餐厅，它既迎合了那些喜欢跳舞的时尚人群，又满足了那些寻求时尚放松生活方式的年轻专业人士。

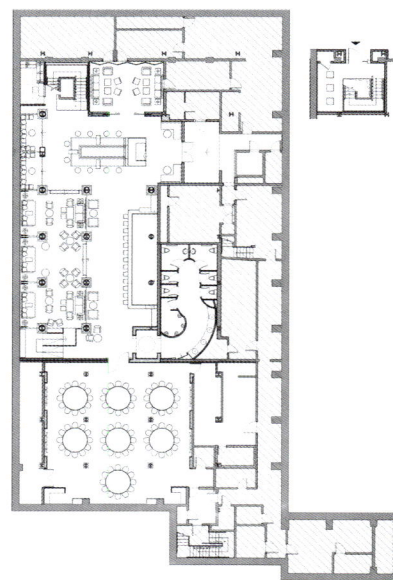

Restaurant & Bar

USA

Florida

2010

Photo: Jeff Dow

Mr. Important Design

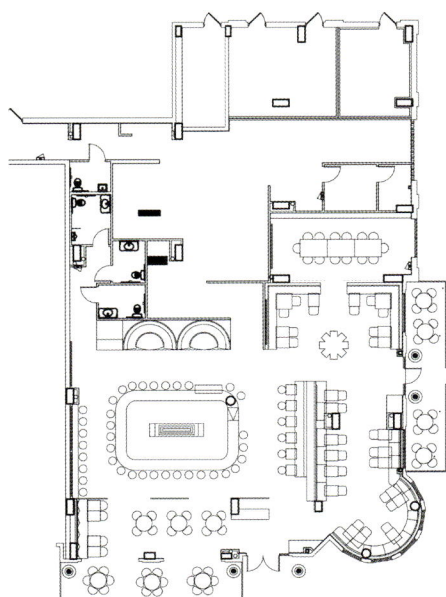

Pure Urban Oasis Restaurant and Bar

Pure Urban Oasis™ will serve reasonably priced American style food, familiar ingredients but with a new twist. In the centre of The Mercato Shops there is an oasis, a refuge from everyday life that will relax your mind, rejuvenate your spirit, and replenish your body. It's simple, genuine, faultless, perfect. It's pure. It's in the midst of a vibrant, centric community and is a culture unto itself. It will quench your thirst for an experience that's fresh and shimmering with life. Rest assured it's not a mirage, but a finely crafted restaurant introducing a unique concept in dining and lifestyle that promises to be a feast for the senses, and a rejuvenation for both body and soul.

纯粹城市绿洲餐吧

纯粹城市绿洲餐吧提供价格合理的美式美食，以全新的方式演绎熟悉的食材。餐厅位于莫卡托购物中心，让人们远离都市的喧嚣，放松心情，重获鼓舞。这里简单、纯真、无瑕、完美，是一个纯粹的空间。这里充满生机，具有自己独特的文化。这里能够满足你对新鲜生活的渴望。这里并不是海市蜃楼，而是一家崇尚独特餐饮理念和生活方式的精致餐厅。顾客可以在这里享受感官的盛宴，让身心都获新生。

95

Restaurant

USA

San Francisco

2009

Photo: Jeff Dow

Mr. Important Design

Gitane

Named after the freewheeling gypsy, Gitane is a modern, funky, and artistically bold restaurant located in downtown San Francisco between the financial district and Union Square. Chef Lisa Eyherabide creates simple, approachable bistro fare emblematic of the Basque region, drawing inspiration from neighbouring Spain, France, and Portugal. An elevated bar programme pairs these exotic tastes and sensations with unique Sherries, Cavas, Madeiras, Spanish brandies, innovative hand-crafted cocktails and small-estate wines. The interior décor from Mr. Important Design unabashedly integrates vibes from three separate decades — the Euro-themed 1950's, Hippie-Driven 1960's and Big Bling 1970's, giving the space a traditional yet modern, eclectic feel.

Gitane餐厅

坐落于旧金山金融区与联合广场的中间位置，Gitane 以自由奔放的吉普赛人的名字命名，是一家现代、时尚、艺术上大胆追求的餐厅。受西班牙、法国和葡萄牙等邻国美食文化的启发，主厨丽莎烹饪出简单而又可口的巴斯克地区特色美食。高层次的酒吧规划将这些异国风味与情调同雪利酒、CAVAS、马德拉白葡萄酒、西班牙白兰地、人工酿造的别致的鸡尾酒和SMALL-ESTATE WINES等独特的酒文化完美融合。

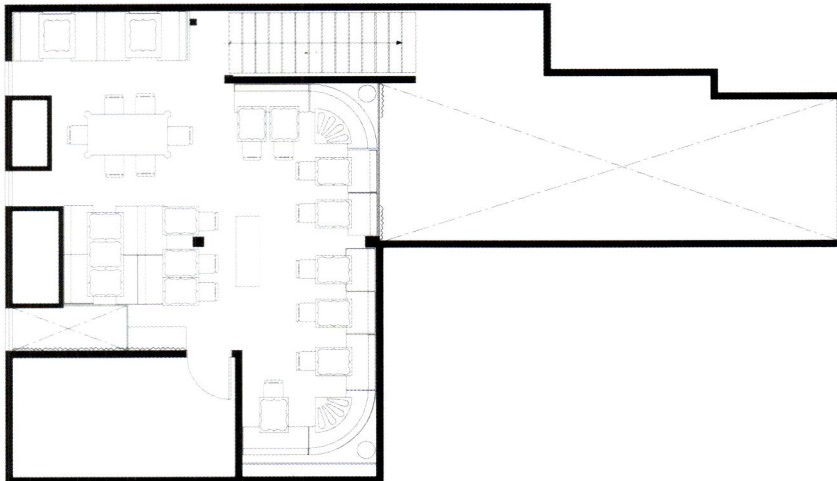

Restaurant & Bar

USA

Las Vegas

2010

Photo: GRAFT

GRAFT

City Centre's Aria Pool Deck, Restaurant and Bar

City Centre is a mixed use, 18 million square feet, development by MGM Mirage that features buildings by several world renowned architects. The unique seventy-six acre urban resort is located in the heart of the famed Las Vegas Strip and is a collection of luxury hotels, condominiums, casinos, shopping facilities, and entertainment venues. To date, it is the largest privately financed development in the history of the United States and is also the largest LEED Certified project in the world. City Centre consists of a collection of spectacular hotels and residences, casinos, spas and retail areas, including Aria, Vdara, the Mandarin Oriental, Veer Towers and Crystals. Located at the base of Aria, the main resort and casino, the pool deck is a lush tropical lagoon. It is an intimate sanctuary of cabanas around a luscious pool area in the midst of the larger vision of the City Centre.

市中心艾丽娅甲板湖餐厅与酒吧

此项目是美高梅集团旗下的一个面积达1800万平方英尺的综合性项目，由几个世界著名设计师联合设计。这个别致的城市度假村面积达76英亩，位于著名的拉斯维加斯海岸的中心地带，是一个包括奢华酒店、公寓、购物设施和娱乐场馆在内的综合性度假村。迄今为止，它是美国历史上最大的个人出资建成的开发项目，也是世界上最大的获得能源与环境建筑认证的项目。市中心度假村项目包括奢华酒店与住宅，娱乐场、水疗中心和艾丽娅（Aria），Vdara,文华东方酒店（the Mandarin Oriental），Veer Towers and Crystals 在内的零售区等。

Bar & Lounge

USA

Atlanta

2009

Photo: Kris Tamburello; New York, NY

CCS Architecture (NY)

Drinkshop at the W Hotel

Drink Shop is an artisanal bar on the second floor of the new, 28-storey, W hotel and residence development in downtown Atlanta. The concept expresses the craft of drink making and guests can watch the drink being made through each step. A large ice block is featured at the point of the bar and used by all bartenders. Macerated fruits and fresh ingredients, served with chipped ice from the giant block, are all on full display. Three ingredient rails are built into Drink Shop's bar. The back bar is stainless steel with glass-door freezers displaying frozen glasses and liquor. The whole bar becomes a production stage. The ice concept runs throughout the lounge. The ceiling is composed of striated panels in cool whites, greys and blues.

W 酒店内的酒吧

Drink Shop 是一家以人工酿酒为特色的酒吧，位于亚特兰大市中心28层的新W酒店与公寓开发项目的二层。酒吧设计展现了酿酒的工艺技术，在这里，顾客可以观赏到酿酒的全部过程。酒吧最大的特色就是位于中心位置的大冰柱，所有酒吧服务员都从冰柱上取冰块。用冰柱上切下的冰块浸泡的各种新鲜水果全部供顾客享用。酒吧内有三种原材料混合而成的栏杆，酒吧背墙选用不锈钢原材料，前面是玻璃门冰箱，里面冷藏着各种酒和玻璃酒杯。整个酒吧成了一个展示酿酒工艺的舞台。

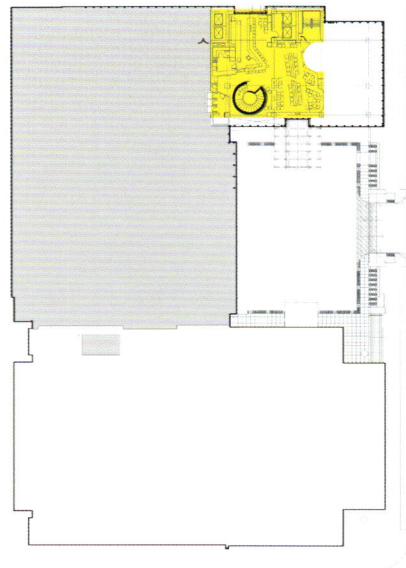

Canteen

USA

Illinois

2008

Photo: Barbara Karant + Associates

De Stefano Partners

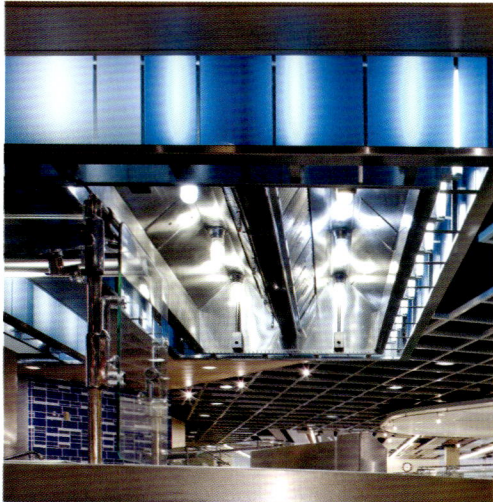

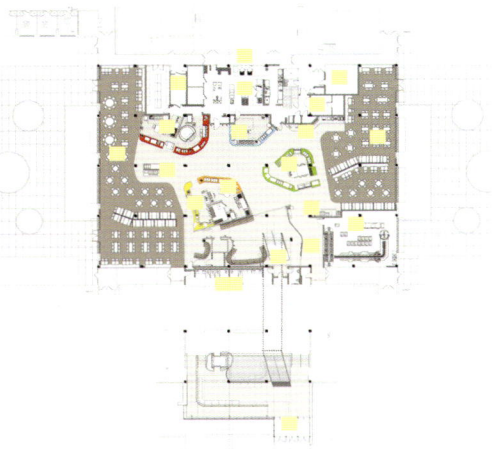

University of Illinois PAR Dining Hall Remodel

De Stefano Partners provided full architectural services to implement the comprehensive renovation of a student dining services centre for the Pennsylvania Avenue Residence Hall on the University of Illinois Urbana-Champaign campus. The scope of work for this 28,000 square feet project was to reconfigure, upgrade and modernise the dining rooms, dining services facilities and related support spaces, such as offices. The work also included renovation of the street level building entrance, lobby and reception area, as well as new building sprinkler system infrastructure, new air handling and ventilation system, new restrooms and main entrance reconfiguration to address accessibility and security issues.

伊利诺伊大学宾夕法尼亚大道学生公寓餐厅重建项目

De Stefano 建筑事物所为伊利诺伊大学乌尔班纳分校宾夕法尼亚大道学生公寓餐厅的重建项目提供了全方位的设计方案与手段。面积达2.8万平方英尺的重建项目包括对现有用餐大厅、各独立餐厅、用餐设施及诸如办公室等空间的重构、升级与更新，还包括街道标高的建筑入口、休息大厅、接待区及新建筑的自动喷水灭火装置、新空调与排风系统、新休息室和主入口在内的一些列重建项目来实现整个项目的可达性与安全性。

Restaurant

USA

Mill Valley

2007

Photo: Rien van Rijthoven

BCV Architects

Piatti Mill Valley

Piatti is a contemporary upscale restaurant overlooking Shelter Bay next to Mt. Tamalpais, featuring the rustic cuisine of a traditional Italian Trattoria. Piatti Restaurant in Mill Valley continues its successful Italian trattoria atmosphere in this contemporary remodel of an existing building. Featuring a wall of windows and an alfresco patio overlooking Shelter Bay, the restaurant is imbued with a resort-like feel. Recycled teak furniture, a community table, and classic Carrara stone reinforce the ethos of timelessness and sustainability reflected in the restaurant's use of seasonal food from local producers. The restaurant, completed in 2007, has 3,790 square feet of interior space and 435 square feet of exterior patio. With the open kitchens and stone pizza hearths, Piatti Restaurant reflects the warm charm and welcoming atmosphere of a traditional Italian trattoria.

皮亚蒂米尔谷餐厅

皮亚蒂餐厅是一家以传统意式乡村风味肴为特色的现代高级餐厅。在这里，可以环视塔玛佩斯山旁的肖特湾。通过改造，皮亚蒂餐厅成功地沿袭了意式餐厅的特有风格。以玻璃墙和可环视肖特湾的室外露台为特色，整个餐厅都有一种度假村的感觉。餐厅四季所用食物都从当地食品商那里购买，室内装饰选用可再生的柚木家具，交流桌和古典白理石，增添了餐厅的时尚性与永恒性。

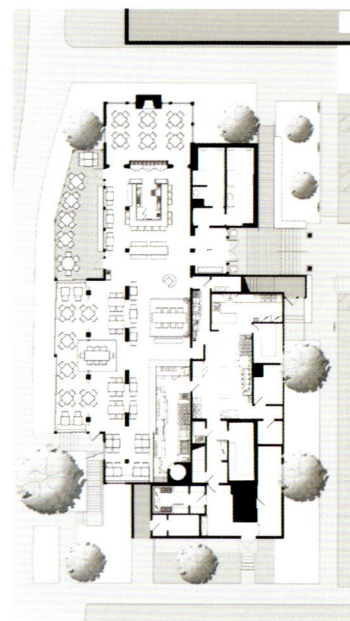

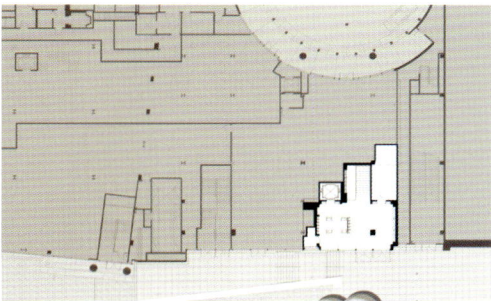

Bar

USA

San Francisco

2008

Photo: Rien van Rijthoven

BCV Architects

Press Club

The charge for this venue was to create a sophisticated urban identity for this innovative wine collective / lounge business model while simultaneously capturing a modern expression of the 'spirit' of California Wine Country. Within a distinctly modern sculptural expression, the architects strived to capture the casual sophistication of the wine country and create a space that is both 'timely' and 'timeless'. Taking cues from Napa's unique blend of industry and natural beauty, the design celebrates the juxtaposition of the 'industrial' against the 'organic' as a foundation for the project's expression.

Press 俱乐部酒吧

此俱乐部的设计宗旨是要为这个酒吧业的典范——创新型酒业集团创建一个高级酒吧品牌，同时弘扬加利福尼亚酒国度的精神文化。建筑师用独特的现代雕刻艺术制品，力图展示酒故乡质朴的酿酒工艺，打造时尚而经久不衰的完美空间。受纳帕独特的工业美与自然美的融合艺术激发，设计表达了工业艺术与有机事物自然融合的核心理念。

Restaurant

USA

Los Angeles

2010

Photo: Eric Laignel

MARKZEFF

Delphine

Delphine is a Mediterranean eatery that combines the charm of the Mediterranean with Hollywood glamour. Nestled inside the W Hotel and Residences in West Hollywood, CA Delphine conveys the feel of a casual yet elegant eatery reminiscent of those found along the European Riviera, and is evident in its indoor-outdoor layout and romantic decor. The expansive 6,000-square-foot space is located in the hub of the legendary district of Hollywood at Hollywood Boulevard and Vine Street, Delphine seamlessly blends the glamour and glitz of its surrounding locale with the relaxed sophistication of the Côte d'Azur.

戴尔芬餐厅

戴尔芬餐厅是融合了地中海风情与好莱坞韵味的地中海式小餐馆。位于西好莱坞W酒店及公寓内。戴尔芬拥有轻松自在而又高雅的餐厅环境，酷似欧洲里维埃拉沿岸那些优雅的小餐馆，而又有着其独特的室内外布局与浪漫的装饰布置。 广阔的达尔芬餐厅，面积6000平方英尺，位于好莱坞大街与藤街交汇处传说中的好莱坞中心商业区,其将好莱坞地区的魅力浮华与地中海蔚蓝海岸的自在豪放绝妙地融合为一体。

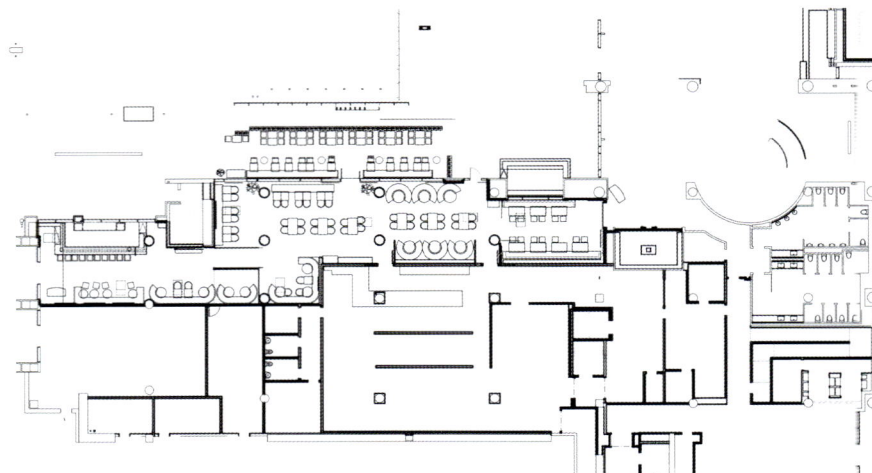

Knuckles

The Knuckles Sports Bar at the Hyatt Fisherman's Wharf was remodelled to create a 'crossover' restaurant that would appeal to both the regular sports bar clientele and tourist/business guest. The previous concept was an adaptation of a past Italian restaurant, called the Marbleworks after the original use of the historic building. The main entry lobby had also been recently remodelled, and there was a significant aesthetic split between the old restaurant and new lobby. The project also included the renovation of a 'games room' adjacent to the restaurant into a wine lounge. The design included an update of the overall dining room seating strategy, dividing the room by low partition walls to create a central dining area, the introduction of booths for large parties, and a variety of seating types, including banquettes, 4-tops, booths, casual lounge furniture for continental breakfasts and pre-dinner cocktails, and a community table in the bar area.

音速小子酒吧

位于渔人码头凯悦酒店的音速小子运动酒吧被改建成那种所谓的"跨界"餐厅——既迎合长期光顾运动酒吧的客人口味，又满足了过往游客与商人的用餐需求。在老餐厅与新大厅中间修了一个十分唯美的分隔带。而且，通往葡萄酒廊的餐厅旁的游戏厅也进行了改建。布置设计包括餐厅整体座椅的更新、中心用餐区周围的矮隔断墙、展台、休息大厅的欧式早餐餐桌和饭前鸡尾酒桌及酒吧区的聚餐桌等，还增加了一个新的接待台和黑胡桃木制成的等候台。

Restaurant

USA

San Francisco

2009

Photo: Kris Tamburello

CCS Architecture

Alto Ristorante E Bar & SWB Hyatt Resort & Spa at Gainey Ranch

CCS Architecture has created two energetic venues at Hyatt Resort & Spa at Gainey Ranch in Scottsdale. The new restaurants replace two previous concepts at the resort, with the goal of attracting the local Scottsdale market as well as resort guests. Gainey Ranch is set amidst flowering cactus and framed against the majestic McDowell Mountains. SWB is a casual, three-meal bistro featuring indigenous Southwest cuisine and materials. Alto Ristorante E Bar, an eclectic, upscale, dinner destination, brings to life the flavours and design sensibilities of Italy. Both spaces are distinctive and modern, offering new indoor-outdoor connections and a current, 21st century perspective on the restaurant experience. The restaurants have separate entries, meeting at the lively wine and liquor bar.

盖尼牧场艾尔托餐厅E酒吧与SWB 海特度假村与水疗中心

CCS 建筑事物所曾在斯科茨代尔的盖尼牧场凯悦度假村与温泉浴场建造了两个富有活力的餐馆。本着抓住SCOTTSDALE市场商机并吸引度假村来客的目标，将之前两个项目改建为如今这家新餐厅。盖尼牧场背靠雄伟的麦克道尔山脉，四周仙人掌花围绕。SWB 是一家提供一日三餐的简朴小餐馆，以西南本土特色菜肴与烹饪技术为特色。艾尔托餐厅E酒吧是一个折中风格的高级用餐场所，酒吧设计运用了意大利设计风格与韵味。运用新型室内外串联式设计与21世纪前卫的餐厅设计理念，无论餐厅还是酒吧都别致新颖、极具现代风格。餐厅拥有独立的入口，各个入口都通向活力动感的葡萄酒与烈性酒酒吧。

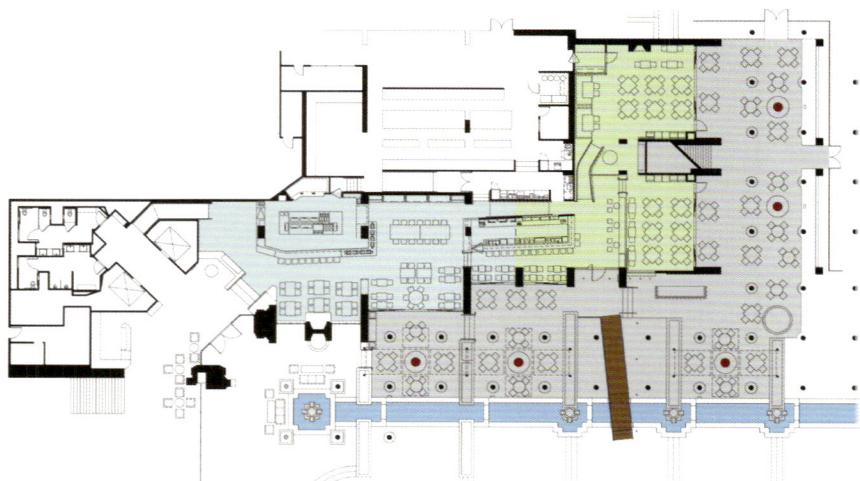

Restaurant

USA

Scottsdale

2008

Photo: Eric Laignel

CCS Architecture

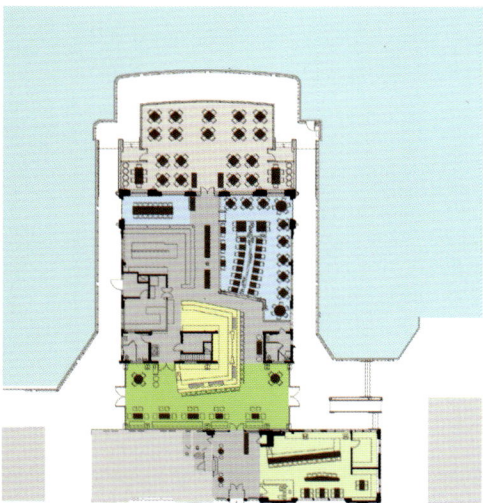

La Mar

The design of La Mar began with a tour of Lima, Peru, and an introduction to the cuisine of celebrated Peruvian chef Gaston Acurio. La Mar in Lima is Acurio's flagship; its modern interior, range of colours and materials, and vibrant cebiche bar all add up to a casual hot spot with unforgettable food. This became the starting point for the design of La Mar San Francisco. The challenge was to bring the Lima experience to life in one of the great historic piers on the San Francisco waterfront, which is just to the north of the renowned Ferry Building Marketplace. The pier buildings front Embarcadero Boulevard to the west and open up to San Francisco Bay on the east. The recently redeveloped waterfront extends from the Ferry Building northward to Pier 5 and is fast becoming San Francisco's food and dining Mecca on the water.

La Mar 餐厅

La Mar 餐厅的设计灵感来源于秘鲁首都利马的一次旅行。餐厅引进了著名的秘鲁厨艺大师Gaston Acurio 烹饪技术。La Mar 餐厅是Acurio创建的旗舰店，店内的现代化设计、材料与颜色的合理搭配、充满生机的时尚酒吧——所有这些元素造就了拥有别致菜肴而又轻松自在的用餐环境。这就是旧金山La Mar 餐厅设计理念的来源。设计所面临的挑战是要在旧金山海滨渡轮大楼市集北面的码头这样一个历史性区域设计Lima餐厅。码头大楼西临内河码头大街，东朝旧金山海湾。新扩建的海滨从北面的渡轮大厦一直延伸到5号码头，并且很快成为旧金山广受欢迎的水上餐厅。

Bar

USA

San Francisco

2008

Photo: Eric Laignel

CCS Architecture

Restauran & Bar

USA

San Jose

2010

Photo: Francois Frossard

Francois Frossard

The Forge Restaurant / Wine Bar

The restaurant's makeover was designed and executed by Francois Frossard, who replaced its dark woods, stained-glass murals and gilded-framed art with walls of hand-carved blonde wood and antique smoked mirror, octopus-like lilac and white crystal chandeliers, and an eclectic mix of upholstered and metal furniture. Frossard also eliminated more than 100 seats, to create an open floor plan with the ambience of a large private home where every seat is 'the best in the house'. The Forge Restaurant is a monumental design challenge, a 'redux' of a cherished venue, known for impeccable design, beloved by the owner, Shareef Malnik, and its A-list celebrity and socialite clientele.

锻造餐厅&酒吧

设计师弗朗索瓦·弗洛沙德负责餐厅的设计与改造：手工雕刻的白木幕墙、古色古香的烟熏镜子及章鱼形状的丁香花和白色水晶吊灯，还有装饰家具与金属家具的时尚混搭，取代了原来的黑木头、钢化玻璃壁画、镀金镶边的艺术品。弗洛沙德还淘汰了100张座椅，开放式地板上只有一个巨大雅间，似乎里面的每一个座位都是这个房间里的绝佳位置。锻造餐厅是一项不朽的设计，是珍惜设计的重现。因其美妙绝伦的设计而受到店主、当地名流和社会名流等顾客群体的爱戴。

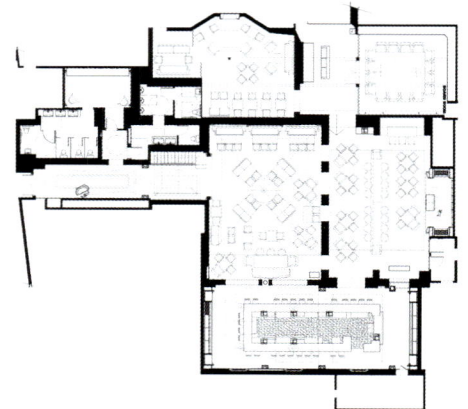

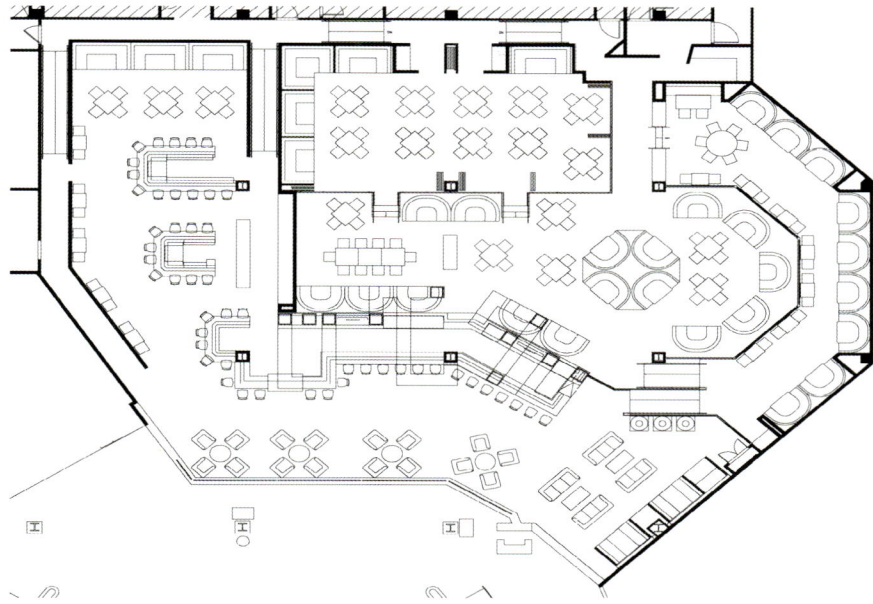

Fin Fish

Sharing a central lounge with Charlie Palmer Steak, Fin Fish has a sea blue, sage and sand coloured shoreline environment that features a 'wall' wave created by a suspended collage of natural driftwood, offset by white penny tile flooring reminiscent of a New England-style oyster. The seafood restaurant takes inspiration from the historic Oyster Bar in Grand Central terminal wit bar height counter dining, signs with daily oyster offerings and a open chowder kitchen. The space has a slightly industrial feel with 'low bay' lighting fixtures and ceiling fans. To reinforce the seafood theme a custom-designed hanging screen wall of driftwood logs was created. Tons of driftwood was imported from British Columbia Canada, cleaned and hung on wires.

鳍鱼餐厅

鳍鱼餐厅同查理·帕默牛排餐厅共用一个中心休息室，拥有海蓝色、鼠尾草和沙滩颜色的海滨式用餐环境。悬挂空中的天然浮木抽象拼贴画形成一堵富有特色的波浪墙，在白色瓷砖地板的映衬下，使人想起新英格兰的牡蛎。这家海鲜餐厅的设计灵感来源于中央车站历史著名的吧台进餐理念，以开放式的海鲜杂烩厨房和每日的牡蛎大餐为特色。低天井灯和吊式风扇为整个空间增添了一丝工业化气息。

Restaurant

USA

Reno, Nevada

2009

Photo: Projects Design Associates

Projects Design Associates

Restaurant

USA

San Jose

2008

Photo: Mr. Important Design

Mr. Important Design

Motif

The designer mixed things up by using futuristic materials such as stretched glossy polymer film to provide an undulating ceiling of reflected colour and pattern in the DJ area. The designer used cutting edge Asian furnishings with traditional materials from Kenneth Cobonpue coupled with the modern sm of Konstantin Grcic all mixed up with the 1960's and 1970's glam of Tommi Parzinger style pendants and sleek black linear silk shades. Hundreds of feet of black and silver chain hand worked into a bold floral tapestry (art provided courtesy of Amy Butler) that envelopes the space without suffocating it. Yards and yards of thick black silk rope to soften and provide pattern to the institutional cinder block construction of the building's interior.

主题餐厅

设计师通过使用极具现代未来感的材料将各种设计元素混搭在一起——扩散性光滑高分子膜材料制成的曲线型天花板在DJ区反射出五颜六色的图案。设计大量使用了天然原材料。设计师将7000块黑色玻璃组成的两个叶子形状的饰物悬挂在楼下餐厅与休息厅上空。上千英尺的黑色刺魁材料手工制花纹挂毯挂满了整个室内空间，却丝毫不会给人压抑感。

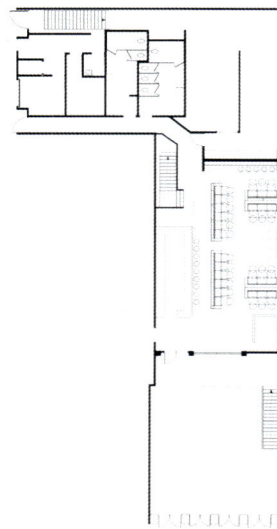

Restaurant & Bar

USA

Oklahoma

2010

Photo: Scott McDonald, Hedrich Blessing

Elliott + Associates Architects, Rand Elliott, Brad Buser

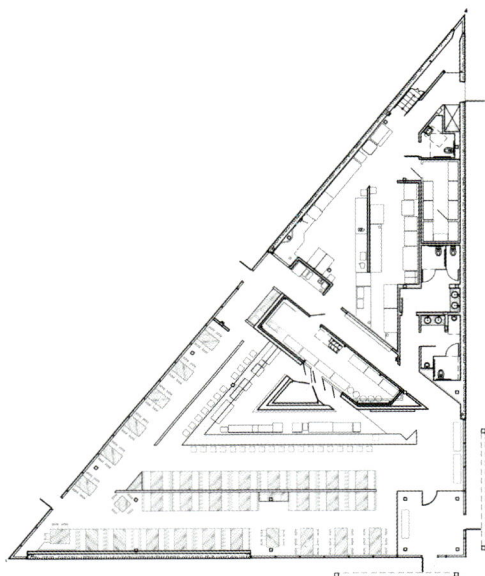

Republic Gastropub

Republic is a contemporary American public house bridging the gap between sports bar and upscale eatery. With an emphasis on American craft beer and an imaginative menu built specifically to complement this beverage. Republic Gastropub is located in the new Classen Curve shopping district. With 100 unique beers drawn from custom taps, close to 250 hand-selected bottles from around the world, a zippy wine list, and unique cocktails, there is no shortage of thirst quenching options to pair with their modern American takes on classic pub food favourites. It's a neighbourhood gathering place with a high energy atmosphere, sleek decor and one of the best audio visual systems in the country to watch your favourite sporting event.

共和酒吧

共和酒吧是一家兼具啤酒吧与高档餐厅的现代美国酒吧。酒吧以美国工艺啤酒及与酒搭配的独创菜单为特色。共和酒吧位于新克拉森曲线购物区。酒吧内100种特制啤酒、近250个全球范围内手工挑选的酒瓶、生气盎然的酒水单和独特的鸡尾酒等，与现代美国经典酒吧菜肴相匹配的酒水应有尽有。酒吧位于一个住宅集中区，拥有繁华鼎沸的餐厅氛围、时尚的装饰，还有全美最好的试听系统，供客人观看喜欢的体育赛事。

Café

USA

New York

2010

Photo: David Joseph

Nema Workshop

D'espresso

The D'espresso Café is a small café on Madison Avenue in New York City wich has an interior concept of a library which turned sideways. The interior designed by Nema Workshop to build a unique espresso brand and to develop a creative environment that connects to its location on Madison Avenue near Grand Central Station. In Conceptually and literally, the espresso bar turns a normal room sideways, creating a striking identity for the emerging brand. The design inspired by the nearby Bryant Park Library with the booklined shelves become the floor and ceilings and wood floor ends up on the walls. The pendants lamp installed sideways from the wall to make the room really turned sideways. For the 'shelves floor' they lined it with sepia-toned full size photograph of books printed on custom tiles, installed along the floor, up the 15 foot wall and across the ceiling.

意式D咖啡厅

D酒吧是纽约麦迪逊大街的一家小咖啡厅，咖啡厅由原来的图书馆改建而成。担任室内设计的Nema设计事务所旨在创建一个独特的意式咖啡品牌，营造一种适合中央车站麦迪逊大街地区特色而又充满活力的餐厅环境。无论从字面上还是设计理念上，这个意式咖啡厅都将普通的室内空间进行了成功转型，彰显了这个新兴品牌的独特身份。设计灵感来源于附近的布莱恩特图书馆，用书架作地板和天花板的原材料，木制地板直铺到墙底。墙面装饰的吊灯为空间增加了独特韵味，使空间设计实现了真正转型。

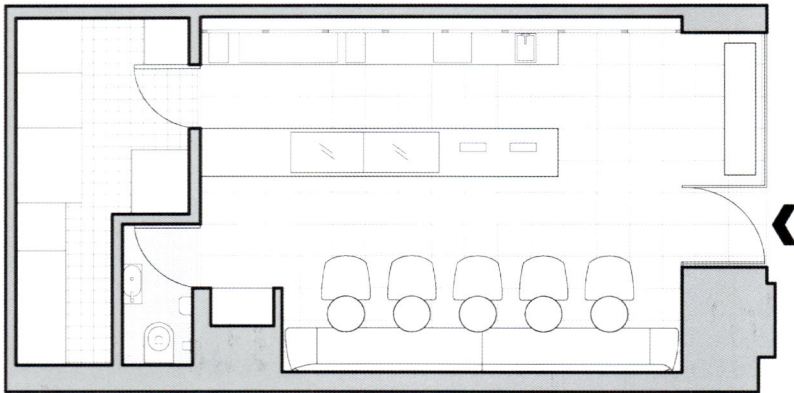

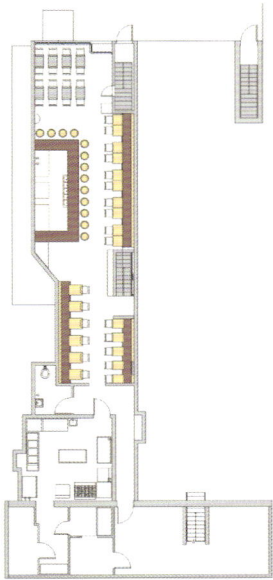

Highpoint Bistro

Highpoint is an American Bistro located on 7th avenue between 22nd and 23rd streets. The ground level is 2000 square feet with a sidewalk café and a backyard patio and the cellar is 750 sf with bathrooms, storage and an office. Market Design envisioned this project including all aspects of a full-kitchen restaurant; dining hall, 18-foot-long bar, expanded kitchen, storefront, custom furniture and banquettes. An amazing 32-foot-long mural designed by the prominent British illustrator Miles Donovan was commissioned to be installed on the north wall. Market Design created a unique atmosphere by modernising the historical American Bar typology with oak bar top and classical wood details into a clean wood-panelled, geometrically gridded eatery. The gridded structure which was called 'the trellis' not only created a canopy defining the bar, but also inhabited the backlit mural boxes.

海波因特酒吧

海波因特是位于22号与23号街道间7号大街的一家美国酒吧。Market Design 设计事务所旨在设计一个包含餐厅、长酒吧、开放式厨房、储物间、特制家具及宴会桌在内的拥有全套厨房设施的餐厅。北面墙上张贴着杰出的英国插图画家梅尔斯·多诺万的长达32英尺的壁画。设计选用橡木板作为屋顶材料，用古典木制品装饰格子形状的几何木质空间，营造了独特的空间氛围。被称作"格子"的隔断间不仅有覆盖酒吧区的顶棚而且将背光壁画囊括其中。

USA

New York

2010

Photo: Market Design

Market Design

Restaurant

USA

Las Vegas

2009

Photo: Karim Rashid Inc.

Karim Rashid, Project architects: Camila Tariki, Karim Rashid Inc. Evan McCullough, Karim Rashid Inc. Architect of Record: Leo A. Daily; Local Project management: MGM Mirage Kitchen Consultant: GEM

Silk Road Restaurant

Silk Road is an elaborate multi-cultural vision where Mediterranean Spice & Trade Market encounters the plush & intimate opulence of the Merchant Route. Silk Road's seductive bar invites one to lounge in its sculptural fibreglass seating that flows to the outside space. The seating separates the relaxed yet elegant bistro style seating, intended to create the atmosphere of a Merchant Meeting House, from the private booths of the intimate dining room. During the day, natural light will illuminate the vibrant & vivid colours of the room; while at night, subtle ambient light encourages an intimate yet vivacious ambiance. Silk Road invites guests to find themselves immerged in a seductive experience that invites to socialise freely in its sensuous surroundings from breakfast to diner.

丝绸之路餐厅

丝绸之路餐厅是一家精致的多元文化餐厅。这里地中海香料贸易市场的喧闹与商人之路的富贵奢华相互交融。在丝绸之路的一个诱人酒吧里，顾客可以仰躺在一直延伸到室外的玻璃雕刻座椅上。雕刻座椅的一面摆放着简单随意而又优雅别致的酒馆式座椅，给人一种商人会客室的感觉；另一面摆放着雅间会客室的私人座椅。白天，在太阳光的照耀下，整个餐厅色彩艳丽、明亮动人；夜晚微妙柔和的光线又营造出亲密而又活泼的氛围。在丝绸之路，顾客可以一日三餐在亮丽迷人的环境口自由随意地交流，感受极富魅力的用餐体验。

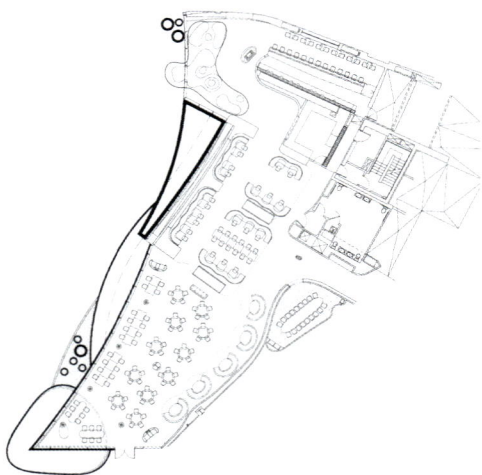

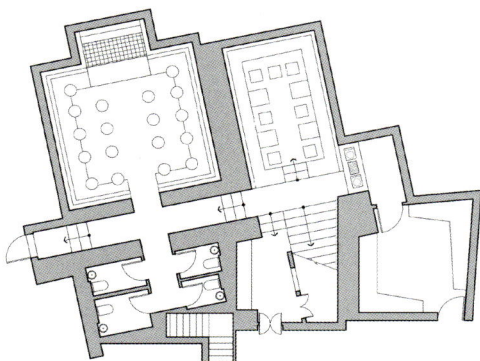

Restaurant & Bar

USA

New York

2008

Photo: David Joseph

Nema Workshop

Delicatessen

For the design of Delicatessen, Nema Workshop explored the concept of urban identity, namely the vibrant SoHo neighbourhood and more specifically the New York newsstand. Traditional boundaries between street and restaurant evaporate, allowing the main dining space to be loosely defined and astonishingly inviting. Underneath the bar stands a collection of clear soda pop bottles while behind the bar the wall is clad in leather subway tiles. The concept of informal urbanity develops complexity as one travels downstairs into the more intimate spaces below street level. The existing boiler room has been transformed into a glass-roofed lounge lined with an oversized mural by local artist Juan Jose Heredia. Just a few steps down from the courtyard lounge is minibar.

精品餐厅

Nema 设计事务所力求设计一家与周围环境相匹配的都市餐厅，比如考虑到整个生机勃勃的SoHo街区甚至细化到旁边的纽约报亭。餐厅与街道间没有明显的分界，主餐区环境宽敞，盛邀来访顾客。吧台下方摆放着一排排干净的苏打水瓶子，后面是皮革材料的地铁砖幕墙。来到楼下更亲密的半地下餐厅，那种不拘礼节的都市气息渐渐地变得错综复杂。现存的锅炉房已被改建成玻璃屋顶的休息大厅，里面张贴着当地著名画家约瑟·胡安（Juan Jose Heredia）巨幅作品。

Bar

USA

West Hollywood

2008

Photo: Grey Crawford and Witt Preston

Franklin Studios,Inc.

Foxtail Supperclub

Foxtail was conceived of as a place where its larger-than-life Hollywood investors could have some drinks, do some work, and bring their celebrity friends. Like the movies that might be made over its tables, Foxtail's rich environment has the effect of transporting its guests away from their everyday lives. The predominant patterns and details have Art Deco and Art Nouveau origins but have been modified and blended together with an eye on the seventies disco scene. The classic, white-plaster fluting of Art Deco was re-imagined as a glossy black skin on the walls, accented by copper stripes. The classic dome supported by columns is another Art Deco move that was tweaked by making the dome segmented, imprinting it with a Bibasque relief pattern and illuminating it with discoesque LED lights and runners.

狐尾夜总会

狐尾夜总会是其好莱坞投资人都来喝酒、工作、与名流界朋友聚餐的高级会所。餐厅装修模式与细节主要采用古典装饰艺术，也混杂着20世纪70年代的迪斯科酒吧装修风格。经典装饰艺术与新艺术风格的白石膏墙面覆盖着带有黄铜条子图案的光滑黑色表层。另一个装饰艺术风格的体现是柱子支撑的古典圆屋顶，屋顶被分成几部分，上面印有巴斯克浮雕图案花纹，灯光采用迪斯科专用的LED灯具与滑道。

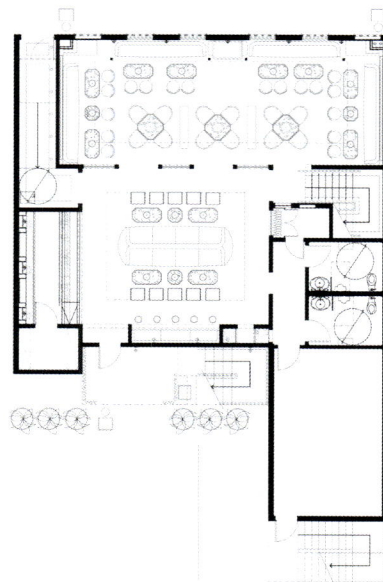

The Deuce Lounge in Aria Casino at City Centre – Las Vegas

Franklin Studios was asked to design a new type of venue - a bar/lounge which incorporates gaming. Guests can simultaneously enjoy the service and atmosphere of an intimate club while enjoying high-stakes gaming. The design was inspired by the surrounding desert with its sweeping geometries and rich textures. For instance, the main walls of the lounge are formed as a lattice of wood, modelled after the cactus skeletons that lie on the desert floor. The function of these open wood walls is to create a sense of enclosure while allowing the energy of the casino and the lounge to pass through and create a visceral connection. The ceiling is also a light, skeletal form defined by a loose grid of wood members which define a sprawling topography akin to the bare mountains surrounding Las Vegas. The bar front and custom carpet pattern reflect the scalloped forms of wind-blown sand. All of these organic references are formalised and contained within a framework modernism.

拉斯维加斯市中心Aria 赌场的The Deuce 鸡尾酒沙发酒吧

弗兰克林设计事务所受邀设计——兼具赌场的酒吧。设计灵感来源于周围沙漠被风吹过的自然图形和丰富质地。比如休息室的木格子墙壁就是仿照沙漠的仙人掌轮廓制成的。这些开放的木制墙营造了一种围墙式的感觉，又使赌场和餐厅的能量相互交融，打造一种内脏式的连接体系。天花板也是由松散的木格子组成的明亮框架，木格子的不规则图形很像拉斯维加斯周围起伏的群山。吧台正面的图案和特制的挂毯图案呈风沙德扇形图案。

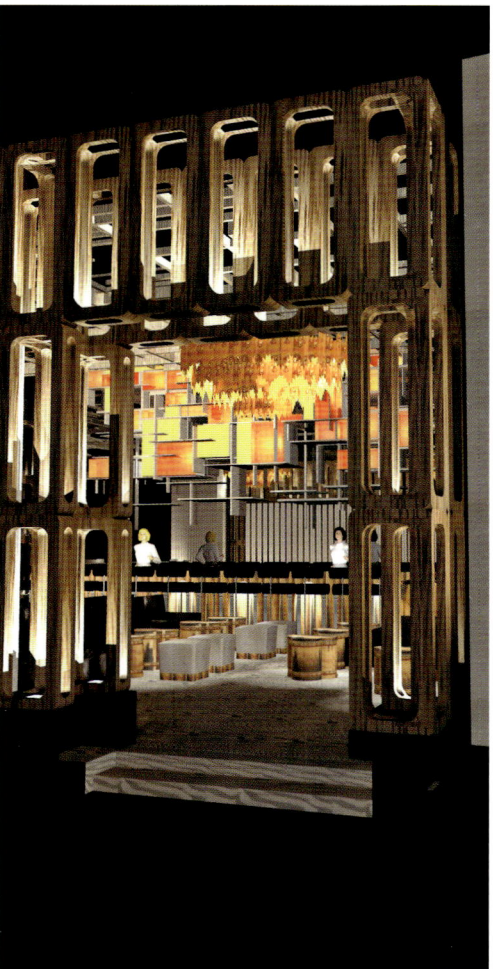

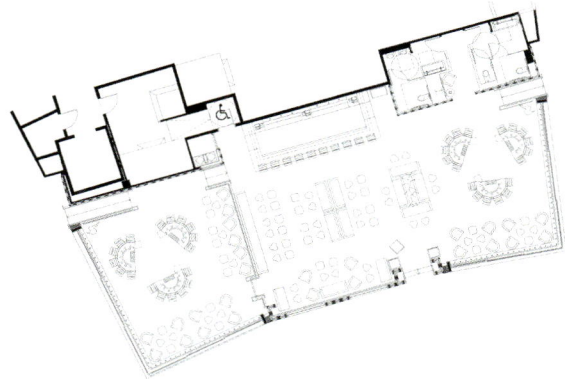

Photo: Franklin Studios, Inc.

Franklin Studios, Inc.

Américas

The new Américas Restaurant and 1492 Lounge in the Woodlands blends distinctive architecture with progressive Latin cuisine to create a captivating culinary experience. Michael Cordúa teamed up with Studio Gaia, a renowned New York City architectural firm, to create sleek interiors and bold, imaginative elements that complement the romantically provocative menu. From the hand-crafted wood tables to the slow tumble of water down three water walls to the oversized banana leaf murals overhead, guests are transported to the tranquility of a tropical rainforest. Fare Américas offers an adventurous menu of Progressive Latin cuisine taking ingredients that span the Americas, from Central America to Argentina. Executive Chef Jonathan Jones routinely ventures into unexpected culinary territory, exploring and adapting South American regional dishes in imaginative new ways.

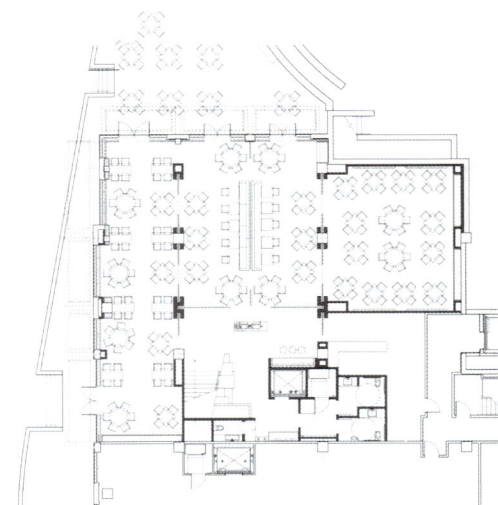

美洲餐厅

新美洲餐厅与林地地区的1492餐厅将杰出建筑与高级的拉丁厨艺相融合，为顾客创造了精美绝伦的美食体验。迈克尔（Michael Cordúa）与纽约著名的建筑事务所Studio Gaia 合作，创作了优美的室内设计作品，并在浪漫诱人的菜肴中注入了前卫的创新性元素。沿着楼梯下行就来到了餐厅。这里活力四射的红色、橘色与石头、木头等自然元素交相辉映，还可以环视林地航道酒店。从手工木制桌子到三面水墙上流下的涓涓细流再到头上面的巨幅香蕉叶子壁画，每一种元素都使顾客感觉到热带雨林般的宁静。

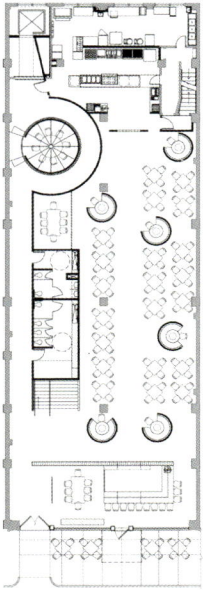

Red Primesteak

The architectural design takes full advantage of 18-foot ceilings, skylights and sheer volume to create spectacular urban beauty and drama. Each table offers an exciting vantage point for memorable dining experiences. The spectacular Wine Wall, some 55 bottles tall by 130 bottles wide, for a total capacity of 7,150 bottles, separates the RED bar from the Main Dining Room. Sweeping the room's grand dimension with a warm glow are suspended 'rays' of red neon. These create energy and light that refract off the building's rustic walls. The rays frame a dramatic procession for diners entering the Main Dining Room. The focal point is the Exhibition Kitchen where red portal highlights the activity and a glowing grill.

特级红牛排餐厅

餐厅设计充分利用18英尺高的天花板、天窗及整个室内空间，创造了奇妙的而戏剧般的城市美景。每一张餐桌就餐都可以感受到刺激而难忘的用餐体验。壮观的酒墙，横着能摆放130个瓶子，竖着摆放55个瓶子，总面积7150个瓶子大小，是红色酒吧与主餐厅的分界线。悬挂空中的红色霓虹灯放射出温暖的光线，照耀整个房间。这些光线沿着手工幕墙墙面向四周折射，释放出巨大能量。这种设计理念叫做"赤色风能"红色光线一直照射到主餐区，勾勒出奇妙的进餐路线。

Bar

USA

Oklahoma

2007

Photo: Scott McDonald, Hedrich Blessing

Elliott+Associates Architects

Restaurant

USA

New York

2008

Photo: Moon Lee Photography

Studio Gaia

Crisp

In designing Crisp, vibrant orange and green colours were selected to represent and enhance the freshness of Crisp's ingredients and owners' philosophy. Warm wood and light flooring brighten this petite eatery. Custom seating kiosks with stainless reflectors create playful unique ambience. The special designed desks and chairs look like the music studio, the metals above the desks are shinning brightly, in the meantime, the metals can reflex the orange colour of the desks, and so the whole dinning area is full of lovely bright. The large picture windows were designed all around the restaurant; the wall was covered by the yellow brickl. On one side, some large mirrors were designed to enlarge the whole interior area.

克里斯普餐厅

设计师选用艳丽的橙色和绿色作为主色调，彰显了餐厅食材的新鲜感。暖色的木材和亮色的地板使得狭小的空间变得开阔；特制的座椅带有不锈钢反射器，营造了趣味十足的氛围；桌子上方的金属结构闪闪发光、高大的落地窗环绕在空间四周、墙壁采用黄色砖石覆面，共同打造了一个特色十足的就餐环境。

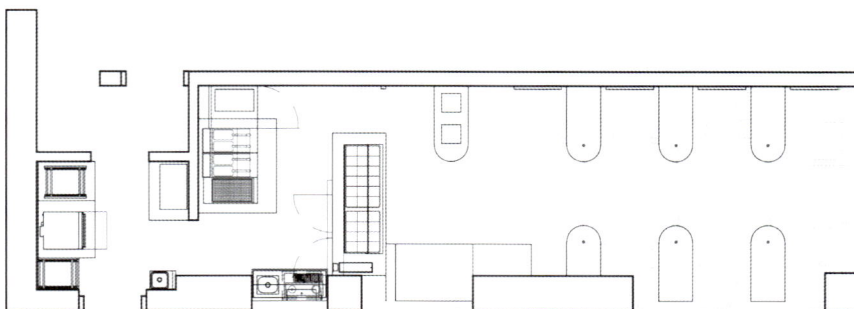

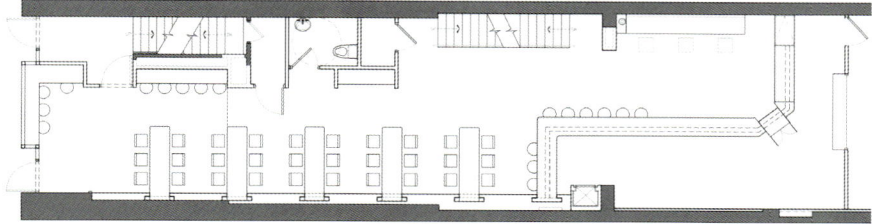

Aroma Espresso Bar

The Israeli café chain Aroma Espresso Bar recently opened its second location in New York, on Manhattan's Upper West Side. With vibrant colours, playful wall graphics and designer chairs styled for lingering, the two-level coffeehouse translates the concept for the company's original New York venue—a long, narrow Soho space with a huge window facade—to a completely different neighbourhood vibe. The material template starts with red wall tiles that arch down from the ceiling and extend the brand's modern-yet-homey ambiance. "We took the colours that are characteristic of Aroma in Israel, and created a more comfortable feel, like a kitchen at home," said Ilan Waisbrod.

香气咖啡吧

香气咖啡吧是一家以色列咖啡连锁店，于2009年在纽约曼哈顿开设了第二家分店，共为两层。设计师将原有的细长家庭式办公空间转变成另一种韵味的休闲咖啡厅，醒目的色彩、趣味十足的墙饰图案以及特别定制的椅子更添特色。红色的瓷砖墙壁从天花板处蜿蜒而下，形成拱形结构，营造了现代而舒适的氛围。

Bar

USA

New York

2009

Photo: Moon Lee Photography

Studio Gaia

Restaurant

USA

New York

2008

Photo: N/A

Studio Gaia

Mc Donalds Urban Living Prototype

The design focuses and celebrates on the freshness of the ingredients and promotes a fresh and healthy outlook to the household brand name. Upon entering, guests are greeted with a full height custom transparent divider panel with oversized produce graphics. The main inspirations behind Urban Living design are: Inviting, Lounging, Intimate and Relaxation. There are various main design features within the Urban Living design: a. The yellow horizontal continues 'band' that wraps the interiors from ceiling to the vertical high table (TV) wall then transitioning to the floor. b. Light wood ceiling panels with oversized pendants above 'lounging banquets' creating a warm and inviting feel. c. Custom black and white graphic wall covering and semi-see thru panels creates privacy and graphic ideas based on store localisation.

麦当劳—城市生活设计原型

设计师在入口处设置了与天花板等高的定制玻璃隔断板，上面绘制着店内食物的巨幅图片。'城市生活'理念由四个要素构成：热情、休闲、亲切、舒适，特色主要表现在：a.黄色包裹着整个空间，从屋顶延续到墙壁再延伸到地面上；b.休闲宴会厅内木质天花板以及悬挂着的巨型吊灯增添了温暖舒适的氛围；c.定制的黑白相间墙面装饰以及半透明的板条营造了私密感。

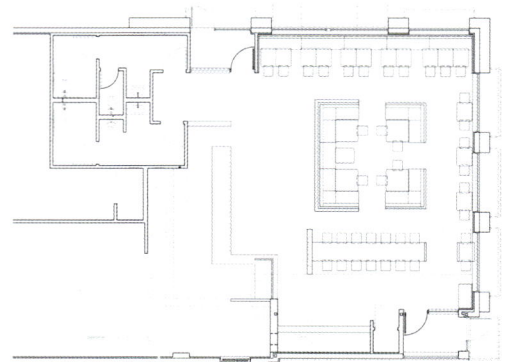

Restaurant in Marriott Marquis

The $138 million renovation of this Atlanta landmark hotel respects the original Portman design while enhancing existing features and relocating all food and beverage outlets off the lobby bar creating a hub of activity. Phase 1 of the renovation was completed in August 2007 and included the opening of four new food and beverage facilities: 'Sear' fine dining restaurant, 'High Velocity' sports bar, 'Pulse' lobby bar and a new Pool Bar. The 'Pulse' lobby bar is the focal point and brings new life into the hub of the hotel. The fifty foot structure is clad in resin panels that are backlit and change colour throughout the day to reflect the 'pulse' of the property.

万豪伯爵酒店餐厅

一期的改造于2007年8月完成，四个新的食品和饮料卖场设施开始营业：美式"煎炸"餐厅、"高速运动"酒吧、"脉动"大堂酒吧和台球酒吧。"脉动"大堂酒吧是项目的重点，它将新的活力带入酒店核心部分。15.24米高的结构用树脂板建成，利用背光照明，让它在一天中变换色彩来反映酒店的"脉动"。

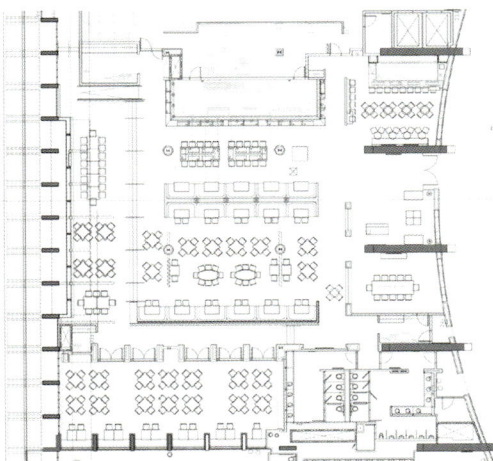

Restaurant

USA

Atlanta

2008

Photo: Thompson, Ventulett, Stainback & Associates

Thompson, Ventulett, Stainback & Associates

Café

USA

Brooklyn

2008

Photo: Seong Kwon

Giancola Contracting

Root Hill Café

The design presents something new to the streetscape, while referencing the archeology of the space within. The salvaged tin panels were stripped down to the bare metal and recomposed into a series of bands that shift across the ceiling, intersecting the two consistent strips of fluorescent lights. Rotten floor joists provided the motivation and economic justification for creating an entirely new multilevel floor structure within the café. Finished with reclaimed lumber and dominated by a gradually sloping ramp, the floor is the most dynamic surface within the space. At the lower level, the counter area slowly transforms into upholstered recessed seating as one moves towards the rear of the café.

根山咖啡厅

根山咖啡厅为这一区增添了一抹全新的精致。内部空间设计遵循历史特色，天花板上原有的锡板被拆除之后重新组合，荧光照明设备穿插其中，更加引人注目。看上去有些破旧的地板接缝营造了层次感的同时，带来不同的韵味。一层，柜台区缓慢延展渐渐形成了一排凹陷的座椅，特色十足。

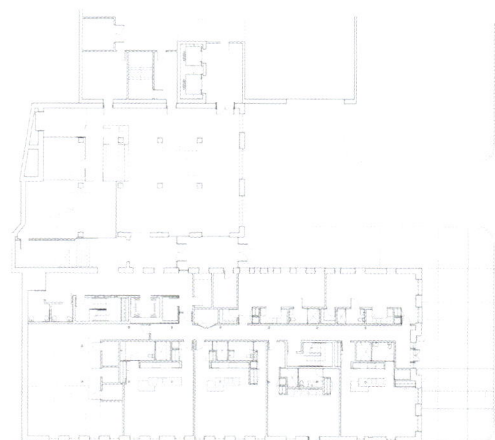

Mixt Greens

William Duff Architects designed Mixt Greens' first retail space, creating a distinctive, flexible retail concept that the up-and-coming restaurant has rolled out at additional locations since. The focal point of the design is the long transaction counter where customers cue, and where ingredients are on display. Like the bench seating around the perimeter of the restaurant, the front of the counter is wrapped in Kirei Board. The countertop is made from a locally harvested slate, a natural stone that provides a dark accent. Other important aspects of the sustainable concept include the specification of all FSC-certified wood products, and the protection of indoor air quality through the use of formaldehyde-free plywood and zero-VOC paints.

米克斯特绿色食品店

WDA建筑师事务所负责打造了第一家米克斯特绿色食品店，强调独特及灵活的零售空间。长长的餐台是整个餐厅的核心元素，上面展示着各种菜品，同时也为顾客点餐提供了场所。餐台的表面采用天然石材打造。此外，设计师格外注重环保理念，采用FSC认证的木质产品以及不含甲醛的胶合板等。

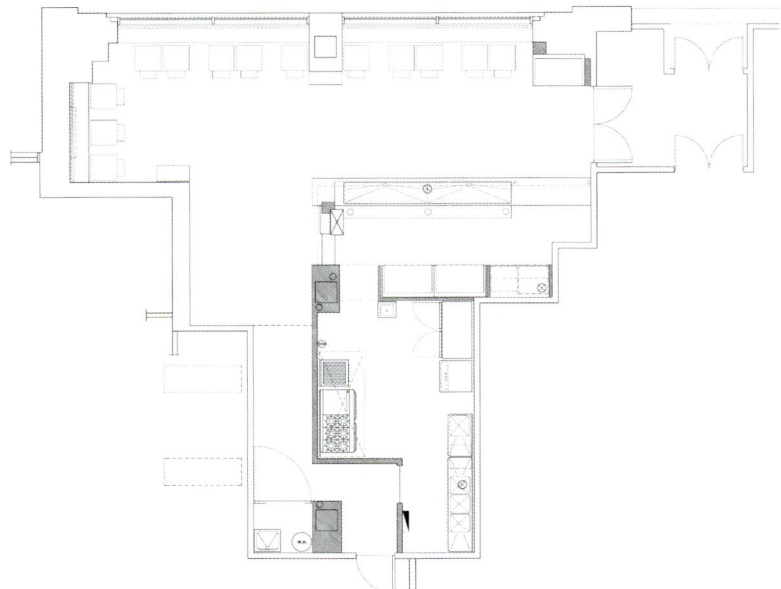

Restaurant

USA

San Francisco

2007

Photo: Lucas Fladzinski

William Duff Architects

Restaurant

USA

New York

2008

Photo: Studio Gaia

Studio Gaia

McDonalds

Studio Gaia was commissioned by the multi-billion corporation that helped shape American culture as well as the international food marketing in designing their brand new wardrobes that is not restricted by the industry standards but to create a brand experience that reflects fresh, modern and warm interior elements. The design focuses and celebrates on the freshness of the ingredients and promotes a fresh and healthy outlook to the household brand name. Upon entering, guests are greeted with a full height custom transparent divider panel with oversized produce graphics. Red wall tiling, light and wooden tile flooring as well as vibrant colours seating help create an open but warm dining and café experience.

麦当劳

GAIA工作室受邀负责麦当劳室内空间设计，要求摆脱行业标准的制约，打造一个全新的现代化氛围，旨在突出麦当劳清新健康的形象。为满足这一需求，设计师在入口处设置了与天花板等高的定制玻璃隔断板，上面绘制着店内食物的巨幅图片。红色瓷砖墙壁、木质地板配以活力十足的颜色营造了愉悦而温暖的就餐氛围。

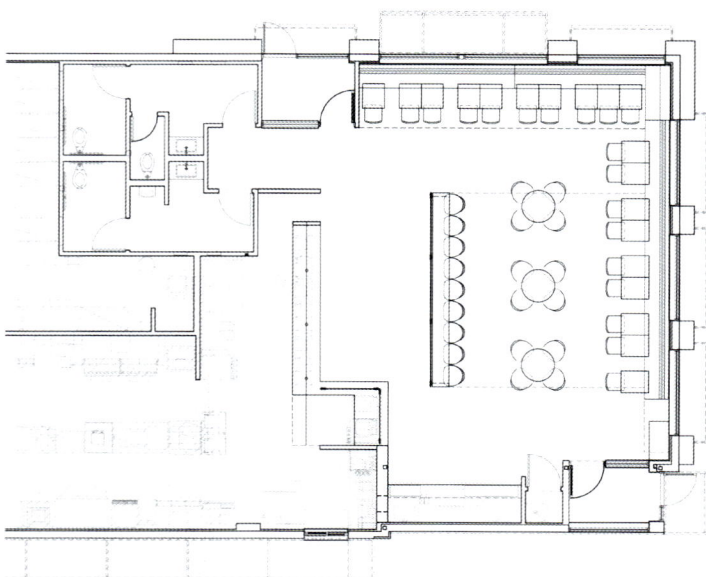

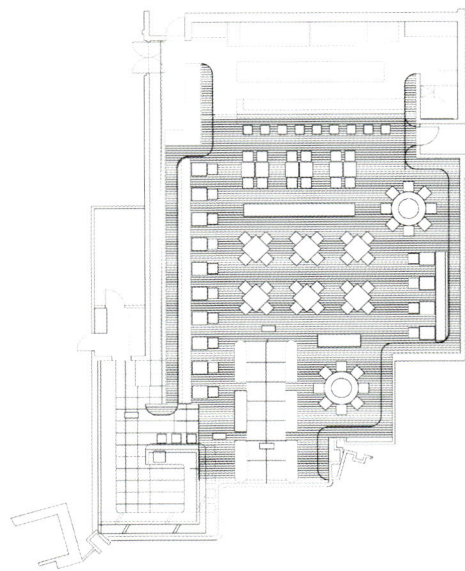

Restaurant

USA

Las Vegas

2009

Photo: Barry Johnson

Design Spirits Co., Ltd.

Beijing Noodle

A space, gently covered with light through a silk-like skin. The challenge was to break the preconceived ideas that make up the elements of an interior, by governing the space into 'one coordinated element' through the creation of a seamless border. This contemporary Chinese restaurant, Beijing Noodle No. 9, is located within the huge casino hotel holding over 3,300 rooms in Las Vegas. The restaurant is adjacent to the casino; consequently, the excitement, gaming machine sounds, and neon lights are naturally overflowed into the space. In general, a space usually consists of various interior elements, materials, series of products, and patterns placed appropriately.

北京九号面馆

此餐厅是一个拥有丝绸般表面，柔和灯光设计的室内空间。设计师要打破固有的室内元素设计理念，通过无缝式的空间设计将餐厅设计为一个统一的和谐元素。北京九号面馆这家现代中式餐厅位于拉斯维加斯拥有3300个房间的赌城酒店内。餐厅与赌场毗邻，因此赌场的喧闹声、赌博机的噪音和霓虹灯的闪烁都能映射到餐厅内。空间通常由不同的室内元素、材料、产品和相对应的图案组成。

Restaurant

Mexico

Mexico City

2009

Photo: Simone Micheli

Simone Micheli

La Casa Italiana

The way that the designer manages the distribution of spaces and places the furnishing elements tunes up according to the increasingly rapid changes that characterise the contemporary lifestyle. The stereotype linked with the traditional will be overcome, to outline noteworthy living scenarios. This is the context of 'The Italian Home' project: a chameleonic metropolitan creation where to celebrate the spirit of the Italian design in modern times.

拉卡莎意式餐厅

拉卡莎意式餐厅整个空间的布局精巧、装饰灵活、风格清新而富于变化，彰显当今生活方式的日新月异。设计风格突破常规设计，生动的应用场景设计分外引人注目，充分展现当今意大利餐厅设计的多姿多彩。

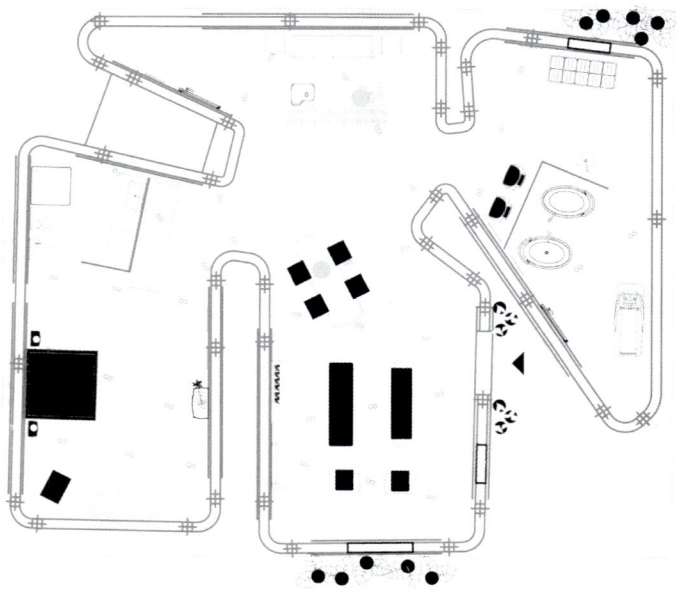

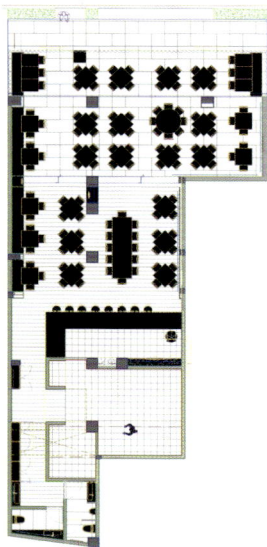

El Charro

This Mexican restaurant bar El Charro located in La Condesa, Mexico City, allowed the designer to make an intervention in an existing place, taking care of the functional requirements while satisfying and expressing the client's character and personality. This project was done in collaboration with a Mexican artist called Jeronimo Hagerman who designed the ceiling and called it: 'Ando volando bajo' ('I am flying low'). The ceiling is covered by a black perforated shield. Purple and pink dissected flowers are hung in the ceiling creating different forms, which evoke Mexican landscapes of clouds and flowers. The moment you go inside the restaurant the smell of wood and flowers fill your lungs. The floor is made of dark stone and carries a wood platform that takes you into the restaurant.

埃尔·查罗餐厅

设计师在原空间内加入新设计，既满足了餐厅的功能要求又尊重和表现了客户的品质与个性。这项工程和一位叫做Jeronimo Hagerman 的墨西哥艺术家共同完成。天花板由Jeronimo Hagerman 设计并给它取名为"Ando volando bajo"（意思是我在低处飞）。天花板外附有一层黑色打孔防护层，上面挂着一片片紫色和粉色相间的花丛，象征着墨西哥白云和花朵。顾客一走进餐厅，就会感受到沁人心脾的花香与木香。地面材料由黑石组成，上面还有一个木台，顾客可以沿着木台走进餐厅。

Restaurant

Mexico

Mexico City

2009

Photo: Cheremserrano

Cheremserrano

127

Restaurant

Mexico

Mexico City

2009

Photo: CHEREMSERRANO

CHEREMSERRANO

Capicua

The Capicua bar and restaurant is designed by the studio of CHERESERRANO and located in the capital of The United Mexican States, Mexico City. The word Capicua means head and tail, beginning and end, a place to gather together with friends without time or formalities. In the Capicua bar and restaurant, the specialty is the traditional tapas, which are a part of the Spanish culinary culture. Located on the top of Ligaya Restaurant, the Capicua bar and restaurant required an architectural intervention of an existing restaurant in order to provide an open space that would allow clients to smoke outside.

CAPICUA 餐厅

Capicua 酒吧与餐厅由CHERESERRANO 建筑事务所设计，位于墨西哥合众国首都墨西哥城。"Capicua" 的意思是头和尾，开端和结局，是一个摆脱正规礼仪束缚、和朋友聚会的绝佳场所。Capicua 酒吧与餐厅的特色就是西班牙餐厅文化中传统的餐前小吃。Capicua坐落在Ligaya 餐厅的顶楼。在现有的餐厅中增加另一个餐厅设计是为了给顾客提供一个到室外吸烟的开放式空间。

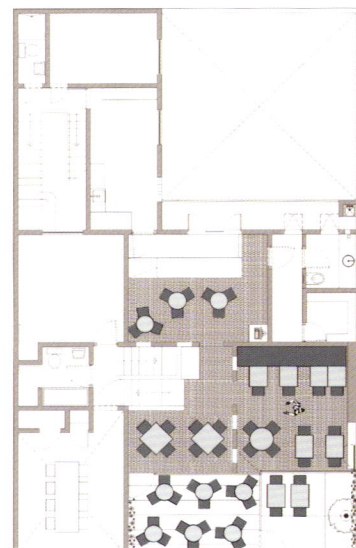

Distrito Capital

The Enrique Olvera-curated restaurant on the fifth floor is one of Mexico City's newest hotspots. The giant windows offer a great view to the outside scenery. The interior design keeps a low profile, yet is very luxurious. The seemingly simple design is actually very detailed, which would offer guests comfortable top-grade experiences.

Distrito Capital商务酒店餐厅

位于酒店5层的餐厅已经成为了墨西哥城的一个最新的热点，由Enrique Olvera 操刀设计。大面积的开窗给了餐厅很好的视野，室内设计是一种低调而华丽的风格，简约而细致的设计给顾客闲适而高档的享受。

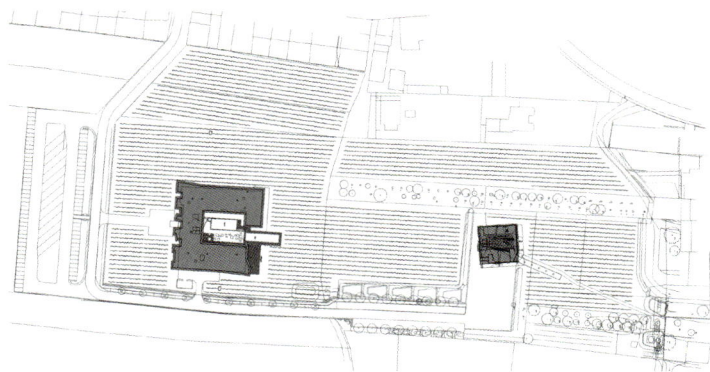

Restaurant

Mexico

Mexico City

2009

Photo: Enrique Olvera

Enrique Olvera

Desserts

Mexico

Nuevo Leon

2009

Photo: Caroga Foto (Carlos Rodriguez)

Anagrama

Theurel & Thomas

Theurel & Thomas is the first patisserie in Mexico specialised in French macarons, the most popular dessert of the French pastries. For this project it was very important to create an imposing brand that would emphasise the unique value, elegance and detail of this delicate dessert. One of the most important extensions of a brand, which is business based in store selling, is the design and ambiance of the stores. The patisserie of Theurel & Thomas has an enlightened space with an exclusive and elegant atmosphere. The store location is found in San Pedro, MX, Latin America's most affluent suburb. White is a central part of the design and it plays as a contrast with the colours of the French macarons.

Theurel &Thomas餐厅

Theurel & Thomas 是墨西哥首个制作广受欢迎的法国甜点——法式小圆饼的美食店。项目设计的核心就是要设计一个有影响力的品牌，来突出这道精致甜点的独特价值、韵味与工艺。一个店面销售品牌最重要的宣传手段就是店面设计与店面氛围。Theurel & Thomas 拥有法式甜点的专业权和优雅的室内氛围。商店位于拉美地区最有名的富人区——墨西哥的圣佩德罗。

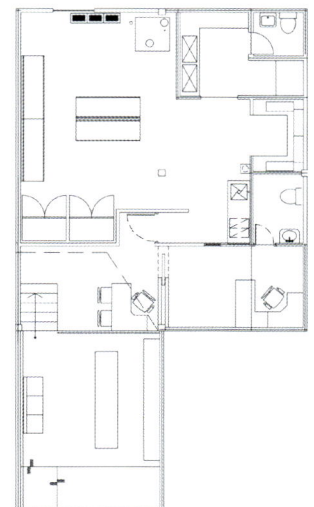

Rubik, Restaurant & Bar

Nested on the upper level of a shopping plaza, Rubik was conceived as a young but elegant eating place. Design was to develop an experience that built upon Merida's new wave of high quality gourmet restaurants; To attract clientele and make them feel comfortable, the space is outfitted with dark painted walls, carpet and porcelain floors and aluminium compound elements. The project maximised the environment by using the full width of the area and placing the bar as a focal point at the end corner concealing the service corridor and pulling the view to the back of the restaurant. Muted shades of white and orange were used to colour the dining area, while the bar is a bit brighter in tone.

鲁比克餐厅酒吧

鲁比克坐落于一家购物广场的二层，虽建成不久却是一个极为高雅的用餐场所。设计旨在响应梅里达地区高级美食餐厅建设的新浪潮。为了吸引顾客并给顾客一种温暖舒适的感觉，空间配有黑漆墙、地毯、瓷质地面和铝制配件。设计极大地利用了空间宽度，并将设计核心——酒吧吧台设计在房间的后角，遮挡了工作通道而且将顾客的视线吸引到餐厅后面。白色与橘色是餐厅的主色调，而酒吧色调略显明亮。

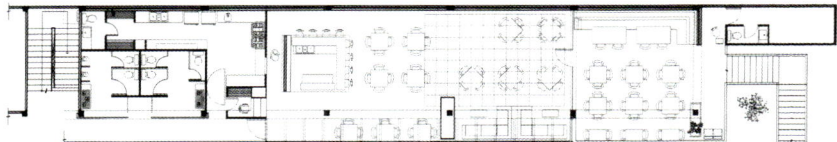

Photo: Eduardo Cervantes

Arquitectos Interiores, S.C.P.
Mauricio Ramirez, Guadalupe Avila
Collaborators: Fabiola Solis, Aniela Mendiburu

Lounge & Bar

México

Cabo San Lucas

2009

Photo: Romana Lilic + Onairam Saira / LA76 Strategic
Design

A10studio

Re.Evolution Lounge+Bar

This is a project that you'd most probably be seeing in Miami or Latin America than in a touristy Mexican beach resort; it provides some fresh air to the city nightlife, and has also unintentionally triggered a micro urban renewal in the neighbourhood. Cabo San Lucas, a popular vacation spot and one of the places in Mexico with the highest density of high-end resorts, faces two public images; on one side, a place of highest luxury and style, visible through resort catalogues and sponsored promotions, while creating a 'vision' of 'traditional' Mexico with palapas, colourful taco stands and mariachis. On the other side the town of Cabo San Lucas in its mere heart reveals dusty streets, poverty, cheep local labour and poor urban image due to the negligence of urban planners.

演化酒吧

这样的酒吧相对墨西哥海滩度假区来说好像在迈阿密或拉丁地区更常见。它为都市夜生活注入了新鲜的空气，也在冥冥中为整个街区掀起了小范围的城市革新。汇集了众多高端度假村的卡波圣卢卡斯，是墨西哥最受欢迎的度假地点。它的对面是代表墨西哥城市形象的两个区域：一面是极为奢华时尚的度假村；墨西哥草棚、五颜六色的玉米卷亭子和墨西哥流浪乐队等打造出了一片传统墨西哥的城市景象；而另一面则由于城市规划者的忽视成了满街灰尘、贫穷落后、到处都是廉价劳动力的穷人区。

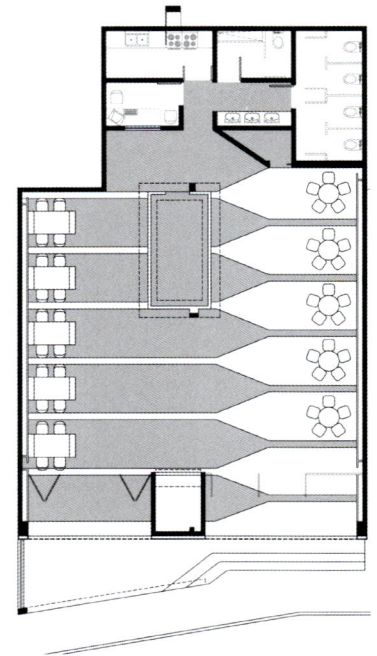

Tea Vana

The use of wood defines the elegance of the place. Found in a second ceiling, above the central sitting area, is a platform raised 15 centimetres from the entrance level. By separating the false ceiling from the walls, giving it air and independence, this element gains a stable and even character. The right wall made out of wood with a staple form, contains the area for buying infusions, teacups and teapots. The rest of the walls are painted in white, which enlarges the space and makes an interesting contrast between the wood and the different colours of the furniture. The service area is completely hidden, which makes the space more compact and makes it whole and understandable.

瓦纳茶坊

木材的运用奠定了茶坊典雅的基调。设计师将墙壁和假吊顶分离开来，赋予空间平缓的特色。右侧的墙壁由木材打造，U形的结构区用于出售茶杯、茶壶等。其余的墙壁全部粉饰成白色，不仅在视觉上扩大了空间的面积，同时与彩色的家具形成对比。服务区完全隐藏起来，增添了空间紧凑感。

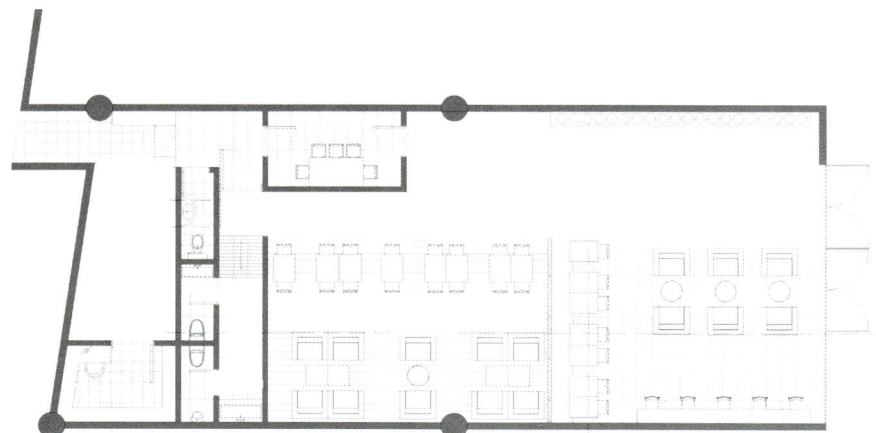

Lounge

Mexico

Mexico City

2008

Photo: CHEREMSERRANO

CHEREMSERRANO

Restaurant

Mexico

Mexico City

2009

Photo: Pedro Hiriart

Serrano Monjaraz Arquitectos

Jaso Restaurant

Mexico City is very well known for its restaurant culture. The Polanco area where most of the five and more stars hotels are located has seen the blossoming of very interesting epicurean proposals. In this same area Serrano Monjaraz Arquitectos were asked to develop the interior design for a new restaurant. The Chefs and owners of Jaso come from New York and they wanted to imprint their expertise and savoire flare not only through the food, but also from the atmosphere of this global restaurant. Step by step the concept was designed based on the sensorial elements: fire, air, water and sound. The materials play a very important role in this project; wood, marble and iron are mixed in different percentages and combinations in search for a very natural atmosphere. The clients are greeted at the reception with a big tree trunk that was rescued and incorporated into the layout.

Jaso 餐厅

墨西哥的餐厅文化闻名遐迩。五家星级宾馆驻扎的波兰科地区已经见证了世人瞩目的美食规划项目的成功运作。就在此地，Serrano Monjaraz Arquitectos 受委托负责一家新餐厅的室内设计。Jaso 的老板和主厨都来自纽约。他们不仅想要通过食物而且通过这家全球连锁餐厅的氛围来展示他们高超的厨艺。设计理念通过一系列感官元素得以展现：火、空气、水和声音等。原材料在项目设计中发挥着重要作用，木头、大理石和铁等材料的不同比例组合烘托出极为自然的气氛。接待处有一棵大树，热情地欢迎顾客。

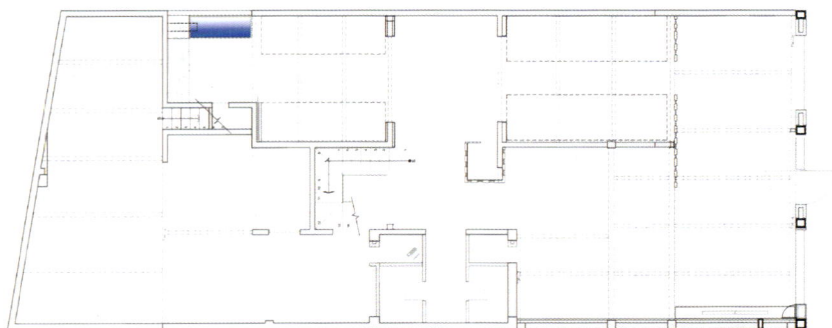

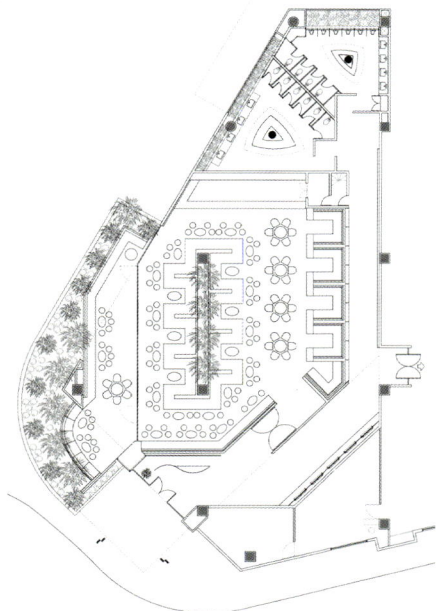

Nisha Acapulco Bar Lounge

This lounge bar is an entertainment place dedicated to the senses. Expressed through the architecture, the people, the music and the images, an alternative and virtual atmosphere is created. The bar's interior is a contemporary interpretation of a nautical theme. Upon entering, visitors are met with curved, wood-clad ceilings and walls punctuated by porthole-shaped screens displaying underwater landscapes. Five high definition screens framed as oval windows that may show either a clouded sky travelling at high speed or bottom of the sea images transforming it into a submarine. A dark foyer leads to a wooden lounge decorated with small groups of sofas that have a view of an outside smoking area. At one end there are several screens framed as pictures in a living room and on the other a 15 metre long bar counter and further up, at the back, a same size by 3 metres height hi-res video screen.

妮莎•阿卡普尔科酒吧

这家酒吧是一个愉悦身心的娱乐场所。设计师通过建筑、人群、音乐及酒吧形象表达了设计理念，打造了全新的空间氛围。酒吧的室内设计是对航海主题的现代诠释。顾客刚一进入酒吧，映入眼帘的就是木制包层的曲线天花板和装有舷窗形状屏幕的墙壁，屏幕上显示着各种海底景象。五个卵圆窗边框的高清屏幕有的显示着天空中高速流动的乌云，有的显示着潜水艇潜入海底的图像。黑色门厅直通向木制的休息厅。休息厅内摆放着成排的沙发，坐在沙发上可以看到吸烟区。休息厅一端是饰有图画屏幕的起居室，另一端则是15米长的吧台，最后面是15米长、3米高的高清显示屏。

Bar & Lounge

Mexico

Acapulco

2009

Photo: Sófocles Hernández

Carlos Pascal, Gerard Pascal / Pascal Arquitectos

Desserts

Mexico

Colonia Centro

2008

Photo: Sófocles Hernández and Jaime Navarro

Pascal Arquitectos, Carlosy Gerard Pascal

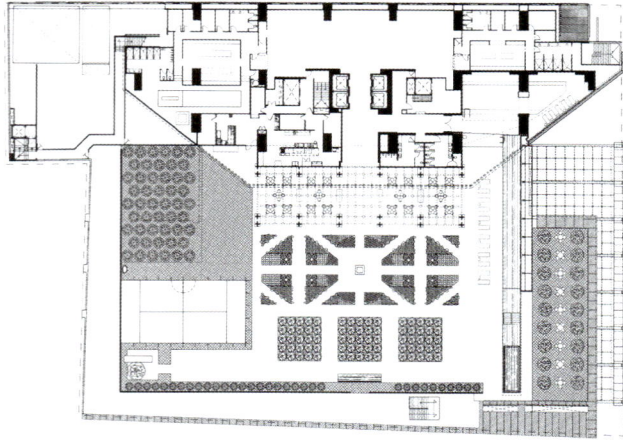

'Terraza Alameda' Cafeteria – Hilton Reforma Hotel

As part of the whole project a special area for guests who stay over the weekend in the city or more days was designed: a fully equipped and modern Fitness Centre & Spa, a 3,000 square metres and Roof Garden and the carefully-designed 4500 square metres' cafeteria. This gardened area match in a certain way with the original design pattern of the old Alameda Park. It also sourrounds the 'Terraza Alameda' cafeteria. This wonderful cafeteria locates under the roof topped pergola, and opens to the the lap pool and paddle tennis court, with an open kitchen and bar that allow to have a view of the whole roof garden, and share and participate in all the events from a climate protected environment, and at the same time enjoying the fresh air and food.

泰拉莎•拉梅达餐厅

为了顾客能更好地享受城市的周末时光，酒店项目又增添了一处特殊区域：设备齐全的现代健身中心和水疗馆、一个3000平方米的室内公园和这个精心设计的4500平方米的餐厅。花园中间就是泰拉莎•拉梅达餐厅。这家绝妙的自助餐厅位于藤架屋顶之下，直通小型健身游泳池和平台网球馆。开放式的厨房和吧台设计使餐厅拥有广阔的视野，可以环视整个花园。顶棚设计使顾客可以不受天气的影响参加各种活动，并享受新鲜的空气与美食。

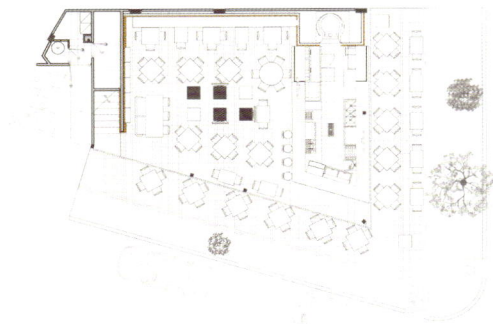

La Nonna

La Nonna is an Italian restaurant located in La Condesa. This project was made in collaboration with DMG architects. The design premise was to liberate the plan so the restaurant could take the most advantage of the 200 square metres local. The restaurant fuses its environment with simplicity. Incorporating the use of local materials, the floor is made of dark stone and is built unto the sidewalk, while the walls and ceiling are surrounded by red brick with special cutting on top of mirrors. The mirror was placed in order to enlarge the space and create a game of light and shadows. The furniture is elegant and simple. At the centre is placed a bar with a pizza stove and a wood counter to sit and enjoy a nice wine.

La Nonna 餐厅

La Nonna 餐厅是位于拉康迪萨的一家意式餐厅。这项设计与DMG architects 建筑事务所共同完成。设计的前提是要自由发挥空间规划，因此餐厅可以充分利用当地200平方米的空间。简约是整个餐厅设计的主线。设计中大量使用了当地的原材料：地板选用黑石材料，一直铺到人行道，墙面和天花板外附有红砖包层，在镜子顶端形成特殊的分界线。镜子可以扩大空间感并且可以实现美丽的光影效应。家具设计简单优雅。

Restaurant

Mexico

Mexico City

2010

Photo: CHEREMSERRANO Mexico

CHEREMSERRANO Mexico

Photo: José María Sáez, Daniel Moreno, David Barragán, Gabriela Delgado

Jose Maria Sáez, Daniel Moreno

Boca Del Lobo Restaurant

The intervention gives continuity to the restaurant and it is shown as a showcase to see and be seen. The floor works as the same time as platform, scenery and a place to sit. Outwards, a metal box frame the indoor activities that are crossed by an existing tree, a tree that becomes an object with an iconic presence. The project intensifies all the existing, as the tree, or recycles itself almost entirely. The existing building walls are cut with a grinding machine and moved to become the new walls. The wooden roof rises up and prolongs itself strengthened by external tensors. The structure above the roof allows a large and continuous space in the inside just with a single support.

Boca Del Lobo 餐厅

插入式设计增加了餐厅的持久性，使餐厅看起来像是一个可以观赏外部风景也可以供人观赏的陈列柜。地板像是一个平台，一处风景，一个提供座椅的地点。餐厅外部，金属外层将整个室内空间包裹起来，保留下来的横跨餐厅内部的大树已经成为餐厅的标志。就像大树沿自身环绕一样，项目设计也突出所有现有的装饰。原始的建筑墙壁经过研磨机研磨后又建成了新墙壁。木制房顶在外部张量的支撑下一直向上延伸。屋顶上的特制结构使室内空间在只有单一支撑底架的情况下保持宽敞明亮。

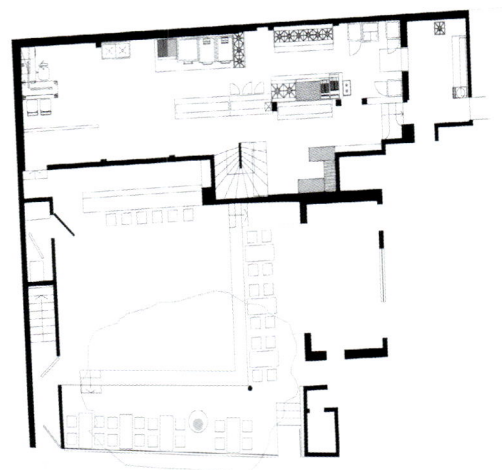

138

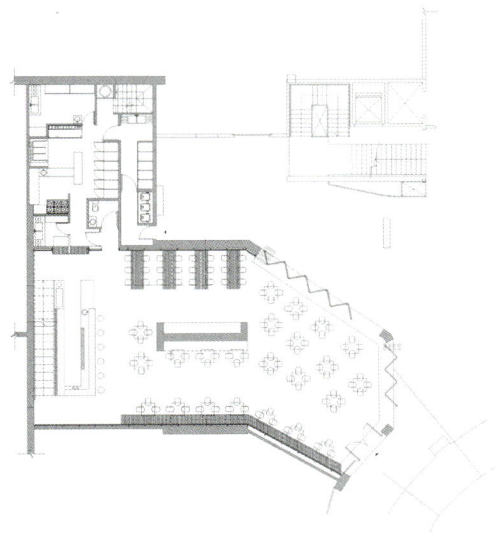

Restaurant Santa Pizza Parque Arauco

Create a new atmosphere, work with low budget materials were some of the conditions for the architectural project. The site, an irregular prism of 350 square metres with an open facade to the Boulevard II in Mall Parque Arauco Santiago de Chile, was to small to accommodate all the tables required for the restaurant. The programmatic distribution allocates the kitchen in a corner of the prism and on top of it the bathrooms and the service area. Three materials were chosen to work with: wood, stone and glass. The ceiling and the laterals walls were coated with a vertical pine skin tinted white, creating a continuous surface suspended in the interior of the restaurant that fold creating tables and benches. The lateral walls folds opening the view to the entrance, the ceiling with the same tactic shows the total height of the space announcing the access to the second level.

阿洛克圣诞比萨餐厅

设计师需要用低成本的原材料创造出新的餐厅氛围。350平方米的不规则棱柱形餐厅正面朝向智利圣地亚哥阿洛克购物中心的二号大街。室内面积很小，很难容纳规定数量的餐桌。餐厅布局规划将厨房设计在一个角落，上方是浴室和服务区。设计搭配使用木头、石头和玻璃三种材料。天花板和外侧墙外部采用松树皮质的白色涂布包层。餐厅室内空间的褶皱外层将餐桌和长凳子包活其中。

Photo: © Daniel Corvillon, © Aryeh Kornfeld

01ARQ. Cristian Winckler/ Pablo Saric/ Felipe Fritz

Restaurant

Chile

Santiago

2007

Photo: © Daniel Corvillon, © Aryeh Kornfeld

01ARQ. Cristian Winckler/ Pablo Saric/ Felipe Fritz

Restaurant Mercat

The 277-square-metre restaurant is located in the east side of Santiago de Chile. The area, mostly dedicated to art and design exhibitions, has an intense pedestrian activity. The design strategy focuses on the visual continuity between the street and the interior patio and to achieve this effect the perimeters of the public areas were surrounded by a continuous transparent strip. A fish pond reflects the sun light on the main saloon ceiling and helps to create a fresh atmosphere during the summer season. Material choices were reduced to minimum, wood for the interior deck and main access, black stone for floor covering and white stucco.

梅凯特餐厅

梅凯特餐厅位于圣地亚哥东部，占地277平方米。设计理念强调室内空间与室外街道的视觉连贯性，为此，设计师在空间四周运用了透明条状元素。主餐厅区，光线透过屋顶照射进来，经过鱼池的反射，空间更加明亮。材料选用以简约原则为主，入口处采用木材打造，地面铺设黑色石材。

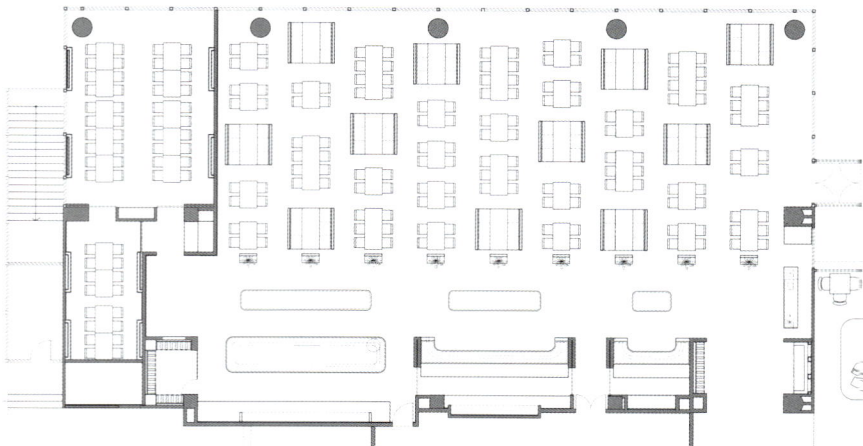

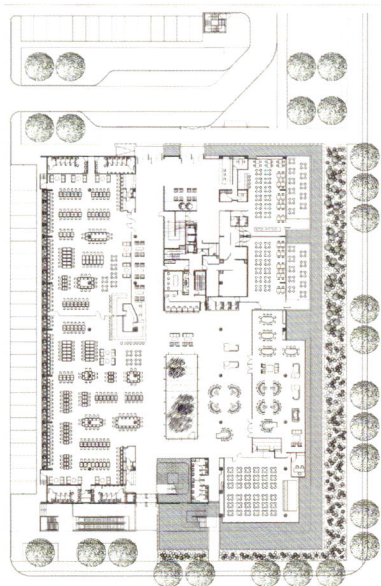

Restaurant in Talca Hotel & Casino

Since the very beginning, this commission was determined by a preexisting disposition of the various parts of the programme (casino, hotel, restaurants, and convention/event's centre), predefining surface areas, position in plan layout, and corresponding levels within the structure. A second important condition was its location sharing the site of a Shopping Mall in a peripheral location to the City of Talca. Though allowing to opening the project towards the further context gaining views of the Andes and Talca itself, the immediate context needed particular attention: the intersection of two high traffic main roads, and the Shopping Mall's desolated parking lot. The third relevant subject to take into account was the thermal conditioning of the building.

塔尔卡酒店及娱乐城餐厅

塔尔卡酒店及娱乐城项目设计主要面临三大挑战：其一，如何根据各种功能空间（娱乐城、酒店、餐厅、会议室）的特色决定室内格局；其二，处于塔尔卡市城郊并与大型购物中心共用一块场地，并需在其中穿插两条高速公路及停车场；其三，建筑内外温度调控。

Restaurant

Chile

Talca

2008

Photo: FVM Decoración y Diseño

FVM Decoración y Diseño

Restaurant

Brazil

Sao Paulo

2009

Photo: Leandro Uhlmann

Forte, Gimenes & Marcondes Ferraz Architects

Rohr's Employee's Refectory

The project for the new employee's refectory at the Rohr Company was based on some client premises and in the fact that this new area was in a previous built warehouses. The space for the installation of the refectory, which represents about half of one of the warehouses of the factory is now occupied Rohr's deposit and was chosen because of factors such as sufficient area, privileged location - next to a disabled parking, ease of removal of the warehouse and logistics satisfactory from the point of view of employees and the operation of the cafeteria itself. The kitchen area was located along the side of the warehouse that enables easy access for trucks and cars that make the delivery of prepared food, raw materials or the trash removal. This kitchen was designed to operate in two different regimes, allowing total flexibility for Rohr and also for the concessionaire that will operate the cafeteria.

Rohr公司员工餐厅

遵照一些客户的要求，Rohr新员工食堂位于公司之前建成的仓库内。员工餐厅占据仓库面积的一半，原是公司的储藏室，由于面积充足、位置优越（位于残疾人停车场附近）、员工可方便地从仓库运送货物及餐厅本身的易操作性等优势，选择在此修建餐厅。厨房位于仓库的一侧，以便卡车和汽车方便进出运送预加工食品、原料和垃圾。厨房的设计既符合Rohr灵活作业的特性，又满足了餐厅经营者的要求。

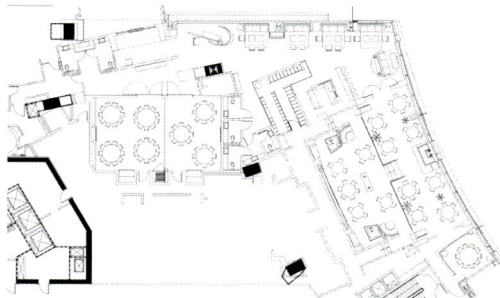

Sushi Restaurant

The Sushi bar was designed to function as a main entrance to the restaurant. A mezzanine floor was located above the Sushi bar, given a lower ceiling to this area. Arthur Casas designed a backlit ceiling at the Sushi bar area to estabilish a relationship with the Japanese culture and the restarurat cousine. At the same time the backlit ceiling signalised the main entrance, catching the public attention and amplifying the lower ceiling space. The comun dining table has received Tom Dixon's pendant lights and directonal lights were placed along the walls, washing the artwork and the plaster finished walls, creating a more dramatic and theatrical effect to the dining area.

寿司餐厅

寿司吧的功能就是作为进入餐厅的主要通道。寿司吧上方有一层中层楼，使寿司吧的空间有了一个较低的天花板。阿瑟•卡萨斯给寿司吧的空间设计了背后照明的天花板，为了在日本文化和餐厅的菜肴之间建立一种关系。公共餐桌有汤姆•迪克森的垂饰照明，另外，艺术品和石膏装饰的墙面上还装了方向照明灯，给就餐区创造出一种戏剧效果。

Restaurant

Brazil

Sau Paulo

2008

Photo: Arthur Casas

Arthur Casas

Restaurant

Brazil

Sao Paulo

2008

Photo: Micro Kogan

Micro Kogan

Fasano Hotel E Restaurant

Located next to luxury shops and refined restaurants in the centre of Sao Paulo, Brazil, is the Fasano Hotel E Restaurante. The Fasano Family, recognised for over a century for their hospitality and fine northern Italian cuisine, runs this new Sao Paulo hotel and restaurant. Under the direction of Rogerio Fasano, proprietor of the leading restaurant in Brazil, the hotel's restaurant serves traditional Italian cuisine based on the Fasano's family Milan origins. National and international jazz stars perform in the hotel's baretto, while the charming Ruggero Fasano room offers a large terrace for breakfast and private meetings.

法萨诺酒店餐厅

法萨诺酒店位于巴西首都圣保罗中心区域，周边遍布奢侈品店和精品餐厅。享有盛名的意大利酒店大亨法萨诺家族经营着这家酒店和餐厅。在酒店经营者罗杰里奥•法萨诺的带领下，酒店的餐厅提供意大利传统美食。知名的爵士明星会在酒店里表演，酒店的客房提供可享用早餐和进行私人会晤的巨大平台。

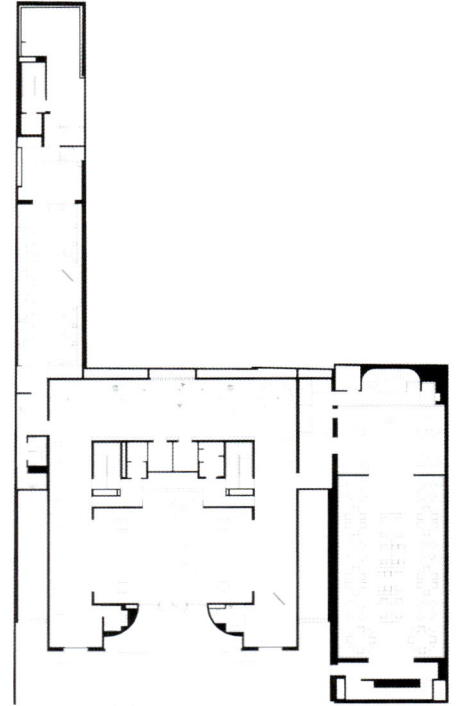

Urban Station

Urban Station has been conceived as a hybrid space, combining café and temporary office, specially designed for nomad workers. The use of the café as a working space has been a long-lasting tradition in Argentina, but the development of connectivity over the last years has boosted this habit. Urban Station is organised on two floors. The ground level accommodates almost classical bar tables and living areas for small informal meetings. On the upper floor a number of private rooms equipped with projection equipment and video conferencing are located. Next to them, a flexible open space can serve as a small auditorium area for conferences. Connecting both floors, a group of three extra-large lamps – specifically designed by the studio – floats over the tables.

城市车站

城市车站是一个综合空间，结合了咖啡厅和临时办公室，特别适合自由职业工人。在阿根廷，将咖啡厅作为工作场所是一个传统，近几年更加流行。城市车站分为两层。一楼设有经典的吧台和非正式会客区。二楼由一系列的独立房间组成，里面配有投影和录像等会议设施。这些房间旁边是一个灵活的开放空间，可以作为小礼堂。一组特别定制的巨型灯具悬在桌子之上，将两层空间联系了起来。

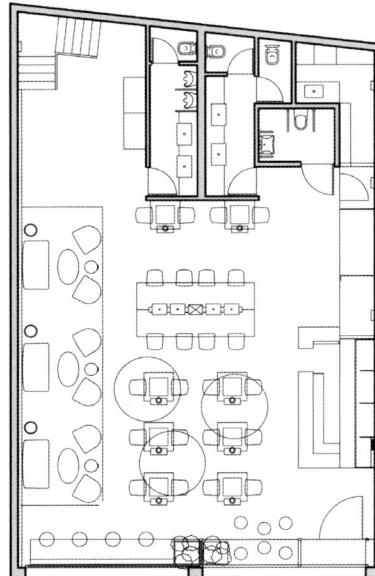

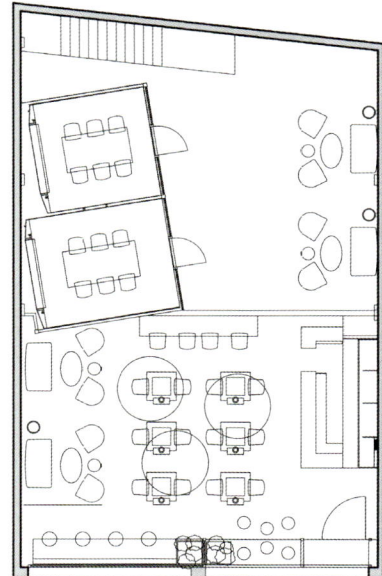

Café

Argentina

Buenos Aires

2010

Photo: Sergio Esmoris

TOTAL TOOL BA

Restaurant

Peru

Lima

2009

Photo: Michelle Llona R

LLONA + ZAMORA Arquitectos (Michelle Llona + Rafael Zamora) +
Fernando Mosquera

El Camion Restaurant

On the Panamerican Highway, nineteen kilometres south of Lima, there is a mandatory rest stop for truck drivers, which include a gas station, rest areas and food services. EL CAMION is located on one of the corners of this rest stop, in an area of 22.5metresx10metres, where it is unfeasible to park. Surrounded by trucks, with constant movement, there is always a new landscape. The restaurant is locked up between trucks; hence the project looked for the creation of an 'interior', providing patios in which truck drivers could rest after extended working hours. The materiality distinguishes two systems. First, a system of reinforced masonry painted white for the smaller volumes. Second, a system of Guayaquil cane for the container volume.

卡车餐厅

在距离利马19千米的帕纳莫里坎高速公路上，有一个卡车司机休息站，包含加油站、休息区和餐饮服务。卡车餐厅位于休息站的角落一块22.5米x10米大小的地方，不便停车。来往的卡车为餐厅营造了不断变换的景观。餐厅被卡车夹在中间，形成了一个供卡车司机休息的天井。两种不同的材料将两个区域分开：白色的钢筋结构是小空间，厄瓜多尔藤条则装饰着较大的空间。

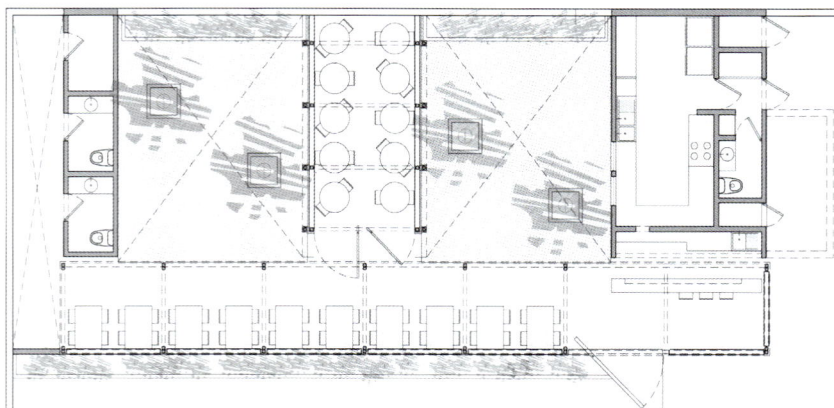

Zebar

The idea looks complex but actually is very simple and was born naturally from the digital 3D modelling environments. The space was subdivided into slices to bring it back from the digital into the real world; to give a real shape to each of the infinite sections of the fluid rhino nurbs surfaces. Using a projector the workers placed all the sections the designers drew on the plasterboards and then cut each of them by hands. The construction is finished soon.

泽吧

项目的设计理念看似复杂，实则简单，其灵感来源于数码3D模型环境。空间被划分为小块，从数码世界带入现实，成为现实的造型。工人们利用投影仪将设计师的图纸投到石膏板上，进行手工切割。整个工程并没有耗费过多的时间。

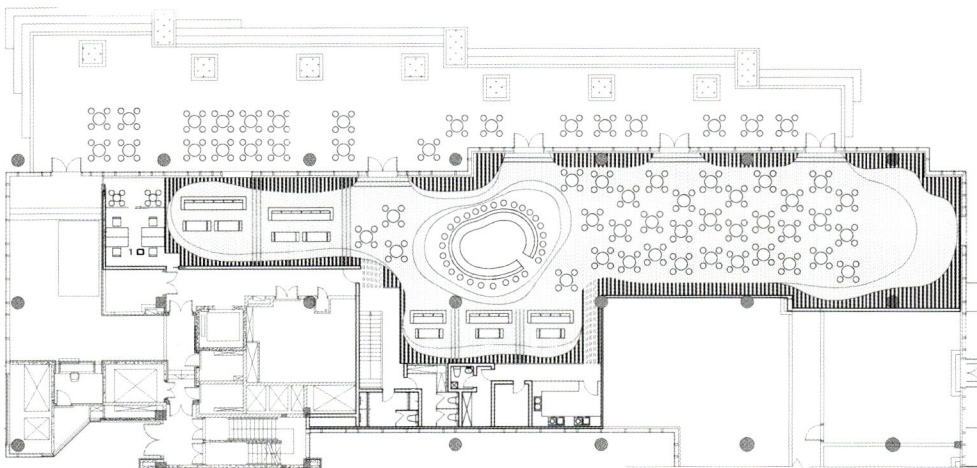

Restaurant

China

Shanghai

2010

Photo: Daniele Mattioli

3GATTI

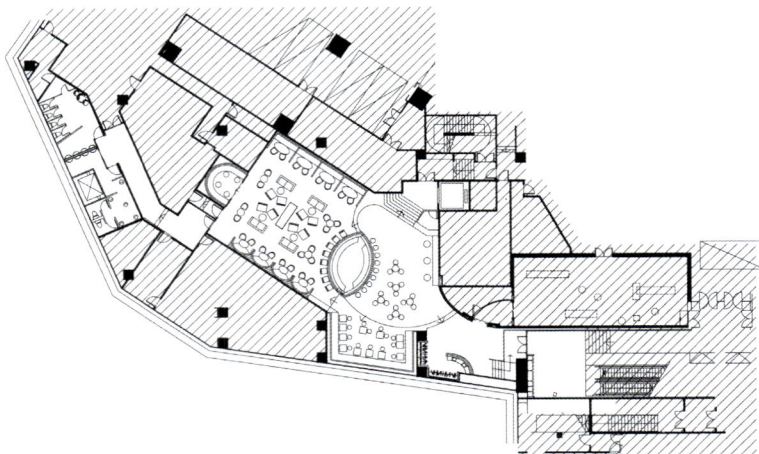

Aurora Restaurant Absolute Icebar

The scheme houses two parts, one warm and the other a frozen cellar. One wall flanked by a row of log booths with fluorescent amber glow creates a warm fireside atmosphere in a snowy environment. The wooden hut and the Eskimo's dome are another featuring area with inside walls covered with flannel and the outside in shining marble. A sparkling dancing area with lazy blue nights enjoys the delightfully comfortable scene from inside the bar. The separate lavatory facilities are situated behind the communal washroom. It is disguised as a snow-laden forest with urinals fitted into the cavities of large hollowed-out trees, which brings visitors close to the nature.

北极光北欧餐厅冰吧

整个酒吧的设计分成冷暖不同的区域。装饰方面用许多木条纵横交错地砌成柴枝感觉的特色墙，暗红的灯光营造出在寒冷的天气下人们围在火炉边取暖的气氛。另外一个座位区则设计成爱斯基摩人的圆顶冰屋形状，内里墙身铺上绒布，外面墙身由闪亮的大理石堆砌成，很有冰天雪地的感觉。跳舞区是暖吧内闪亮的一角，以灰蓝色调为主，结晶闪亮的效果正配合冰吧整体的感觉。

Boca & Ashanti Dome

Ashanti Dome, elegant with a similar style to that of the church, occupies the ground floor while on the first floor lies a Boca Pub. In addition to enjoying French-style food, more people are obsessed with its unique environment. Ashanti Dome was in careful repair of St. Nicholas Cathedral in Russian Orthodox Church, which was built in 1928. The interior design makes full use of the 1920s' original architectural details. Decoration has not taken great changes with elegant murals and lighting. Simple colour and illumination add a dreamy mood for the environment. Just imagine what kind of feeling it should be while sitting in it dining and seeing Our Lady's smile at you?

阿香蒂&博卡酒吧

阿香蒂餐厅很雅致，有着和教堂一样的风格，位于二楼，而一楼是博卡酒吧。在这里，除了品尝法国大餐，更多的人是迷恋这里特殊的环境。餐厅设在经过精心修复的俄罗斯东正教圣尼古拉大教堂里。餐厅内部设计充分利用教堂建于1928年的原有的20年代建筑细节，装修基本遵从当年的格局，壁画、灯光都很优雅。

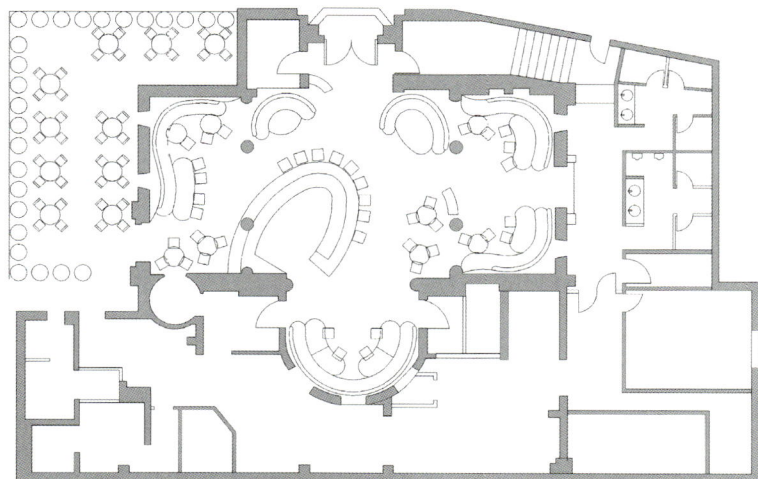

Bar

China

Shanghai

2007

Photo: Kinney Chan

Kinney Chan

Restaurant

China

Tangshan

2007

Photo: Gaohan

ISPACE & Studio 63

Green House

This is a renovation project. The original building has two floors and a backyard garden which locates in the main street in Tangshan city, China. Client wants to transfer it as a business restaurant. Considering the business proposal, designers use 'Around', which has been extracted from Su Shi's poetry 'Die Lianhua' as design concept. According to the layout of original building, designers change old garage door as main entrance to let guests experience 'Around' when they walk in. Interior space was extended to backyard. Wicker partition wall and large scale wicker flower container divide dining area.

绿水人家商务餐厅

本项目是将一处两层楼的老建筑和后院改造成为商务餐厅的项目。根据原有建筑和庭院的布局,设计师把苏轼的《蝶恋花》作为设计概念,巧妙的融入进了新餐厅的平面布置和空间装饰中。餐厅的名字也是取自词中的上阕:"花褪残红青杏小。燕子飞时,绿水人家绕。"餐厅空间和流程的设计注重"绕"这一概念,使得室内和庭院的空间变得更舒展神秘。

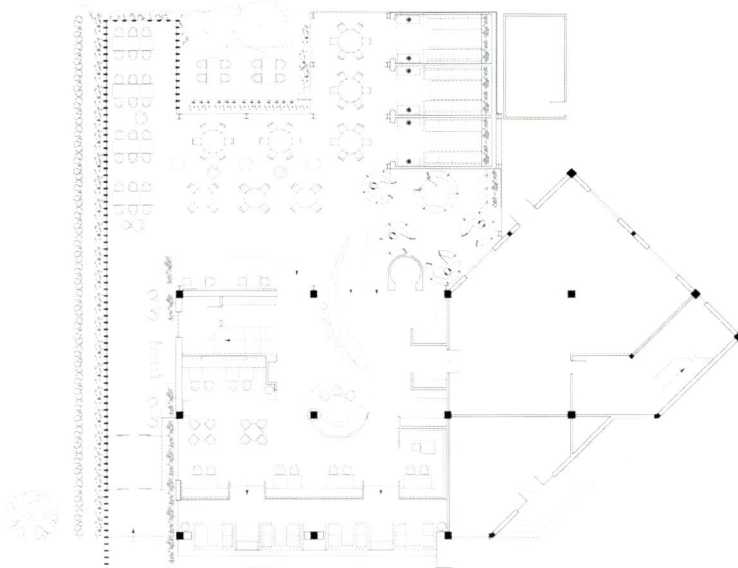

Slice Jinqiao

The space is divided into three distinct 'jewel boxes', a two-storey marketplace and two separate casual dining rooms, one on each floor – creating a series of intimate and humane spaces. In the dining rooms, one is enveloped in wooden screens that filter natural light. These screens are very much an integral part of Chinese domestic spaces and by abstracting them designers create a comfortable, modern atmosphere. The market is a more dramatic double height circular space, creating a mundane sense of luxury in a routine activity of shopping for food. The design of the whole space gives customers unlimited inspiration and imagination.

思莱仕金桥

该空间由三个各具特色的"珠宝盒"组成，其中包括二层的购物区及两间独立的休闲餐厅。整个设计给人一种亲切感。两间餐厅中，一间四周用绿色木质屏风装饰并遮挡强光。屏风是中国传统家庭中不可或缺的陈设，通过这种设计，更添加了舒适和现代感。购物区双倍高度的环形空间，更为其注入一种现代奢华感。整个空间设计给人无限的灵感和想象空间。

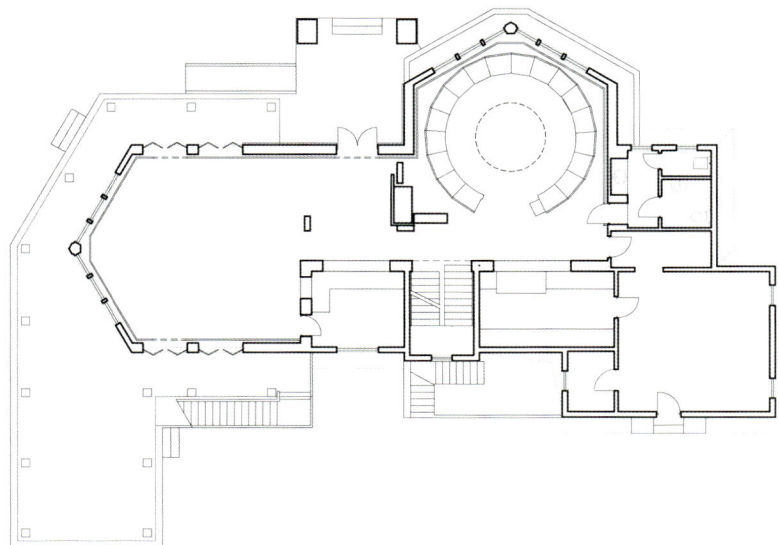

Restaurant

China

Shanghai

2007

Photo: Lyndon Neri & Rossana Hu

Lyndon Neri & Rossana Hu

151

Restaurant

China

Shenzhen

2009

Photo: Ray Lau

Joey Ho

Shiki

Shiki Japanese Restaurant is a spacious elegant dining room featuring delectable Japanese cuisine. The design reflects the theme as implied by its name 'Shiki', meaning the 'four seasons' in Japanese. Being an important concept in Japanese culture which demonstrates their strong appreciation of nature and celebration of every change of season, Shiki was designed to reflect this, which took inspiration from its picturesque surroundings, maximising the lush green views and using natural materials like stone, wood, wallpaper and bamboo with a contemporary Japanese aesthetic.

四季餐厅

四季日式餐厅是一家以赏心悦目的日式美食为特色的优雅宽敞的餐厅。餐厅名字"Shiki"，在日语里的意思是 "四季"，这也反映了餐厅的设计理念。四季是日本文化中一个重要的概念，反映了日本人对自然的欣赏及四季轮回的纪念。四季餐厅的设计灵感来源于生动如画的周围景色，将树木的苍翠繁茂发挥到了极致，并用当代美学的装饰手段对石头、木头、墙纸和竹子等自然材料加以发挥利用。

Xiao Nan Guo Restaurant in Suzhou

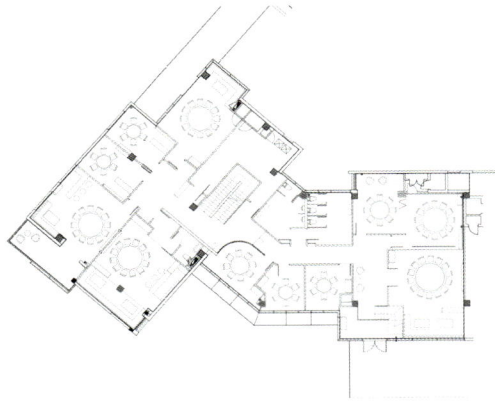

Facing the spectacular view of the Jinji Lake in Suzhou, the designer for the Xiao Nan Guo Restaurant takes advantage of its local context and creates a bright, vivid and elegant environment for the diners. Unlike other typical Chinese restaurants, minimal emphasis was given to the private dining area, letting the eclectic Western and Chinese concept reflects the sense of community. In contrast to the vast majority of Chinese restaurants with self-contained dinning rooms, this restaurant was conceived as a transparent volume with a series of layers that allowed diners to enjoy an open and roomy space. Each chosen element is traditional in form, yet modern in treatment. An open floor plan on the first level draws inspiration from the Chinese courtyard gardens, with floating baubles of glass as spirited birds in flight against free form tree motifs that were captured as wall paintings and freesias.

苏州小南国餐厅

小南国餐厅正对着苏州优美的金鸡湖。设计师充分利用了餐厅的地理优势，为用餐者打造了一个明亮、生动而又高雅的用餐环境。和其他典型的中式餐厅不同，设计着重彩于私人用餐区。中西合璧的折衷设计理念反映了较强的社区意识。不同于大多数中式餐厅拥有独立的用餐区，小南国各个楼层间开放透明，顾客可以在开放宽敞的空间内用餐。

Restaurant

China

Suzhou

2008

Photo: Chai Zhi Cheng, Bao Shi Wang

Joey Ho

Xiao Nan Guo Restaurant in Shanghai

Located at the prestigious cultural heritage Shanghai Xintiandi, the interior design of the Xiao Nan Guo Restaurant celebrates the historical architecture in a modern expression. The restaurant design reflects the monumental cityscape of Shanghai, which is a wonderful mix of old and new, East and West. It maintains this unique quality and spirit by preserving the existing architectural form and all the delicate detailing, such as the brick wall, the circular-pattern window frame as well as other old fabrics of the original structure, leaving a succinct and clean backdrop in contrast to the old accents. Taking inspiration from the Chinese garden, classic Chinese garden elements are applied in its interior to express the natural rhythm and so link different spaces within the restaurant.

上海小南国餐厅

小南国餐厅位于著名的文化遗产建筑——上海新天地，室内设计以现代的表现手法来赞美这栋历史性的建筑。餐厅设计反映了上海新老结合、中西合璧，灿烂不朽的城市风光。设计通过对现有建筑形式和砖墙、圆形窗框和原始建筑结构的旧纤维织物等精致装饰的保留，保持了原始建筑的独特品质与设计精神，而又以简洁、干净的背景与古老设计形成鲜明对比。设计灵感来源于中式花园。应用古典的中式花园元素于室内设计中，以自然节奏将餐厅的不同空间进行连接。

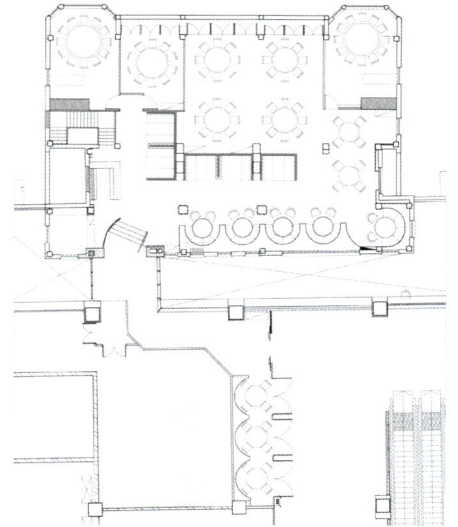

Restaurant

China

Shanghai

2008

Photo: Chai Zhi Cheng, Bao Shi Wang

Joey Ho

Tiandi Yijia – Hangzhou

The project has a fascinating history. It is a long narration that interprets with contemporary elements of Chinese culture and tradition and is set against the conception of a poetic text as opposed to the typical restaurant concept. The project takes up an area of 1,800 square metres, including wide common areas and VIP rooms. The ground floor is dedicated to the category Fusion: Eastern & Western, and the merging of oriental and Western flavours; the first floor features Hot Pot & Fusion Western, with a light Western note immerged in oriental flavours and perfumes, while the second floor is devoted to the VIP area offering purely Chinese fare.

杭州天地一家

这家餐厅拥有令人神往的历史文化。和典型餐厅设计理念不同，此餐厅以诗文为设计理念，用现代设计元素解释中国传统文化。包括宽阔的共享大厅和贵宾室在内，项目占地面积达1800平方米。地下一层以中西餐相互结合、东西方混搭式菜肴为特色；一层主要是中西式火锅——传统中式火锅中加入一点西方口味，而二层主要是以中国菜为特色的贵宾区。

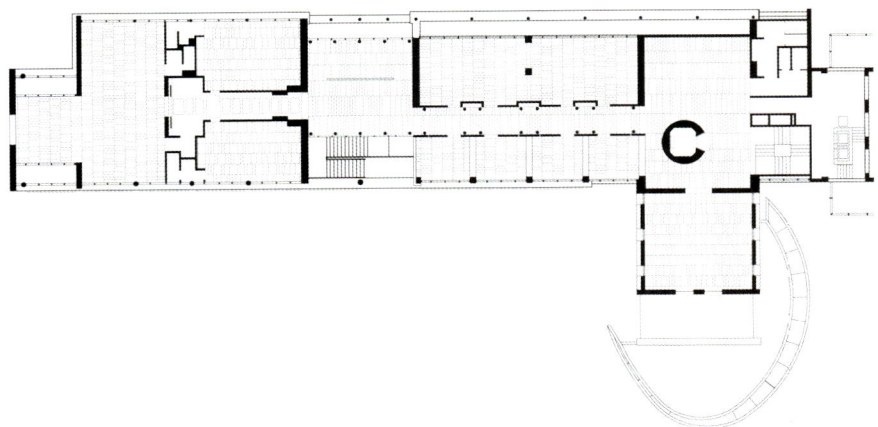

Restaurant

China

Hangzhou

2006

Photo: Gionata Xerra

Mauro Lipparini

Photo: Keiichiro Sako, Takeshi Ishizaka, Keigo Miyaichi/ Sako Architects

Keiichiro Sako, Takeshi Ishizaka, Keigo Miyaichi/ Sako Architects

Honeycomb in Shenzhen

Shenzhen Honeycomb is a restaurant in Shenzhen City, Guangdong Province. The partition walls formed by the free form curved surfaces convolute around to form six Taiko drum-shaped nests which take up the whole restaurant space. There are nearly 1000 oval holes sticking out inwards or outwards along with the change of the curved surfaces on the partition walls. Each of the six nests separates into two floors, which forms 12 VIP rooms. In the night, there is light penetrating through the oval holes on the curved surface of the six Taiko drum-shaped nests. Looking from outside, the organic gesture really seems as if a nest full of life as a lift body. The acrylics partition walls is arranged from the hall towards inside.

深圳蜂巢餐厅

深圳蜂巢餐厅位于广东省深圳市。自由形态的弯曲表面围成了六个日本太鼓型蜂巢式隔断墙，将整个空间分隔开来。沿曲线型隔断墙内外有近1000个椭圆形洞孔。每个椭圆形蜂巢都有两层，这样就形成了12个贵宾室。夜晚光线从洞孔穿出，这六个有机体仿佛是一个充满生机的鸟巢。

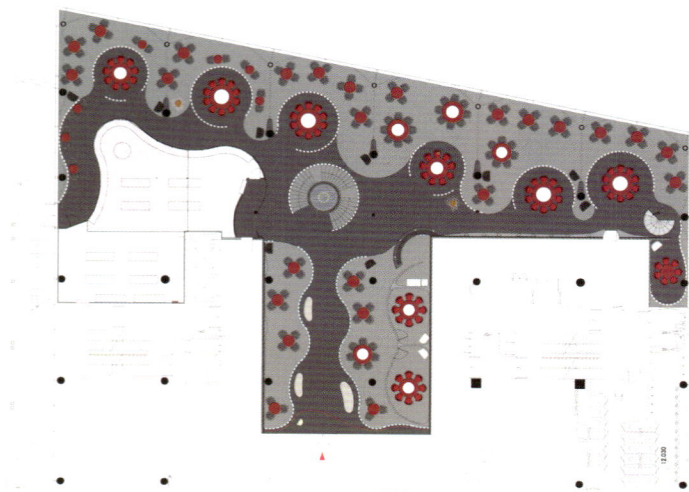

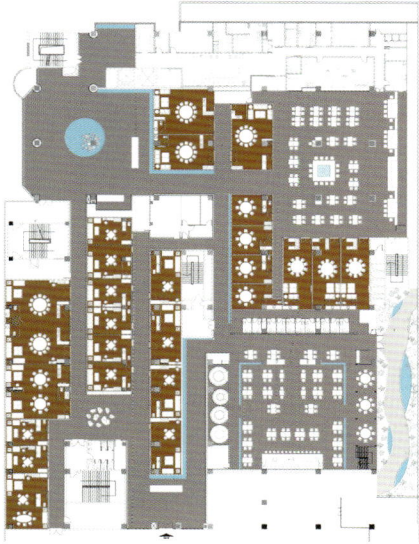

Made in Kitchen II

Located within the down town area in the city of Wuhu, China, Made in Kitchen II is the newest roll-out of this high-end F&B brand, serving contemporary Chinese cuisine. The site is facing a beautiful lake in the city centre with a total floor area of 4,000 square metres. The design strategy aimed at creating a unique dining experience by reinterpreting various beautiful scenes of a 'lake'. The resulting environment incorporated narrative elements in different zones: Entrance Lobby & Corridor – Motifs of 'pool' + 'ripples' + 'butterflies' + 'rocks' + 'falling water' + 'flying lotus' created a unique sense of arrival to the restaurant.

厨房制造II

厨房制造II坐落于中国芜湖市繁华的商业区，是首家提供当代中国美食的高端餐厅品牌。设计规划旨在通过对湖边各种美丽景色的重新利用，来打造独特的用餐环境。餐厅环境在不同区域融入了叙事元素。门厅与走廊的池塘、波纹、蝴蝶、岩石、瀑布和莲花灯装饰图案打造了餐厅的独特氛围。

Restaurant

China

Wuhu

2010

Photo: Ng Siu Fung

Horace Pan, Alan Tse, Nick Wong

Restaurant

China

Shanghai

2008

Photo: Jia Fang

Norio Ogawa, Zou Zhonghe
Shanghai Infix Design

Ninsei Jinmao Branch

Japanese cuisine emphasises on the accordance with seasons. In order to express the ingredients'
freshness and provide them with creativity, the arrangement of the dishes and selection of tableware
are all combined with seasonal flavours. Japanese cuisine cannot be separated from seasonal elements.
The designers expanded the open space in depth and in height, fulfilling the whole space with
openness. Guests will feel likebeing in the infinite sky. They could enjoy both the seasonal food and the
openness of the upward space.

仁清 金茂店

作为日餐的表现手法，极其注重与季节的呼应。最大限度地发挥应季食材的鲜味儿，并赋予其创
意，在菜品的拼摆以及餐具的选用上，都融入季节的气息。日餐，就是这样一种与季节密不可分的
料理。 在原本就已非常宽敞的平面上，随着向深处的不断推进，上方也打开的空间构成，使得整个
空间充满开放感，让人仿佛置身于无限空旷的天空。在享受来自于四季的恩惠的同时，也体验着由
此向上无限延伸的开放空间带来的美妙感受。

Jin Xiu Jing Ya

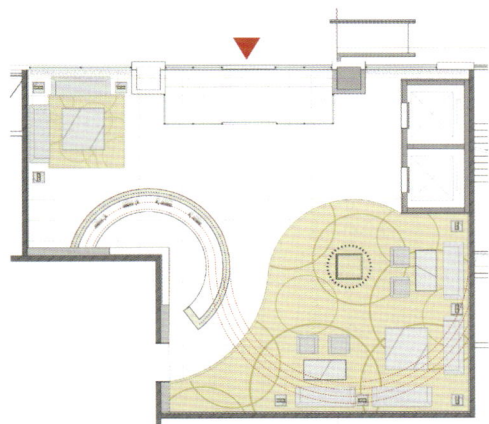

The golden entrance and arc patterns on the ceiling compliments each other, creating an illusion of light and shadow in the sea. Besides, through the idea to insert small spaces in the open room, the restaurant looks like an underwater world. In order to enable the guests to share a happy time with the one they loved, there are plenty of private rooms. The elegant light on the table are waiting guests' coming. The designe¯s hope that guests could enjoy an unforgotten time in this special space.

锦绣净雅

被金色包围的正门入口，和天花上的弧形花纹互相映衬，仿佛海水中美好的光影交汇一般让人神往。另外，通过在开放空间中故意穿插狭小空间的构思，表现出了时而宽广时而幽深的海底世界。为了能使客人与特别的人共享最为美好的时光，店内几乎完全采用包房的构成形式。而每张餐桌上那束安静而优雅的灯光，也仿佛正在期待客人的光临。设计师希望客人可以在这特别的空间里，悠然享受佳肴美酒，度过一段难忘的时光。

Restaurant

China

Hangzhou

2010

Photo: Shu He

Atelier Feichang Jianzhu

Tang Palace

The restaurant is located on the top floor of a superstore in the new town area of Hangzhou, with nine metres of storey height and a broad view to the south. Composite bamboo boards were selected as the main material, conveying the design theme of combining tradition and modernity. In the hall, to take advantage of the storey height, some of the private rooms are suspended from the roof, creating an interactive atmosphere between the upper and lower levels, thus enriching the visual enjoyments. The original building had a core column and several semi-oval blocks which essentially disorganised the space. Hence, the design wanted to reshape the space with a large hollowed-out ceiling which was made from interweaved thin bamboo boards, and extended from the wall to the ceiling.

唐宫餐厅

餐厅坐落于杭州新城区一家超市的顶层，楼层9米高，南面拥有广阔的视角。设计选用复合竹板作为主要原材料，体现了传统与现代结合的设计理念。为充分利用楼层高度，大厅里一些雅间悬挂于棚顶，营造了上下层的互动气氛，丰富了视觉感官。原建筑的一个芯柱和几个半椭圆形木块使空间显得布局紊乱，因此设计师用薄竹板编织成的空心网状木板从墙面一直铺到天花板，来打造一个全新的空间。波浪式的天花板为整个用餐大厅打造了丰富的视觉盛宴。

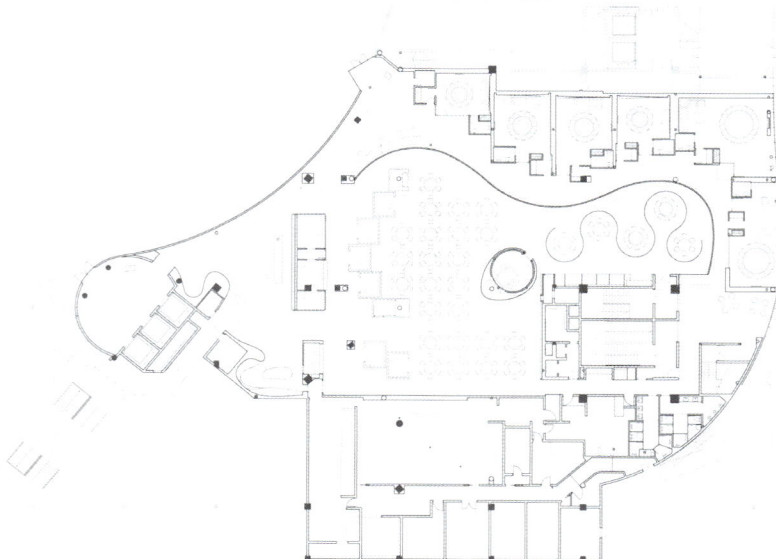

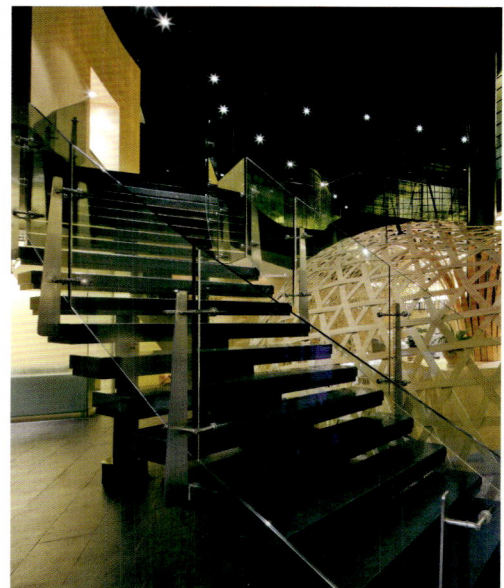

Qiandaohu Chinese Restaurant

The design concept is to evoke a traditional Chinese dining atmosphere, albeit in a subtle, modern manner. The whole space was divided into several zones – namely the main hall, VIP dining room and an outdoor dining area. The main hall is similar to Chinese banquet hall with round table setting to create a more social, joyous dining ambiance. Circle and square patterns are carried through and take on a more prominent presence here as wooden screens, delineating space while offering visual privacy. The use of dry flowers and branches that accompanied by an earthy palette with rich textures and inviting coffee colours, endow the space with an understated opulence – a welcoming departure from the stereotypical Chinese restaurants.

千岛湖中式餐厅

千岛湖餐厅的设计理念是运用精致的现代方式打造传统的中式就餐环境。餐厅由不同的区域构成，包括主厅、VIP厅及室外就餐区。其中，主厅模仿中式宴会厅风格，大圆桌的选用营造了欢快愉悦的氛围。圆形及方形图案遍布其间，格外引人注目；干花及树枝等装饰元素与丰富的纹理及咖啡色调相得益彰。

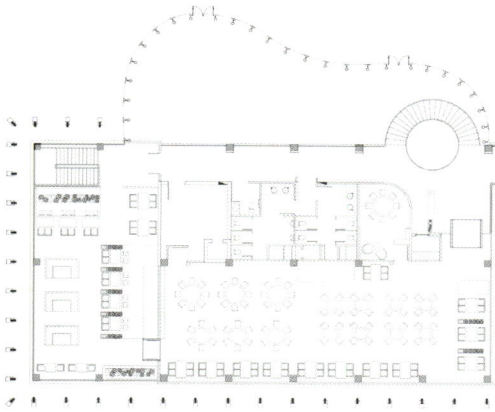

Restaurant

China

Hangzhou

2008

Photo: Chai Zhi Cheng, Bao Shi Wang

Joey Ho / Joey Ho Design Ltd.

Restaurant & Bar

China

Shanghai

2009

Photo: Jia Fang

Shanghai Infix Design

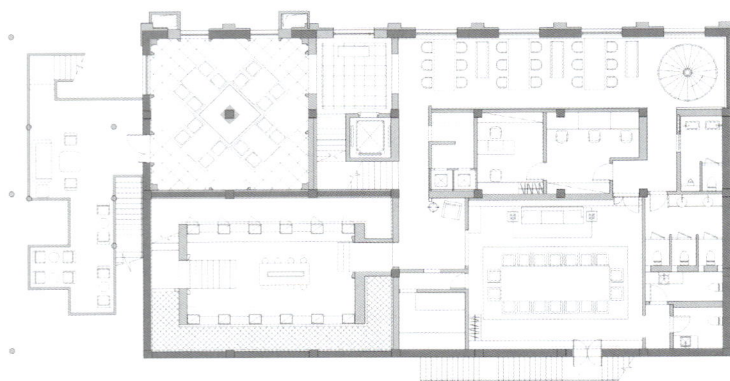

KAYLA1158 Restaurant and Bar

In order to meet the guests' need, there are both bar and restaurant in KAYLA1158 Club. The guests can enjoy a fashion and lively atmosphere in the bar, and a tradition and decent atmosphere in the restaurant. It is a colourful world. In the main restaurant, guests can order Western and Eastern food with appropriate wine. The glass wall shows different views with the sunlight's change in a day: at daytime, the sunlight come through the glass and shines beautifully. At night, wine glasses run into each other, creating an inviting sound and rendering a steady and thick atmosphere. Time flows in the relaxing.

KAYLA1158餐厅&酒吧

为了迎合来自客人的多种需求，店内分别设有酒吧和餐厅。设计方面，既有让客人感受前卫轻快的一面，又有让客人沉浸在比较传统的、具有厚重感氛围内的一面，堪称是一个多彩的空间。在主餐厅，客人能够按照自己的喜好选择西餐或是东方美味，并配上与料理相合的红酒。整面玻璃的墙壁，随着太阳光线的变化，让你享受不同的时间段呈现出的截然不同的景象：白天，阳光透过玻璃照射进来，闪耀着夺目的光辉，时间，就在这份舒适惬意中慢慢流逝；夜晚，酒杯碰撞时发出的诱人声音，在空间中以一种深沉的方式交错回响着，氤氲出沉稳、厚重的氛围。

Restaurant in Nanjing Central Hotel

The Nanjing Central Hotel is located in Xinjiekou — the bustling centre for business and commerce in the city, with convenient transportation and a lively market. In order for all the guests to enjoy the new view of the atrium, the front desk was moved from the first floor to the fourth floor and the buffet to the third floor, so that the best views are left for the guests. This organic restructuring of hotel functions allows practicality to intergrated with quality of service. The mastery of a unique space is to make the building to stand out from other surrouding hotels. Instead of the simplistic, introverted mainstream style, the designer combined culture, aesthetics, market, and operation.

南京中心大酒店餐厅

南京中心大酒店位于南京繁华商业中心——新街口，周围交通便利，商业发达。总台区被从一层转换到了四层，自助餐厅改建到三层，让每个入住的宾客都能够最大限度的尽享中空大堂的景观，使各功能区有机结合，把窗外最好的风景留给客人，真正做到了功能与服务的完美结合。设计师一改现今简约、内敛，强调低调奢华为主流的设计风格，把文化、美学、市场、营运完美结合。

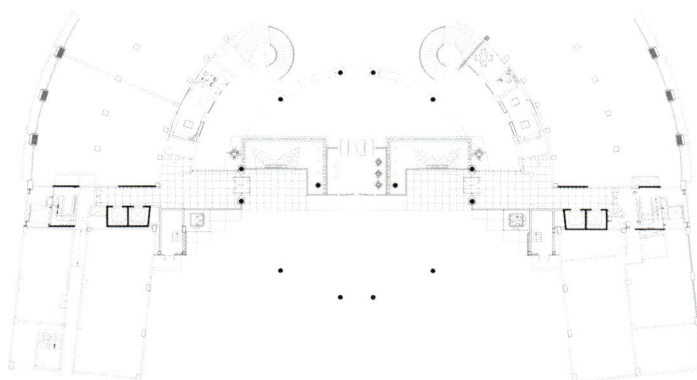

Restaurant

China

Nanjing

2010

Photo: Li Gan

W To The Power Two Architects

Restaurant

China

Beijing

2009

Photo: Misae Hiromatsu

Sako Architects

Feng Wo De in Beijing

'Feng Wo De' downtown CBD in Beijing, is operating boutique Sichuan and Cantonese cuisine in the main restaurant. Here the spaces are, like many other Chinese restaurants, organised by the lobby and large, inter-packet form. Owners wanted the proposed design elements to be placed in the East. According to this requirement, the design in the lobby with oriental traditions, started off the grid and was extended to the whole space. The thin and soft woven was formed into a box-shaped wood panel, horizontally and vertically bending into wave form, and became a mix of softness and hardness of the partition. Thus, sofas near the window seats, cutting off the middle area and surrounding by semi-private room were arranged in the three space forms.

丰沃德餐厅

"丰沃德"位于北京的CBD商业区，是以经营川菜和粤菜为主的一家精品中餐厅。这里也和其他很多的中餐厅一样由大厅和大、小包间组成。在大堂设计带有东方传统色彩的方格隔断展开延伸到整个空间。将又薄又软的木条编织成方格状面板，在水平垂直方向弯折成波浪形状，就变成了兼备柔软度和硬度的隔断。由此，就形成了窗边的沙发席、中间区域以及被隔断所包围的半开放式包间等三种空间形式。考虑到CBD周边的客户群体，在大厅席位中也充分照顾到各桌席位的私人空间。在以黑色为基调的空间中呈现出中国红色彩的蜿蜒而曲折的方格花纹。

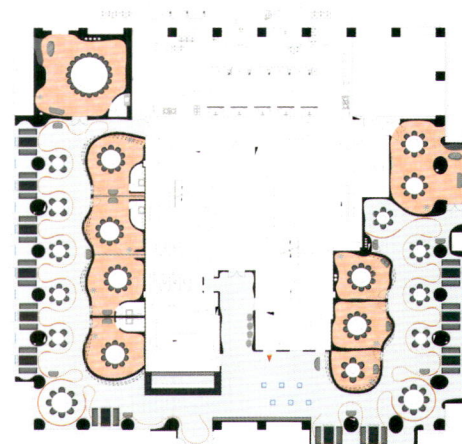

Restaurant

China

Suzhou

2009

Photo: Song Weijian

Song Weijian, Zhang Nan/Yjian Design Office

Suzhou Laodongwu Restaurant, Yadu Branch

The designers redesigned the booths in the upper floor. The space plan applied Suzhou Gardens' layout and created a feeling of 'house in house'. The ground floor's design continued the former one and was added some new ideas, more relaxing in layout. While the upper floor is a 'house in house', the ground floor looks more simple and natural, in a courtyard environment. Hallways and booths are connected and separated by the louvre wood partitions, providing certain privacy and not cutting off the exterior view as well. These partitions created a 'connection in breaking' atmosphere.

苏州老东吴食府雅都店

设计师对该餐馆的二层贵宾包厢区域进行了设计，整体空间布局上引用苏州园林的布局动线，营造出楼中楼的感觉。一层的设计在呼应之前二层的设计理念基础上，有了一些新的理念内容，在布局上也更为自然轻松。如果说二层是楼中楼，有云中楼阁的感觉，那一层大厅的设计则是开宗明义，更加看重自然、灵动的庭院式氛围。廊和厢的关系通过百叶式的木质隔断，既巧妙地保证了一定的私隐性，又没有完全阻隔外界的视野，有一种"笔断意不断"的意境。

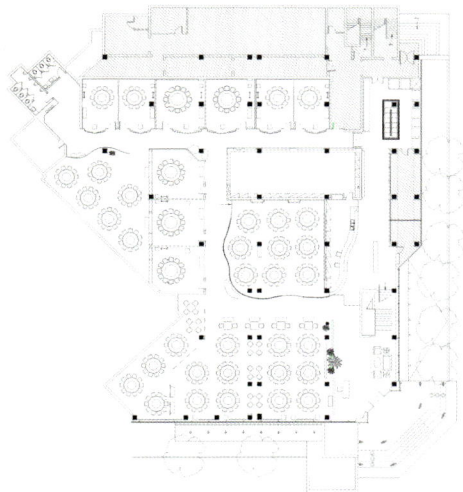

Photo: Kerun Ip

Joseph Wong Design Associates

Jardin de Jade Restaurant, Shanghai

The Jardin de Jade Restaurant was designed and created within the grand structure of a historical building in the heart of Shanghai, which is China's largest city. It is a fusion of historical and modern design, bringing new life into an old and historical building and creating a bold mix of existing and newly imagined elements. It is difficult to believe that this 12,000-square-foot space that now houses Jardin de Jade was once a slaughterhouse. The building, originally built in 1933, offers vast, beautifully-proportioned loft-like spaces, floor-to-ceiling windows with art-deco flair and stately fluted columns at large, regular intervals. The structure of the building and its interior layout helped tremendously in the adaptive reuse of the slaughterhouse that was used more than 70 years ago.

上海翡翠园餐厅

这家翡翠园餐厅位于中国最大的城市——上海市市中心一栋历史建筑里。设计将新旧元素融会贯通，是古典与现代的完美结合，赋予古老的历史建筑以新的生命力。很难想象，翡翠园餐厅所在的12000 平方英尺的建筑原始是一个屠宰厂。建筑始建于1933年，拥有广阔、完美对称的阁楼式空间，装饰艺术风格的落地窗和每个一定大间距的宏伟的槽型圆柱。建筑结构和空间布局为有70年使用历史的屠宰厂的改建发挥着巨大作用。

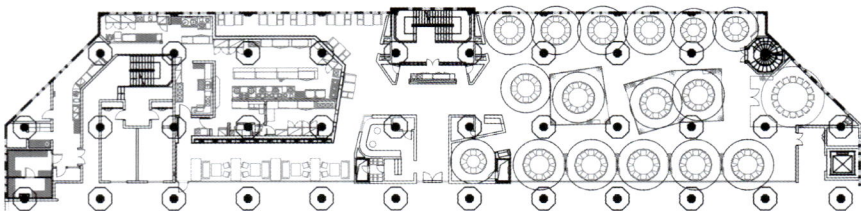

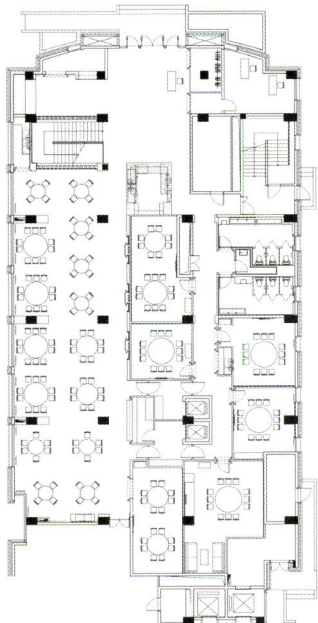

Jardin de Jade Restaurant, Hongqiao

Hongqiao currently stands as a highly coveted and prestigious residential area of Shanghai. After underwent significant reform, it quickly developed into a high-traffic economic development zone and presently serves as a bridge in foreign trade and as a centre for cultural exchange. Hongqiao is well-known around the world for this unique unifying function, and it is where this particular Jardin de Jade restaurant resides. Because of cultural emphasis in this town, the focus and theme of this restaurant design was exploring and promoting the Shanghai culture. The guiding ideology of design was heritage combined with development, so the designer attempted to infuse Shanghai-style historical and cultural elements, using symbolic representation at times. Three specific avenues that they used to bring about were architecture, painting and detailed accents.

虹桥翡翠园餐厅

虹桥是上海十分令人向往的享有声望的住宅区。经过彻底改造后，迅速发展为交通便利的经济开发区，如今已成为国际贸易的桥梁和文化交流中心。虹桥因其强大的综合实力而世界文明，而虹桥翡翠园餐厅就坐落于这个城镇中。与虹桥区独特的文化氛围相协调，餐厅设计的主题就是对上海文化的探索与传播。设计的核心理念为继承与创新相结合，因此设计师力求将象征性元素贯穿于建筑、绘画和细节设计于始终，传播上海式历史与文化元素。建筑是室内设计的艺术展现，虹桥翡翠园餐厅精致的细节设计既传播了虹桥的历史与文化，也展示了城镇多年来的发展变化。

Restaurant

China

Shanghai

2010

Photo: Derryck Menere

Dariel & Arfeuillere - A Lime 388 Company

The Purple Onion Restaurant

Designed by the young talented designers Thomas Dariel & Benoit Arfeuillere, conceptualised by the internationally renowned Chef David Laris, The Purple Onion has been directly inspired from the French Bistrot and Italian Trattoria. The design features all the codes of these typical spaces yet deepening, sharpening and reinterpreting their usual characteristics. For instance, like the Bistrot, a homelike and popular place where you can always meet someone that you know, The Purple Onion exhibits a series of oil-painting portraits representing the venue's ancestors. Witnesses of the time passing by, these characters, all of them sipping a glass of wine or a cocktail, are yet winking at the present by sharing a drink with the guests. The traditional family portrait is here hijacked to keep its familiar feeling while bringing to it humour and modernity.

紫洋葱餐厅

设计灵感由在南欧流行的颇具家庭温暖氛围的法式Bistrot及意大利式Trattoria激发，体现了这些独特空间的密码而又更加深入、鲜明地重新诠释了其通常的特质。餐厅陈列了一系列祖先的油画肖像，目睹时光流逝，这些人物都拿着一杯葡萄酒或者鸡尾酒啜饮，微笑看着来此的客人并与大家共享欢乐休闲时光。

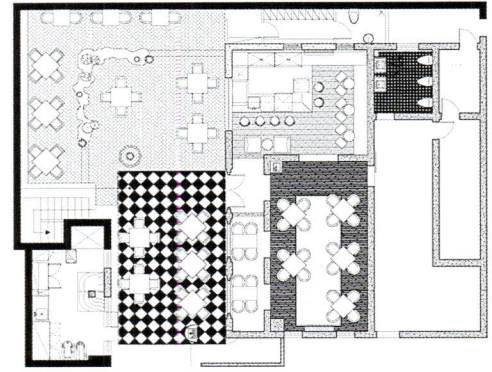

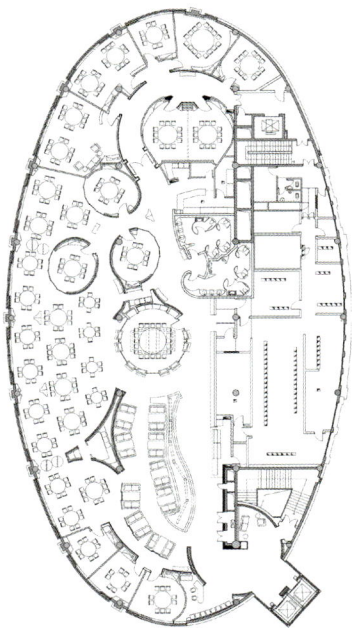

Jardin de Jade Restaurant, Pudong

Much like Jardin de Jade Restaurant in Shanghai, the Jardin de Jade Restaurant in Pudong places a special emphasis on the natural elements. While the one in Shanghai highlights the element of fire, the one in Pudong puts a spotlight on the beautiful element of water. Each aspect of this restaurant was designed with the theme of water in mind — from the wave-like patterns of the curved walls, to the raindrop-like crystals hanging from the ceiling's light fixtures. These crystals shine as one of the eye-catching features of this modern restaurant, each crystal hanging from nearly invisible strings at various lengths to give the impression of actual raindrops. Yet another unique attribute is that these crystals not only glimmer with rainbow-coloured refractions, but they are also suffused with a synthetic changing LED light from within, playfully altering the mood of the restaurant with its greens, blues and purples.

浦东翡翠园餐厅

和上海翡翠园餐厅一样，浦东翡翠园餐厅也极为突出自然元素设计。上海餐厅强调火元素设计，而浦东餐厅则突出美丽的水元素。无论是曲线墙面的波浪式图案，还是天花板灯具上悬挂着的雨点一样的水晶，餐厅的每一部分设计都围绕着水的主题。这些水晶饰物耀眼夺目，是现代餐厅的最大特色。每一颗水晶好像用长度不等的无形线连接，就像雨滴一样。更有趣的是这些水晶不仅散发着耀眼的彩虹光，内置的发光二极管还变幻发出绿、蓝、紫等各种颜色的光芒，为餐厅增加了生动活泼的气氛。

Restaurant

China

Shanghai

2008

Photo: Kerun Ip

Joseph Wong Design Associates

Restaurant

China

Suzhou

2008

Photo: Kerun Ip

Joseph Wong Design Associates

170

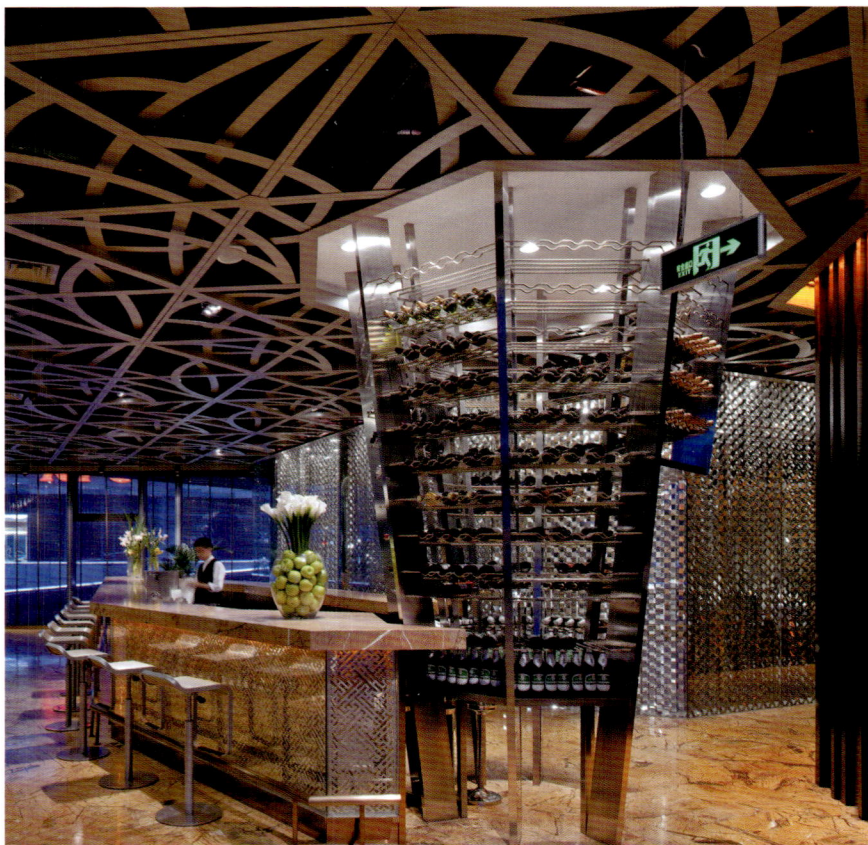

Jardin de Jade Restaurant – Suzhou

The history of the city of Suzhou goes as far back as two thousand years ago, and has much to boast about, particularly in obtaining the highest achievements in both Chinese southern culture and literary art. One example of this is the Suzhou Gardens, which have been deemed the 'world cultural heritage of humanity' by the United Nations. These particular gardens are breathtakingly designed to enhance the unity of nature, drawing together different aspects of natural life and combining them with beautiful and exquisite handmade sculptures. The Jardin de Jade restaurant in Suzhou is designed with the same goal in mind: to cater to the unique and age-old cultural heritage of Suzhou, while blending the old with the beauty of modern interior design and formal architectural language. This particular branch of the Jardin de Jade restaurant series is a matchless marriage of the ancient cultural elements of Suzhou and modern space design.

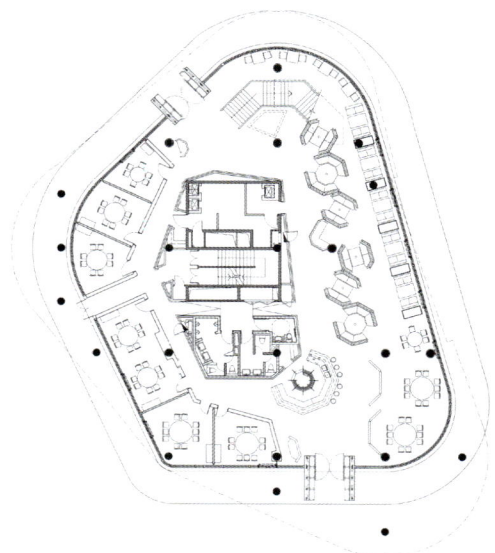

苏州翡翠园餐厅

苏州拥有2000年发展历史，最值得骄傲的就是其在中国南方文化和文学方面的至高成就。其中一个例子就是被联合国列为世界非物质文化遗产称号的苏州园林。这些独特的园林，凝聚了自然生命的点滴精髓，并将其与精致优美的手工雕刻相结合，提升了大自然的和谐美，设计工艺精湛，堪称一绝。翡翠园餐厅的设计也围绕这样一个目标：继承和弘扬独特而又古老的苏州文化遗产，并将古老文化与现代室内设计之美互相结合，用专业的建筑语言来描绘。

Nantong Prince Restaurant

The restaurant has three levels: the ground floor is a two-storey lobby, grand and decent; the first floor includes a large banquet hall and booths of different sizes; the second floor consists of VIP rooms. The designers took Indigo Printing patterns as a thematic element in the design. The patterns were used widely in surfaces of walls and ceilings, textures, furniture, lighting fixtures and decorations. The large glass lamp with Indigo Printing patterns looks gorgeous and sophisticated; the patterns and colours in glass and black mirrors have a modern and magnificent feeling; Indigo Printing patterns are also everywhere in the carpet, the sofas and oil paintings.

南通王子饭店

饭店分为三层：一层为两层挑空的大堂，气势恢宏；二层包含一个大型宴会厅和大小不一的餐饮包厢；三层主要由VIP包厢组成。我们把南通蓝印花作为贯穿设计的一个母题元素。在设计中，"蓝印花"被大量应用在顶面立面造型、材质、家具、灯具和装饰品上：以蓝印花图案定制的大型玻璃水晶灯富贵华丽中透出一份雅致；蓝印花的色彩模式被提取成连续花纹形式，用玻璃与黑镜面不锈钢雕刻，呈现出现代而又绚丽的视觉感；而地毯、沙发以及大幅的油画，随处都可见蓝印花的精彩演绎。

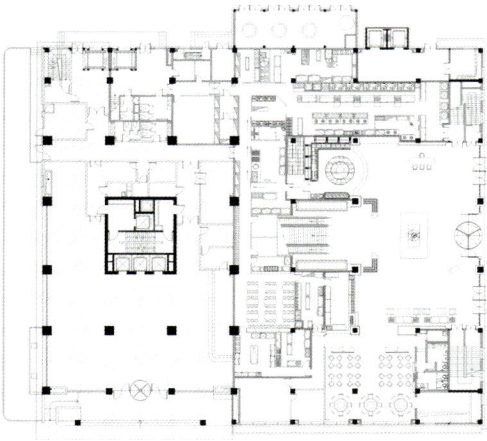

Restaurant

China

Nantong

2008

Photo: Wen Zongbo

Song Weijian, Feining, Lu Ronghua/Vjian Design Office

Restaurant

China

Chengdu

2009

Photo: Still Life exposed works, Kevin Best

Graft Architects

Gingko Bacchus Restaurant

The starting point of the design was a blacked out space. The public spaces were considered as a river or a stream along which one 'floats' through the depth of the 1200-square-metre black space. The stream is entered directly from the elevators which are the only access to this restaurant on the 4th floor of the Gingko Restaurant Building. The undulated wood ceiling and the stainless steel intarsia of the floor pattern create the flowing sensation of this 'stream' which terminates at the open dining and show cooking area. Eight private dining rooms are located like boulders along the stream, and each was colour coded and themed by food and famous Bacchus depictions. Food is used in various layers of abstraction throughout the restaurant. Each private dining room has its own customised wallpaper of simple food products like carrots, mushrooms, walnuts, broccoli, beans, chilies, and artichokes.

银杏餐厅

前期设计是创造一个没有光亮的黑色空间。公共区域就像一条河流或是一条小溪，顾客沿着溪流"漂流"到1200平方米的黑色空间海洋内。人流从银杏餐厅大楼的四楼电梯上来就直接来到了餐厅内。用餐大厅和展示厨房以外的空间内，起伏的木制天花板和不锈钢木制地板嵌花营造出了一种小溪的流畅感。八个雅间好像溪水旁的岩石一样，每一个都有独特的颜色编码，都以食物和著名的酒神巴克斯的图画为主题。食物被应用到餐厅各个层面的抽象装饰中。每个雅间都有其独特的墙纸，上面绘有胡萝卜、蘑菇、核桃、花椰菜、豆子、辣椒和朝鲜蓟等普通蔬菜。

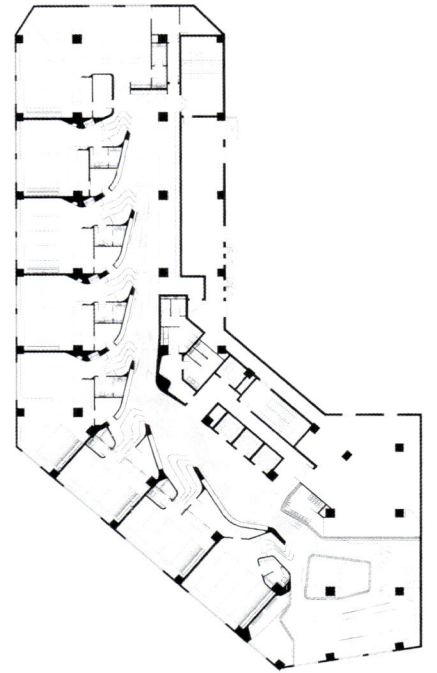

Restaurant in Wynn Macau

A Japanese culinary adventure awaits you at Wynn Macau, also unique for its sumptuous Cantonese cuisine. The open kitchen of Ristorante il Teatro allows you to view the cooking of classic Italian specialties such as traditional pizzas, salads, homemade pasta and other authentic dishes from Southern Italy. You will be entertained while dining by the spectacular views of the Performance lake. An international café with an eclectic menu of Asian, Macanese, Mediterranean and Western dishes, Café Esplanada overlooks the pools and gardens, where you can enjoy savory and irresistible desserts. A chic lobby bar where the warm glow of fuchsia, vivid orange and honey onyx tempt guests to enjoy delectable cocktails with distinctive ingredients.

永利澳门酒店餐厅

帝雅廷意大利餐厅为开放式厨房设计，可让顾客一睹经典大利名菜的烹调过程，如传统意大利比萨、沙拉、自制意大利面及其他意大利南部的正宗菜式。客人可一边用餐，一边欣赏窗外表演湖的精彩演出。 咖啡苑集合各地美食，供应亚洲、澳门、地中海及西方菜式。宾客可一面享用无法抗拒的法式甜品，一面欣赏泳池与花园的优美景致。

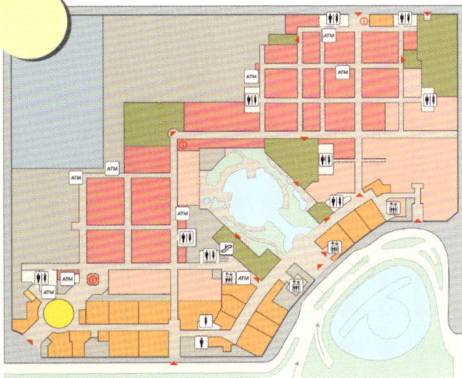

Photo: Wynn Design & Development

Wynn Design & Development

Super Potato, Takashi Sugimoto (Producer) Norihiko Shinya (Director) Yoshifumi Tanaka (Chief Designer) Shinya Maeda (Designer)

Grand Hyatt Macau – Restaurant ' Mezza9 '

Mezza9 is a restaurant on the 3rd floor of Grand Hyatt Macau Hotel which opened in 2009 as a part of the complex facility project from a hotel and commercial facility called City of Dreams Project. There are several show kitchens at the Asian and Western stations which allows for guests to see the cooking in front. The interior is expressed in a modern manner through materials which elaborates naturalism in a strong way such as stones which were cut out from the mountains, timber with natural faces left, and iron with rust. By segregating these materials with usage of floor, wall and counter, the materials do not only become a material for the interior, yet, also elements to constructing an art installation space.

澳门君悦酒店Mezza 9 餐厅

Mezza 9 餐厅位于新濠天地酒店与商业设施一体化项目中2009年开业的综合设施项目——君悦酒店的三层。亚洲和西式餐厅区有几个展示厨房，顾客可以看到前面的厨师做菜。室内设计采用现代设计手法，选用山体采发的石头、带皮的木材和生锈的铁等原材料，以强劲的手法展现了自然主义。通过使用地板、墙和柜台等装饰将这些材料应用到不同区域。因此，它们已不仅仅是普通的室为装饰材料，更是构建艺术空间必不可少的元素。

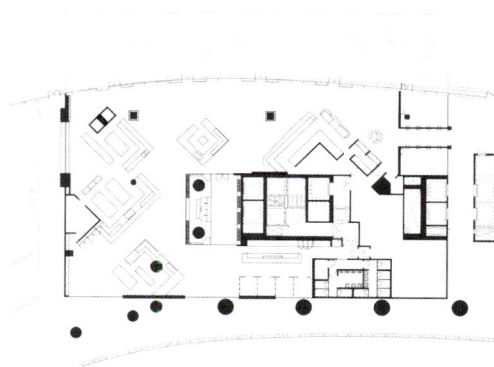

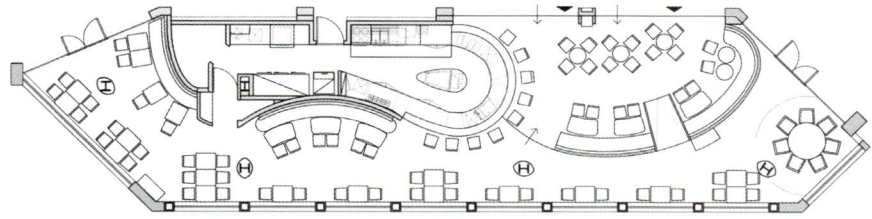

BOCA

With the success of BOCA Tapas & Wine in Hong Kong, Eclipse has created another stylish wine bar and café with a Spanish flavour perched right on the San Luca Canal in The Venetian Macau. Contemporary in nature, curves pervade almost every element of design from the bar and banquette seating to the tabletop setting. Lush colours throughout with modern architecture feel more Barcelona than Madrid. Lounge chairs and soft banquettes will ensure a fine balance of those dining with those simply sipping wine and gazing at the gondolas gliding by. Music from flamenco to acid jazz will round out the cool ambience. As shown by the name, BOCA, which means 'mouth' in Spanish, offers everything that pampers that mouth from delectable tapas and a broad range of wines to good conversation, all in modern and sophisticated surroundings.

博卡酒吧

博卡在西班牙语中是"口"的意思。这是一间西班牙小吃的酒吧，其色彩及风格就如西班牙的舞姿。选用物料：不锈钢，水磨石地面，丝绒布料，胡桃木，金属铁纱窗。弯形沙发及酒吧台面营造出多角度空间及层次感。此外，沙发的流线型与地面家具及天花的流线统一起来。 以不同高度的凳来区分各个区域，加上灯光效果，构造出满有特色的西班牙风采。

Bar

China

Macau

2007

Photo: Chen Dejian

Chen Dejian

Bar

China

Hong Kong

2006

Photo: Axiomoval

Axiomoval

Champion Circle

In order to provide a comfortable, spacious and convenient space, the designers opened up one third of the viewing terrace and inserted a skylight in the middle of the betting hall, which now provides a spectacular unobstructed view of race track. Running parallel to the new view is a 33-metre-long bar, which may be the longest bar in Hong Kong. Here, the customer can effortlessly enjoy a wide selection of high quality foods and beverages. The modern style of the betting counters is another focal point in the betting hall. The computer programmed LED graphic Arch provides a contemporary approach in housing the many audio/visual elements that need to be placed throughout.

骏星汇

为了保证空间的舒适、宽敞和便利，设计师把观景台的三分之一打开，并在博彩大厅设置了天窗，能毫无障碍的观赏壮观的赛场。与景观平行的酒吧长达33米，这大概是香港最长的酒吧。在这里，顾客可以毫不费力地享受各色高档食品和饮料。风格现代的投注区是博彩大厅的另一个重点，电脑控制的LED拱门能显示大量音频和视频内容，十分先进，值得在世界各地推广。

Pissarro

Michael Young has designed Pissarro, a unique dining experience situated in the centre of Hong Kong's night life district. Young's ambition to unite traditional craft ideology with technology-led processes of today pre-empts a typology in design which celebrates progress and the timeless appeal of design placed in its time. The restaurant covers 152 square metres and seats approximately 50 diners at any one time. Furniture was manufactured by Accupunto Indonesia. The Coen chair (winner of Wallpapers 2008 best dining chair award) was created as special edition in replanted teak and black leather.

Pissarro餐厅

Pissarro餐厅由迈克尔•杨亲自操刀设计，营造了独特的用餐体验。他将传统的手工制作工艺同现代的高新技术相结合，符合当今时代风格。餐厅占地152平方米，可容纳50人。玻璃结构渐渐变薄，使得整个空间光线充足，洋溢着温馨的氛围。家具、座椅以及桌子全部由知名设计师打造。

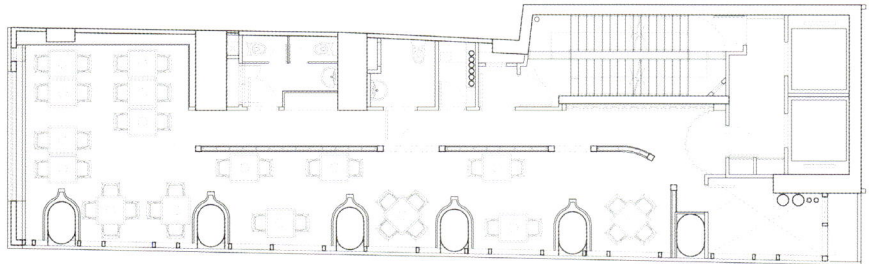

Photo: Hartim Djauhar Winata

Furniture design - Michael Young, Lighting design - Michael Young
Design Engineering -Ken O'Rourke, Architectural Lighting -
Anlighten Design Studio, Project Architect -Sebastian Saint Jean

Café

China

Hong Kong

2008

Photo: Raqe Wan

Alex Choi

Café Habitu Chain Store

It is a chain of contemporary 'Italian looking' coffee shop located in a modern city Hong Kong. This chain of cafés were located in commercial area mainly, and the behind is for a considerate thought. Think about it, in the hustle and bustle of our daily lives, especially of the commercial city, sometimes a sip of coffee, sit back and relax would bring the lives back. That is why the ambience of the café should be cosy and comfortable. To remain the chain café's origin, 'Italiano' style is the axis of the design, and the use of colour black and white keeps the taste of Italian. The uniqueness and cosiness of the shop was created by using abandoned building materials and techniques, like stucco paint and concrete plastering on walls, bear face wrought iron with exposed cap bolt, wired glass, oxidised copper tile, the antique silver coated light bulb, and the like.

Habitu 连锁咖啡馆

此咖啡馆是现代都市香港的一家现代意式连锁咖啡店。项目设计的背后包含着设计师体贴周到的设计思路——人们在紧张而忙碌的现代生活中，偶尔坐下来品尝一杯咖啡，放松身心，可以找到返璞归真的感觉。正是秉承着这种设计理念，才营造出了温暖、舒适的餐厅氛围。为了保持这家连锁咖啡店的原貌，意大利风格贯穿整个设计的主线。为体现意大利式设计风格，餐厅设计全部采用黑白两个色系。通过使用粉刷油漆和混凝土抹灰墙面涂料等废弃建筑材料与技术，实现了独特个性。

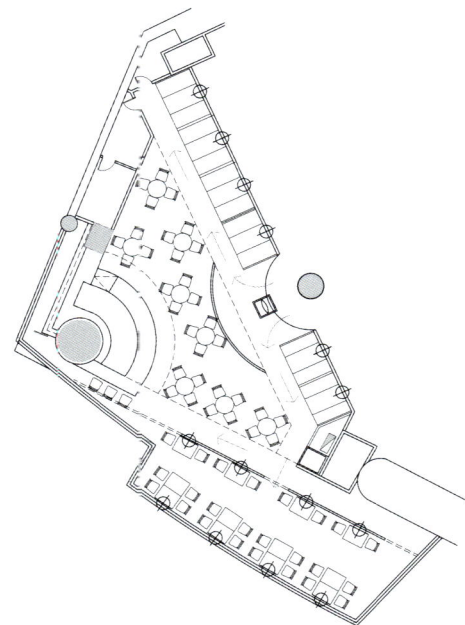

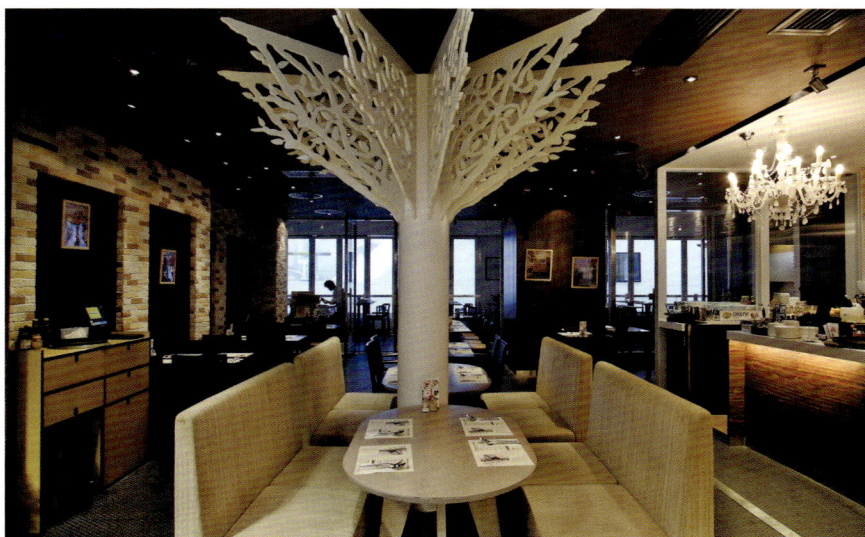

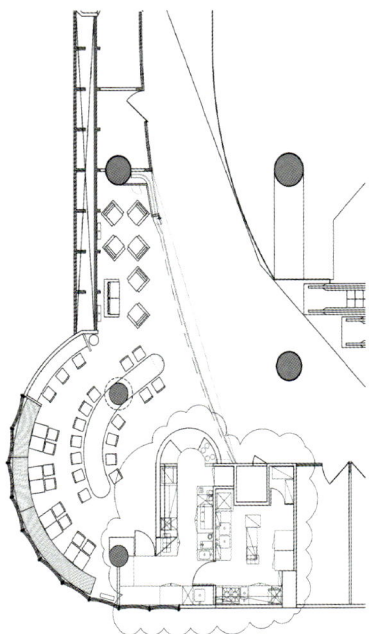

Café

China

Hong Kong

2006

Photo: Rage Wan

Alex Choi

Suzuki Café Chain

Bright spark has been made when a chain of archetypal Japanese café entered Hong Kong. While keeping its own Japanese style, designer concerned how the two different cultures can work together without any confliction. As a chain café, it has to deal with fluctuating situation; location is one of the matters, and how to stand itself out is another one when a new café is no longer an exciting issue in the town. Café is always welcome to the youths; in order to show the young and warmth, pastel colour and wooden wall furnish were exercised in the design. The cross-cultural design happens when creation should not limit to trend, style, shape or anything; mix and match is not limit to material either. The philosophy is also convincing in this chain café design. This time, designer mixed the local culture, and matched with the cafés with their different locations.

铃木连锁咖啡馆

这家日式连锁咖啡店自宣布入驻香港后，就引起了极大反响。然而，要保持日式餐厅的原始风貌，设计师又担心怎样才能实现两种不同文化元素的完美融合。作为一家连锁咖啡店，要考虑怎样处理地理位置变动带来的影响，因此位置是要考虑的头等大事，而另一个重要问题就是在人们已经对新餐厅的出现不足为奇的情况下，怎样使这家咖啡厅有别于其他同类餐厅，出类拔萃。咖啡厅总是受到年轻人的青睐。因此，为了体现时尚与温暖，设计采用了暖色材料和木制的墙面装饰材料。当创作不在拘泥于潮流、风格或外形等的局限，混合搭配不仅限于材料本身时，就实现了跨文化设计。

Aropa Mediterranean Restaurant

The idea to bring up the restaurant was based on the spirit of Anutans – 'Aropa', which means giving, sharing, respect to nature. Located at the Solomon Islands, the island of Anuta stands out in one of almost a thousand islands, which is as pure as it has never changed since the creation of the world. The viewpoint from above Aunta gave a stunning panorama of the whole island. The wonderful panoramic view of the island of Anuta and the artistic handcrafts of the tribes become the inspiration of theme of the design. With the impression of characteristics of Anuta together with Aropa, the restaurant is shaped with plenty curves filling around from the side to the ceiling to create rhythmic rise and fall of the ocean. As the island is in the breast of the ocean, the curvy open frames around the restaurant embrace the whole in an open way instead of isolation. The water jelly lamps and cloud chandelier are the highlights that stand out the concept of being natural.

阿罗帕地中海餐厅

餐厅的设计理念来源于阿奴塔精神——"Aropa"。这个词的意思是给予、分享、尊重自然。阿奴塔岛屿位于所罗门群岛海域，由约上千个小岛组成，这里如同原始大自然一样的纯净。从阿奴塔上空鸟瞰，可以看到整个岛屿的全景。阿奴塔的全景图和当地部落的手工艺品是此餐厅设计的主线。为突出阿奴塔的地区特色，发扬阿罗帕精神，餐厅一侧和天花板都呈波浪式图纹，营造出了海洋波浪起伏的节奏感。就像阿奴塔岛是整个地中海的心脏一样，这些开放式曲线框架以开阔的臂膀将一切揽入怀中。水胶灯和云形状的枝形吊灯突出了回归自然的设计理念，是整个餐厅设计的亮点。

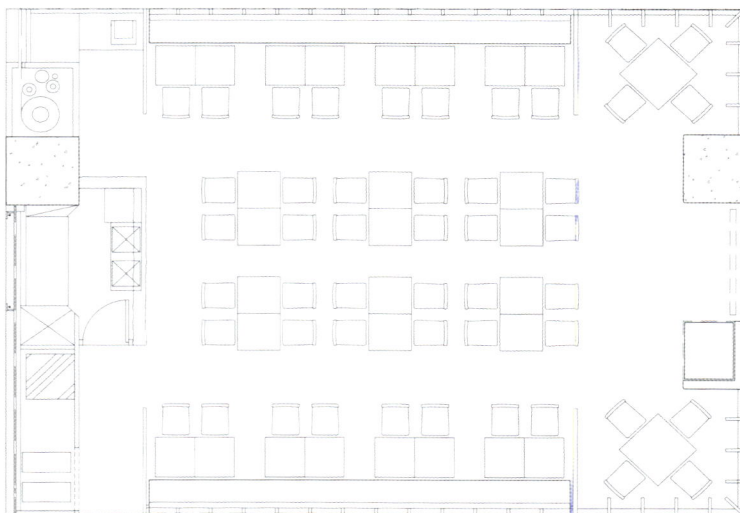

Restaurant

China

Hong Kong

2009

Photo: Lam Ka Ho

Alexchoi Design & Partners Ltd.

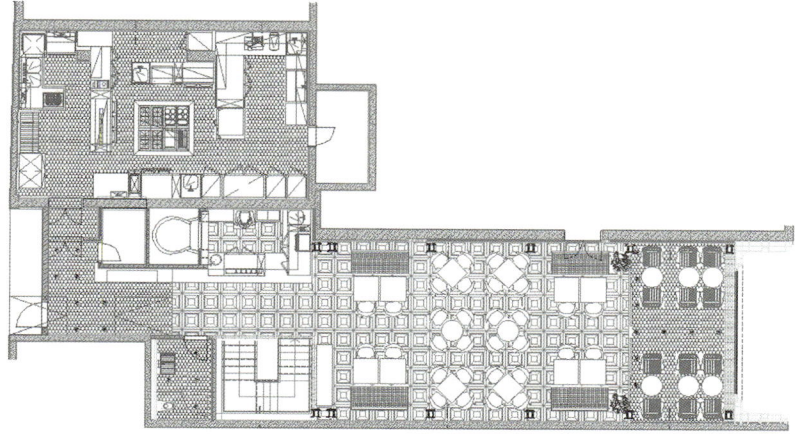

208 Due Cento Otto Restaurant

Autoban's first overseas project, 208 Duecento Otto Restaurant has opened in Sheung Wan, Hong Kong. Inspired by the district surroundings and the New York-style Italian menu, the design story was created to bring the authentic cuisine to Hong Kong. Art and design play a large part in the make-up of the restaurant's surroundings, creating a fresh Bohemian energy. A striking rustic iron facade welcomes visitors along with the two Nest Chair, the latest item of the Autoban's collection. Two-storey restaurant seats up to 90 people, bar on the ground floor with a warm private dining room for up to 18 people, which leads up with an interior staircase to the dining room on the first floor.

208号铺餐厅

Autoban 设计事务所的首个海外项目——208号铺餐厅在香港上环开业。餐厅位于208号铺地区，以纽约式的意大利菜肴为特色，设计旨在带给香港纯正的意式风味菜肴。艺术与设计对于餐厅环境设计发挥着重要作用，带来了新鲜的波西米亚能量。显著的乡村风格铁制面墙与Autoban的最新设计作品——两张鸟巢座椅，将顾客邀请到餐厅内。两层餐厅共容纳90人。一层是可以容纳18人用餐的私人餐厅，沿室内楼梯就来到二楼餐厅。

Restaurant

China

Hong Kong

2010

Photo: George Mitchell

Seyhan Ozdemir, Sefer Caglar

Photo: HEAD Architecture and Design Limited

HEAD Architecture and Design Limited

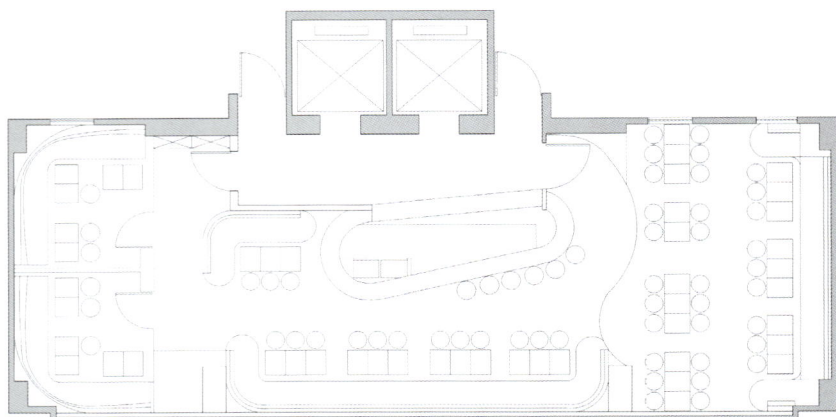

Muse Bar

The brief was to design a bar on the 26th floor of an office building in Causeway Bay that could accommodate 92 people in an open bar area with two VIP rooms. According to the dictionary 'muse' means 'to think deeply' and 'to forget about the world around one'. This word formed the starting point of an interior which was to be relaxing, stylish and provide a retreat from the implications of daily life in Hong Kong. A soft colour scheme of purple and black provides an 'other-worldly' backdrop to a long curving bar running almost the entire length of the building core. This bar forms the focus of the interior and is fronted with a three dimensional sculptural facing.

沉思酒吧

客户要求在香港铜锣湾一栋办公大楼的26层设计一家酒吧。设计包含两部分：两个贵宾室和能容纳92人的开放酒吧。按照字典解释，MUSE 的意思是"沉思"、"忘掉周围发生的一切"。室内设计就以这个词为出发点，打造一个轻松、时尚、使人远离香港平日喧嚣的雅致餐厅。紫黑相间的柔和色系为占据大楼核心位置的长长的曲线形酒吧提供了远离世俗的空间背景。吧台是酒吧的核心，吧台前面是一面三维的雕刻面墙。

The French Window

This high-end French restaurant combines retro and modern, which is a surprisingly good place. The delicious tradition food provides the guests with both a good appetite and a casual emotion. The design principle of this restaurant is noble but not overcautious. Since the neighbourhood is filled with well-known high-end brand shops, the designers' main task is to create a distinct image for the entrance. The designers create a pair of luminous lighting columns, made of wood and cast iron. The guests will confront a long hallway shared with the neighbour shop when they come into the space. The hallway is decorated with modern lanterns and lush vertical plant panels.

法国之窗餐厅

这间高级法国餐厅的设计糅合了怀旧与摩登，令人喜出望外。这里优质的传统佳肴让客人得享时尚美膳之余，又令人感觉自在。法国之窗餐厅的设计原则是：高尚而不拘谨。设计师的首要任务，是为餐厅入口位置营造一个鲜明形象，因为其隔邻商店皆是为人熟悉的高级品牌，拥有显赫名声。因此，设计师设计了一对极具标志性的发光灯柱，以木材及锻铁造成。宾客一踏进餐厅，迎面的便是与隔邻商店共用的修长通道，沿途尽是现代灯笼及茂盛的植物。

Photo: Chester Ong, AB Concept

Ed Ng, Terence Ngan/AB Concept

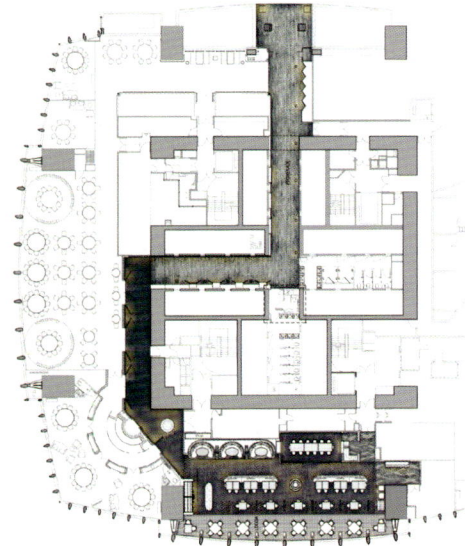

Photo: Lim.Teo+Wilkes Design Works Pte Ltd.

Lim.Teo+Wilkes Design Works Pte Ltd.

Restaurant & Bar in Mandarin Oriental Hotel

The captain's bar is just off the lobby, and the predominance of the red walls will also remain. Instead , the designers have altered the navy accents, and brought in the sophisticated taupe colours into the carpet and leather upholsteries, along with some new updated tables so that rejuvenation has occurred to an old familiar face. The same goes upstairs in the Chinnery bar, where all is intact. The new Cake and Coffee Shop is a 21st century interpretation of the colonial era... not literally, but in a contemporary sense. Up on the 25th level, the former home to Vongs Restaurant will be reborn as Pierre, with stunning views of Hong Kong Harbour. Mirror and glass and reflective surfaces will make the room sparkle and allow Pierre to take centre stage.

文华东方酒店餐厅酒吧咖啡厅

船长酒吧邻接大堂，红墙的主色调没有更改。我们改变了海军（深蓝）的主色调，地毯和皮沙发运用了有一定深度的褐灰色，新换了一些桌子，是"旧貌换新颜"。楼上的Chinnery酒吧也和这里一样，所有的一切都互相呼应。新建的蛋糕、咖啡店是对殖民地时期的现代诠释，但并不夸张。在蛋糕售卖区，这里的椅子看上去就好像盒装的巧克力。25层上，原来的Vongs餐厅被新改造为Pierre餐厅，并总览香港港口的壮观景色。

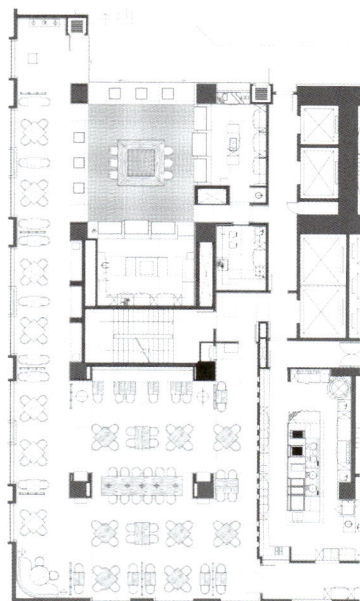

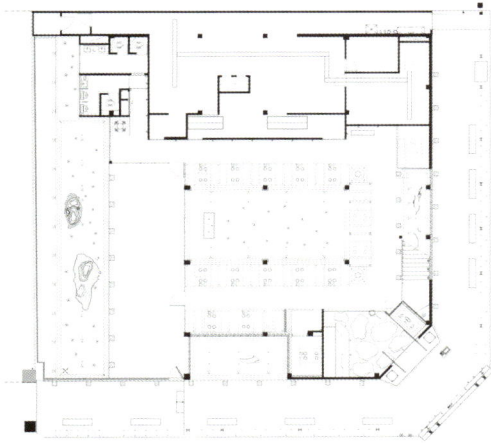

Karuisawa Restaurant

The project gets a simple and clean look due to the designers' good job. The designers change the original appearance and express the restaurant's extension with an 11-metre-high cast iron grilling. The interior continued the previous theme and was considerately add some impressive experience. Full of Japanese style, the whole building has various visual images, natural and rich in meanings. Minimalist Japanese style is everywhere in this design. From the direction of lines to the strength of the textures, from the materials' grain to the use of clean lines and natural materials, the iron detail and glass created a perfect combination of colours and elements.

轻井泽锅物

本案利落的室内外肌理焕然一新，设计师改变了原结构外观，以挑高11米的铸铁格栅展现张力，内部方面除了延续先前一贯的风格外，更精心营造让人印象深刻的体验。 整体建筑充满浓浓日式风格的轻井泽，拥有多样的视觉印象，不假雕琢的自然意涵，渗透以"极简日式"意象，从传递动线方向，排列出富肌理的力道，从讲求素材的细腻纹路，运用简单利落的线条与自然材质，搭配铁件和玻璃，呈现色彩与元素的完美结合，借由物件之间串联出的独特语汇，织构递进的空间层次。

Restaurant

China

Taiwan

2010

Photo: Lou-Kwuo-Chi

Chou-Yi

Restaurant

China

Taiwan

2008

Photo: Nacasa and Partners Inc.

Hashimoto Yukio Design Studio

Silks Palace

'Silks Palace' locates in one of the top four museums in the world, National Palace Museum in Taipei. It is a complex of 1st floor dining, 2nd floor private rooms, 3rd floor ballroom and basement 2nd floor food court. The designer achieved to fuse various exhibit motifs of National Palace Museum into traditional and expressive Chinese space by using wood and bricks as basic elements. Large version of 'Ts'ung Tube' (important ritual) continuously line up at dynamic double height ceiling between 1st floor and 2nd floor. Open kitchen at the centre back, 'frosting' pattern wooden partition in both sides which is traditional pottery crack motif. Bronze bell-shape acrylic pendant lights hanging from the wood-panelled ceiling.

故宫晶华

故宫晶华就在世界上顶级的四个博物馆之一——台北国立故宫博物院的院内。它的一层是餐厅，二层是私密房间，三层是舞厅，地下二层是美食广场。我们将国立故宫博物院的展览特色融于传统的、富有表现力的、用木和砖作为基本设计元素的中式空间内。仿制的大型琮管（用于宗教仪式中的重要饰品）有序地排列在一层和二层不断变化的空间里。开敞的厨房位于中偏后的位置，两侧"磨砂"图案的木隔断是模仿陶瓷瓦片的特征。铜钟状的树脂吊灯从木天花板上垂下。

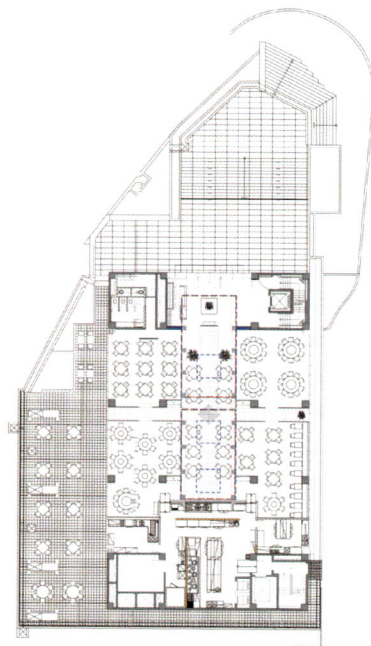

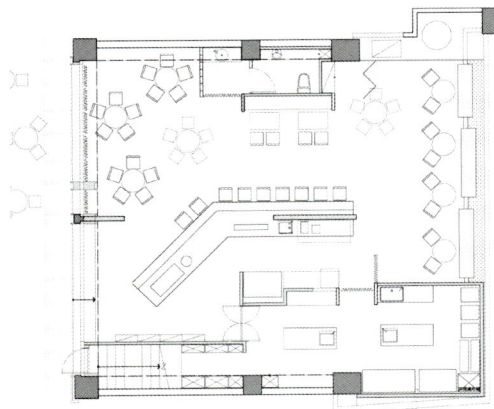

Restaurant

China

Taipei

2009

Photo: Black

Janus Huang/Taiper Base Design Centre

Oden Studio

Located in Inner Lake of Taibei, Taikobann Restaurant is a Japanese Oden restaurant. The designers used clean and simple lines and raw wood with bamboos, vines, fibres and other natural material colours to convey a concise aesthetics. Guests are welcomed by the large barrels and red curtains at the door, full of Japanese styles. To cover up the disadvantage of orientation and location, the designers apply a slanted surface at the entrance, to get the passers-by and guests influenced by the interior atmosphere. In order to echo with this line, the floor uses slanted paving too. Wood in different colours and sizes are matched together, providing the space with simple and raw traditional atmosphere.

太鼓判餐厅

坐落于台北内湖的太鼓判，为一家日式关东煮餐厅。设计师运用简洁利落的线条和自然原木材质，以竹、藤、麻和其他天然材料颜色，形成朴素的自然风格，用设计来传达简约的美感。走进餐厅，浓浓的日本风情映入眼帘，门前大型酒桶与红色布幔更是揭开了迎宾意象。由于方位与设点位置的不利，设计师将餐厅的门口以斜面牵引出主题，让店内的氛围顺势感染路人与宾客；而呼应动线的转折，地板质材同样也以斜面铺陈，并搜集各式深浅新旧不一的木料拼接，让空间内部洋溢着古朴的传统气息。

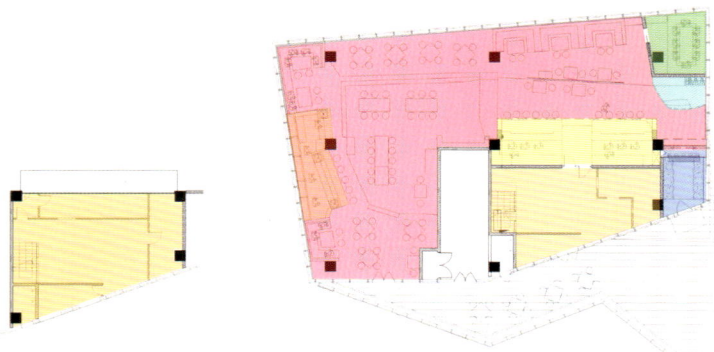

Hatsune Japanese Restaurant

An entrance without any signboards, the restaurant keeps itself to its most sensitive guests. The water curtain is opened by a secret button. A large rock reception, ceilings and walls like waves and fish schools coming create a colourful underwater world for the guests. Submarine contours simply divide the space into counter area, banquette booths, separate tables, VIP rooms, bar area. Stainless steel bars, stone, wood floor, concrete floor and carpet enrich the texture of this underwater ground. The figure shadows behind the sushi bar look like a nonstop performance. The owner exquisitely uses some modern methods to show Japanese traditional patterns, such as interior wall paintings, films on the glass and sofa design.

隐泉日本料理

没有任何招牌的入口，将这块圣殿保留给嗅觉最灵敏的时尚贵客：秘密的机关打开了水帘幕，一块大岩石的接待台，犹如海浪波涛的天花板及墙面，跟着由远处逼近的鱼群，仿佛置身于绚丽的海底世界，优游其中。海底等高线简单地区分空间为吧台区、卡座区、散客区、VIP包房、酒吧区，层层叠峦，错落有致。透过不锈钢条、石材、木地板、水泥地及地毯的质感来表达海底地质的丰富性。寿司吧上方的磨砂玻璃里，虚虚实实的人影，如同永不停息的表演秀。老板细腻的心思，利用现代的一些手法来全释日式传统的图腾，如室内的墙彩绘、玻璃的贴膜与沙发等等。

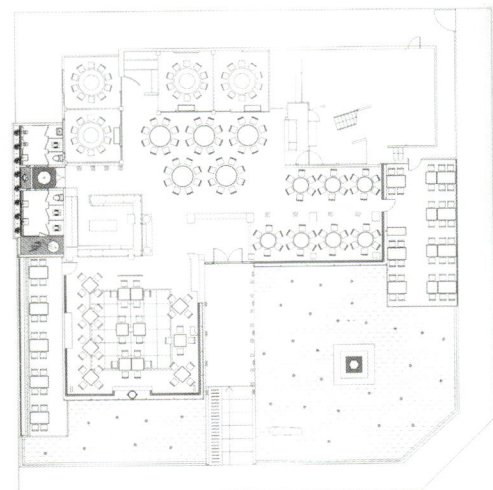

Restaurant

China

Taichung

2009

Photo: Black

Janus Huang/Taiper Base Design Centre

Natural Hakka Restaurant, Taichung

The designe jumped out of tradition elements to express this space in multiplex Eastern styles. The architecture is simple and clean with façade wrapped with black flowered windows, creating a sense of mystery. Matched with a pool in the front yard, it becomes the focus of this busy crossroad. The entrance gallery is expressed in torn peddles and a heavy cypress gate is the main feature in the end. The foyer leads guests into a reception on the left side. The black flowered window element was also used in interior area. The space is divided by seat's numbers and each area has a distinct character. Lighting is soul of the space, on which the designers pay more attention than other design features. A birdcage of Chinese style is in the centre of space arrangement.

台中客家本色餐厅

设计师发挥所长并跳脱传统元素，以多元的东方调性来呈现。建筑造型简单利落，外墙以黑色花窗包覆，带着若隐若现的神秘感，搭配前庭水池所营造的气氛，成为往来频繁的十字路口处焦点所在。入口长廊以复古磨石子材质表现，而厚实的桧木大门则成为长廊上的主要端景，迂回的大门玄关引导用餐客人往左方的接待柜。外墙黑色花窗仍引用至内部，空间大致以客人的用餐人数座位来区分，在不偏离主轴下每个区皆呈现各自的风格。灯光是空间的灵魂，设计师在灯饰设计及光线的营造上颇费心思，当中加入中国风的鸟笼做为空间布置的主角。

Café

South Korea

Seoul

2009

Photo: Courtesy LDV Hospitality

Crème Design

Lugo, Seoul

As a continuation of the first Lugo Caffe in New York City, this location in Seoul, South Korea is a smaller more formal version of the original. At the entrance, a wood-grain texture concrete wall wraps the front of the building and a backlit brushed nickel sign, oak door and brushed nickel door handle, and Vespa scooter greet patrons as they enter. Inside the front door, the courtyard style dining space is perfect for more casual dining with its red metal chairs, metal tables, a red and white awning and a lattice wall of ivy. Inside, the walls of the main dining room are lined with white subway tile and wooden mill work. Tin coffered ceiling panels, inseted in a wooden framework, cap the double-height space. Vintage black and white photographs line the top of the walls and are up-lit for emphasis.

首尔卢戈咖啡厅

继纽约首家卢戈咖啡厅之后，位于韩国首尔的这家卢戈咖啡厅空间比前一个小但却更加正式。入口处，木制纹理质地的混凝土墙面覆盖了整个建筑正面。背光式拉斯镍标志、橡木门和拉斯镍门把手，还有黄蜂牌摩托车，使顾客一进门就有一种宾至如归的感觉。走进前门，庭院式的用餐空间内摆放着红色金属椅子、金属桌子、红白相间的遮阳篷和常春藤格子墙，是休闲餐饮的极佳场所。室内主餐厅的墙壁上镶有白色地铁瓦和木制品。镶嵌在木子框架内的锡格子天花板将整个双层空间覆盖住。

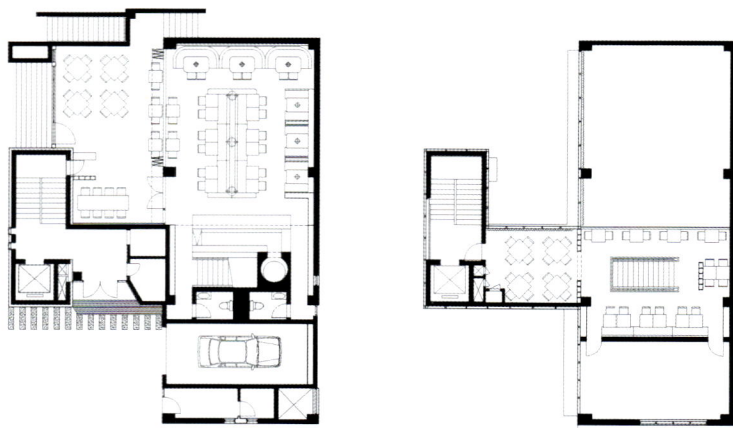

Café de Matinee

The space is divided into three distinct 'jewel boxes', a two-storey marketplace and two separate casual dining rooms, one on each floor – creating a series of intimate and humane spaces. In the dining rooms, one is enveloped in wooden screens that filter natural light. These screens are very much an integral part of Chinese domestic spaces and by abstracting them designers create a comfortable, modern atmosphere. The market is a more dramatic double height circular space, creating a mundane sense of luxury in a routine activity of shopping for food. The design of the whole space gives customers unlimited inspiration and imagination.

日场咖啡厅

室内空间由三个别致的"珠宝盒"一样的空间组成———个两层的市场和楼上楼下两个独立的休闲餐厅，打造了一系列亲密而又人性化的用餐空间。走进餐厅，顾客就被包围在洒满自然光线的木制屏风中。这些屏风更符合中国式室内装饰的整体风格，而经设计师利用后，则营造出舒适而又现代的气氛。市场是一个更别致的双层圆形空间，打造了日常食品购物空间的极度奢华感。整个空间的设计能勾起顾客无限的激情与遐想。

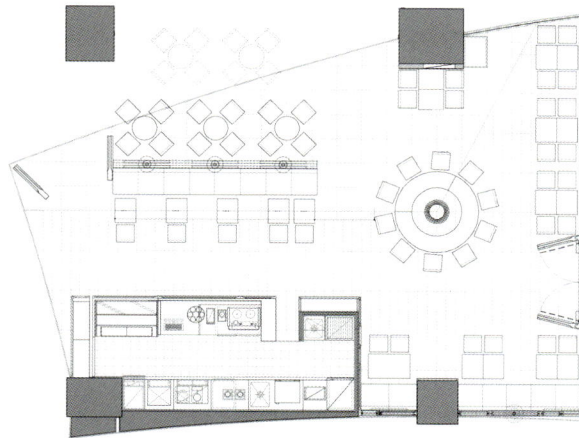

Café

South Korea

Gyeonggi-do

2009

Photo: Park, Woo jin

Studiovase

Café

South Korea

Seoul

2009

Photo: CP Group, Seoul, South Korea

Beyond The Void   Seoul . Hamburg

Hollys Brand Identity

Simplification, abstraction, and at the same time upgrading of actually modest materials to a precious design pearl describe the further approach of this design. In this sense, a band of grey felt is attached to the open wall in a rhythmical order - binding the space together by spatia. framing. Following this rhythmical order, the HOLLYS COFFEE logo is punched into the felt panels by laser cut, whereby the actual wall becomes visible and, thus, an architectural ornament in its elevation. On the one hand, the architects' goal to apply authentic materials, which is anything but comm on in South Korea, and, on the other hand, keeping the low construction budget, led to a concentration on the essentials. While abandoning complete cladding of spatial elements such as ceiling and floor, the focus is on a sophisticated execution of space modules, furniture, and illumination.

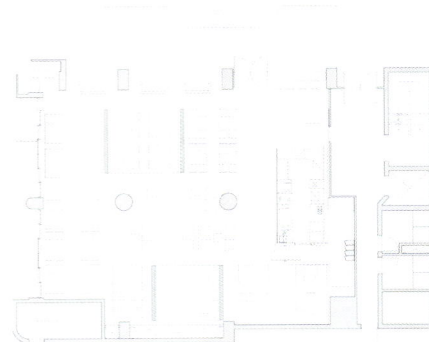

霍丽斯品牌餐厅

设计简单化、抽象化、普通材料的重新利用及采用珍贵材料设计的珍珠等体现了餐厅设计方法的别样性。灰色毛毡以极强的节奏感包裹着开放式墙壁，用空间框架将空间连接。采用激光切割技术，餐厅标志"HOLLY COFFE"以这种节奏感镶嵌在毛毡式嵌板上，这样，一面实体墙就变得醒目突出，实现了墙立面的建筑装饰。忽略了天花板和地板等空间元素的包层设计，设计师将设计重点转移到空间构件、家具和照明设计上。

Buffet Restaurant in Busan Paradise Hotel

The buffet restaurant is in a way an extension of the lobby where the emphasis is on grand food display. In order to break the monotonic space, unique, stylish banquette booths were placed all around in order for the space not to appear as seas of tables and chairs. Warm colours, lightings and luxurious wood as well as stone flooring contribute to the overall feel of modernity, comfort and luxury.

釜山天堂酒店自助餐厅

自助餐厅是大厅的扩展部分，重点展示了丰盛的美食。为了打破单一的空间，餐厅内部设置了独特而时髦的卡座，免得整个餐厅看起来像是桌椅的海洋。温和的色彩、灯光、木材以及大理石地面的使用让整体氛围更现代、舒适、奢华。

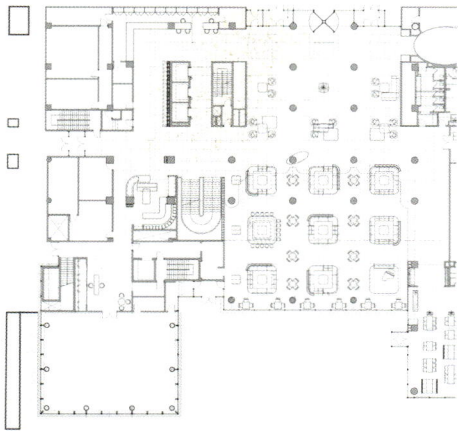

Restaurant

South Korea

Busan

2007

Photo: Studio Gaia

Studio Gaia

CGV Gold Class Lounge

CGV Yongsan Gold Class Lounge is the space to be provided to the CGV theatre users. It is made of three different themed spaces that address specific needs to three different client targets. Café Zone is accessible to all incoming CGV theatre users. Gold Zone is accessible to all the Gold Class clients. Private Zone is the celebrity and VIP space for publicity events and junkets. These three spaces are divided specifically according to the thematic materials used for each individual space. The designers emphasised the creative direction on to the exterior design. Exterior design is the first thing that the CGV users notice as they climb up the escalator to get to the entrance of the theatre where the designers wanted to deliver the most impactful space design concept for all to see.

大宁影城黄金级休闲吧

龙山大宁影城黄金级休闲吧是特别为前来观影的观众所提供的空间，由三个不同主题的空间组成：咖啡区适宜全体观影者进入；金色区适用于黄金级客人；私人区专为名人和贵宾举办活动。这三个空间分别采用了不同的材料。设计师将重点放在了外部设计。休息厅外部是观影者走上扶梯、到达影城入口时最先看到的，设计师试图在这里传达最抢眼的空间设计。

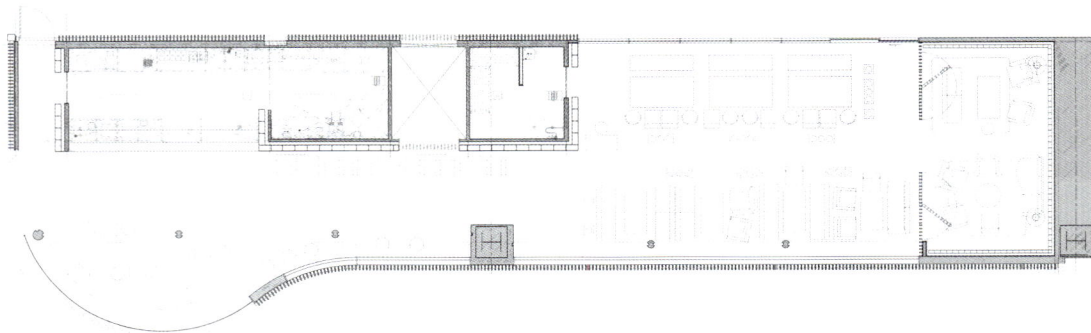

Lounge

South Korea

Seoul

2010

Photo: Urbantainer, Namgoong Sun

Urbantainer Co., Ltd

Restaurant

Japan

Tokyo

2009

Photo: HBA

HBA/Hirsch Bedner Associates

Lobby Lounge-Shangri-La Hotel Tokyo

The lounge consists of an eclectic mix of seating groups, functioning for both high tea service and evening cocktails. The area has windows to Tokyo on three sides, giving the guest a 180 degree view of the Tokyo skyline. High windows accentuate the view which is framed with dramatic drapery. The rich palette in the hand made area rug along with the silk, velvet and other rich fabrics give the guest a rich tactile experience not found in any other hotel. There is a mix of seating, from dining to cocktail to lounge to bar seating which creates a wide selection for the guest's use. Rich materials at the bar of onyx and gold leaf along with a signature custom designed chandelier consisting of Czechoslovakian crystal finishes off this amazing area.

东京香格里拉酒店大堂餐厅&休闲吧

休闲区由各种风格的座位区组成，既能高雅地品茶，也能在傍晚来一杯鸡尾酒。这个空间三面都是窗子，让酒店客人有180度的视野来欣赏东京的天际线。高高的窗子配上窗帘，仿佛为窗外的美景加上了画框。手工制作的小地毯色彩丰富，此外还有丝绸、天鹅绒等其他质地丰富的织物，给客人带来在其他酒店中从未有过的丰富的触觉体验。座位区分为很多种，有用餐座位、品酒座位、休闲座位、酒吧座位等，客人使用时可以有多种选择。酒吧区采用了丰富的材料，如编玛瑙和金叶子装饰，还有订制的捷克斯洛伐克水晶吊灯，共同打造了这个令人叹为观止的空间。

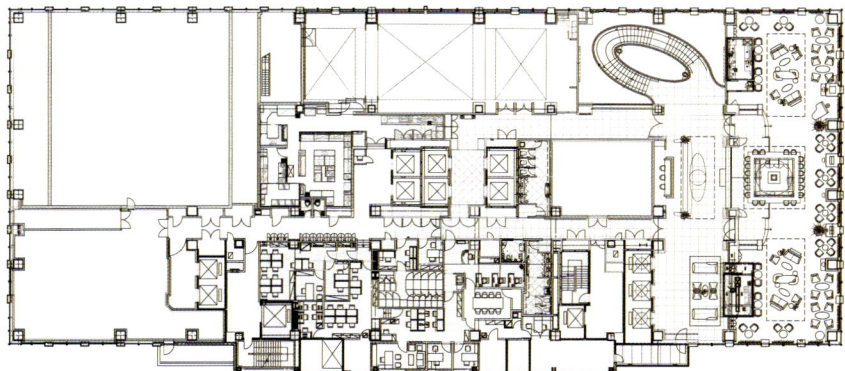

Kamura

The designers were asked to design a traditional Janpanese restaurant with simple style. 'Less is more' can express the essence of the design vividly. So the main design concept of this restaurant is to create a natural feeling by featuring lots of natural wood. The panels connected to the ceiling have a gentle curve and gracefully envelop the whole counter, which is the charming character of this design. The cooks can show their dishes in front of the customers, and people can taste the fresh food which is also the atrractive point of this resturant. People can get a tranquil and relaxed atmosphere in the restaurant.

Kamura餐厅

设计师打造了这个简约风格的传统日式餐厅——"少即是多"的理念形象地体现了设计的本质。餐厅的主要理念是通过使用大量的自然色木料营造天然的感觉。天棚由带有弧度曲线结构连接,优雅地"蜿蜒"在吧台的上方,成为了整个设计的亮点。厨师可以在客人的面前展示他们的厨艺和最新鲜的食品,这也是餐厅的特色。此外,餐厅内环境优雅、气氛和谐,让人备感轻松。

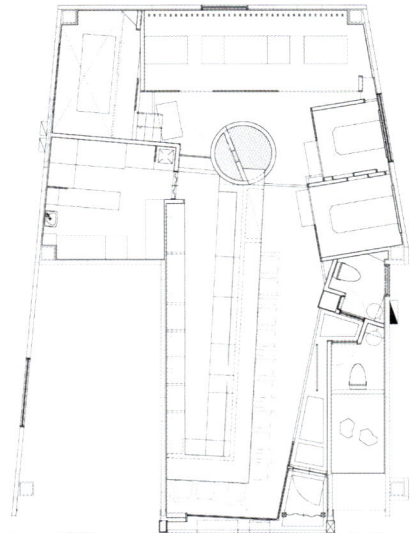

Photo: Masahiro Ishibashi

Yusaku Kaneshiro+zokei-syudan Co., ltd

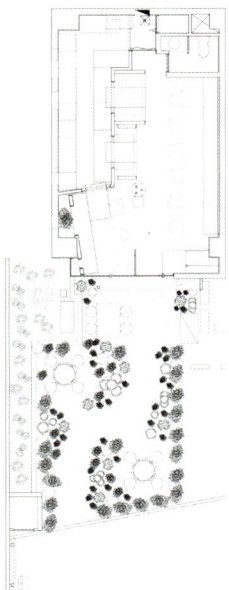

Photo: Masahiro Ishibashi

Yusaku Kaneshiro+zokei-syudan Co., Ltd

Moutrachet

This space has been designed to look a bit decadent with a near future concept. Stylish elements and decadent images were cleverly mixed. This chic Japanese style pub restaurant features design themes of Lights and Gardens and is targeted at young women. The lamps installed on the wall behind the seats mimic fireflies, producing a comfortable hideaway with shafts of light scattered throughout the garden. There are some openings in the kitchen wall and the dining tables extend into the kitchen. Customers can see the chef at work and also experience the unique opportunity to converse with the chef while their meal is being prepared.

Moutrachet餐厅

空间的设计目标是颓废的未来感。时尚与颓废的元素巧妙地融合在一起。这家日式酒吧餐厅的目标顾客是年轻女孩，它的灯光和花园设计别具特色。座位后面的墙上安装着萤火虫一样的灯，灯光洒在花园中，营造了一个世外桃源。厨师也充当侍者的角色。厨房墙壁的开口与餐桌相连，顾客可以看到厨师的工作，体验由厨师上菜的独特感受。

Restaurant

Japan

Osaka

2007

Photo: Nacasa & Partners

Space Planning-LAR

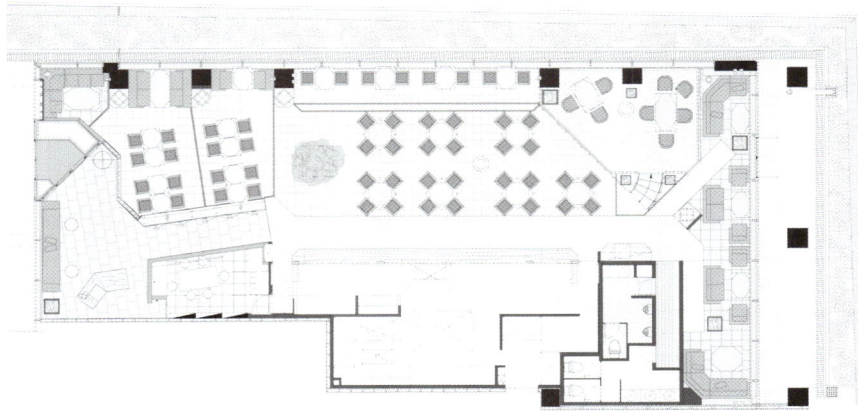

Grand Monde Classy

The dining room is divided into two sections; a main dining room with two-or-four-seat tables arranged with generous spacing, and a semi-private room separated from the other diners by a sheer curtain. In the main dining room, guests can dine with a view of the scenery outside, visible through a vast window. Designer Masaru Shimizu made good use of mirroring to cover the walls and give the restaurant an even more spacious feel. The light stand in the centre of the room and orange lamp shades suspended from the ceiling provide the accents in this simple and modern interior.

大世界精品餐厅

餐厅包括两个部分：设有两个或四个座位的主餐厅以及半私人包间。在主餐厅就餐，客人可以在享受美味佳肴的同时，饱览窗外美景。室内特殊的镜墙设计令空间时刻宽敞明亮。来自中央照明设备的橘色灯光将空间烘托得分外时尚。

Restaurant

Japan

Ikoma

2006

Photo: Kaori Ichikawa

Ryuichi Ashizawa

Temp

This Chinese restaurant is in front of the station at suburb in Nara prefecture. The designer considered to create a space like 'Garden' where people can come together while thinking of a forest in the mountains destroyed by typical development. A movable bamboo partition wall is on the ceiling over seats for corresponding to the change of the number of customers. Moreover, the floor, the tile made by resin shaped like border gives the space an impact of a space profound. The designer arranged the common showcases in a straight line, which can be one part of design for the façade in the same area.

节奏中餐馆

这个中餐馆位于城郊奈良管区的火车站前面。设计师想把这个空间设计得如同"花园"一样,当人们来用餐时,就可以联想到毁于社会发展的山间丛林。座位上方的顶棚上有一面可移动的竹制隔断墙,以相应地配合顾客人数的变化。此外,地面铺有树脂制成的瓷砖,如同边框一般,使空间更通透。设计师将通常的展柜排布方式改为直线式排列,使一般意义上的展柜成为门脸的一部分。

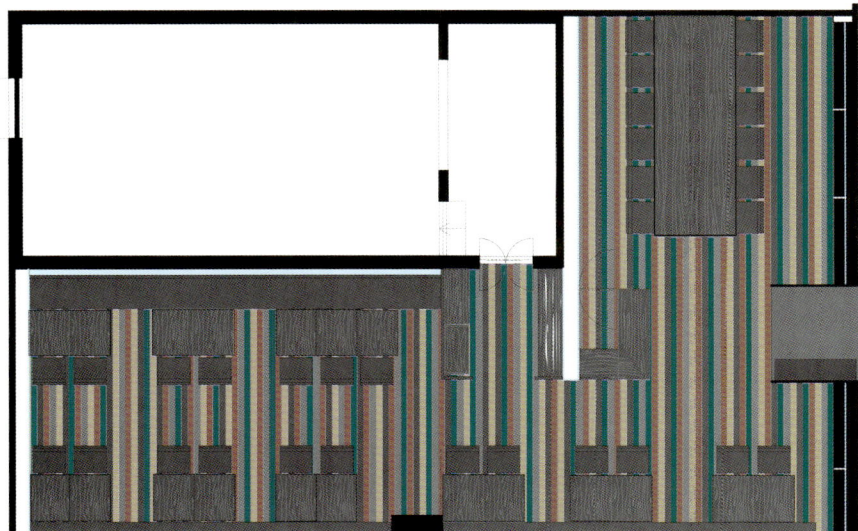

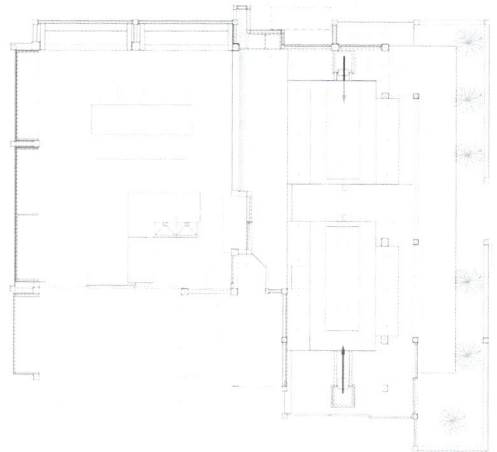

Bar Anzu

The highlights of this bar are based on four seasons, such as cherry blossom in the spring and full moon in the fall. A reflective sheet of holographic film covers the bar counter tops and the wall. Sakura petals give the illusion of an endless space which is also intensified in the lower-level room when one is immersed in a palette of red and black. The walls were painted in vibrant red and the lighting enhances the sensual mood to a subtle erotic level.

日本京都山酒吧

酒吧以四季的更替为设计理念，力图打造出春日樱花烂漫，秋日月满如盘的浪漫空间。全息薄膜的反光板覆盖了吧台台面及墙壁，从视觉上将空间无限延伸。同时，低层房间中的黑红色的经典搭配使空间更加宽敞。亮红色的墙壁和精致的照明设备为整个空间烘托出微妙、神秘的气息。

Restaurant

Japan

Tokyo

2007

Photo: Shinichi Sato

Akemi Katsuno & Takashi Yagi (Love the Life)

Alternative

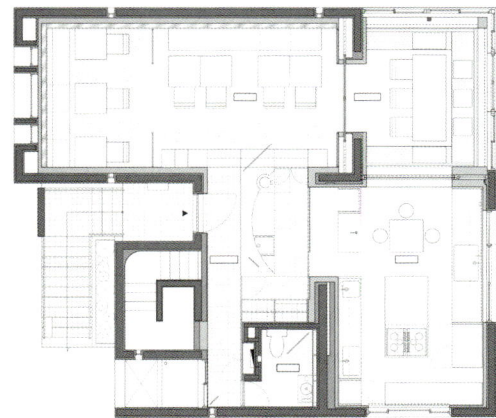

'Alternative' is a restaurant of contemporary Japanese-style foods. The reception and the kitchen are on the one side, and the hall on the other side. The visitors enter the reception and descend the steps and go to the hall. The partitions of frosted glass divide the hall vaguely. The indirect lightings are on top and centre of the wall. Those aspects of absentminded colours influence the whole space as 'blank with high density'. The mat black reception counter has sculptural form. The design idea of those fixtures is from motif of 'Waves at Matsushima' drawn by Korin Ogata. This place is isolated from the babel of surroundings, and filled in stillness.

竞择

"竞择"是一家现代日本料理餐厅。前台与厨房在一侧，大厅在另一侧。客人通过前台后下台阶进入大厅。磨砂玻璃将大厅朦胧地分开，射灯分别安置在顶部和墙壁中间。朦胧的颜色使整个空间产生"高密度留白"感。黑席子前台带有雕刻元素。这些设计灵感源自尾形光林画作"松岛湾之波"的主题。整个空间从周围的喧嚣中独立出来，显得宁静温馨。

Café Tetote

Environmental issues are not issues of someone else, but yours. The whole concept of the Café Tetote is to use leftovers for most of the finishing materials. Producing leftover materials whose quality is too good to be dumped as garbage or smashed to be disposed cannot be avoided. The designer used these leftover materials after fabricating them for suitable and the best use depending on the original materials' characteristics and the craftsmen's work in order to give them new values. Then all these details are forming a unified whole. This concept itself is already beyond the design details.

泰托蒂咖啡厅

环境问题不是其他人的，而是每个人的。泰托蒂咖啡厅的设计理念就是充分利用装修建材的剩余材料。这些弃之可惜的剩余材料的产生是在每次施工之后不可避免的。在泰托蒂咖啡厅的设计上，让这些剩余材料得到适当且最充分地利用之后，设计师再依据材料自身的特点，以先进的技艺赋予材料新的利用价值。然后，这些所有细节慢慢造就成一个完整的设计。

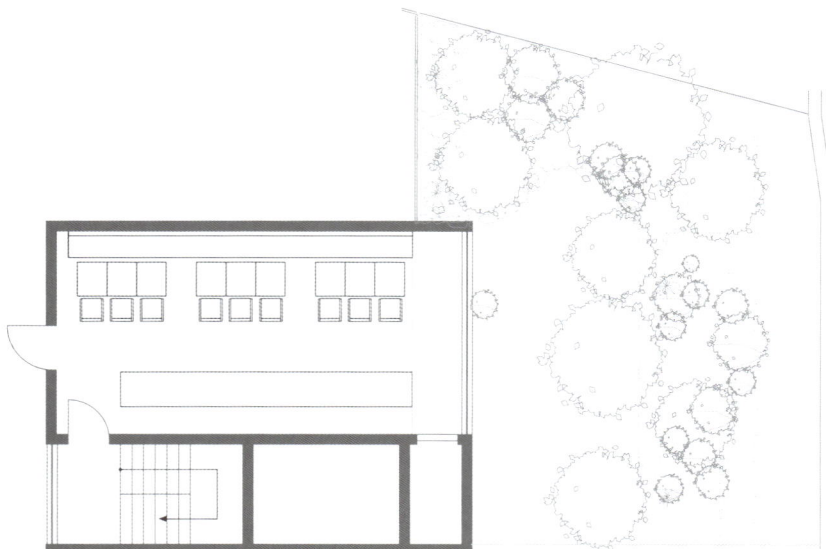

Beer-Hall

This beer hall presents the retro atmosphere of the Showa period. Food samples are displayed in a glass case where the plates are set in a reverse direction so customers can see them from underneath, which present the owner's care of customers and their business concept. Clever placement of mirrors reflects those plates, producing an impact of space missing new and old. The light is not to bright, and people can enjoy their time in a peaceful atmosphere. People can sit along the bar desk and also in a private space which avoids disturbing others. The photos on the wall and on the turning of the stairs offer a kind of cultural taste.

啤酒屋

这个啤酒屋给人轻松的感觉. 摆盘以相反的方向陈列在玻璃柜里, 这样通过镜面的反射顾客就可以在下面很清晰的看到食物, 这充分体现了餐厅经营者的用心和他们的经营理念, 镜子的巧妙使用, 通过反射可以让这些摆盘看起来很新鲜, 人们可以坐在吧台上, 也可以坐在私密的空间,避免别人打扰, 墙面上和楼梯间的挂画让啤酒屋更有文化的气息。

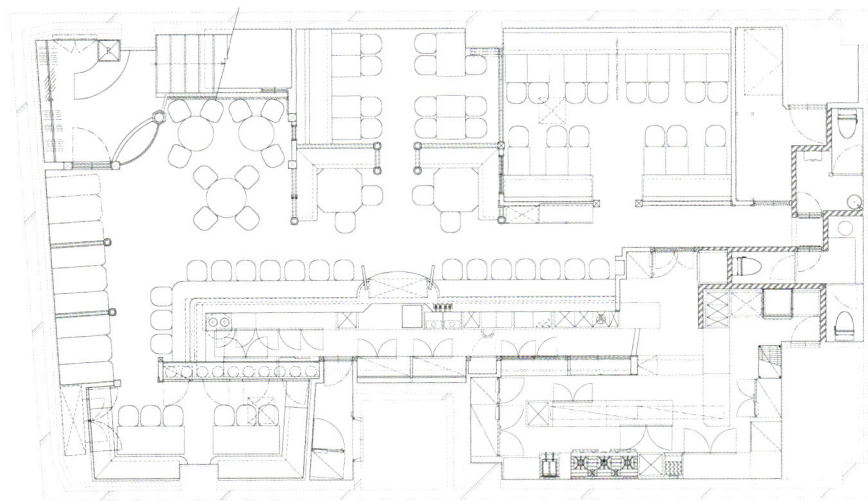

Bar

Japan

Tokyo

2007

Photo: Masahiro Ishibashi

Yusaku Kaneshiro+Zokei-Syudan Co., Ltd

Restaurant

Japan

Tokyo

2007

Photo: Masahiro Ishibashi

Yusaku Kaneshiro+Zokei-Syudan Co., Ltd

Ajima-Ashib

To differentiate this restaurant from other Okinawan restaurants, unique and unusual interior designs have been utilised throughout. For example, a 'flower hat' worn in traditional Okinawan dances is used as a lamp shade, and instead of just displaying bottles of the local spirit called awamori as done by other area spots, awamori bottles are used as servers by attaching pouring spouts. This restaurant has a variety of dynamic and positive design elements.

Ajima—Ashibi餐馆

与冲绳的其他餐馆不同，这家餐馆的室内设计非常与众不同，独特的设计贯穿始终。举例来说，冲绳人喜欢佩戴的传统花帽子被用作灯罩，当地人信奉的awamori瓶子并不单单作为展示之用，而是被店员与浇铸槽连接在一起使用。这家餐馆有很多鲜活、富有朝气的设计元素。

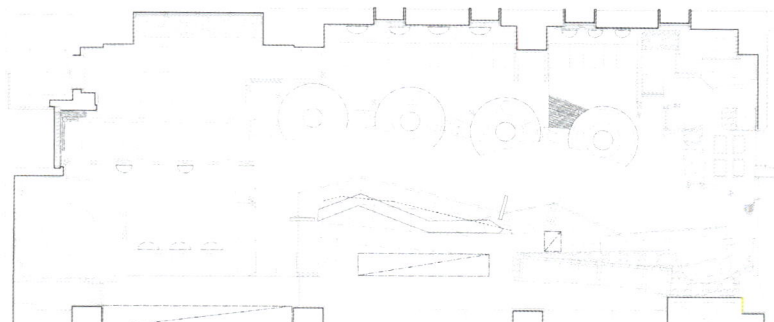

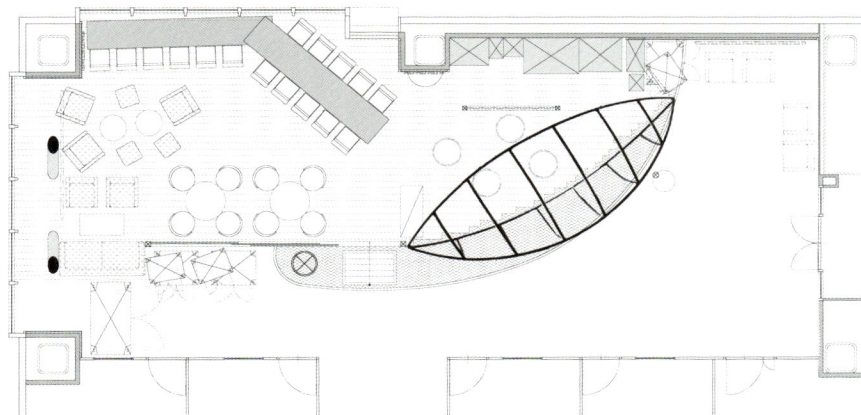

AJITO Private Bar & Lounge

This space has been designed with the theme 'Pirate's Hideout in the Future'. To surprise customers, pirate objects (a pirate is hung upside down) are displayed that serve as external walls and partitions. Steel and wood were used throughout, creating an enigmatic atmosphere.

AJITO酒吧休闲所

空间设计围绕这个主题——"未来海盗的藏匿处",为了给顾客制造惊喜,餐厅里布置了很多海盗物品,如一个海盗玩偶就被颠倒着放在店里。钢制和木制的材料在多处使用,制造了一种迷离的气氛。

Bar & Lounge

Japan

Tokyo

2007

Photo: Masahiro Ishibashi

Yusaku Kaneshiro+Zokei-Syudan Co., Ltd

Restaurant

Japan

Tsushima-Shi

2007

Photo: Shinichi Sato

Akemi Katsuno & Takashi Yagi (Love the Life)

Mottainai Tsushima

'Mottainai' is a life miscellaneous goods brand which assumes 3R (Recycle, Reuse, Reduce) as the concept of the goods development. It is the first examination store of 'Mottainai', and constructed temporarily in an atrium of a shopping centre. Hinoki (Japanese Cypress) was used as the main material. The frame (120mm x 120mm) is made of Hinoki and constitutes a random beam. The floor and the display wall were finished with the panel board of Hinoki. Low stages and round tables are made in S-wood (board molded the tip of the Hinoki). The top surface of a low stage is fitted with glass, and the inside is filled in tip of Hinoki.

Mottainai对马岛店

Mottainai是一个生活用品品牌，再循环、再利用及再降低是其主导理念。对马岛店作为第一家试验店，临时开设在一家购物中心的中庭内。设计师选用日本柳杉作为主要材料——120毫米X120毫米的柳杉框架构成随意的梁柱，地面以及展示墙面采用柳杉板装饰，诠释了品牌的主旨理念。

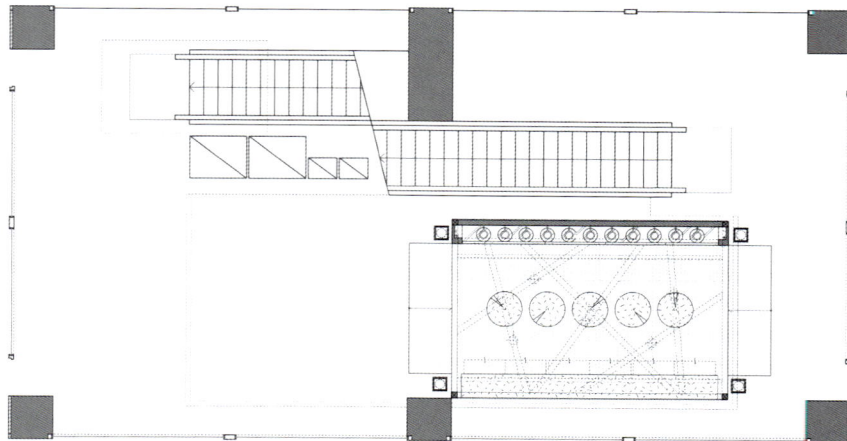

Bar

Japan

Osaka

2006

Photo: Yoshihisa Araki

Shimizu, Masaru

Zodiac

Zodiac is located in Osaka (a port city on southern Honshu on Osaka Bay, an important commercial and industrial centre of Japan), and its total area is only 49 square metres. Zodiac's furniture and interior is kept simple, so graphics on the ceiling become even more captivating in contrast. As the bar's name might suggest, the graphics are actually depictions of the twelve astrological constellations. On the first sight, they appear to be luminous, but there are actually just reflections. The tabletops below them are mirrors covered with opaque foil that has reflective cutouts in the form of graphics. When light is shone upon, these graphics seem gleaming against the ceiling.

黄道带酒吧

酒吧"黄道带"坐落于大阪（本州岛北部大阪湾的海港城市，日本重要的工商业中心），总建筑面积为49平方米。为了彰显顶棚迷人且富有魅力的图案，酒吧的室内陈设被设计的十分简单。其顶棚图案来源于酒吧的名字"黄道带"（源于十二星座占卜术）；而那些图案的形成，实际上只是吧台的反光效果产生的倒影。

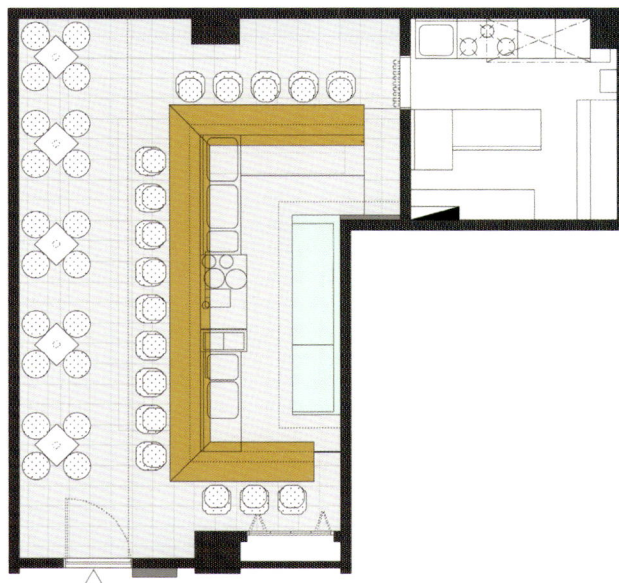

Bar

Japan

Chiba-ken

2007

Photo: Masahiro Ishibashi

Yusaku Kaneshiro+Zokei-Syudan Co., Ltd

Yoi-No-Myojo

The whole façade is glass so that the inside of the bar can be viewed as a stage for visual effects from outside. As the name indicates, the bar's concept is based on fish. The restaurant features contemporary Japanese taste throughout. A folding screen with a Japanese motif has been installed on the wall behind the counter. Huge geometry pictures hang on the wall. The vivid motifs imbue the atmosphere with a feeling of oneness. Lines are used as the main motif. The chandelier accent inspires a romantic mood and the many theme characters displayed throughout give additional dream to the space. People can enjoy themselves and relax in this limited space.

优之明星酒吧

酒吧的外墙全部由玻璃制成，从外面看，酒吧内就像一个舞台。酒吧的设计理念是鱼。餐厅云集了日本各地的风味菜色。柜台后面装着日式屏风。墙上挂着几何图案的大挂画。所有生动的细节构成了一个整体。线条是餐厅中主要的图案。吊灯等浪漫的装饰赋予餐厅梦幻的感觉。人们可以在有限的空间内放松身心，享受快乐时光。

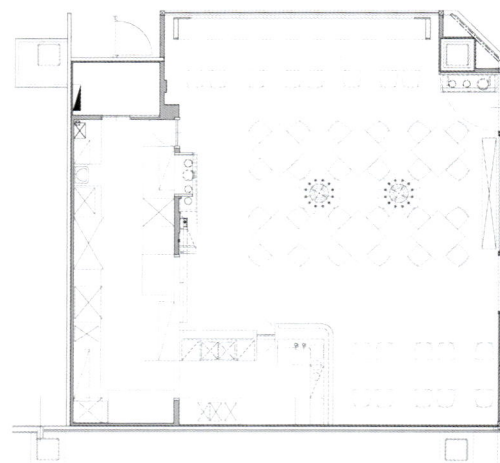

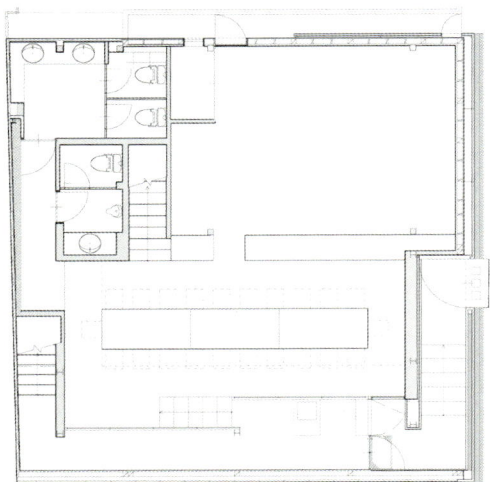

Photo: Syuich Aida

Yusaku Kaneshiro+Zokei-Syudan Co., Ltd

Soya

'Soya' is a restaurant which serves cuisine of various beans. The planning site was situated in the quiet area surrounded by commercially developed areas; the building was a warehouse of timbers that was planned to be demolished five years later. The facility would exist for only five years as there is the coming disassembling planning. The main object of this project aims at the alteration of the neglected building that is determined to be disassembled into a landmark building that gradually fades out as the surrounding economic situation matures. This is the disassembling project which is phased into several steps in a determined period.

索亚餐厅

索亚是一家以豆类菜肴为主的餐厅。餐厅的选址位于商业开发区内一处安静的位置。这里原是木材仓库，并将在五年后拆除。拆除计划只给这里五年的寿命。设计着眼于建筑拆除后这里将建起的地标式大厦，在周边的经济形势成熟后，它终将淡出人们的视野。设计实施前经过几个步骤的决策过程，最终在该地区创造了活跃的氛围。

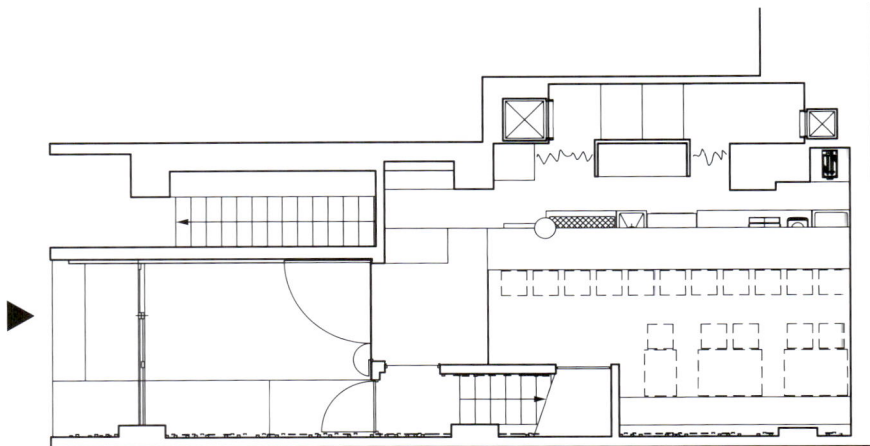

Bar

Japan

Tokyo

2006

Photo: Guen Bertheau-Suzuki Co., Ltd.

Guen Bertheau-Suzuki Co., Ltd.

Gomitori Yakitori

In order to distinguish itself from many competitors nearby, the client asked for an arty and modern design. The ground floor features a bar, the first floor has table seatings, the second a Japanese style floor seatings, and kitchen is situated at the top floor. Vertical lines are used on the first floor in relation to its European style of seatings, while horizontal lines are used on the second floor in relation to its Japanese style. Also, red and black colour scheme evokes an appealing glamour for young ladies while a traditional yakitori-shop would attract only male businessmen.

五味鸟居酒屋

为了凸显自身的个性而与竞争对手有所区别，按照客户的要求采用了兼具艺术感和现代感的设计：一楼设有特色酒吧，二楼配有桌椅，三楼则提供了日式的可坐式地板，顶楼设有厨房。二楼用垂直线条的设计与其欧洲风格的座椅相呼应，而具有日式风格特色的三楼则用水平线条的设计相衬托。此外，以红黑两色为主的区域对年轻女士极具吸引力，而传统串烧区吸引的主要为男性顾客。

Feast Village (Starhill Gallery, KL)

This project named Feast Village looked like a real feast place. The entire floor of a shopping centre houses 12 different restaurants of different styles, including Malay, Hong Kong, Korean, Thai, Lebanese, Norwegian, Indian, and Chinese style. It is not simply 'modern'; it is the Ethnic and Rustic Village for the contemporary lifestyle. It being a multi-ethnic village, brought into an exclusive shopping centre, the special design vision offers people with opportunities for communication.

盛宴村庄

这个名为盛宴村庄的餐厅看起来就是一个真正的盛宴之所。整个购物中心有12家风格不同的餐馆。设计师在"异国和乡村"的概念下进行设计，这些餐馆包括马来餐厅、港式茶餐厅以及韩国、泰国、黎巴嫩餐厅，挪威面包房、海鲜坊、印度菜和中国菜。设计师并不只是简单地遵循"现代感"，而是考虑到现代的生活方式而设计出"异国民族村"的概念。

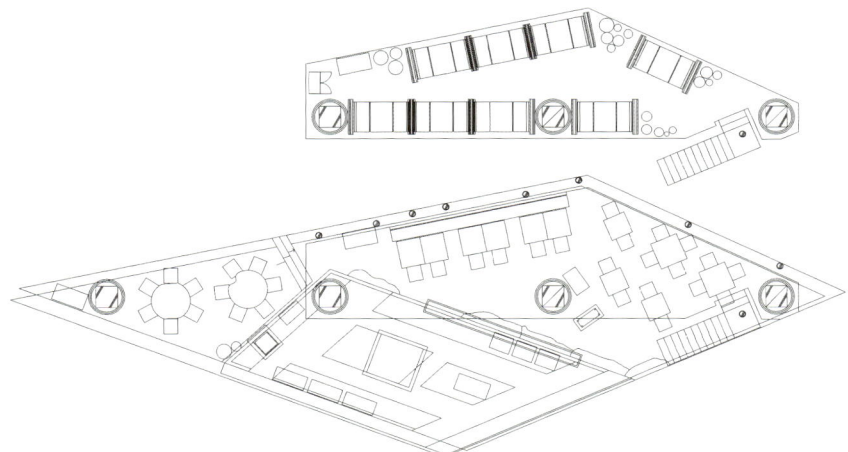

Photo: Koichi Torimura

Yuhkichi Kawai (Design Spirits Co., Ltd.)

Restaurant

Japan

Osaka

2006

Photo: Yoshihisa Araki

Shimizu Masaru

Kura

'Kura', a typical Japanese restaurant with a total area of 155 square metres and containing 55 seats, is located in Osaka. In Japanese, Kura means big warehouse for rice. Kura is a Japanese restaurant because the designers used the traditional Japanese chairs, desks and windows to decorate the whole interior; and also, the whole interior is surrounded by a warm and fragrant atmosphere due to the utilisation of warm colour! The designers wanted to express their style for shop design. It exudes both modern and trational feeling.

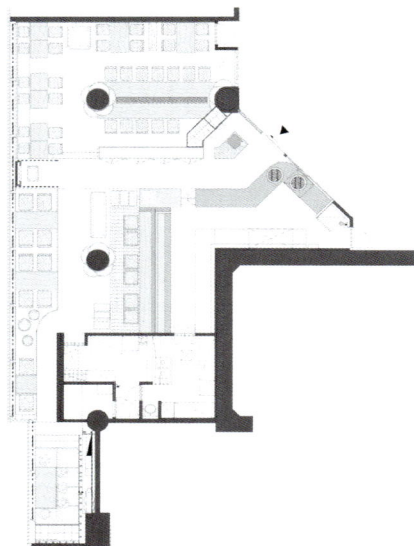

库拉餐厅

"库拉"是位于大阪的一家典型日式餐馆,总面积为155平方米,可容纳55位客人。"库拉",日语意为"装大米的仓库"。作为一家传统的日式餐馆,其室内设计采用了传统的座椅、窗户;同时,其暖色系的色调令整个空间都给人一种温馨的感觉。设计师旨在通过对店铺的设计,让肯花时间在这里坐一坐的客人发现:本餐馆具有现代与传统兼得的设计方式!

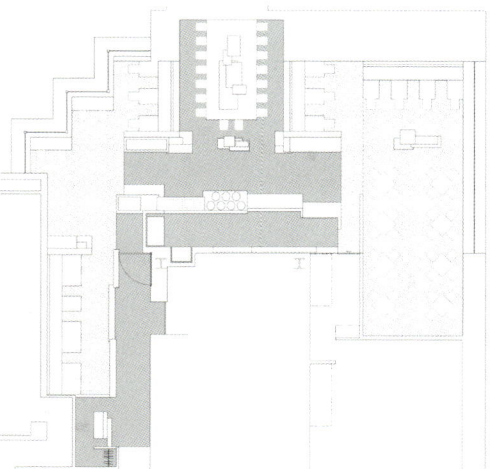

Café in the Park

The space was created around two large windows. The 'greige' (beige+grey) tone of interior combined natural stones and washed wood, highlighted with a metallic sculptural lighting in the ceiling; thus a sense of modern luxury was created with the combination of lighting and materials. The contrast between the different zone, open space and intimate area, created the ideal setting for different customers. In the centre presented on large stones, a generous buffet is the highlight of the space, and the special lighting will emphasise the colour of the food on display. Carefully designed flower arrangements were placed at different areas.

公园咖啡厅

空间围绕着两扇大窗户。天然石材和水洗木将室内空间装饰成灰褐色。天花板上金属雕刻吊灯格外突出。灯光和材料的搭配营造了一种现代奢华感。开放空间和亲密空间等不同区域的对比为不同顾客群体打造了理想的用餐场所。种类繁多、品种齐全的自助餐在灯光的照耀下，色彩诱人、激发食欲。精心设计的花束遍布餐厅的不同区域。

Café

Japan

Osaka

2008

Photo: Nacasa & Partners

Gwenael Nicolas, Curiosity

Restaurant

Japan

Tokyo

2006

Photo: Nacasa & Partners, Inc.

Glamorous Co., Ltd.

Chateau Restaurant Joel Robuchon

Respecting the classic design of the chateau, the designers have provided champagne gold glass in front of wall to create classic and modern mood as well as warm atmosphere. With the Baccarat chandelier and embedded crystals, guests will have a graceful time in the luxurious environment. Rouge Bar on the first floor has a distinctive character apart from the restaurant; it is not a merely waiting bar but an independent, proper bar. Based on red colour, which is Robuchon's theme colour, the space has been added spicy black colouration effectively such as the impressive chandelier of black crystals and oil lamp holders of black crocodile leather. This is the bar for the elegant times in your life.

罗布松餐厅

为体现餐厅的特色，设计师精心设置了香槟色玻璃墙壁以烘托出现代、温馨的气氛。百家乐吊灯和镶嵌式水晶为空间注入了豪华时尚气息。酒店二楼的"丹红酒吧"风格独特，既可以作为休息区，同时可以作为独立的休闲酒吧，结构灵活。充满神秘色彩的红色基调和黑水晶以及黑鳄鱼皮油灯的巧妙搭配将空间衬托得分外高雅深邃。

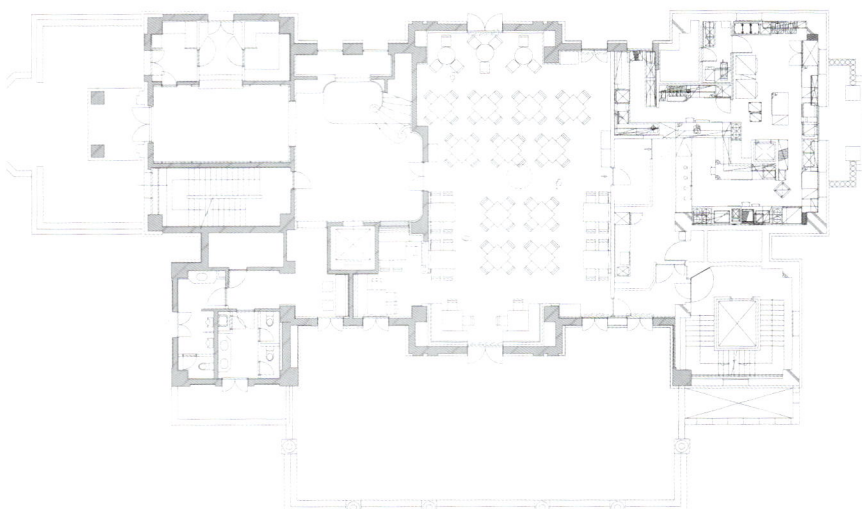

Peter Restaurant

Occupying the entire 24 th floor of The Peninsula Tokyo, Peter Restaurant offers 128 seats for main dining, two semi-private dining areas, one private dining room and a banquet room accommodating up to 60 guests. Peter has been carefully designed with extraordinary backdrops, both by day and night. The designers incorporate the visual stimulation of the splendid view into Peter's design with specially commissioned artwork to enhance the scene. Curved windows surrounding the walls were interspersed with special acrylic panels by artist Marc Littlejohn, featuring embedded mirror strips in an abstract grid pattern which capture the cityscape lights.

彼得餐厅

高居于东京半岛酒店的24层楼上，彼得餐厅拥有128个座位的主就餐区、两个半私密就餐区、一个私密餐室和一个可容纳60人的宴会厅。餐厅四面都可以看到东京的景色和著名的皇居。彼得餐厅的背景设计是很用心的，无论是日景和夜景都很特别。环绕墙面的曲面窗是用艺术家马克小约翰设计的丙烯塑料板点缀的，镶嵌上去的镜面条以一种抽象的格栅样式布局，捕捉城市的风景线。

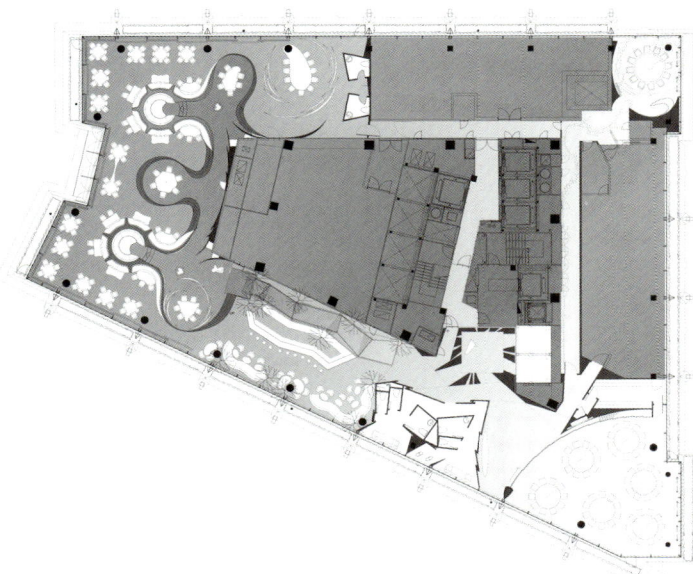

Restaurant

Japan

Tokyo

2007

Photo: KKS Group

Yabu Pushelberg

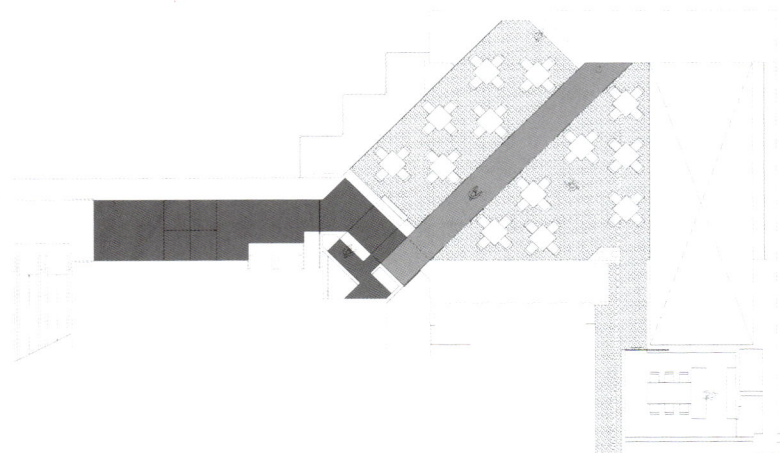

Restaurant

Japan

Osaka

2008

Photo: Nacasa & Partners

Gwenael Nicolas, Curiosity

The Dining in ANA Crowne Plaza Osaka

This restaurant locates in ANA Crowne Plaza Osaka and there are two different aspects of the restaurant, the day time and the night view. The designers used various kinds of materials to interpret these two aspects: A casual setting day time where generous natural lighting enter the large windows. The combination of light colour materials, wood ceiling and soft curtains creates a soft feeling. At night the space becomes more dramatic with a reflective floor and dynamic interior elements, emphasis ing a more focus lighting, creating a more intimate and private space. The large wine cellar at the entrance suggests enjoying a nice conversation around a glass of wine.

大阪全日空皇冠广场餐馆

餐馆坐落于大阪全日空皇冠广场，白天和夜晚呈现不同的风景。设计师使用许多不同的材料诠释这两个方面。白天，充足的阳光透过巨大的玻璃窗射入室内。其浅颜色的原料、木质天花板、柔软的窗帘，为整个室内带来了温馨的感觉。夜晚，整个空间闪耀着夺目的光辉，创造了更加私密的个人空间。入口处巨大的酒窖给客人带来想在此开怀畅饮的感觉。

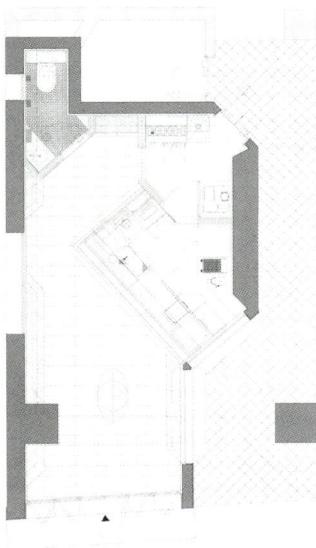

'Caffe Da Gino' Wine Bar

The concept was to recreate a typical Italian ambient but with a modern vision, with adult men as main target customers. Designers decided to approach the design in two different ways: at first they designed a modern space, linear and essential, and then they chose Italian traditional materials such as basaltina stone for the front of the counter, walnut wood for the top, flooring and entrance windows. The ceiling was unusually high here in Tokyo, so designers decided to build up a louvre false ceiling allowing to take lower the perspective which otherwise would be too high.

"咖啡达吉诺"酒吧

设计的宗旨是要营造一个典型的意大利风情却又不失时尚的视觉冲击的氛围，将消费的群体定位为成年的男性。设计师决定将设计分为两个方面：首先他们勾画了一个时尚的空间，流线感强可谓设计的精华所在。之后他们采用意大利传统的材料为主材，如黑陶石作为前台的点缀，胡桃木用以装饰天花板、地板和进气窗。设计师修建一个假的天窗，从视觉上增大了空间感。

Bar

Japan

Tokyo

2007

Photo: Anonimo Design Corp

Luigi Campanale

Photo: 45g Photography-Junji Kojima

Kidosaki Architects Studio- Hirotaka Kidosaki, Principal-in-charge; Satoshi Itasaka, project team

AG Café

AG Café is planned for a gallery. The design aims to offer the artists at Osu Shopping street in the centre of Nagoya City a comfortable, artistic space to meet. The owner requested a space where people can enjoy art, with pictures showing the scenes of daily life, but not an inorganic space far away from everyday life. Also he requested to harmonise the two contrary factors: the café's lively and brisk ambience and the gallery's dignified atmosphere. The space is composed of an extremely warm tone and the 'quality' which is necessary to the gallery by sticking to the design proportion and detail. The designers designed the abstract patterns of birds' nests, which contribute to the warm feeling and spread throughout the café.

AG咖啡厅

为画廊而设计的AG 咖啡厅旨在为名古屋市Osu购物街的画家们提供简单舒适的艺术空间。画廊里展示的并非是无机空间的图片,而是体现日常生活场景的图画。因此,客户要求空间设计要使顾客能在用餐时欣赏艺术之美,还要求将两个对立因素——咖啡厅的生动活泼与画廊的庄严高雅相协调。空间充满了暖意融融的设计与基调,也严格遵守设计比例与细节,兼具画廊的的品质。为增加咖啡厅的暖意,空间内布满了鸟巢的抽象图案。

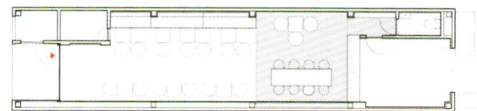

Restaurant

Japan

Yamagata

2006

Photo: Hirohisa Sako

Yasutaka Yoshimura Architects + Emiko Tsuno / Troom

Kameya Ryuguden

The dining hall called the 'Ryuguden' is renovation of a large hall in an old Japanese style hotel. The designers minimised the walls that were disturbing the services, but when thinking of creating a feel of a private room, they decided to build an extremely low ceiling that was two metres high. In this low ceiling, they made a hole, and placed the table right below the hole. From the hole, you can see the existing structure. For the hole and the food to stand out, the other areas, like the floor, walls, ceiling, furniture, and even a few parts of the tableware were made with the same material.

昭和畦餐厅

餐厅位于一家传统日式风格酒店内，由一个大厅翻修而来。设计师尽量减少墙壁的设置，为营造私密感，专门打造了2米高的低矮屋顶。之后，他们又在屋顶上开了一个大洞，将餐桌摆放在其正下方。从这里望上去，可以看到原有的结构。为突出餐厅内的食物，地面、墙壁、屋顶及家具等全部采用同种材料装饰。

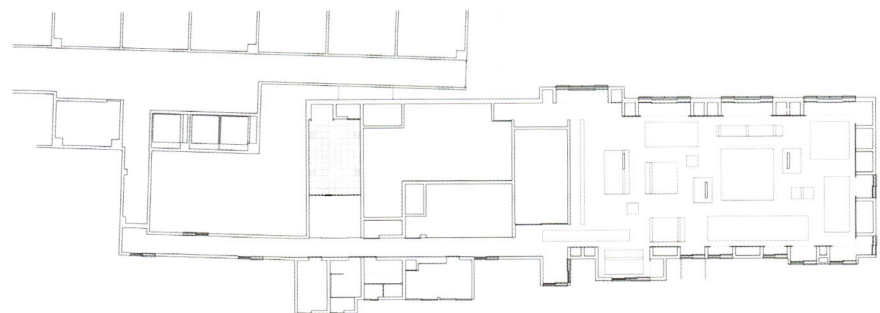

Bar

Japan

Tokyo

2006

Photo: Hironori Tomino

Tange Associates

Jwc 2ⁿᵈ Avenue Ginza

'Ginza' in Tokyo is well known as one of the most luxurious shopping districts in the world. Sleek, modern buildings grace the main streets of Ginza, while smaller buildings cluster on the back streets. This unbalanced sense of scale creates the unique character of the area. Walking through a narrow, spotlighted lane from the main street of Ginza, one could find the private lounge bar 'JWC 2ⁿᵈ Avenue Ginza' designed by Tange Associates. The owner was inspired by his love of jazz, wine and cigars to open a private lounge where people like himself can gather to indulge in their favourite pastimes. Upon entering the bar on the 2ⁿᵈ floor, as their eyes adjust to the dimness, guests are struck by the glow emanating from the long blue glass counter, which seems to be floating in the darkness.

银座二号街Jwc 酒吧

东京的银座是世界上最奢华的购物街之一。时尚前卫的建筑群装点着银座的主街道，而小巷内则低矮建筑簇拥。这种不平衡的规划感是当地最突出的特点。从银座主街道穿过聚光灯闪耀的窄胡同，就可以看到由丹下健三都市建筑设计研究所设计的银座二号街私人酒吧。酒吧老板出于对爵士乐、葡萄酒及香烟的喜爱开的这家酒吧。希望和他有相同嗜好的人能够在这里相聚，共度最喜爱的休闲时光。进入二楼的酒吧，由于习惯了昏暗的灯光，顾客会觉得眼前一亮，长长的蓝色玻璃柜台里放射出的耀眼光芒仿佛在黑暗中漂浮。

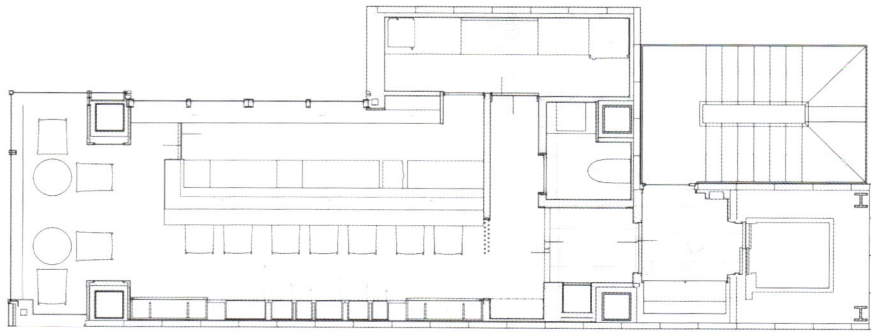

Rigoletto Bar and Grill

This is the 5th restaurant of Rigoletto brand. The theme is contemporary America and makes the space both avant-garde and classic. Avant-garde bar counters of 13 metres in length is made by metal, mirror and glass. The designers install light inside counter. It is like floating. The carved patterns on the counter is delicate and beautiful with light inside. There are seats dashes from wall and projected image on TV monitor. The designers add sense of fun by such items. The classic feeling is emphasised for wine cellar and dining space as if they exist since the opening of Roppongi Hills. The designers made luxury space by using flooring and door made by antique wood, marble lamp shade and 930 pieces of mirror that installed between bar and hall in theme colour gold and purple.

里戈莱托餐吧

这是里戈莱托餐吧系列的第五个餐厅。以现代美国为主题，餐厅兼具现代与古典特色。13米长的前卫吧台由金属、镜子和玻璃等材料制成。柜台上的图案在灯光的照耀下美丽而精致。电视监视器的投影图像为酒吧增添了娱乐感，增加了餐厅的古典魅力。吧台与大厅中间930面金色与紫色等主题颜色的镜子、大理石灯罩和古木门与地板等装饰打造了一个奢华空间。餐厅的紫色椅子在灯光的照耀下，把整个室内空间都笼罩在紫色的海洋中。

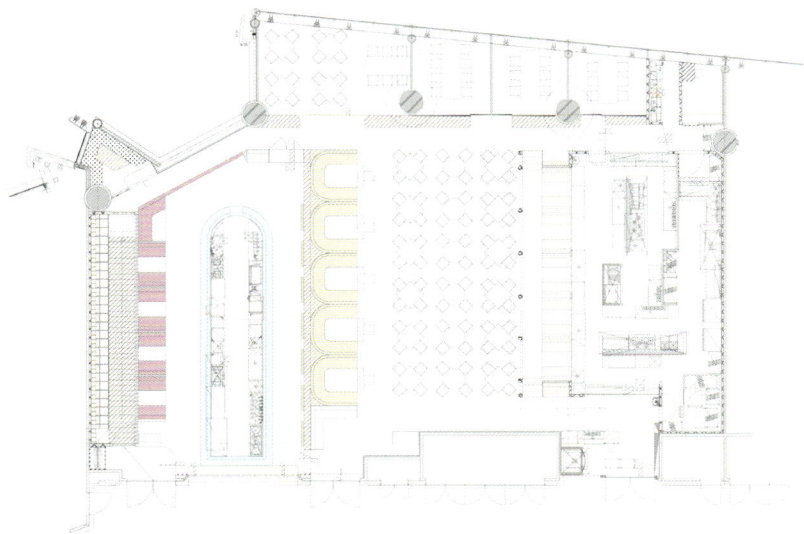

Restaurant

Japan

Tokyo

2006

Photo: Nacasa & Partners Inc.

Takeshi Sano / SSDesign

Restaurant

Japan

Tokyo

2009

Photo: Nacasa & Partners Inc.

Takeshi Sano /SSDesign

Arata

The design theme is Edo Japanese Modern and has two points. The first point is 'SARAKU' art in main dining with the size of 3500*2500mm made by SUS frame and 18,000 pieces of 20mm-diameter crystal bolls. It is installed light inside and like flowing in dark. By expressing popular art in Edo Era through contemporary art filter, it will succeed in bringing Japanese traditional culture to future. The second point is bar space. The designers made bar space in the place which used to be common part in the buildings. Normally Japanese dining does not have bar in entrance part. This made the restaurant oneness with international place 'Roppongi Hills common part'.

新Arata餐厅

项目的设计主题为江户时代的日式现代感，共有两个特色。特色之一是主餐厅的东洲斋写乐浮世绘，长3.5米，宽2.5米，由18,000个20毫米的水晶球组成。其内部的灯光宛如在黑暗中流动。这种以现代艺术来表达江户时代传统的形式将日本传统文化带到了未来。另一个特色是吧台，设计师将吧台设在了门厅的位置。传统的日式餐厅并不这样做，这让餐厅看起来更具国际感。

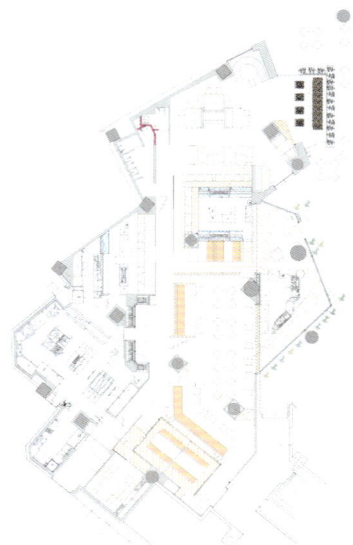

Restaurant

Japan

Tokyo

2009

Photo: Toshiyuki Yano

Chikara Ohno / Sinato

Salon Des Saluts

For the connection between the outside street and inside space, considering more worthy of ambiguous and deep transparency, the designers decided to adopt a way crowding an opening of the building with small glass boxes running across over the outface. Except one box used for private room in inside, others which were used for a court with single olive tree, approach and terrace are outside. So it's difficult to know which is a real boundary between inside and outside because this line runs very complicatedly. Besides, each corner of boxes with curved surfaces brought visual distortion and blurred transparency to the view from the outside. Another important component: grand plants filled with asparagus, rosemary, ivy and the like run free curve with floor tile and cross over the glass wall swinging like shore and wave.

致意沙龙

设计师在内外空间交界处设计了一些错落的小型玻璃间，增添了室内外的联系和透明度。其中只有一个盒子在室内作为包间，其他的都设在室外，与橄榄树、小径和露台在一起，让人很难分清室内外的界限。此外，玻璃间的弯曲的边角让空间从外面看起来扭曲而模糊。设计的另一主要元素是植物：芦荟、迷迭香、常春藤等遍布店内，在地板和玻璃墙上卷曲摇曳，宛如沙滩上的波浪。

Desserts

Japan

Fukuoka

2008

Photo: Hiroshi Mizusaki

Koichi Futatsumata, CASE-REAL

Suzukake Honten

'A Japanese pastry shop with Japanese black in silence' – The Japanese pastry shop locates in Kami-Kawabata, Hakata Ward, Fukuoka City, a historical merchant town. The flagship shop has Japanese 'Kaho', a pastry shop, and Japanese 'Saho', a tearoom. Under the idea that the place where pastries are treated is the holiest of holies, the 'Kaho', a pastry shop is independent from the crowd, even though the shop is on the main street. To make conspicuous the existence of pure, simple pastries, there is only one half-floating, nine-metre display case in Japanese black. By using Japanese black plastered wall to all the materials, details and all of the walls, the designer successfully abused the existence of the display case.

铃挂点心店

"一个有着日式黑色沉默的日式点心店"——这家日式点心店位于日本福冈有着悠久商业历史的一个小镇上，店内分为日式点心和日式茶室两部分。在这个店的经营理念中，点心是被虔诚的对待的。尽管它位于小镇的主干道上，点心店仍然被定位为独立存在于外界的熙熙攘攘之外。 同现在的那些干净、简洁、光鲜的点心店不同，这里只有一个一半悬浮的、9米长的点心柜台，还是日式阴沉的黑色。但是通过将这种日式黑色墙材运用到周围所有的墙面、材料、细节中，我们成功的将这个柜台的存在感做到了极致。

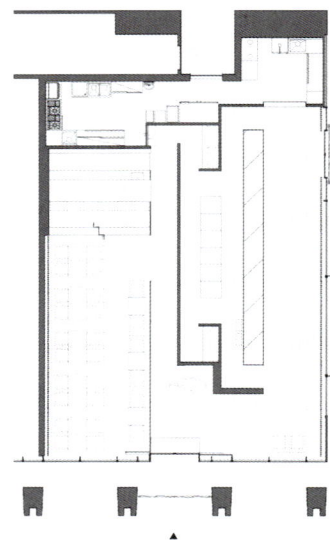

Bar & Lounge

Japan

Tokyo

2010

Photo: Nacasa & Partners

Curiosity

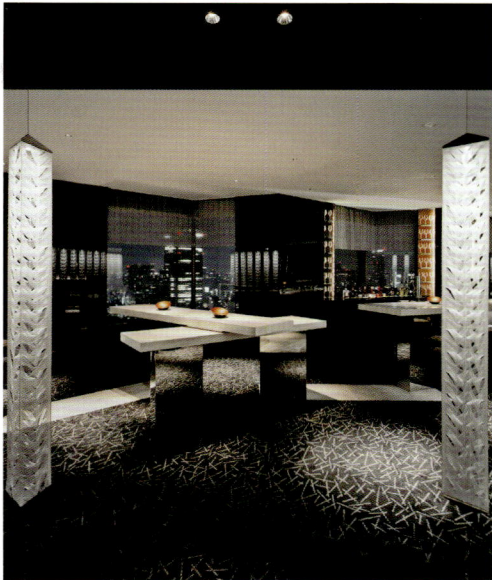

The MIXX Bar & Lounge

Located at the Intercontinental Hotel in Tokyo and spanning 600 square metres at the top of the hotel, MIXX is a playful exploration of light and shadows offering an atmospheric window over the city of Tokyo. Against a palette of neutral grey beige are multilayered materials of washed wood, ray skin covering, hammered bronze and stone complemented by textured fabrics. The space fuses traditional craftsmanship and Japanese art with modern design to create an invisible connection with the nation's rich cultural heritage. The overall effect? One of softness, designed to highlight and enhance the sharp digital portrait of modern Tokyo that unfolds through the windows. The entrance is marked by a 'gate of light': a wall of floating white fabric sculptures reflected through a play of mirrors, their delicacy balanced against the dynamism of the lighting.

MIXX休闲吧

MIXX休闲吧位于东京洲际酒店的顶楼，总面积600平方米，其独特的光影设计为东京城打造出一片耀眼的风景。米灰色基调上是多层次的材料：水洗木、反光面板、铸铜和石材与手工织物相得益彰。空间将传统手工业和日本艺术与现代设计融为一体，巧妙地展示了日本悠久的文化。柔和感凸显了现代东京锐利的数字形象。入口处的灯光大门十分抢眼：流动的白色织物雕塑通过镜子反射，与动感的灯光达成了微妙的平衡。

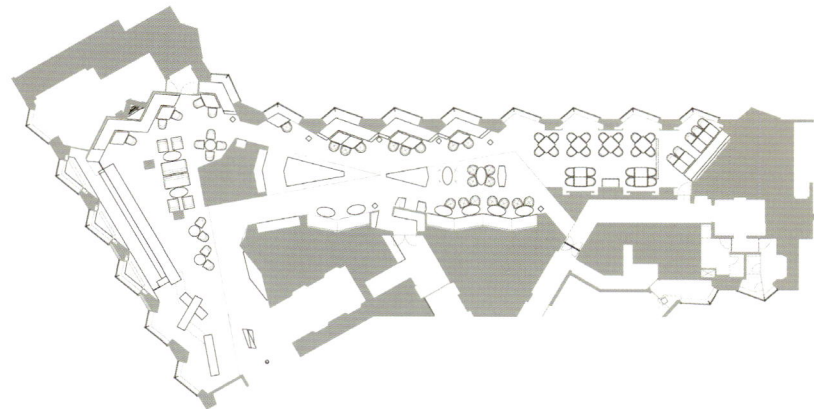

Restanrant

Malaysia

Kuala Lumpur

2009

Photo: Toshihide Kajiwara

Design Spirits Co., Ltd.

The Hutong

The project involves redecoration of a food court located in the basement of a shopping centre in Kuala Lumpur. As the designer has designed for the food court named 'FOOD LOuVER' in Jakarta, Indonesia before and it leads him more experience of designing food court and the flow of human traffic. Of course, one of the main objectives of the redecoration was to draw more customers to the food court. The designer aimed to create the environment of a 'street' and 'area' controlled by invisible chaos that is at the same time nostalgic. However, the space was somewhat futuristic with round exposed air ducts on the ceiling to give the space a retro-future ambiance.

胡同美食广场

该项目是位于吉隆坡某购物中心地下室的美食广场的翻新工程。设计师曾为印尼首都雅加达的"美食天窗"做过设计，积累了很多美食广场设计和人流疏导的经验。翻新工程的重要目标之一就是吸引更多的顾客。设计师致力将美食广场打造成街道和区域环境，既熙熙攘攘又怀旧。然而，天花板上裸露的通风道又为整个空间增添了未来感，有种回到未来的感觉。

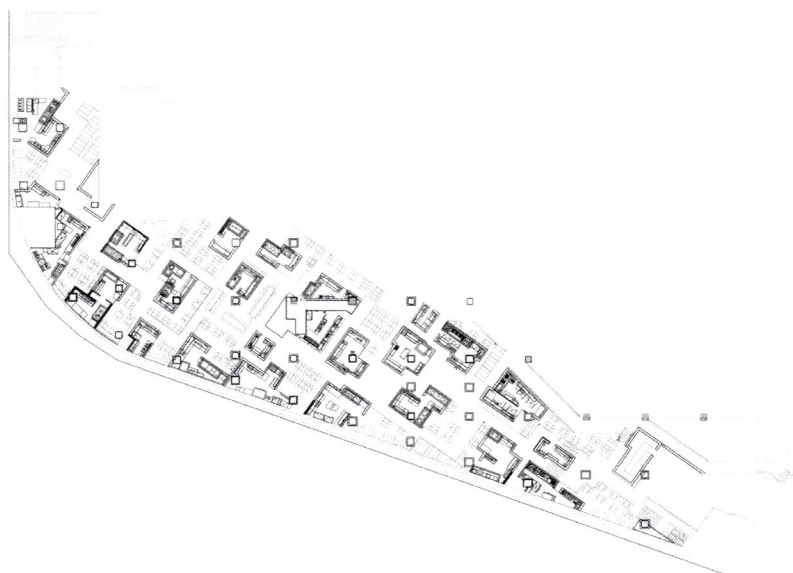

The Ark Café

The Ark Café is conceived as part tree house, part café, viewing platform and urban landmark in Sibu. Sibu is a riverine town in Sarawak with its roots in the timber industry. The catalyst for this project grew from a 60-year-old rain tree on the Sibu's riverside esplanade. The tree provides a quiet backdrop to all the spaces; a lush green wall paper to all the rooms that change hue and character through the day. The Ark is designed to be a permeable building to minimise the flow of foot traffic along the esplanade and the street. The structure of the Ark is deliberately open with glazed walls that slide open to allow breezes and reveal glimpses of the rain tree. The formal riverside entrance uses the esplanade as a fore court whilst by contrast, the street entrance is a non-descript break in the screen fence. Both entrances bring one into a sanctuary of calm and repose; foliage filtered sunlight painting shadow murals on the floor.

方舟咖啡馆

方舟咖啡馆一半是树屋，一半是咖啡馆、观景台和诗巫都市坐标。诗巫是马来西亚沙捞越州的一个河畔小镇，以木材工业闻名。这一项目的灵感来自于滨水公园一棵60年的雨树，大树为咖啡馆提供了一个幽静的背景，房间里翠绿的壁纸奠定了空间的整体基调。方舟咖啡馆是一幢可穿越的建筑，以减少街道和滨水公园的人流。方舟的结构开放，滑动玻璃门可以保证通风。咖啡馆的正门是滨水公园，另一侧的门则临街。咖啡馆为人们提供了一个宁静休闲的避难所，阳光透过树叶在地板上留下斑驳的阴影。

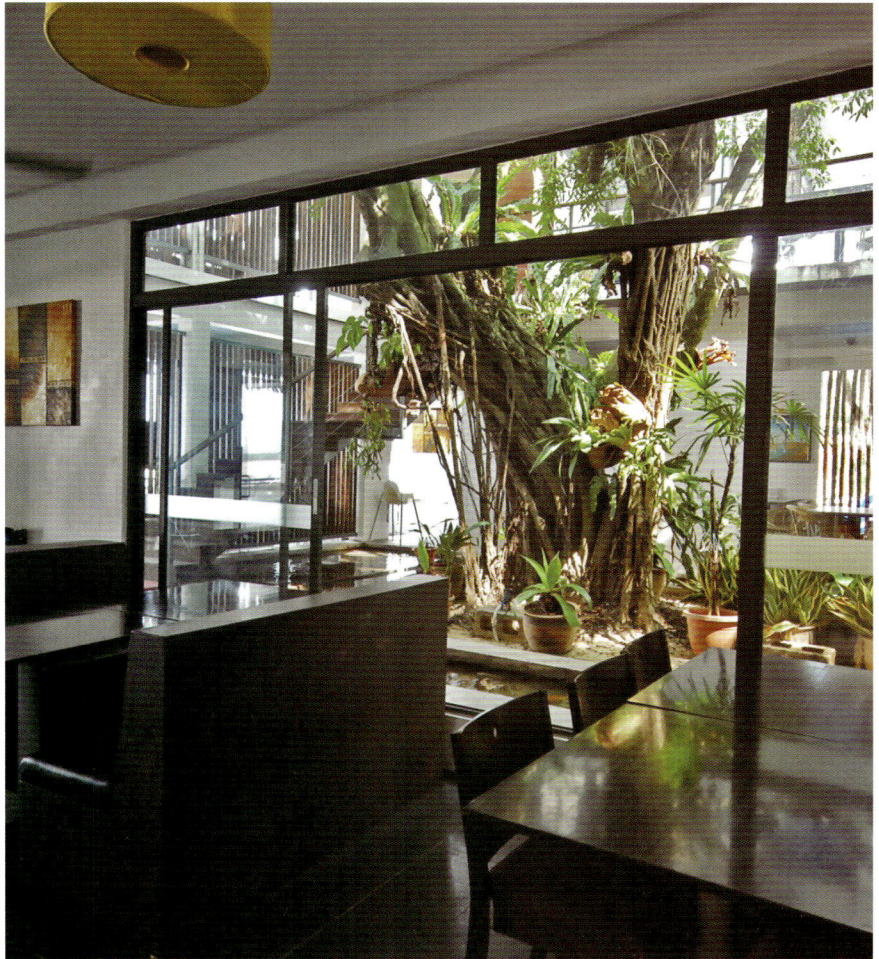

Café

Malaysia

Sibu

2008

Photo: Design Network Architects

Design Network Architects

Restanrant

Malaysia

Kuala Lumpur

2009

Photo: Zainudin

Design Spirits Co., Ltd

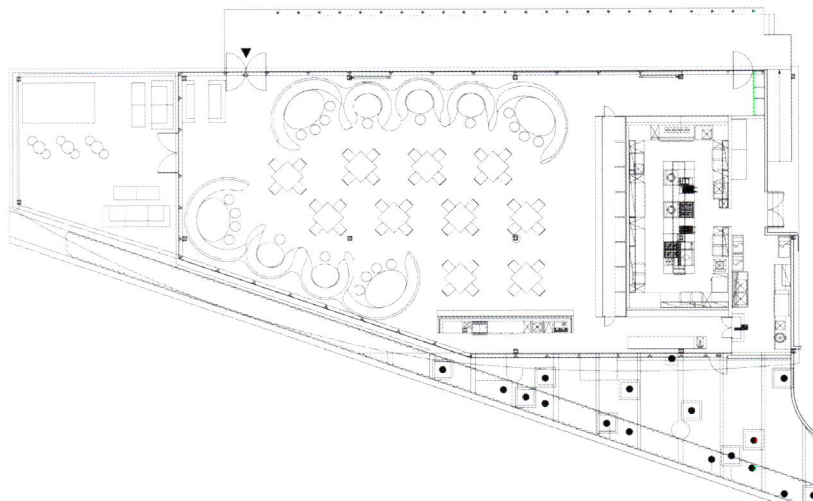

Shopping Centre Rooftop Teeq

This restaurant, Teeq, was planned in Kuala Lumpur Malaysia, at the roof parking space of the existing shopping centre. The designer chose the location to see the scenery. As a result, the designer concentratedon the ceiling design because only ceiling design was not left to surrounded restaurant with glass keeping scenery. Teeq becomes a restaurant at noon, and the lounge where guests go up and dance on a sofa at midnight and weekend. The designer aimed at the ceiling said to be wonderful by a visitor without disturbance, without being too beautiful, and without being too showy. The execution repeated trial and error many times because the designer wanted to let the thin board of the tree undulate fluidly. After all, the designer hung the strip of the wood board with wire directly and attached LED to the board.

提克屋顶餐厅

提克餐厅位于马来西亚吉隆坡某商业中心停车场的屋顶，可以俯瞰风景。设计师将设计的重点放在天花板上，因为只有天花板不是玻璃做的。提克是一家午后餐厅，顾客可以在午夜和周末进行狂欢。设计师的目标是让天花板既不过分花哨，又能吸引顾客的眼球，他试图将薄木板自然形成流畅的波纹。最后，设计师在木板的条纹上直接缠上了金属线并且添加了LED灯饰。

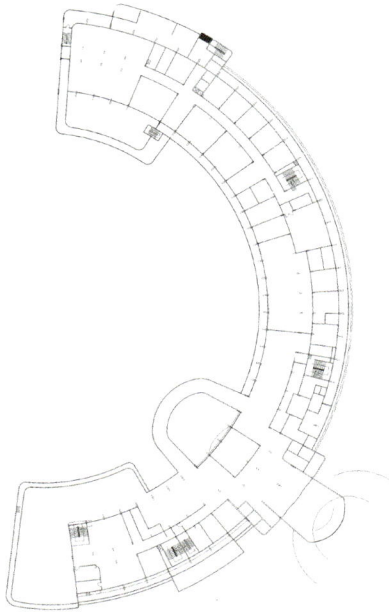

Restaurant in Golden Sands Resort by Shangri-La , Penang

Golden Sands Resort by Shangri-La, located on Penang's popular Batu Feringgi beach, is a veritable tropical paradise for vacationers and families. Cool Lounge, the resort's newest addition, pampers guests with the comforts of home and acts as a transit lounge for early arrivals and late departures. It was designed with an emphasis on space and comfort, and the open-air concept allows guests an expansive view of clear blue sky and sea blending in harmony with the lush tropical greenery. Garden Café, the resort's all-day dining outlet, Garden Café and the Lobby Lounge were designed with open white shutters to both are frame unparallelled views of tranquil gardens, pools and beach.

槟城香格里拉金沙度假村餐厅

香格里拉金沙度假村位于槟城的巴都宜非宁海滩，是一个度假的热带天堂。清凉休息室是度假村最新的设施，给予游客们家一般的舒适体验，让他们在到达和离开时可以做一个短暂的停留。其设计注重空间和舒适感，露天设计让游客们可以尽享蓝天碧海与迷人的热带风光。花园咖啡厅提供全天的餐饮服务，与休息室都采用了开放式的白色百叶窗，框住了花园、泳池和沙滩的美景。

Restaurant

Malaysia

Penang

2009

Photo: One Plus Partnership Limited

One Plus Partnership Limited

7 Atenine

Poole Associates redesigned the restaurant space. Selected areas were totally re-built such as the Ice Bar and railway sleeper flooring. A new metal screen was added to create a foyer, and furniture was refurbished. The new circular bar features a double-storey gold back-bar, resembling a champagne cork on the lower floor, and a flute on the mezzanine. LED lighting was brought up to date and the sound system energised.

7安特耐恩餐厅

普尔设计公司对餐厅进行了重新设计，一些区域进行了完全的重建，如冰酒吧和枕木地板。设计师通过一个新的金属屏风而隔出了门厅，也对家具进行了翻新。新的圆形有一个双层金色后部吧台，在一楼向香槟酒的软木塞，在二楼则像一个长笛。餐厅还采用了最先进的LED灯光设施和音响系统。

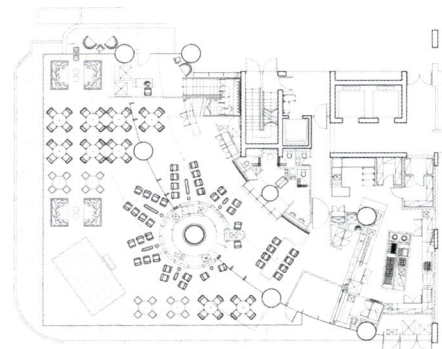

Restaurant

Malaysia

Kuala Lumpur

2010

Photo: Untold images.com.my

Poole Associates Private Limited

Restaurant

Singapore

Marina Bay Sands

2010

Photo: Courtesy of JZA+D

JZA+D

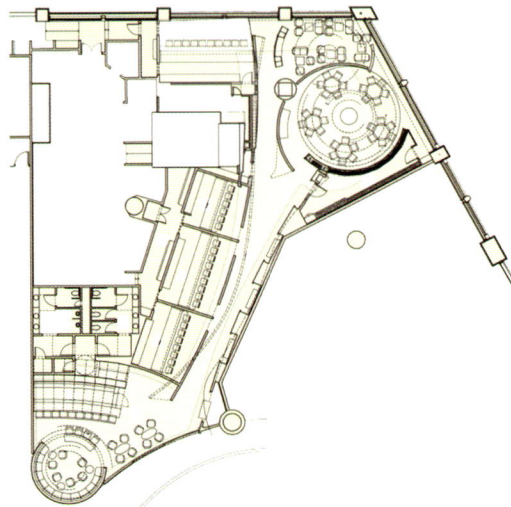

Waku Ghin Restaurant

World-renowned Australian Chef Tetsuya Wakuda brings his culinary vision to restaurant Waku Ghin (Silver), an 8,000-square-foot space located at the Marina Bay Sands in Singapore. The programme consists of a sake bar/caviar lounge, four teppanyaki rooms, a main dining room, drawing room and support spaces. Sitting on the edge of the casino floor the translucent glass-panelled façade hints at the warm lighting and cool forms of the interior. Diners enter the sake bar/caviar lounge and are welcomed by the sinuous walls of the Grand Hall. Manipulated against each other to take advantage of the space's compressed nature, the walls create an elegant gallery which leads to the intimate teppanyaki rooms and the main dining room. A 3,000-bottle wine room, together with another sweeping wall, enfold the main dining space while framing Singapore's skyline.

瓦库银餐厅

澳大利亚名厨久田哲也所开设的瓦库银餐厅位于新加坡滨海湾金沙娱乐场。餐厅由一个清酒吧/鱼子酱休息室、四个铁板烧室、一间主餐厅、一间会客厅和其他辅助设施组成。餐厅位于赌场楼层的边缘位置，透过透明玻璃墙可以看到内部温暖的灯光和出色的室内设计。用餐者走进清酒吧/鱼子酱休息室，便会看到大厅蜿蜒的墙壁。墙壁形成了一个优雅的画廊，引领人们走向铁板烧室和主餐厅。收藏着3,000瓶葡萄酒的藏酒室和一面墙壁包围着主餐厅的空间。

Restaurant & Bar

Singapore

Chin Swee Road

2007

Photo: JP Concept Pte. Ltd.

JP Concept Pte. Ltd.

Wood Restaurant & Bar

Central to the theme of Wood Restaurant and Bar is the use of traditional wood and charcoal to achieve culinary perfection. The design of this new upmarket restaurant combines contemporary elements with a dark and earthy beauty. Greeting the patrons, the entrance is a pillar signage that resembles a tree. The decorative ceiling feature that extends from the pillar is akin to a branch that provides subtle direction that leads patrons into the restaurant. The very stylish, elegant and chic interior is a distinctive feature of the facility. Exuding a modern and classy look, colour changing glass panels mounted behind the reception desk hint at the dark and sensual interiors. Through a timber-floored passage with rough texture that evokes an earthly and mysterious feel, patrons will come to an opening that is a beautiful Octagon-shaped wine cellar. Lined with gleaming glass and steel, the cellar presents a cool and airy contrast like an oasis to the dark earthly feel of the entrance passage.

木餐吧

木餐吧的中心主题是使用传统木材和木炭来达到至善至美的料理水准。这一高档餐厅的设计在现代元素中融入了暗夜和朴素的美。餐厅入口的柱子宛如一棵树，天花板上的装饰物是树的枝叶，指引着顾客走进餐厅。这种时髦而优雅的室内设计是餐厅的主要特征。不断变换颜色的玻璃板在前台后方闪耀，活跃了室内的气氛。质地粗糙的木地板有种自然而神秘的感觉，穿过走廊，顾客便可以到达一个漂亮的八角形酒窖。酒窖四周是闪闪发光的玻璃和钢材，感觉轻盈而凉爽，与走廊入口处的黑暗形成了鲜明的对比。

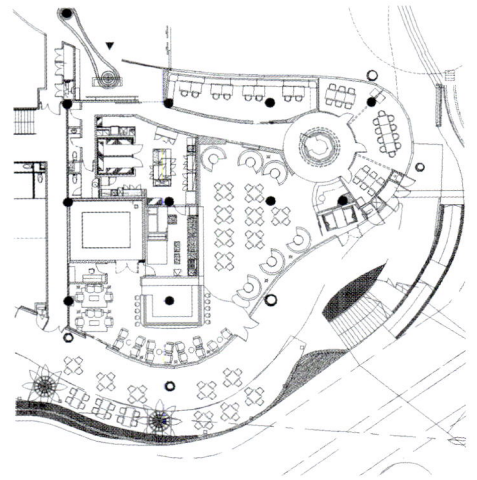

Restaurant

Singapore

Bukit Manis Road

2008

Photo: Jeremy San

Takenouchi Webb

The White Rabbit

The brief called for a single dining space and bar with an outdoor terrace at the back. The designers began this design by looking at traditional British public school dining halls and church buildings. In the main dining space they kept the format of the original raised platform and decided to make it the bar area. The designers added a large double-sided marble counter that naturally separated the main dining area from the lounge bar space. The different heights on either side of the bar meant that guests could sit at the bar at the upper level looking out over the restaurant whilst the lower level could be used more for serving. For the main dining areas the designers differentiated the seating into three areas: long communal tables with pew-like bench seating, plush grey semi-circular banquette seating contained within two oxidised steel containers on either side of the main space, and at the far end a raised natural oak platform with high-backed shaker-inspired chairs.

白兔餐吧

项目集餐厅、酒吧和露天平台于一体。设计师参考了传统英式公立学校食堂和教堂的建筑结构。他们在主就餐区保留原有的上升平台，使其成为了吧台区。设计师运用一块巨大的双层大理石台面将主就餐区和休闲吧自然而然地隔开。吧台两侧的高度差意味着顾客可以坐在上层吧台边俯视下方的餐厅。设计师将主用餐区分为3个部分：带有长靠背椅的长桌区、奢华的灰绒面半圆形卡座区和带调酒椅的天然橡木平台区。

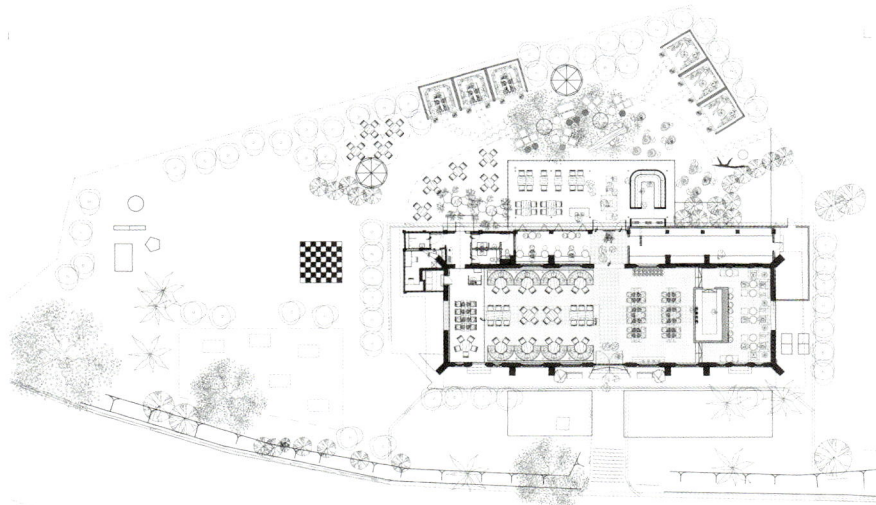

Restanrant

Singapore

Singapore City

2009

Photo: Design Spirits Co., Ltd.

Design Spirits Co., Ltd. / Yuhkichi Kawai

Nautilus Project

The Nautilus Project is located on the third floor of the ION Shopping Centre, which opened recently on the Orchard Road, Singapore. This floor is not only the restaurant floor, but also there are shops as well. The owner is the president of a cargo company and, is a surprisingly beautiful woman boss, so the designer decided to reflect her sophisticate, elegant, and tender characteristic to this restaurant project. The chef was from New Zealand, and was popular among New Zealanders like a celebrity. The designer was unable to eat his dish until the completion of the project because he was very busy. However, the designer could feel his confidence, so he stopped making the unnecessary highlight. Cooking is the leading role.

鹦鹉螺餐厅

鹦鹉螺餐厅位于新加坡ION商业中心的4楼，这一楼层不仅是餐饮区，而且还有许多店铺。餐厅的主人是一家货运公司的女董事长，设计师决定在餐厅设计中体现她的成熟、优雅、温和的个性。餐厅的主厨来自新西兰，非常有名，也非常忙，设计师在项目将近完成之时才品尝到他的手艺。但是从与他的交往中，设计师早已了解了他的自信。在餐厅中，最重要的还是主厨的手艺。

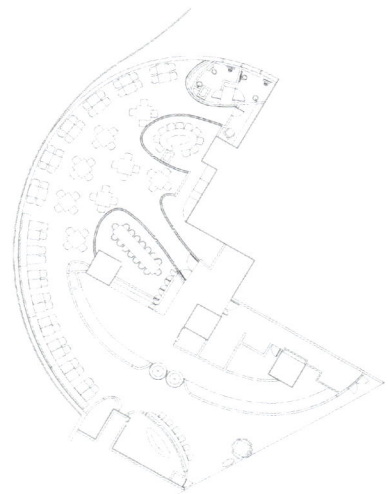

Il-Lido Italian Dining + Lounge Bar

Nestled in the tropical greenery of the Sentosa Golf Club, a sense of grandeur and luxury permeates this restaurant. The colour concept juxtaposes an off white colour scheme against a black background to create spatial depth and visually arresting contrast. Selected furniture by iconic French designer, Philippe Starck, were used to infuse a dash of contemporary vibe to the lavish setting. In the dining area, brown tinted mirrors on dark blackish colour timber veneer serve to provide a 'scattering effect' to reflect the outdoor greenery into the room. A long suspended tinted glass top counter provides storage solutions while maintaining the elegance of the setting.

丽都意大利餐厅+休闲酒吧

项目坐落于圣淘沙高尔夫俱乐部之中，因其高贵和豪华而著称。空间黑白的经典搭配营造出强烈的视觉冲击效果。由著名的法国设计师菲利浦·斯塔克设计的家具为空间注入了时尚高雅气息。餐饮区，深黑色胶合板外部的棕色套色玻璃形成了神奇的"散射效应"，完美地将室外的"绿色"引入到室内。覆以套色玻璃的加长型吧台在盛放商品的同时，与整个空间氛围相得益彰。

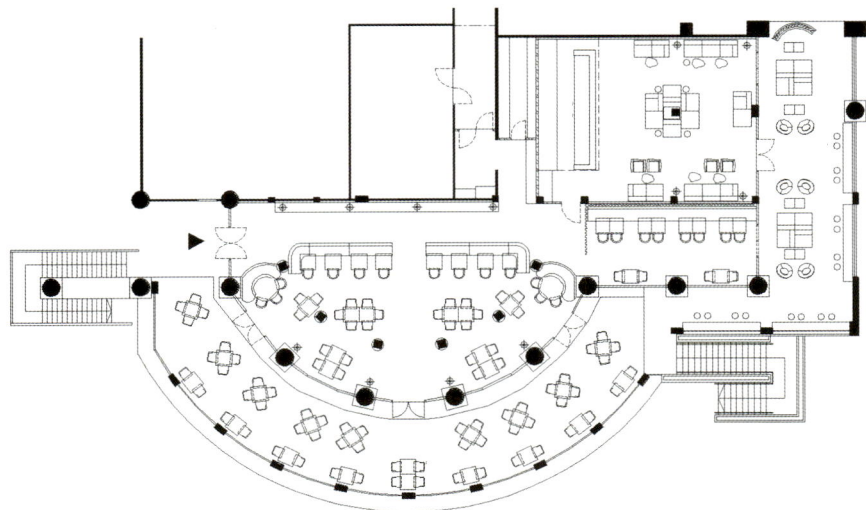

Restaurant

Singapore

Bukit Manis Road

2008

Photo: Jerlyn Poh/JP Concept Pte Ltd

Philippe Starck

Restaurant

Singapore

Resort World Sentosa

2009

Photo: See Chee Keong

ONG&ONG Pte. Ltd.

Rang Mahal Pavilion

Nestled in the luxurious Resorts World Sentosa, Rang Mahal Pavilion serves up authentic Indian cuisine in a traditional setting albeit with a modern twist. The restaurant comprises of a main dining and banquet area, two private function rooms, a kitchen and a restroom. Musical instruments like sitars and flutes make for unique light fittings, adding cultural flavour to these otherwise ubiquitous fixtures. A wall of bronze traditional pots are stacked together at the entrance, forming an interesting entrance screen. Beyond the screen, diners are welcomed by the statue of Lord Ganesh, the Indian elephant diety, perched over a reflective pond. Fabric panels line the restaurant's doors and sections of wall and, together with the plush furniture, projects a dignified yet inviting atmosphere for a meal.

兰玛哈餐厅

兰玛哈餐厅位于新加坡圣淘沙名胜世界，是一家正宗的印度传统餐厅。餐厅由主就餐区和宴会区、2间独立功能室、1间厨房和1个洗手间组成。西塔琴和长笛形状的灯饰为餐厅增添了民族风情。门口由传统铜壶堆叠而成的墙壁形成了有趣的屏风。除了屏风，门口还有一尊象头神伽内什雕像栖息于倒影池之上。餐厅的门和墙壁截面上的纤维板和绒毛家具为就餐提供了优雅而舒适的环境。

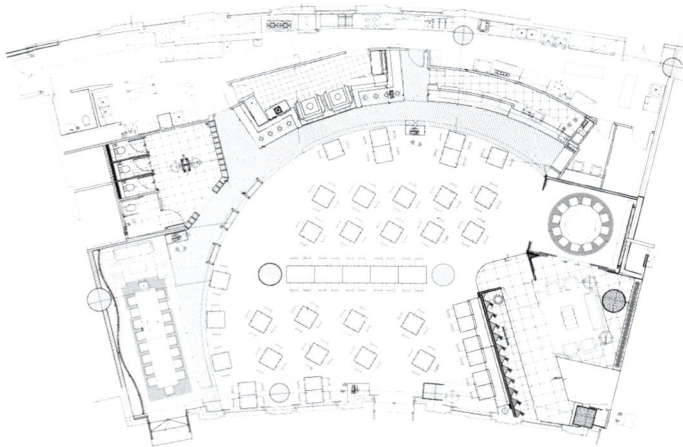

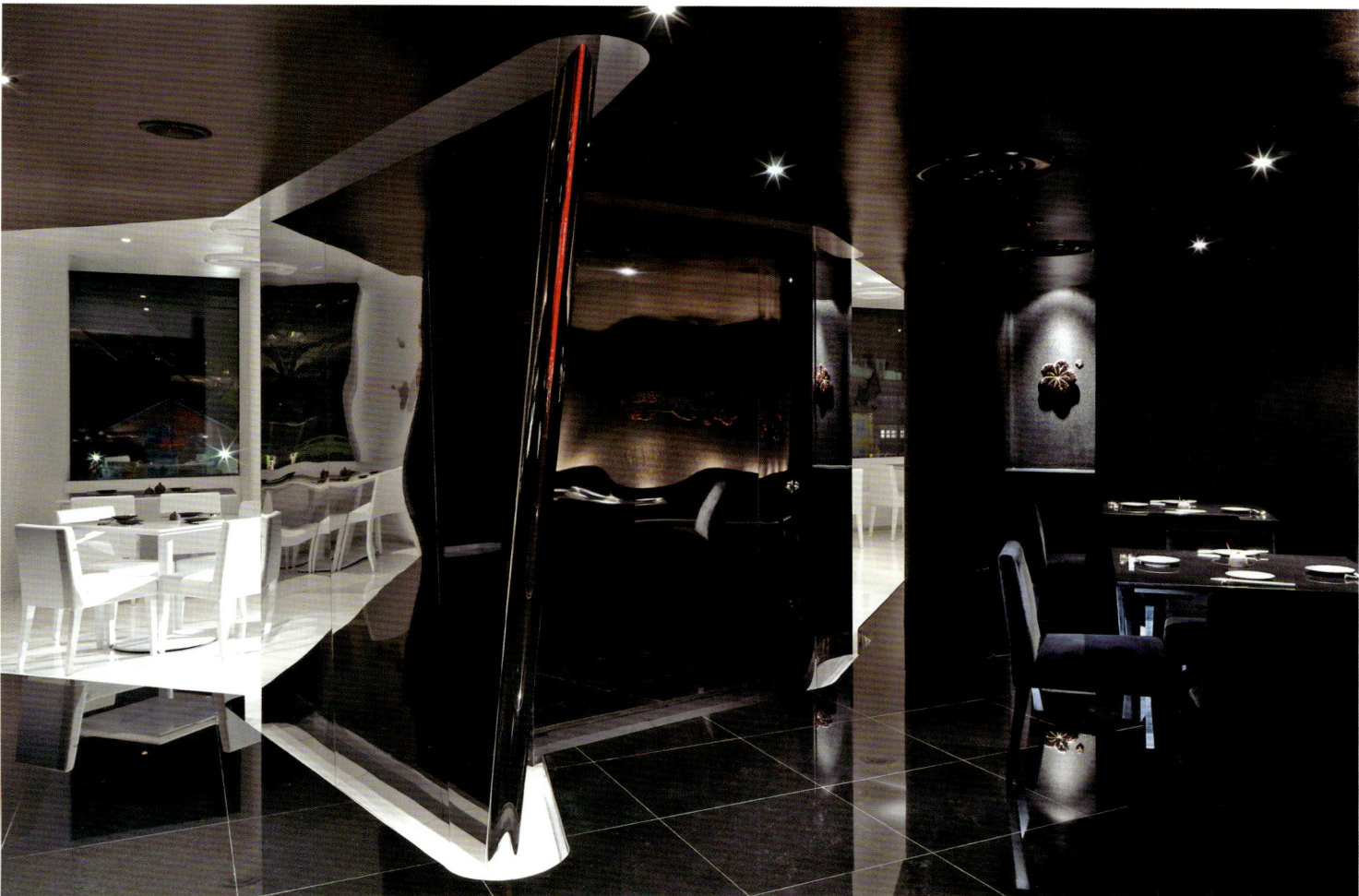

Sho U

Inspired by the dramatic intensity and vivid hues of Japanese Kabuki theatre and also by the spatial quality of Peter Greenaway's 'The Cook, The Thief, His Wife & Her Lover', the design for Sho U provides a dining experience that celebrates the unconventional amidst a surprising aesthetic. Eschewing the typical symbols of Japanese dining, the design consciously avoids the ubiquitous shoji screen, an excessive use of natural materials or obvious Japanese ornamentation of any kind. Celebrating sudden and dramatic spatial and experiential twists, the design begins with an entirely red entry vestibule and unfolds into an all black series of dining nooks. These nooks ring along a full height white dining hall where the spatial experience culminates in an expression of natural light and views to the river beyond.

首屋餐厅

由于受到日本歌舞伎剧院中浓郁鲜艳的色调装饰以及彼得·格林纳威执导的电影《厨师、窃贼、他的妻子和她的情人》中的餐厅氛围的影响，设计师在首屋餐厅中打造了一种非同寻常的就餐体验——在令人惊讶的美感中寻求新鲜感。障子、日式装饰物以及自然材料等典型的日式餐厅标识被完全摒弃。设计从大红色的入口前厅开始，紧接着便是一系列的黑色隐蔽就餐区——环绕着高大的白色开放式就餐区展开，柔和的自然光线以及小河美景营造了极强的空间感。

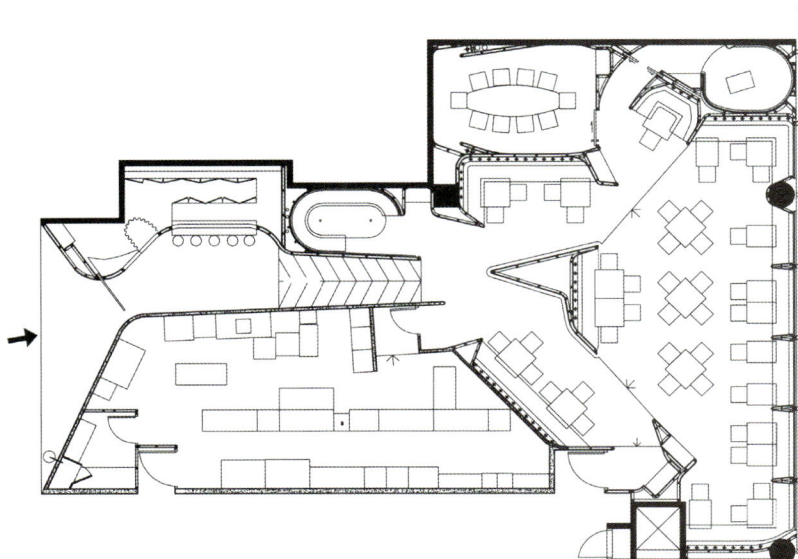

Restanrant & Bar

Singapore

Chinatown

2009

Photo: ONG&ONG Pte. Ltd.

ONG&ONG Pte. Ltd.

Screening Room

Screening Room is an Art Deco Shophouse style conservation building. Previously the historical Damenlou Hotel, it is located within the Telok Ayer Historic District. The design concept was simply to create different expressions of space bearing the same concept of 'film cuisine'. The Shophouse consists of three floors with a roof terrace and a basement. The ground floor was converted into a bistro with indoor dining. Walls were touched up and cleverly covered with intrinsic designed wallpaper to suit the modern yet classic Mediterranean décor. Existing window panels were restored and repainted to retain that rustic old world charm and clever use of large mirrors created an illusion of an open and large space.

映像酒吧间

映像酒吧间是一座装饰艺术风格的古建筑，前身是达门鲁酒店，位于直落亚逸历史街区。设计理念很简单——提供"电影料理"。空间共3层，带有屋顶平台和地下室。一楼被改造成小酒馆，脱漆的墙壁上巧妙地贴上了设计过的墙纸，与经典地中海装饰风格相符。原有的玻璃板经过修复和重漆保持了锈迹斑斑的做旧风格，巨大的镜子让空间看起来更宽敞。

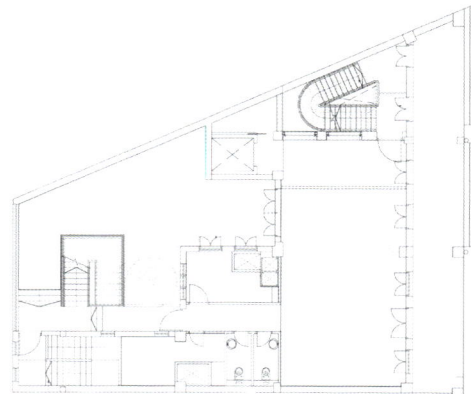

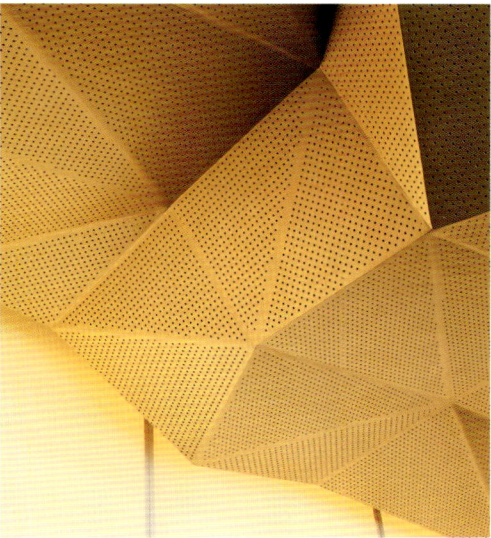

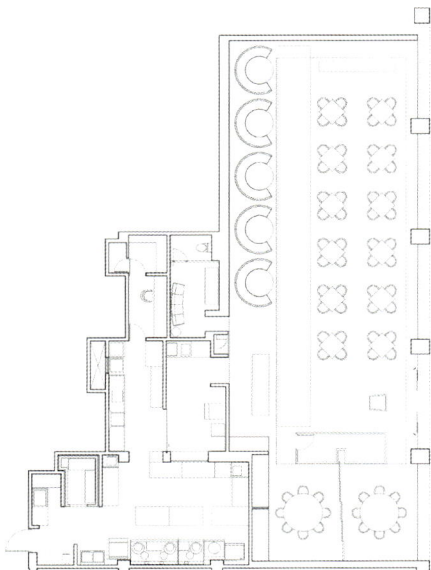

Jing Restaurant

Jing is a contemporary Chinese restaurant situated in a vibrant heart of downtown Singapore. From the very beginning, the restaurant's architecture was informed by an impressive location to enjoy the ever-changing scenery of this city, which is quite beautiful. The first logical step in the design process was to lock in the design layout and technical considerations of the kitchen space, which actually is the heart of any restaurant. This was done with rigorous considerations to the preparatory process of Chinese cuisine. It was only after the kitchen's design resolution that the planning and design focus zoomed in onto the dining area.

景餐厅

景餐厅是一家现代中餐厅，位于新加坡繁华的市中心。餐厅的所在地能够欣赏不断变换的城市风景。餐厅的设计中首先考虑的因素便是规划厨房空间，也就是餐厅的核心。这涉及到许多中餐的准备工作。在厨房设计完全完成之后，设计师才开始考虑就餐区的设计。

Restaurant

Thailand

Bangkok

2009

Photo: Wison Tungthunya

Department of Architecture Co., Ltd

Zense Restaurant

As a part of 'Zen', Bangkok's trend-setting department store, Zense Restaurant possesses a unique design approach by fusing together four design disciplines: fashion design, interior design, landscape architecture and architecture. The project introduces the dynamic world of fashion in the static domain of architecture; combining aesthetic of pleating fabric with architectural elements and function, and at the same time, illustrates the essence of landscape architecture into interior space. The restaurant is comprised of four zones: elevator lobby, vestibule, indoor and outdoor dining area.

禅意餐厅

禅意餐厅位于曼谷潮流百货店"禅"里，其独特的设计融合了4种设计方式：时装设计、室内设计、景观设计和建筑设计。项目在静态的建筑中融合了动态的时尚；将褶皱布料的美学与建筑元素和功能融合在一起，在室内空间中体现了景观设计的精髓。餐厅由四个区域组成：电梯大厅、门廊、室内就餐区和室外就餐区。

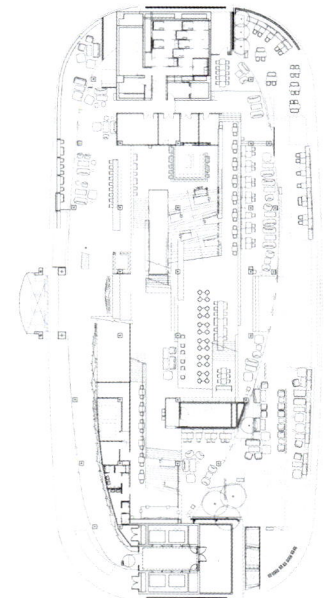

Angelini Bangkok

Angelini is a reinvention of an existing fine-dining Italian restaurant and bar within a five-star resort hotel on the banks of Bangkok's Chao Phraya River. Central to the new design is the relocation of its main entry to the hotel's lower level to create connection to other venues at the riverside level, and to distinguish Angelini as its own destination. The lower level is transformed into a new bar and lounge space, with curved seating arrangements allowing views to a live performance stage beneath the upper balcony. A series of dramatic new window bays were inserted between existing columns to define the newly important riverside façade of the hotel. The two most central of these window bays house striking twin bars, with soaring decorative bottle displays.

安格里尼曼谷餐厅

安格里尼餐厅是对原有的意大利餐厅和酒吧的重建，位于曼谷湄南河畔的一家五星级度假酒店里。新设计的中心是将餐厅的主入口重置在酒店的底层，与酒店内的其他场所相连。底层被改造成一个新的酒吧和休息室，人们可以在阳台下方观看现场表演。柱子之间新增了一系列夸张的飘窗，可以欣赏到河畔的风景。中心的两个飘窗处是一对双子吧台，高耸的酒架引人注目。

Restaurant

Thailand

Bangkok

2008

Photo: Courtesy of Shangri-La Hotels and Resorts

BCV Architects

241

The Maccaroni Club Restaurant

The aim of the project was to make people happy while eating in a comfortable place. Space is on a human scale: chairs, tables and lighting are comfortable. This building material was enriched with timber, white concrete and stones in order to make the surfaces of the restaurant precious. There are huge flower boxes 4metres x 4metres hanging inside the restaurant that mark the natural aspect of the building. It's an image which represents a greenhouse of orchids, which are often in the houses of the native people. There is usually a bower of thousand bamboos outside that, like a sunshade, shade the split-level terraces in the garden.

通心粉俱乐部餐厅

项目旨在让人们在舒适的环境中愉快的就餐。空间极具人性尺度，桌椅、灯光都十分舒适。木材、水泥和石材等材料让餐厅看起来奢华而珍贵。餐厅内4米x 4米的巨大花坛凸显了自然氛围。整个餐厅宛如一个平常人家的兰花温室。外面的竹林仿佛遮阳棚，让花园内错落有致的平台变得阴凉。

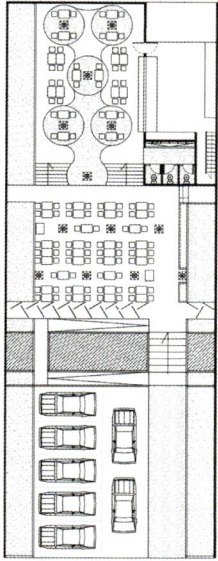

Restaurant at Greenville

The owner of this restaurant dreamed of having an outdoor Japanese Noodle restaurant. But he realised that it is not usual, and for preventing a great loss, he commissioned the designers to design an outdoor and temporary restaurant that hopefully would be easily to be built/assembled and also easily to be disassembled. Those requests directly inspired the designers to take bamboo as the main material for the restaurant. The form of restaurant basically was inspired by an umbrella. The giant form would function as a shelter. A group of shelter then become a building, which hopefully could protect the restaurant guest from sun, wind, and rain. The designers made the giant bamboo umbrellas in different sizes (the width and the height) for arranging the giant umbrellas in such a way that each umbrella could overlap each other to become a giant roof.

格林威尔餐厅

餐厅的主人幻想拥有一个露天日本拉面馆，但是这似乎并不现实，所以他委托设计师设计一个临时的露天餐厅，便于建造，也便于拆除。设计师利用竹子作为主要建筑材料，用竹子做主要结构和外墙。餐厅的造型宛如一把巨大的雨伞，使顾客免于日晒雨淋。设计师将巨大的竹伞打造成不同的尺寸，让它们相互重叠，形成巨型屋顶。

Principal Architect: Suwardana Winata, Susan Soetanto
Design Team: Robby Soetanto

243

Restaurant

Indonesia

Banten

2008

Photo: Rikin Junaedi

Principal Architect: Suwardana Winata, Susan Soetanto
Design Team: Johan Gozali, Gregorius Don Pieto, Ade Ikhwanri

Restaurant at Serpong

This building wants to exploit three elements that restaurants usually must have. First are customers. Customers, who are the heart of restaurant, must be shown from many angles from either outside or inside. Second element is kitchen. Kitchen, where food is made and served to customers, is the soul of restaurant. It also must be shown to customers in order to make them confident to choose this restaurant as their favourite place. The third element is environment. Environment has important position for increasing appetite, relaxing the mind and comforting the body of customers. The exploitation translates into openness of the building. The building must show what inside and also what outside without any barrier. In all manners the designers tried to manage all the spaces to be seen through directly or indirectly and physically or none physically.

瑟邦餐厅

项目设计涉及了餐厅所需的3个必要元素。一是顾客，顾客是餐厅的核心，设计的内外部都要吸引他们。二是厨房，厨房作为食品的加工地，是整个餐厅的灵魂。顾客要能看见厨房才能对餐厅充满信心。三是环境，环境对增进食欲、放松身心起到至关重要的作用。因此，建筑的空间极其开阔，内外部没有明确的界限。设计师力求让建筑保证视觉上的通透性。

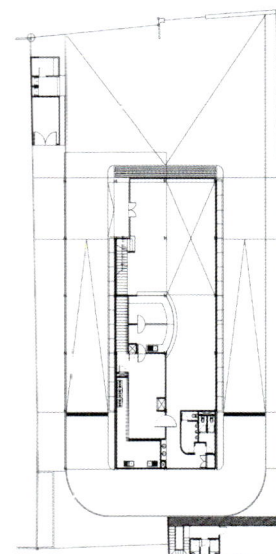

Alila Jakarta

The hotel in the middle of Jakarta's central business district stays true to its name, a Sanskrit word meaning 'surprise'. A welcome retreat from the energy of its surrounding streets, the Alila Jakarta exemplifies simple elegance and skips all unnecessary ornamentation or excess throughout its twenty-seven floors. A Shanghai-style modern bistro was designed with open kitchen behind the glass wall. An extensive variety of Shanghainese cuisine, from the simple to the sumptuous – this is the land of the best soup, noodle, rice, and more dumplings.

艾里拉酒店餐厅&酒吧

酒店坐落在雅加达中央商业区，共27层。简约的装饰风格彰显典雅，使其成为这一喧嚣区域中的"避风港"。餐厅酒吧：上海风格的小餐馆带有开放式厨房。这里供应菜肴品种齐全，靓汤、面条、饺子一应俱全。酒吧坐落在二层，阳光透过开阔的大窗洒落进来，明亮而舒适。在柔和的音乐和美酒的陪伴下，约上好友畅饮或是看看报纸或是浏览一下网页，这里无疑是最好的选择！

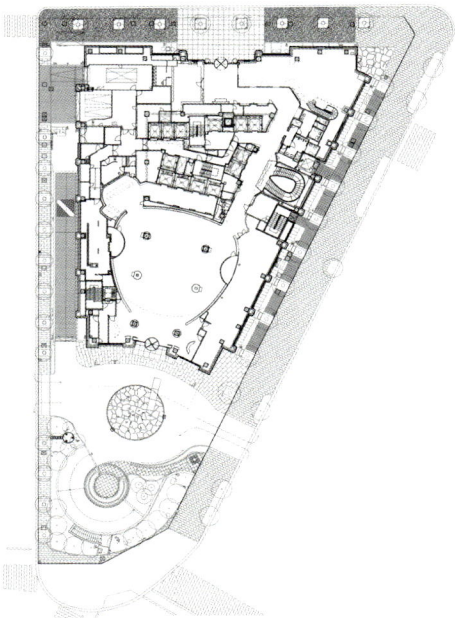

Restaurant

Indonesia

Jakarta

2006

Photo: Denton Cork Marschall

Denton Cork Marschall

Hospitality Destination, Banglore

The programme brief defined the requirement of four different dining experiences and offerings in a single hospitality destination. These were accommodated on two levels, a floor and a terrace, each having a 4,000 square feet footprint. On the lower level, a multi-cuisine vegetarian eatery with open kitchens, on the upper, a lounge bar, fine dining and an open air grill. The challenge for designers was to first manage the requirements within the given space, second offer dismountable construction solution to cover the terrace in conformity with local byelaws, and third, create experientially rich and varied interior experience to address different patron need-states for different spaces within the singular entity.

班加罗尔景点餐厅

项目在同一地点提供四种截然不同的餐饮理念，分为底层和平台两层，每层面积为370多平方米。底层是素食餐馆和开放式厨房；上层是休闲吧、精致餐厅和露天烧烤。设计师所面临的挑战一是在有限的空间里满足项目要求，二是在当地法律允许之内为平台设计一个华盖，三是根据顾客和菜式的需求尽可能创造更多的室内环境体验。

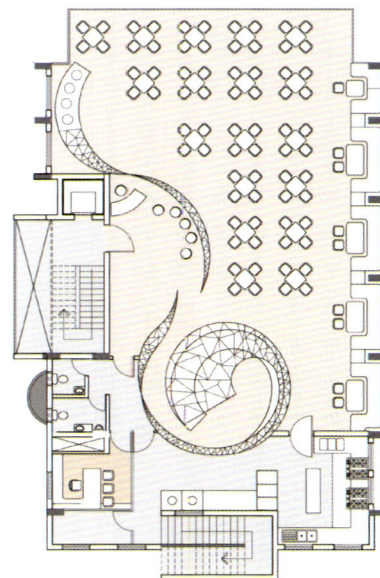

Bar

India

Mumbai

2007

Photo: Fram Petit

Chris Lee / Kapil Gupta

Blue Frog Acoustic Lounge and Studios

A large north lit industrial warehouse within the old mill district in Mumbai was converted into a complex of sound recording studios and an acoustic lounge. This lounge consists of a restaurant, bar and a live stage. Beyond this amalgamation of provisions, Blue Frog seeks to stage an acoustic experience par excellence. The deep structure that was employed is of a cellular organisation composed of circles of varying sizes in plan approximating a horse-shoe configuration. The differential extrusions of these circles encapsulated at different levels allow the diners and standing patrons to be distributed across staggered levels.

蓝蛙录音休闲吧及工作室

这个项目是将位于孟买古老的磨坊区的一座大工业厂房转变成录音休闲吧及一系列工作室。录音休闲吧包括餐厅、酒吧以及一个现场表演舞台，以备不时之用。在这个舞台之上，蓝蛙想要录制最好的声音效果。设计采用的纵深结构是由多个小单元组成，每个小单元由不同大小的马蹄形状的圆圈组成。这些圆圈在不同层面上如分级台阶般突出来，成为圆柱形的座位间，使坐着就餐的人和站着的人的视线分别看到错列的不同层次，高度从舞台起逐渐升高。

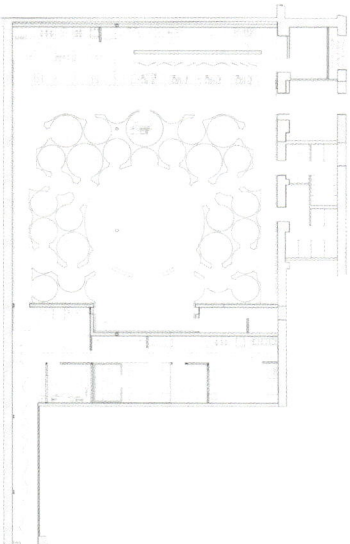

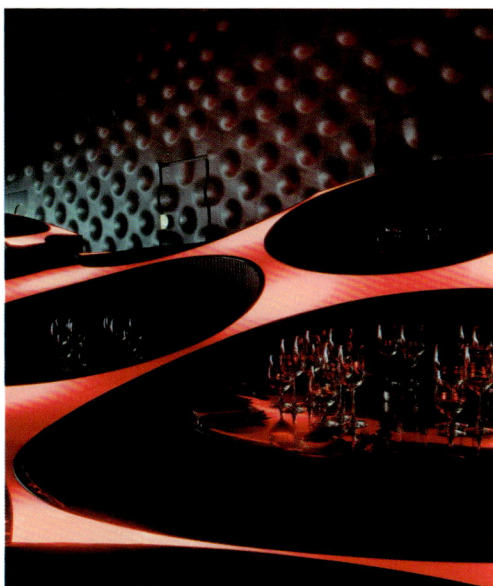

Ville Chaumiere

This restored farmhouse, surrounded by a magnificent garden with a swimming pool, offers all the comfort and charm you need. Here you can enjoy peace and quietness, refined and tasty cuisine prepared for you by head cook, Joel Cesari. Many amusements and activities are available, including a fitness trail directly behind. A large parking lot and five private garages are available for guest use. The hotel – restaurant La Chaumière du Lac is located in the middle of the natural park 'Hautes Fagnes – Eifel'. You will enjoy nature and the tranquillity all around away from the daily stress.

肖米埃庄园

修复后的农舍，在花园与游泳池的衬托下分外妖娆，为客人打造了舒适惬意的休闲空间。在这里可以尽情享受和平与宁静以及手艺精湛的厨师为您提供的各种美味佳肴。酒店后身的众多娱乐及活动场所定会令您身心愉悦。同时，新建的大型停车场和5个私人车库可供住客使用，方便安全。这里位于"高等法涅—埃菲尔"自然公园的中间地段，客人来到此地犹如进入世外桃源之境，能够忘却一切凡尘琐事。

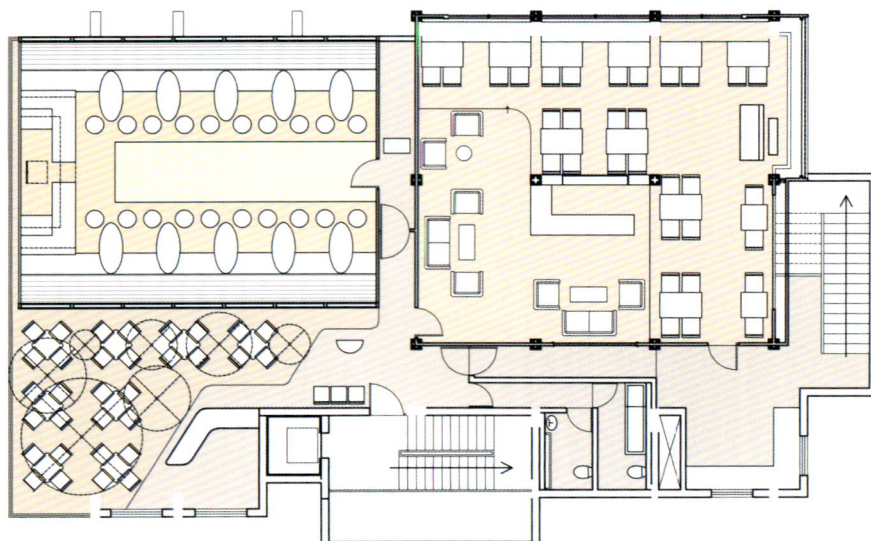

Photo: Planet 3 Studio Architecture Pvt. Ltd

Planet 3 Studio Architecture Pvt. Ltd

Café

India

New Delhi

2009

Photo: Andre J Fanthome

Morphogenesis

Sequel, United Coffee House

Sequel, by United Coffee House is located in a busy market square in the heart of one of the up market colonies of New Delhi. United Coffee House is one of the old and renowned restaurants of Delhi that has deep-rooted memories for most Delhiites who frequented the restaurant over decades. It is a new venture by the people who own United Coffee House; almost a sequel to the existing legend in Connaught Place. The market square is extremely busy and an identity for the restaurant was a part of the design brief – both for the passers-by and for the restaurant customers.

续集联合咖啡厅

续集联合咖啡厅位于新德里市中心一处繁华的广场上。联合咖啡厅是德里最具盛名的餐厅之一，几十年来一直深受德里人欢迎。新建的续集店是位于康诺特广场联合咖啡厅传奇的延续，广场异常忙碌，因此，为餐厅树立一个引人注目的形象至关重要。

Restanrant

India

New Delhi

2009

Photo: Ali Rangoonwala

Project Orange

I-Talia

The design of the new I-Talia restaurant and café was inspired by the collision of rustic simplicity and catwalk glamour. Within the very classical shopping centre environment at Vasant Kunj, the designers wanted to create two very different dining area experiences that were true to the roots of their Italian cuisine. The designers were minded of the historic connections between India and Italy, which saw Roman ships making use of the monsoon to cross the Indian Ocean in order to trade spices, gems and ivory with gold in the first century. The designers felt that this dialogue between aromatic and exotic spices with the allure of gold and silver created a narrative that ties these two countries together and generated the brief for the restaurant.

艾特利亚餐厅

艾特利亚餐厅的设计灵感来自于朴素的乡村风情与T台的奢华魅力所形成的冲突。设计师试图在经典的购物中心中为顾客提供两种截然不同的意大利美食体验。设计师考虑了印度与意大利的历史联系：公园一世纪罗马商船利用季风横跨印度洋来进行香料、珠宝和象牙交易。设计师认为具有金银诱惑的香料交换为两国建立了自然的联系，并以此为餐厅的设计主题。

Punjabi By Nature

Punjabi By Nature caters to the large number of offices in the vicinity with 24/7 operation. It is also catering to the evening dinners. Due to the target client, a sobriety was to be maintained in design in house to many high profile working lunches, business meetings and office celebrations, etc. Given the brief for a 150-seat restaurant with the desired ambience, the space made available to design this contemporary Indian restaurant, was less. Furthermore it was profusely littered with columns and services. The design approach was to carve out space in between this chaos. The area was divided by a series of articulated partitions to expand notional space and offer privacy simultaneously.

彭加比自然餐厅

彭加比自然餐厅为其附近的办公人员提供24小时餐饮服务，也提供正式晚餐。为了适应目标客户群体，餐厅的设计简约时尚，适合工作午餐、商务会议和公司庆典。由于餐厅需要设定150个坐席，留给设计的空间并不多。此外，餐厅里还遍布梁柱和服务区。设计需要在这片混乱中打造一片天地。餐厅空间被一系列的隔断分割开来，扩展了概念空间和私密空间。

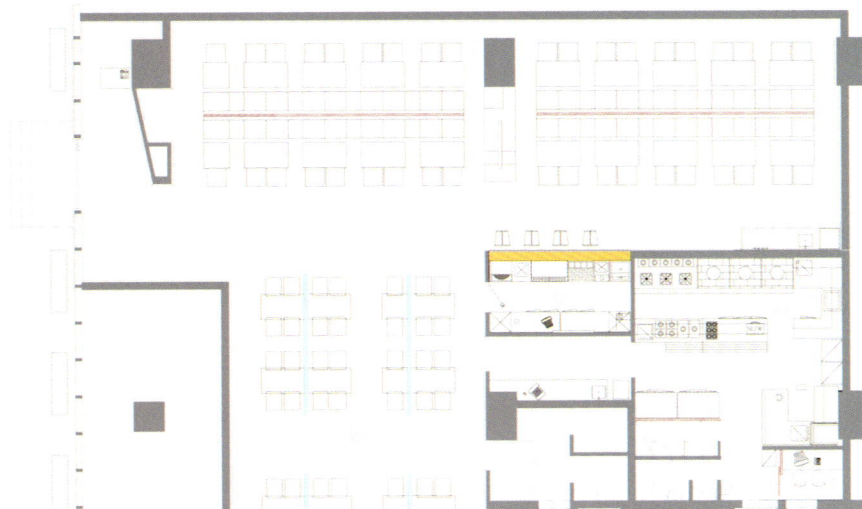

Restaurant

India

Gurgaon

2006

Photo: Amit Mehra

Morphogenesis

Restaurant

India

Bangalore

2009

Photo: Sanjay Ramchandran

Khosla Associates

Shiro Bangalore

Shiro, the lounge bar and restaurant, that significantly upped the ante in Mumbai's entertainment and nightlife scene, now comes to Bangalore. Winner of several design awards and voted as arguably India's finest lounge bar, Shiro makes its entry into a dramatic new space in Bangalore's UB city complex. 'Shiro', which in Japanese means 'castle', has a mysterious aura as you enter its lofty proportions. An unassuming antique acid finished metal door leads you into the space. Crane your head up, and you encounter the awesome monumentality of the area. At a height of 50 feet is a vast ceiling of carefully proportioned bamboo blocks staggered at varying heights. The bamboo is washed with gentle glows of light creating an exciting and complex interplay of textures.

希罗餐厅

曾显著提升了孟买的娱乐业和夜生活水平的希罗休闲吧和餐厅来到了班加罗尔，希罗荣获了许多设计奖项，获得了印度最佳休闲吧的称号，为班加罗尔注入了新活力。"希罗"在日语中的意思是"城堡"，餐厅有一种神秘的氛围。一扇做旧的金属大门引领你进入餐厅，抬起头，你会发现置身于一个宏大的空间。15米高的天花板上装饰着错落有致的竹块。灯光映射在竹子上，形成了复杂纷繁的纹理效果。

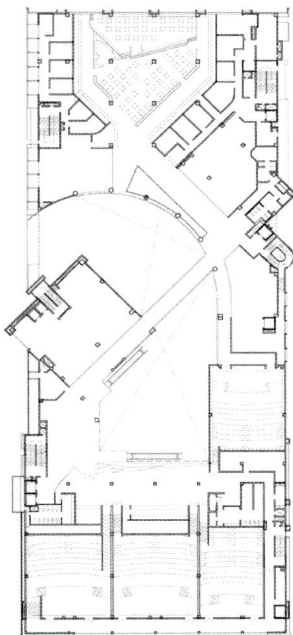

Fun Foodcourt, Lucknow

The play of Tectonic planes evident on the second floor level of Fun Republic Mall continues to be the main theme for the design of the Food-court and its vicinity. Focal point of the design was the prominent end user – the neo-urban youth population of the city who has been exposed to glimpses with newer lifestyle trends around the capital waiting to experience it firsthand. Relatively narrow access leading to the main entry of the food-court prompted the designers to thoroughly work on schemes to capture the patrons' attention to the food joints. To aid this transition which has the guest passing through wide corridor with restaurant on one side and atrium on the other, a vibrant element was introduced in form on a seating around water body. The décor's themes inside are more on the lines of restaurant – with seating spread across at two levels. Both the levels are consciously treated differently – lower level as a more formal zone whereas the upper level as a casual hang-out space.

勒克瑙乐趣美食广场

美食广场的设计延续了乐趣购物中心的层次构造。设计的焦点在于方便终端用户——不断追求更新生活方式、勇于尝试的都市年轻人。美食广场较窄的通道让设计师必须全方位考虑如何吸引顾客。他们运用了一个周围可供人休息的水景装置来装饰通往美食广场的大走廊。内部的装饰主题随着餐厅的变换而变换，座椅遍布上下两层。下层更正式，而上层则比较休闲。

Bar

India

Uttar Pradesh

2007

Photo: Amit Bhandhary, Mumbai

Arris Architects

Restanrant

India

Ranchi

2007

Photo: Vinesh Gandhi

Sanjay Puri

Capitol Hill Residency

A structure designed as a residential building was converted into this small 80-room business hotel, with the challenge of bringing a coherent order into the small dissimilar spaces and low heights within. A small void was created in the ground floor slab allowing a feeling of expanse as the main entrance leads into it. Further it creates a visual connectivity between the public spaces spread over two levels. The reception lounge and conference rooms occupy the ground floor with the restaurant and bar on either side of small atrium at the upper level. Curvilinear fins of particle board create a sculptural effect while unifying varying heights and lend continuity to the main circulation space and large area. The bar at the upper level is made of undulating stacked plywood strips that curve in both planes creating a fluid continuous space with a distinct character while the restaurant is defined by floating planes of particle board at varying levels folding vertically and horizontally.

国会山酒店餐厅

国会山商务酒店共有80套客房，由一座住宅建筑改造而来，内部空间比较狭小，层高也比较低。一楼的一块小空间被划分做酒店的主入口，连接了上面的两层。一楼除了前台休息处和会议室，两侧是带有中庭的餐厅和酒吧。弧形刨花板所形成了雕塑感，让主流通区域和大面积区域形成了一致性。上层酒吧的波纹胶合板条装饰向两侧延伸，十分流畅。而餐厅不同层次的悬浮刨花板在垂直和水平方向叠加在一起。

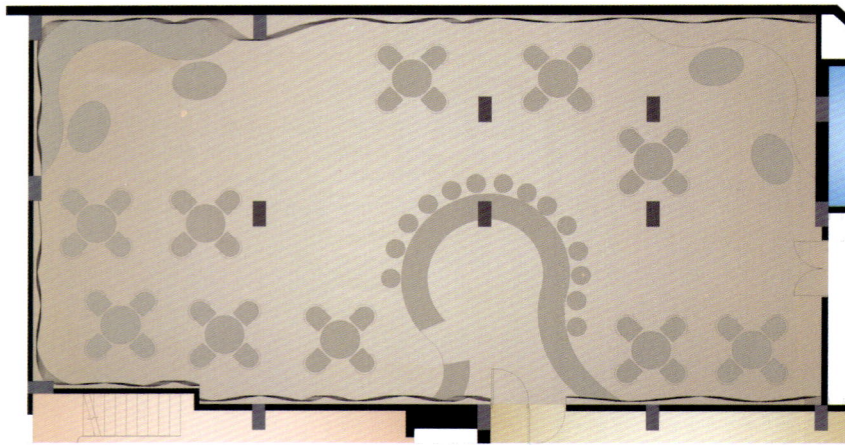

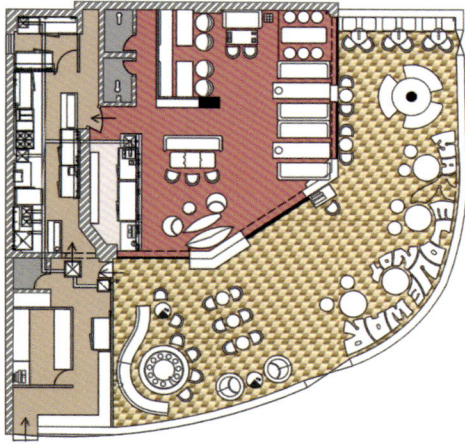

Mocha Mojo

A well-known chain of coffee shops distinguished by a distinct, eclectic kind of interiors wanted to up-end the experience for its patrons at one of the outlets. Except for an interesting ceiling in one section of the existing restaurant, everything else was stripped bare. Within this empty shell, the designers added a retro inspired bar counter and backdrop in signal red with frosted acrylic cutouts and faint backlighting. A collage of Roy Lichtenstein pop art clad on adjacent walls defined the access to a faux fur lined alcove that exuded excess and decadence of the age. Sofas, seats and even a floor rug took the crimson hue. A high recognition pattern was specially recreated and printed on plain wallpaper to create a striking backdrop for an iconic object of the age—an Ambassador car.

摩卡魔力咖啡店

摩卡魔力是一家知名连锁咖啡店，其独特的室内设计给顾客带来与众不同的体验。除了店内一处有趣的天花板，其他所有的部分都需要重新装修。设计师在这个空壳内加入了怀旧感吧台、红色的剪贴画背景和昏暗的逆光照明。旁边墙壁上的罗伊·李奇登斯坦流行艺术拼贴画围绕着一个由人造毛绒装饰的隔间，洋溢着颓废主义时代的气息。沙发、座椅和地毯都是深红色的。具有高识别度的图案被印刷在墙纸上，凸显出那个年代的标志——大使牌汽车。

Photo: Mrigank Sharma, India Sutra

Kalhan Mattoo, Santha Gour Mattoo, Dimple Toraskar, Mansee Jain

Bar

India

Bangalore

2008

Photo: Pallon Daruwala

Project Orange

I-Bar

The new I-bar is luxurious, mysterious and decadent. It uses traditional pink sandstone and carved timber screens in an altogether new way fusing established materials with a graphic sense of pattern and colour to create an ambience that is rich and enveloping. In addition the concept of 'patchwork' was introduced both as a motif and as a metaphor for the city. Elements are collaged together from the different pink colours on the floor to the dusky pink mirror on the walls and the ceiling tiles to the entrance. The layout is understood as a box within a box where the inner sanctum is accessed via the sliding timber screens. This area has a lowered ceiling and features a collection of furniture all covered in red leather, silk and cotton. On the floor are a number of individually designed textured rugs by Project Orange. The bar itself is on the end wall standing in front of a huge textured hand carved sandstone relief.

I吧

新I吧奢华、神秘而颓废。它运用传统粉色砂岩和木雕屏风共同打造出图案与色彩的图像感，形成了丰富而隐蔽的氛围。"拼接"这一理念既是项目的主题，也暗喻着整个城市。地板上不同的粉色色块、墙壁上朦胧的粉红镜子和入口的粉色天花板被拼接在一起。项目被规划成盒中盒，内部的包房可通过滑动木屏风进入。这一区域的天花板较低，家具全部由红色皮革、丝绸和棉布覆盖。地板上铺着特别制作的织物地毯。酒吧后方的吧台后面是一个巨大的手工砂岩浮雕。

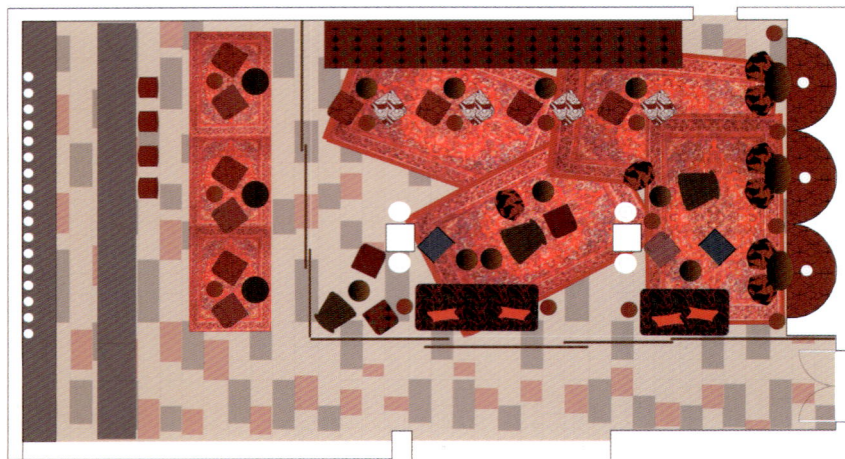

Smoke House Deli

The Smoke House Deli posed a unique challenge to create a fun, informal, buzzing Deli space, which turns into a mood-lit, quieter and more romantic evening space. The designers conceived of an entirely hand-illustrated space, where there is an increasingly eccentric interaction between 2D and 3D elements in the space. The design takes an irreverent, fun look at 'serious' restaurant design: objects exist as a parody of themselves. Every element, from the outdoor signage and Main door, to wallpapers and cabinets, were hand illustrated, over a 33 day period. Currently in India there's this trend towards a very standardised view on 'Luxury', and places are slowly streamlining themselves towards a more maximal, almost ostentatious display of high-budget but disparate materials and processes, what is perceived as 'stylish' is getting into a very mediocre design intent, with a huge disconnect between designer and space.

德里烟房餐厅

德里烟房餐厅是一个挑战，要求打造一个有趣、正式而繁忙的空间。夜晚，餐厅还要气氛优美、安静而浪漫。设计师想打造一个手绘空间，将2D和3D元素融合在一起。设计区别于传统而正式的餐厅设计，物品都像仿制品一样。从户外引导标示、大门、墙纸到橱柜全都是由设计师手绘的，共历时33天。现在，印度有一种奢华的潮流，空间趋于全面化，追求招摇的高预算、迥异的材料和过程，看似时尚，实则平凡，严重缺乏设计师与空间的沟通。

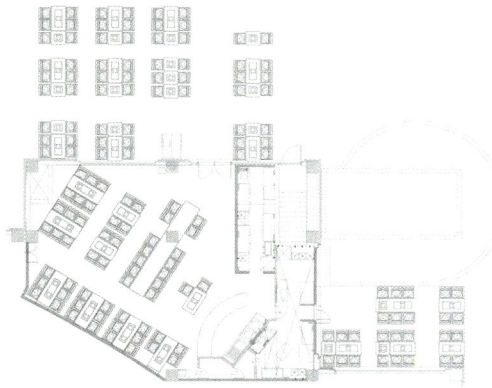

Restaurant

India

Delhi

2009

Photo: The Busride Design Studio, IEHPL

The Busride Design Studio

Bar in Chrome Hotel

The hotel was planned in eight levels with public spaces occupying the first three levels and four levels of rooms above with a rooftop lounge bar on the topmost floor. The bar being small in area was designed in a fluid manner that allows it to be perceived as a larger space while being rendered in a sculptural way. Undulating curved ribbon-shaped panels are suspended from the ceiling with reflected colour-change lighting between them across the length of the bar. The walls and the bar counter are merged fluidly with curvilinear panels of varying widths and projections. Complete white rendering of all the design elements allow the bar to be completely transformed by colour-change lights at intervals, creating different moods.

克罗姆酒店酒吧

酒店设计为8层，下面3层是公共空间，上面是客房层，顶层是个高级酒吧，酒吧外是一个狭长的平台。这个酒吧虽然面积不大，但是设计却具流动性，看起来好像更大了一些，采用了类似雕塑的处理办法。波状的曲线镶板从天花板上悬下来，镶板之间变换颜色的光线创造出一种反光效果。宽度、角度各异的曲线镶板将墙壁和吧台融合在一起。所有设计元素都是白色的，这样整个酒吧就随着不时变换颜色的灯光而有了不同的气氛。

Photo: Vinesh Gandhi

Sanjay Puri & Nimish Shah

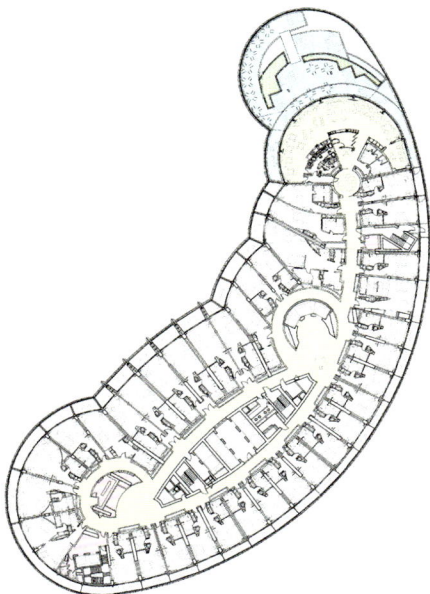

Restaurant

UAE

Dubai

2008

Photo: Atkins Design Studio

Atkins Design Studio

Restaurant in The Address Hotel

Atkins was responsible for the architectural and engineering design of this five-star hotel and serviced apartment building located in the prestigious Business Bay area. The development comprises a business centre, health club and spa, a restaurant and parking bays for 900 cars. The Carbon Critical Design™ ethos ensured features such an air condition condensation recovery process that captures enough water to service the irrigation needs of the whole building.

地址酒店餐厅

位于久负盛名的迪拜商业湾的这幢五星级酒店的建筑和机械设计及酒店式公寓设计由阿特金斯事务所负责。此开发项目包括一个商业中心、一家健身俱乐部和一个温泉浴场、一个餐厅和能停900辆车的停车场。设计师的低碳排放设计回应了向客户保证的（环保）特色，如将空调冷凝水用做整个大楼的（景观）灌溉用水。

Nobu Restaurant

On Friday November 21st, 2008 there was a ceremony in celebration of the first Nobu restaurant in the Middle East in Dubai, U.A.E, in conjunction with the opening of Atlantis – The Palm, Dubai, the resort where Nobu is located. The main restaurant looks like a huge fishing boat. The whole design is so spectacular. It is not only novel, but also can provide unlimited imagination. There also is a small lounge, which is very exquisite in appearance. There are some linked half-circle sofas and full of the walls are decorated by mirrors. The vertical timbers are carved by many delicate drawings. Psychedelic light-fixtures make this space more romantic. The lights twinkle and reflect from the mirrors to make the lounge much brighter.

诺布餐厅

诺布餐厅于2008年11月21日开业。该餐厅位于中东迪拜的朱美拉棕榈岛的亚特兰提斯度假村。主餐厅的设计犹如一个大型的渔船。整体设计很壮观，既新颖又给人无穷的想象空间。同时这里还有一个小型的休闲室，外观上更显精致。这里安置相连的半环形沙发，墙面全部用镜子装饰。屋内的纵梁上有雕花装饰。迷幻的灯饰更为空间增添了一份浪漫的色彩。灯光闪烁并通过镜子反射出的光芒使休闲室更加明亮。

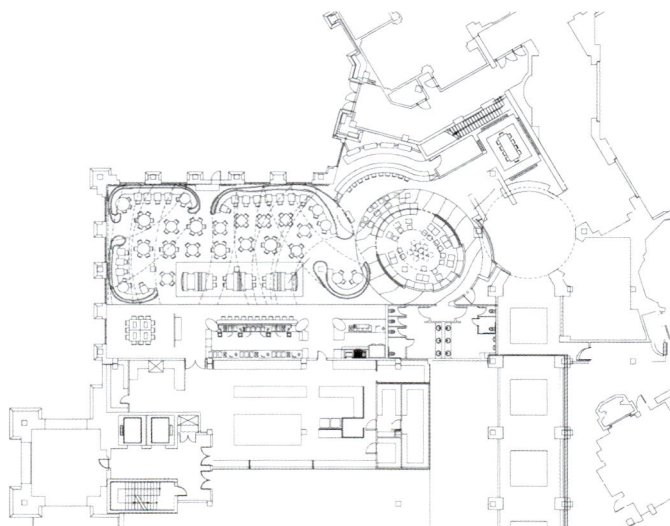

Switch Restaurant

The design created an interesting texture for light and shadow, evoking the sand dunes in the desert. It is a unique environment of symmetry and balance that completely envelops the guests. Every experience is composed of views, smells, tastes and sounds; here, the senses create individual backgrounds for a truly amazing global dining experience. The backlit ceiling artwork consists of stylised inspirational Arabic phrases. The continuous wave seating provides an efficient and dynamic operating system. The designer wanted to create a powerful, clean space that offers a beautiful perspective, an oasis free from chaos.

开关餐厅

设计打造了有趣的光影效果，让人想起了沙漠中的沙丘。这个对称而平衡的空间将顾客完全包裹其中。这里的声、色、味、嗅效果为顾客带来了神奇的全球化餐饮体验。背光天花板上有鼓舞人心的阿拉伯短句。连绵不断的波纹坐席提供了高效而动感的操作系统。设计师力求打造一个有力而简洁，同时又兼具美观的空间，在嘈杂的城市中打造一片绿洲。

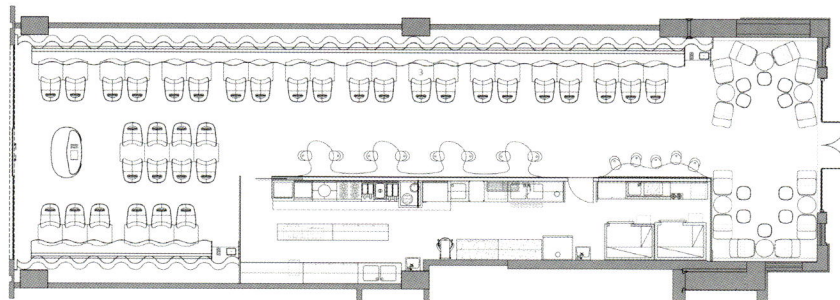

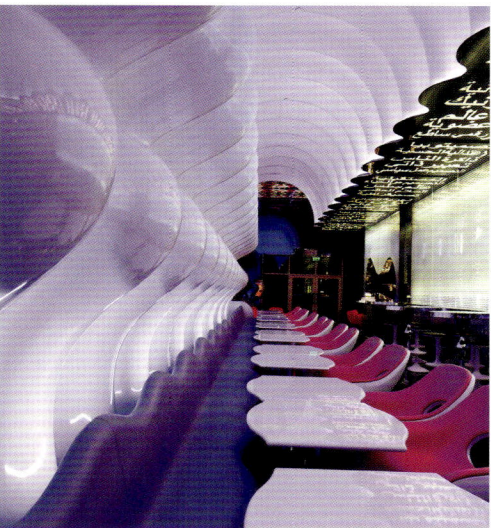

Restaurant

UAE

Dubai

2009

Photo: Karim Rashid Inc.

Karim Rashid, Project architects:Camila Tariki, Karim Rashid Inc.
Kamala Hutauruk, Karim Rashid Inc. Local Project management:
Giane Atallah, Designer Translator. Kitchen Consultant: HICE –
Hospitality Industrial Catering Equipments

Restanrant

UAE

Abu Dhabi

2009

Photo: Gerry O'leary

Jestico + Whiles

Restaurants in Yas Hotel

 The concept of the restaurants was that they should be entirely contemporary, with only very subtle but tantalising references to the culture from which the cuisines originate. The Tandoori Restaurant, for example, takes the saturated colours of traditional saris and actually incorporates panels of fabric into table tops. The Italian restaurant is focused on a heavy, family-sized olive wood refectory table. The noodle restaurant used celadon enamelled volcanic basalt tiles. The speciality fish restaurant is located at water level in the marina wing, and the mood is defined by this marine setting. The entire space is lined by glowing panels of 'ice white' composite, which are pierced, drilled, and partly drilled and stained with undulating, flowing aqua-coloured lights to evoke the serenity of a moonlit coral reef, and a stupendous chandelier makes oblique reference to the translucent, amorphous form of a jelly fish.

雅思酒店餐厅

餐厅的设计理念是，要体现出现代气息，而且要能巧妙又明白地表现出这些菜肴都来自于哪些文化。比如印度泥炉炭火烹饪餐厅，采用了传统的印度莎莉较深的颜色，而且桌面采用织物构造。意大利餐厅则突出一张沉重的橄榄树木材制成的长餐桌。面条餐厅使用了灰绿色的搪瓷黑陶瓷砖。特色鱼餐厅坐落在"水层"，就以海洋背景为特色。整个空间镶着雪白的发光镶板，上面有如水波般流动闪烁的浅绿色灯光，灯光或穿透或部分穿透或者打在镶板上，营造出月光下的珊瑚礁一般的静谧感。头顶巨大的枝形吊灯像是一条半透明的、难以名状的水母的形状。

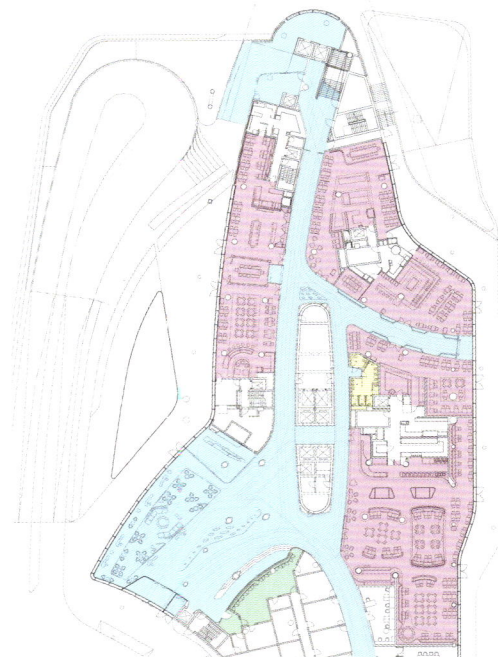

Restaurant in InterContinental Hotel

This hotel will be one of the premier hotels in the Emirates, a unique experience, a relationship informed and enhanced by the surrounding design, catering to both the business and leisure travellers. The sleek, clean, contemporary architecture of the existing building's exterior is the inspiration for the interior design approach to the hotel. The hotel's Three Meal Restaurant is adjacent to the active pool deck with outstanding views of the Gulf. The blue sky and sea surrounding three sides of the glass walled restaurant are screened by a dark wood Arabic screen with soft flowing sheers beneath. A soft, neutral colour palette rich in texture, create a pleasing backdrop for the many food displays and open kitchen.

阿布扎比洲际酒店餐厅

该酒店将会成为阿联酋最重要的酒店之一。酒店的设计拉近了客人和酒店之间的距离，这种非凡的体验，得到各地的商人和游客的好评。现存酒店外部雅致、巧妙的现代建筑风格，为酒店的内部设计提供了灵感。酒店的三餐餐厅泳池，可以远眺海湾的美景。碧海蓝天环绕着餐厅的三面玻璃墙壁，另一面是黑木阿拉伯风情屏风。柔和、中性的色调和丰富的材质为美食和开放式厨房打造了一个愉悦的背景。

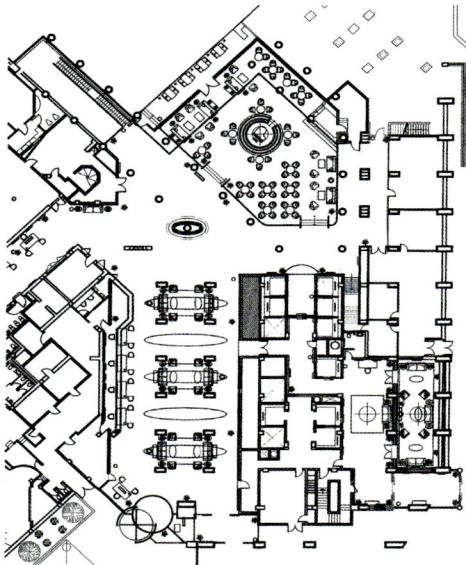

DiLeonardo International

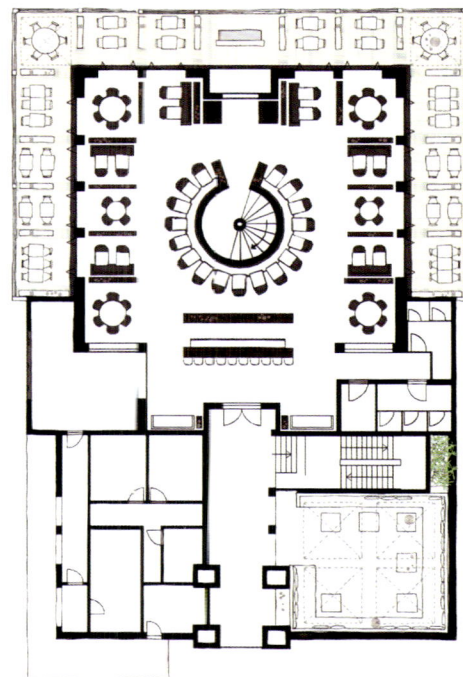

Restaurant

UAE

Abu Dhabi

2008

Photo: Richard Thorn

Concrete Architectural Associates

Pearls & Caviar

'Pearls & Caviar' represents the new Arabian lifestyle, a luxury fusion of East and West, black and white, occident and orient, light and shadow, the extrovert and the introvert, the intimacy and the view. The basic idea of the design of the restaurant has been to create an abstraction of the commonly used oriental forms and materials without loosing their richness. To achieve this the amount and the richness of the oriental patterns and forms which are found in traditional oriental spaces were kept but without colour. All colours were replaced by either shades of black or shades of white - both in combination with silver. Therefore the restaurant was divided into two parts, which also gave its name: 'Pearls' and 'Caviar'.

"珍珠和鱼子酱"餐厅

"珍珠和鱼子酱"餐厅代表了阿拉伯人新的生活方式，一个东与西、黑与白、欧美与东方、光与影、内外兼收，兼容并蓄的奢侈融合。这个餐厅的最基本的设计构思是运用东方形式和材料来打造一个丰富多彩的空间，为了达到这个目的，东方图案和窗体在传统空间中应用的数量和样式被保留，但没有使用颜色。所有的颜色被黑色或白色的阴影代替，所以餐厅被分成两个部分，如同名字一样，珍珠和鱼子酱。

Theodor TLV

In the heart of Tel Aviv, next to Rothchild Street and the world heritage site of the white city, inside a heritage building from the 1930's, a new branch of the Theodor Café-Bistro was put up. Just like the other Theodore Café Bistro in Ramat Ishay, the Tel Aviv one is also a place of culture. It is a café restaurant that, on the one hand exudes a culinary atmosphere, while, on the other, is leaning on an Israeli cultural foundation from the last 60 years, exhibits in literature, song, art and architecture that are displayed on the shelves. This allows one to dine while staying in a very special atmosphere.

特拉维夫特奥多餐厅

这家特奥多连锁餐厅位于特拉维夫市中心,紧邻罗斯切得尔大街和白城世界遗产,设在一座建于20世纪30年代的历史建筑之中。与之前的拉莫依沙特奥多餐厅一样,特拉维夫特奥多餐厅也充满了文化氛围。这是一家一面散发出美食诱惑,一面在文学、歌曲、艺术和建筑多方面展示近60年来以色列文化的餐厅。人们可以在这里享受非同寻常的就餐氛围。

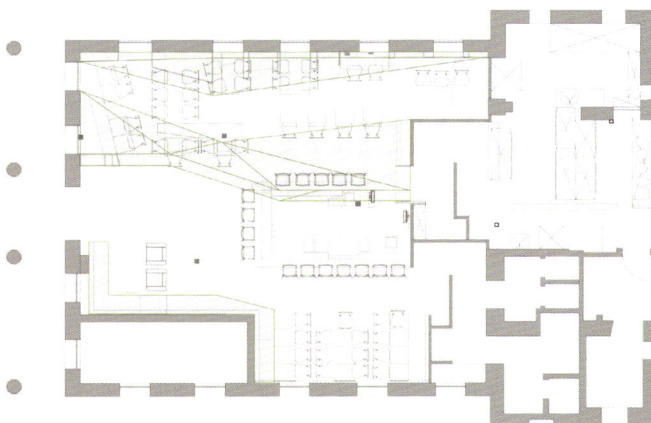

Restaurant

Israel

Tel Aviv

2010

Photo: S0 Architecture

Shachar Lulav, Oded Rozenkier / S0 Architecture

Restaurant

Israel

Ramat Ishay

2007

Photo: Asaf Oren

S0 Architecture - Shachar Lulav, Oded Rozenkier in cooperation with arch. Eran Mebel

Rosso Restaurant

The brief seeked to enlarge the existing restaurant space into a back room with proportion of five to two that should allow the arrangement of intimate functions and parties, as well as being integrated as part of the restaurant. The green hill around the place influenced the design. The planning seeked to explore – how to give to the restaurant space the feeling of its surroundings. The designers checked in depth on how to give the diner a feeling of sitting outside while in the restaurant.

罗素餐厅

罗素餐厅的设计目标是扩展原餐厅空间，使之成为一个密室，长宽比为5比2，这样能够满足密室的私密功能，也可以作为聚会场所，又能融入餐厅，成为它的一部分。餐厅周围的绿色小山影响了餐厅的设计。设计计划探索了这样一个问题：如何将周围环境的感觉赋予餐厅空间。设计验证了怎样才能产生这样一种氛围，那就是坐在室内却感觉好像在室外一样。

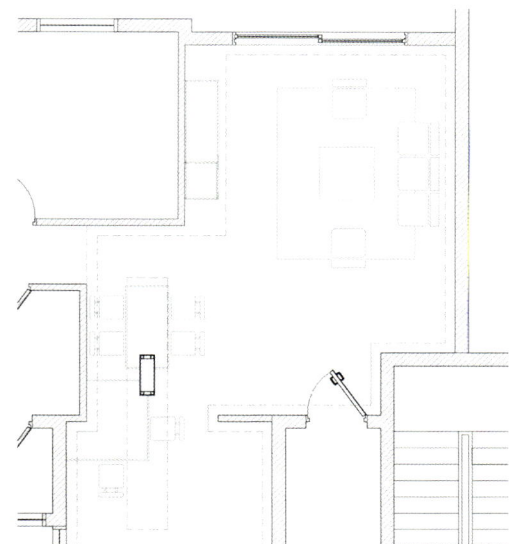

One Out of Four

The project started its conceptional way as a prototype to four different functions in the domestic space as one element. One Out of Four contains in it the bar, dinnig area , kitchen's table and storage area. It functions in different ways on the time line and redefines the space and at the same time answers other needs. It is hung from the ceiling on a vertical shaft with no base and out of it the four elements were created. All these elements were disconnected from the floor and create the levitation effect, in which the project is floating.

四分之一餐厅

设计师的设计理念在于将四种不同功能的空间融入一个单一空间内。 建筑含有酒吧区、就餐区、厨房桌及储藏区。整个空间布局围绕一根直达天花板的立柱展开，立柱没有基底。与此同时，空间内的其他所有元素也都与地板分离，产生一种悬浮效应，使整个空间仿佛浮动一般。

Restaurant

Israel

Gedera

2009

Photo: Felix Spivack

Lior Vaknin + Sabi Aroch

Restaurant

Iran

Tehran

2006

Photo: Ali Daghigh

Arad Office

Touch Restaurant

Touch restaurant is all about searching for different cuisines in a single fine dining restaurant with a different interior ambience in a single volume. It means the answer to a vast range of tastes, appetite and sense of aesthetics. The location is in a commercial centre under a large residential tower. Touch was to fulfill the demands of the commercial centre, residents of the tower above it and also the people who live in this neighbourhood. This restaurant has three different sections in two levels. The kitchen in the lower floor with second service door, the main hall in the same floor with double height ceiling, and the upper floor is like a cosy lounge with more comfortable furnishings in mezzanine. These small parts of the restaurant were designed in different taste with different ambience, lighting and mood but in a specific architectural order.

触感餐厅

触感餐厅试图在同一餐厅中提供各种美食，每个独立的室内空间都有不同的氛围。其设计需要适合不同品味、菜肴风格和审美观。餐厅位于一座大型住宅楼楼下的商业中心，为商业中心、住户和附近的居民提供餐饮服务。餐厅分为2层3个部分，厨房和大厅位于一楼，楼上是休息室，里面摆放着舒适的家具。餐厅中这些不同风格、气氛、灯光和情绪的部分具有特别的建筑结构。

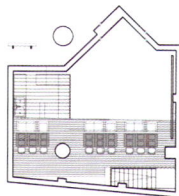

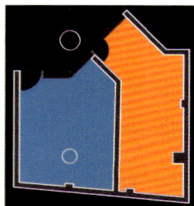

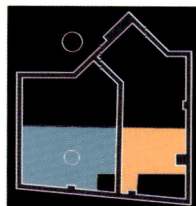

Restaurant

Bahrain

Manama

2008

Photo: SHH

SHH

Nu Asia Restaurant

Internally, the building was arranged with a first kitchen on the lower-ground floor (which was only able to be set one metre down because of the water table), along with a staff area. A glazed reception/ lobby area on the ground floor contains the circulation, with a service core located near the restaurant space. A second, open and more theatrical kitchen is located on the ground floor and is semi-open for diners to see in. A series of six columns break up the ground floor space. From the main entrance, guests pass a planter strip, bringing greenery into the scheme, before arriving at the restaurant reception. A second planter strip was included further into the restaurant itself.

牛扒餐厅

餐厅内部主厨房和员工区的位置较低（由于水案的缘故，只能将厨房降低1米）。接待区、大堂和收银台设在一楼的餐厅旁边，是主要服务区。另一个厨房位于一楼，采用半开放式设计，用餐者可以看到里面的情况。进入正门，顾客穿过绿化带到达餐厅接待处。在餐厅里面还有一个绿化带。入口的楼梯是大理石材质，餐厅前半部的地面也用同样的材料，后半部以深色石材营造不同的感觉。

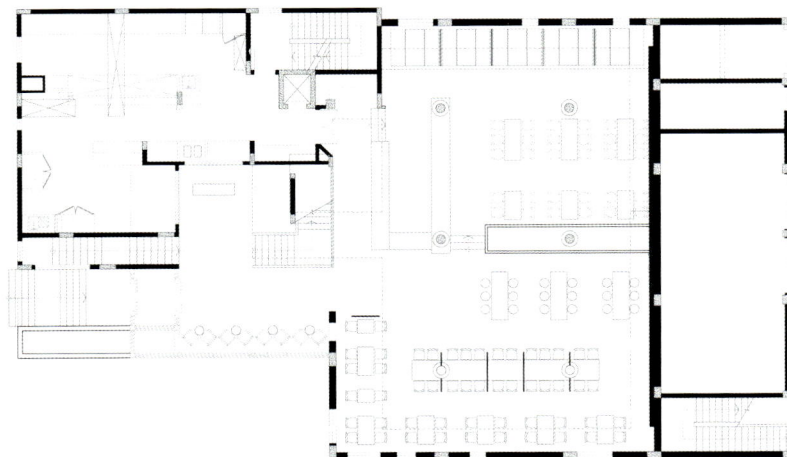

Photo: Nagehan Acimuz

Shimasushi Restaurant

First of all, beside flexible tables that different numbers of people in a group can sit, the client needed to design a drink bar, a sushi bar and a social long and big table that a big number of people can sit together. While the designers were working on the plan, they designed all of these as one single element that starts with the drink bar, transforms into sushi bar, and the table for people eating sushi, and then the big social table. Beside these, the designers placed some of the tables stable, some can be movable so that they can change according to the number of people who want to sit together. They created a lounge table, which is separated from the space with wooden separations; it is a more private table.

岛寿司餐厅

除了可以灵活挪动的桌子之外，客户还需要设计一个饮品吧台、一个寿司吧台和一张可供群体就餐的大型长桌。设计师将这些看成了同一元素进行设计：起点是饮品吧，随后演变成寿司吧和寿司桌，最后是长桌。除此之外，设计师还设计了一些可移动的桌台，以便适应不同数量的团体就餐。他们打造了一个休息桌，通过木隔断与其他空间隔开，更具私密性。灯光设计在空间、色彩中也极为重要，而灯光设计在日式餐厅中起到了重要的作用。

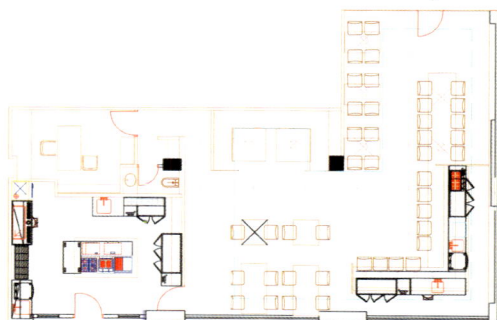

Nagehan Acimuz

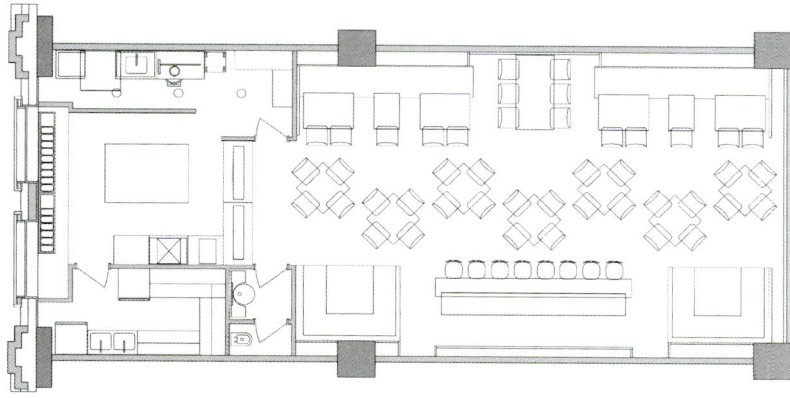

City's Vengeplus

First of all, the designers started this project with the basic concept that has been prepared previously for Capacity Shopping Mall Vengeplus Project and its adaptation on to 160 square metres. The layout plan consisted of disorderly tables and the cedars leaning towards the walls and also lodges and compartments. However, since the space had been in a rectangle geometry, entering it from the short edge and having its end section out of glass made the designers' job quite difficult to have the day-light inside the space. To achieve this, they located the kitchen section to the very end of the space and the parts that are dark become the right places for locating the cupboards, scullery and storage. Moreover, the other daylighted areas left could be used for the open kitchen activities and by this way, daylight had been provided inside the entire restaurant space.

都市温吉加餐厅

餐厅的设计与温吉加购物中心的既定设计理念相一致，总面积为160平方米。设计由随意摆放的桌椅、向墙壁倾斜的雪松木和包房、隔间组成。长方形的空间和狭长的走道让设计师的采光设计工作变得异常困难。他们将厨房设在了空间的后部，昏暗的地方正好用来设置橱柜、洗碗槽和储藏室。这样一来，能够自然采光的区域就被放大了，可以作为开放式厨房，并且将阳光带到整个餐厅。

Restaurant

Turkey

Istanbul

2007

Photo: Nagehan Acimuz

Nagehan Acimuz

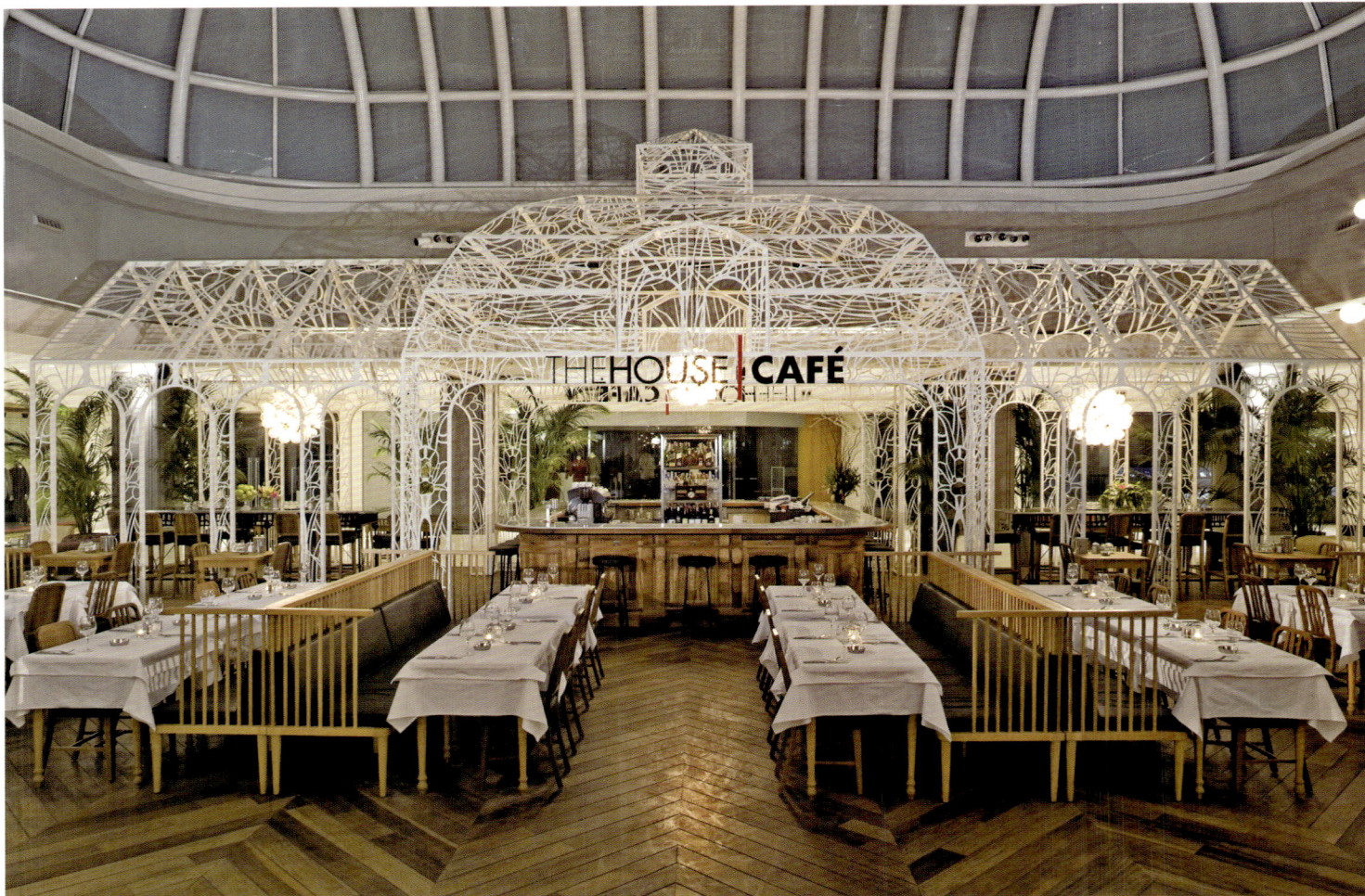

Café

Turkey

Istanbul

2007

Photo: Autoban

Autobahn, Seyhan Ozdemir & Sefer Caglar

The House Café Istinye

This is the widest house café located at Istinye Park Shopping Mall, with a seating capacity of 200. The whole atmosphere is fascinating with the latest special designs accompanying the furniture classics of The House Cafés. The space was just like a city square with no walls or columns surrounding. Autoban created metal towers, just like a clock tower and train station. With the marble tables inside each, these cages become a spectacular space in which to sit. You can feel the relaxing ambiance during day time, with the natural day light coming through the roof.

伊斯汀耶咖啡屋

咖啡屋可容纳200人，是伊斯汀耶商场中最大的咖啡屋。咖啡屋融入了最新潮的特别设计和经典的咖啡屋布置。空间里没有隔断和廊柱，仿佛一个城市广场。设计师打造了一些金属塔，就像钟塔和火车站一样。这些金属塔和内部的大理石桌台形成了壮丽的景观。日光从屋顶上洒下来，形成轻松的氛围。

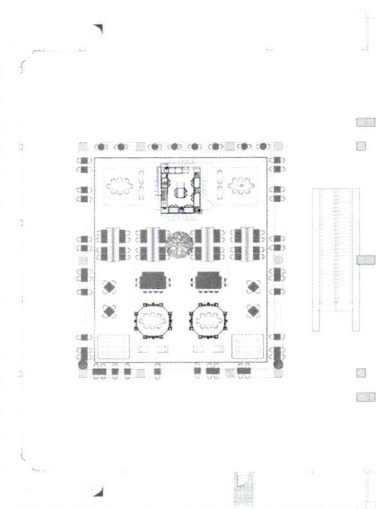

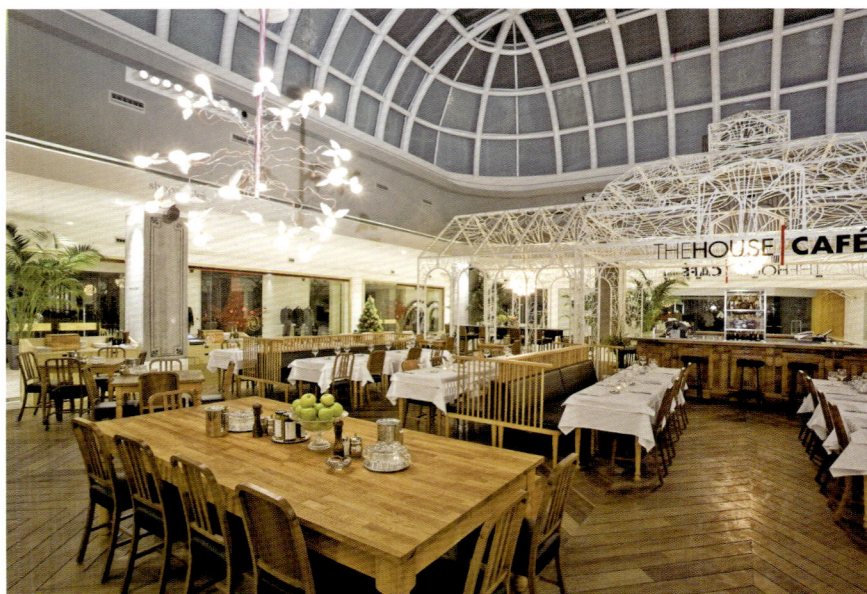

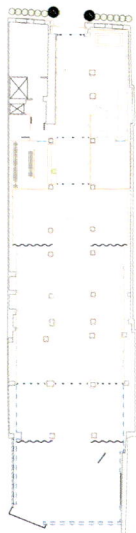

Restaurant

Turkey

Istanbul

2009

Photo: Ali Bekman

Autoban, Seyhan Ozdemir & Sefer Caglar

Anjelique

The award-winning restaurant-bar-club Anjelique is all new, with an interior redesigned by Autoban. Spreading over two floors with jaw-dropping views of the Bosphorus, the new Anjelique boasts a restaurant and lounge bar on the main entrance floor. There are more tables to enjoy a fancy dinner on the upper floor, also where the club is located. The characteristics of Mediterranean style is evident in everything from the navy blue and white colour combination dominating the space, as well as the traditional chintemani tiles in turquoise adorning the columns.

安吉利克餐吧

安吉利克是一家获奖餐吧，其室内设计由Autoban一手打造。餐厅的尽头可以看到波斯普鲁斯海峡的美景，共分为两层，主楼层是一个餐厅和休闲吧。二楼的俱乐部则适合进行更加精彩的晚宴。餐厅设计处处洋溢着地中海风情：海蓝色和白色是主色调，柱子上镶嵌着绿松石色的马赛克。

The House Café Istiklal

This branch of House Café is located on the ground floor of the infamous M.s.r Apartment on the busy street of Istiklal. The glass façade ties in with the beautiful old architecture of the building while the interior carries through Autoban's trademark style of old and new. An elegant metal framed pavilion greets visitors at the entrance behind which the bar runs down the long rectangular space. The suspended ceiling is formed from hexagonal mesh tiles while subtle coloured lighting here and there adds a touch of modernity. For those with an interest in cooking, this particular House Café also invites you through to the kitchen at the end of the restaurant for cooking courses.

伊丝提克拉咖啡屋

这间连锁咖啡屋位于著名的M.s.r公寓的一楼，地处繁华的伊丝提克拉大街。玻璃墙面与古老而美观的建筑相得益彰，而内部装饰采用了Autoban经典的新旧结合式设计。入口处优雅的金属架给予顾客全新视觉感受，后面的吧台一直沿着长方形的空间延伸。悬垂式天花板呈五边形，与彩色的灯光共同为空间增添了现代感。如果有顾客对料理感兴趣，咖啡屋后方的厨房还提供厨艺教学。

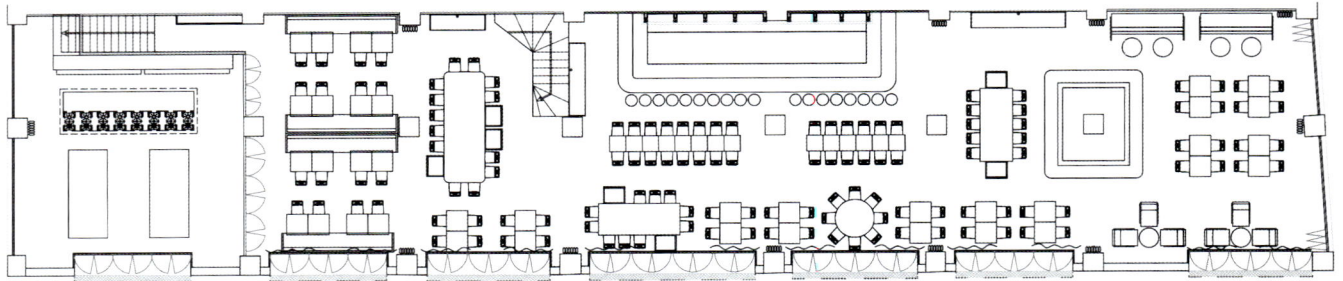

Café

Turkey

Istanbul

2006

Photo: Ali Bekman

Autobahn, Seyhan Ozdemir & Sefer Caglar

Restaurant

Turkey

Istanbul

2010

Photo: Ali Bekman

Autobahn, Seyhan Özdemir & Sefer Caglar

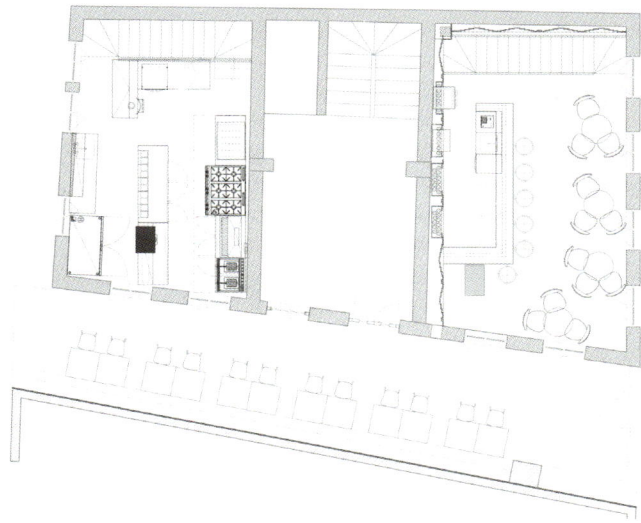

Munferit

Munferit, which is situated in Galatasaray famous for its art galleries, antique shops and Flower Passage (Cité de Péra), is a project that blends traditional Turkish cuisine and Autoban's approach with a modern twist. Metal paneling, Art Deco mirrors and ceramic flooring are found in the two-storey restaurant Munferit, designed with traditional materials in a contemporary fashion, flourishes in its 120 square metres garden and courtyard, where the guests can enjoy a timeless place. Munferit Restaurant has interpreted the traditional Turkish culture in a contemporary way. The designers used very precious 'rosewood' for high wall panelling. Mirrors were placed from the wooden panels to the ceiling to generate a wider atmosphere. The reflection of the ceiling made of metal panels produced an illusion.

木菲利特餐厅

木菲利特餐厅位于加拉塔沙雷，该地区以艺术画廊、古董店和花店而闻名。该项目融合了传统土耳其料理和Autoban的现代风格。两层高的餐厅中包含了金属板、装饰艺术镜子和瓷砖地面等多种元素，融汇了传统材料和现代时尚。在这个120平方米的空间，顾客可以获得经典的享受。餐厅试图以现代方式诠释传统的土耳其文化。设计师运用珍贵的紫檀木来做墙壁镶板。镜子从镶板上方延伸到天花板，扩大了餐厅的空间。网状天花板的倒影造成了视觉错觉。

Restaurant

Turkey

Bodrum

2009

Photo: Ali Bekman & Ozlem Avcioglu

Global Architectural Development

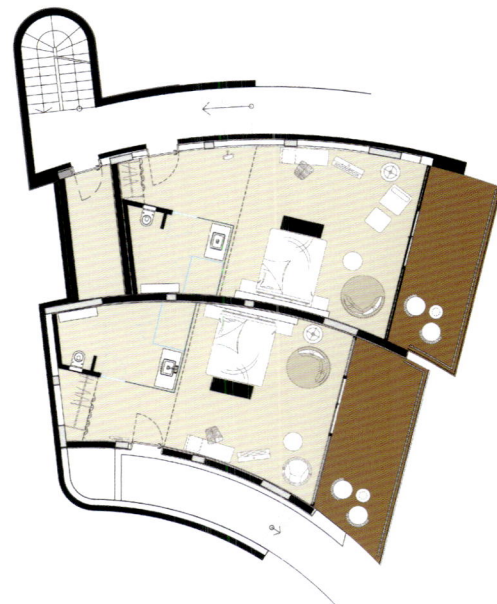

Restaurant in KUUM Hotel Spa & Residences

The restaurant is situated in the part of the hotel that faces the beach. To highlight the feature of a resort, and make full use of the surrounding beach scenery, the designers particularly created the open-air dining area. The giant windows connect the interior and the exterior. In the exterior, you could enjoy the soft breezes while in the interior, spectacular views of the sea are available. The builcing and the decoration in it are all simple and plain, in order to impress guests by the beautiful natural scenery of the beach.

库姆度假胜地

餐厅酒吧:餐厅设计在酒店面向海滩的一面,为了突出度假酒店的特色,充分利用建筑周围的海景,设计师专门设计了室外就餐区。用大尺度的开窗与室内餐厅相联系,在室外可以感受海风的吹拂,在室内也可以看到广阔的海景。建筑与装饰都简单质朴,为了让客人更好的感受度假胜地的自然美丽。

Restaurant in Almyra Hotel

The hotel has undergone an ultra-chic metamorphosis and now features sleek interior elegance inspired from the island's patron goddess of love and beauty, Aphrodite. Splashes of 1970s' boldness – such as white leather sofas and ottomans – are enhanced by a combination of natural and artificial lighting. Seeing as the hotel places as much emphasis on a successful family experience as on good design, the concept focuses on the practical as well.

雅尔蜜拉酒店餐厅

酒店室内以高雅与华贵为特色，这是从岛屿的守护神爱与美的女神阿芙罗狄蒂身上得到的灵感。20世纪70年代的大胆手法运用——例如自然光和人工灯光提升了白色真皮沙发和长脚垫的设计感。酒店如好的设计一样还很重视家居的体验，因此设计理念也很注重实效。

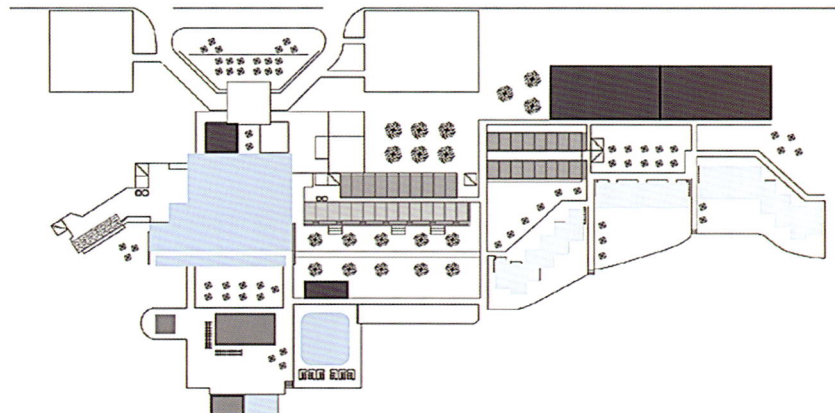

Photo: Joelle Pleot, Tristan Auer

Joelle Pleot, Tristan Auer

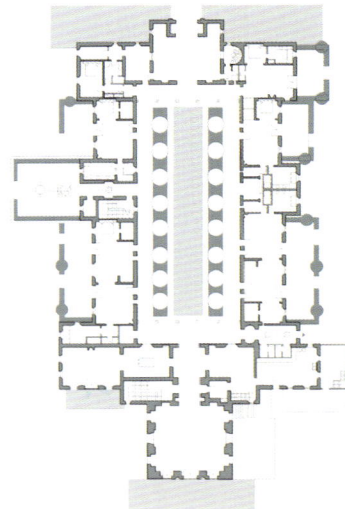

Restaurant in Ksar Char-Bagh

Ksar Char-Bagh is a palace designed in a concept of harmony, a travel through architecture inspired by the Moorish architecture of the 12th to 14th centuries revisited by a contemporary vision of space, volumes and colours, ancestral tradition enlighten by modernism. The exterior aspect as a fortress, and the sophisticated details inside; behind each enormous door, an architectural piece of work.

Ksar Char—Bagh酒店餐厅

Ksar Char—Bagh酒店是用和谐理念设计的一座宫殿，是穿越建筑史的一段旅行，受12至14世纪摩尔人式建筑启发，又以当代人对空间、体积和颜色的视角重新解读"受现代派启发的祖先传统"。外表看像一个堡垒，内部有许多复杂的细节；在每扇大门后都是一个建筑作品。

Restaurant

Morocco

Marrakech

2005

Photo: P de Grandry, M Zublena, J Silveira Ramos

Patrick & Nicole Grandsire-Levillair

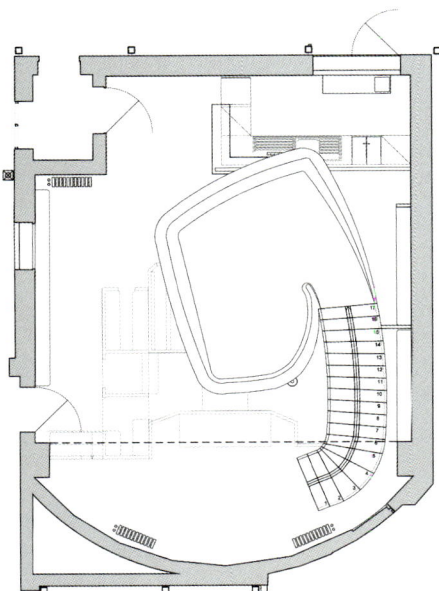

Metropole Luxury Boutique Hotel

Veranda Restaurant, the perfect setting to enjoy an outstanding eclectic dinner menu. It is an ultimate gathering place with stunning style and utterly delectable cuisine. M-Bar & Lounge, shake & stir at one of the hottest nightspots in town. A chic and intimate bar and lounge is home to the hip and young at heart. M Café, opens for breakfast, lunch and casual dining, every day of the week. The buffet lunch is widely favoured amongst the local folk.

都市精品酒店

游廊餐厅是享用经典美食的完美场所，它是一个具有时尚风格和诱人美食的最佳集会场所。M酒吧休息室是城里最炙手可热的夜间活动场所，新潮而私密的酒吧休息室是年轻人的圣地。M咖啡厅提供三餐服务，全年无休，它的自助午餐深受当地人欢迎。

Restaurant & Bar

Denmark

Farum

2010

Photo: Thorbjoern Hansen

Henning Larsen Architects

Restaurant & Bar in Scandinavian Golf Club

Scandinavian Golf Club is located at Farum, Denmark. The concept is fundamental to creating a plateau overlooking the rolling countryside. On this stone plateau placed all primary public functions such as hall, restaurant and meeting rooms in a smooth transition. Norwegian slate is used for both bars and floor and stair surfaces, while tombac used for enclosure of stones and protruding beams. The abundant use of materials in contrast to the large glass windows and the consequent use of daylight make the building a sense both heavy and light all at once.

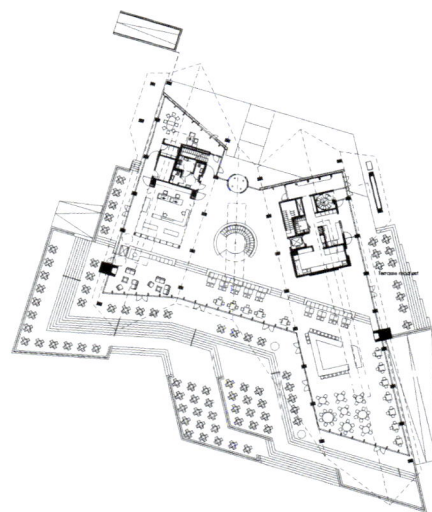

斯堪的纳维亚高尔夫俱乐部餐厅和酒吧

斯堪的纳维亚高尔夫俱乐部位于丹麦法林。其设计理念是打造一个可以俯瞰乡村美景的高原平台。俱乐部里面拥有全套的公共设施，如大厅、餐厅、会议室等。吧台、地板、楼梯广泛应用了挪威岩板，而铜锌合金则用于石材的附件和突出横梁。丰富的材料与巨大的玻璃窗和充足的日光让建筑兼具厚重感和轻盈感。

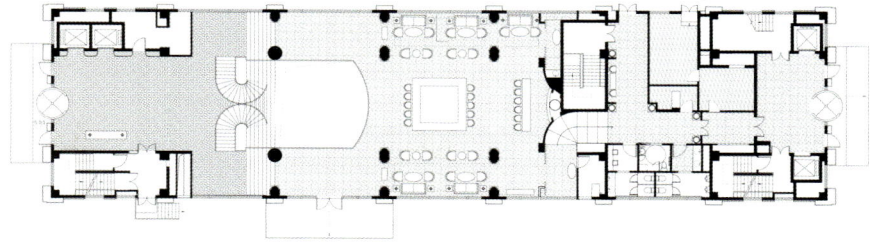

Restaurant in The First Hotel Skt Petri

Situated on a beautiful street in Copenhagen's quaint, trendy Latin Quarter, the First Hotel Skt Petri is a fine example of superior minimalist Scandinavian design – with a warm welcoming glow. Downtown Copenhagen, next to the pedestrians' streets and City Hall Square. One of Denmark's leading visual artists, Per Arnoldi, selected a tricolour scheme for both public and private spaces: bright whites contrasting with his signature cool blues and vivid reds adorn the hotel's interiors down to the smallest accessory.

SK电讯皮特尼酒店餐厅

第一SK电讯皮特尼酒店坐落在哥本哈根古怪时髦的拉丁区，一个美丽的街道上，是简约的斯堪的纳维亚设计风格的杰出代表，带着一种暖意吸引客人。餐厅酒吧：丹麦最重要的视觉艺术家之一佩尔·阿诺迪为酒店的公共区域和私人空间选择了一种三色方案：用明亮的白色与他代表性的酷蓝色形成对照以及生动的红色装饰酒店的内部，小到最小的配件。

Restaurant

Denmark

Copenhagen

2007

Photo: Jim Ellam

Julian Taylor Design Associates

Lounge & Restaurant

Sweden

Malmö

2007

Photo: Abelardo Gonzalez Architect Office AB

Abelardo Gonzalez Architect Office AB

Lounge & Restaurant Caramello

The interior is divided into zones in different colours. The entrance is orange in colour. A black street leads the visitor towards the lounge. The lounge is located in the blue zone, which has a special typology. An atrium serves as the central square. The furniture there is flush with the wall, which is covered with mirrors, emphasising the spatial density of a core, one that represents the entrance to an imaginary world and makes the presence of a spatial infinity visible. Further into the interior, where there is a connecting link to more private areas, the colour becomes green. Arriving there, the visitor finds grey zones for smoking and, in continuing on to the upper level, a white zone, with restrooms.

卡拉梅罗休息室与餐厅

室内的各个区域被划分成不同的色彩。入口处是橙色的。一条黑色的走道将来客引入位于蓝色区域的休息室。中庭是整个空间的中心广场。家具与墙壁持平，墙壁上挂满了镜子，增强了空间的密度感，让空间变得更大、更无限。内部的私人区域是绿色的，其间的灰色区域是吸烟区，楼上的白色区域则是洗手间。

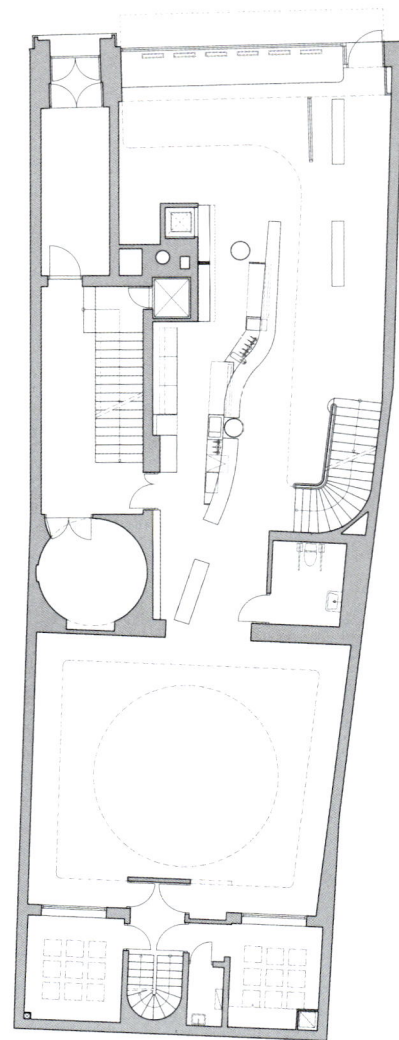

Restaurant

Sweden

Stockholm

2005

Photo: Lars Pihl, Jan Söder and Rolf Löfvenberg

Lars Pihl, Jan Söder and Rolf Löfvenberg

Restaurant in Nordic Light Hotel

Light Restaurant & Bar, serving modern, progressive Californian meals, matched by superb American wines. Complimentary breakfast buffet, 24-hour room service. The aim of the Nordic Light has been to build an arena for the senses – music and light to caress ears and eyes, food and drinks to seduce your taste buds, aesthetics and service that appeal to your soul. The Nordic Light spoils all its guests with a complimentary breakfast buffet. If you prefer breakfast in your room it is available at an extra charge.

北欧光明酒店餐厅

光明餐厅和酒吧提供现代而先进的加州美食和美式美酒。酒店提供免费早餐和24小时的客房服务。北欧光明酒店为客人提供各种感官上的感受——音乐和灯光愉悦你的耳朵和眼睛,美食和美酒诱惑你的味蕾,美感设计和服务吸引着你的灵魂。

Café

Sweden

Stockholm

2010

Photo: Stefano Barozzi

Note Design Studio

Café Foam

The owner wanted to create an interior design that people either love or hate and that nobody is indifferent to. The designers' entrance to the project became the complexity around bull fighting as a phenomenon. They were fascinated by the bull and bullfighter's mutual movements and the directions in the battle. This struggle along with the materials and colours at the stadium were the inspiration when designing the new Café Foam. The designers allowed the Spanish temperament to meet the Scandinavian coolness creating a vivid place that makes meetings easier, protects the integrity of its costumers and enhances the eating experience. A place for everybody to be. The glass lamps were designed especially for this project and hand blown in Kosta glassworks.

泡沫咖啡店

店主需要一个让人们爱憎分明、而不会默然而对的室内设计。设计师觉得斗牛中的复杂关系值得深入研究，牛与斗牛士之间的互动和战斗模式十分有趣，而材料与色彩的斗争与此相似，激发了设计师的灵感。设计师让西班牙的火热与斯堪的纳维亚的冷静共同打造了一个生动的空间，提高了顾客们的就餐水准，是一个人人喜爱的地方。餐厅的玻璃灯是由Kosta玻璃工厂手工制作的。

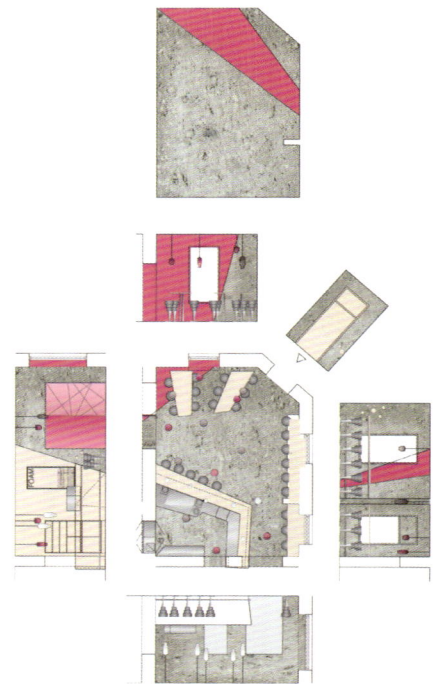

Photo: Martin Adolfsson

Solvalla

In connection with the new office there is a café for everybody involved with the horse races. It is placed with the horserace track on one side and the stables on the other. The designers' task was to create an environment that was both interesting and at the same time durable enough to endure the wear and tear from the drivers' boots. An illuminate large acrylic glass box runs throughout the room, slowly changing colours. Looking up at the café from the track you get a clear view of everything going on inside.

索尔瓦拉咖啡厅

索尔瓦拉咖啡厅紧邻新办公室，为赛马比赛的相关人员和观众提供服务。它的两侧分别是赛马跑道和马厩。设计师的任务是打造一个既趣味十足又能经得住赛马师皮靴的蹂躏。咖啡厅内的有机玻璃箱发出不断变换的灯光。从赛马跑道上可以清晰地看到咖啡厅内的情形。

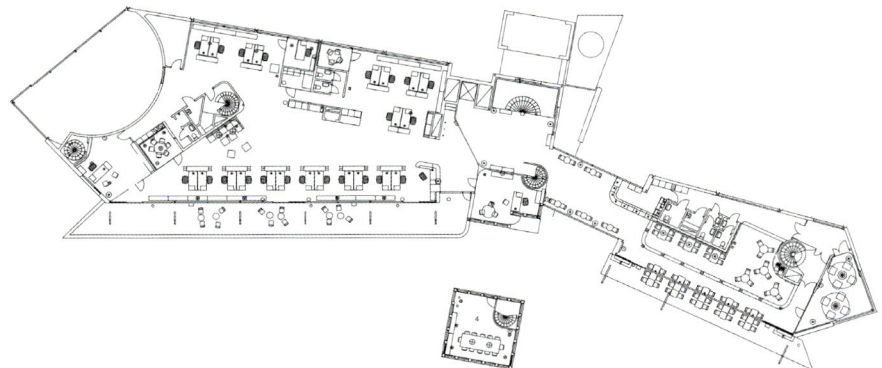

Bar

Sweden

Stockholm

2010

Photo: Jonas Wagell D&A

Jonas Wagell

Design Bar

The bar area will serve light foods and provide a surrounding where you can sit back and enjoy a drink or coffee, while the secluded VIP area will offer an undisturbed seating for meeting and chats. The conceptual theme for the Design Bar and the VIP Lounge is Forest and Industry - a tribute to raw materials, craftsmanship and refinement, which constitutes the backbone of the furniture industry. The area was divided into two parts with the bar labelled 'the industry' and the VIP area labelled 'the forest'. Furnished with Wagells' own furniture and lighting, the two spaces had large balloons suspended from the ceiling representing white clouds over the forest and black smoke for the industrial theme. Cutouts in the shape of mountains and trees were positioned on the forest side, and corresponding buildings and clouds of smoke on the other.

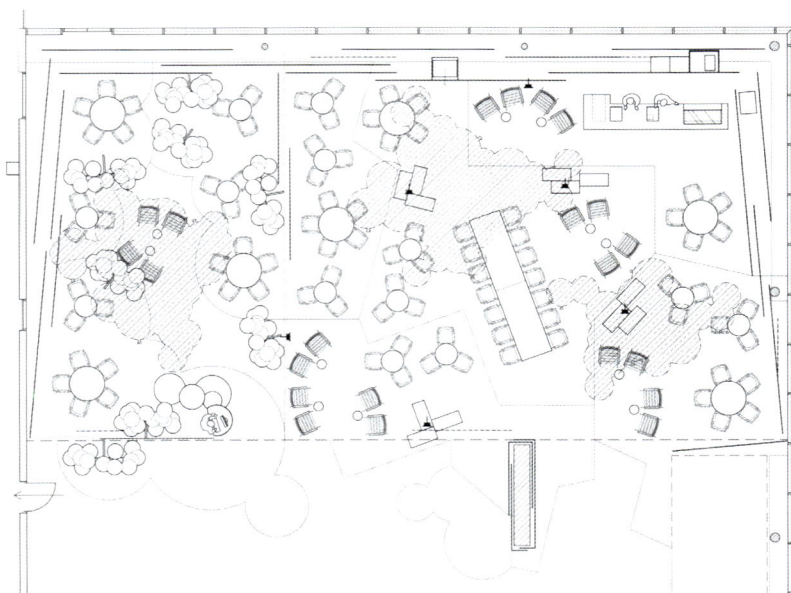

设计吧

吧台区域供应低热量食品，顾客可以环绕而坐，享受饮品或咖啡。特别划分出的贵宾区能够提供一个安静的聊天场所。设计吧和贵宾休息区的设计主题是森林和工业——注重原材料、手工艺和精工细致，正如家具工业一样。吧台区的标签是"工业"，而贵宾区则是"森林"。天花板上的巨型气球宛如空中的白云和工厂排出的黑烟。森林旁边是山脉和树木的剪影，而工业区旁则是建筑和烟云。

Pleasant Bar

As a starting point the designers worked with a fragmented fever dream, in which there was a place where the city meets an enchanted forest. The endless mirror ball ceiling in the bar area permits secret spying on everyone else. They are mirrored acrylic hemispheres of different sizes, originally intended for supermarkets and shops to survey potential shoplifters. Its new context give shy guests the opportunity to take long, undisturbed looks at potential soul mates. The wall framing the bar interior forms a graphic element as well as a bottle stand, DJ stand, and narrow one-cocktail-only tables. The mix of chairs are vintage, and coated with glossy black lacquer to become a family.

愉悦酒吧

设计师试图打造一个狂热的梦境，在城市中打造一个迷幻森林。吧台区天花板上无数个镜面球让人们可以秘密地查看他人的行动。它们由大小不一的镜面亚克力半球组成，原来被超市和商店用于防窃。这种新形式让害羞的顾客可以长时间打量自己感兴趣的人，增加配对几率。吧台既是一个造型元素，又能是酒架、DJ台和鸡尾酒吧。酒吧的座椅是混杂的古董椅，外面涂的黑色亮漆让它们形成了一个整体。

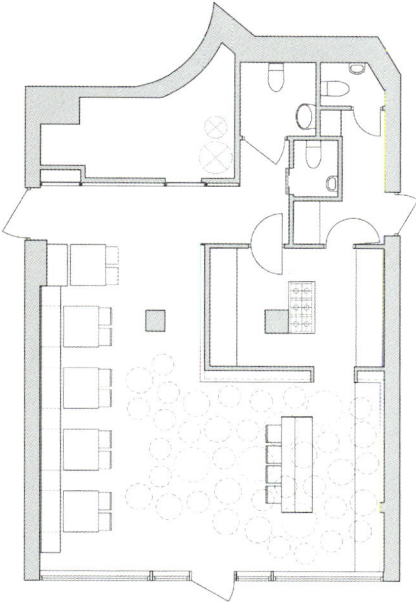

Bar

Sweden

Stockholm

2007.

Photo: Fredrik Sweger

Electric Dreams

Restaurant

Sweden

Stockholm

2008

Photo: Åke E:son Lindman

Wingårdh Arkitektkontor AB

Restaurant in Clarion Sign Hotel

The interior fittings live at a faster pace, although perhaps the tone can also be preserved here. Scandinavian classics have been selected. Some furniture has already been on the market for more than fifty years. These are classics that are comfortable and durable. On the ground floor, the 'Aquavit' restaurant meets Scandinavia with a dash of Manhattan. Globalisation is here.

克莱瑞恩酒店餐厅

在克莱瑞恩，设计师选择了斯堪的纳维亚的传统装饰风格。有些家具已经上市五十几年，但却依旧经典、舒适且耐用。一层，Aquavit餐馆在装饰上以斯堪的纳维亚风格为主，并掺杂了些许的曼哈顿风情。那么在这里，就可以感受到全球化的气息了！

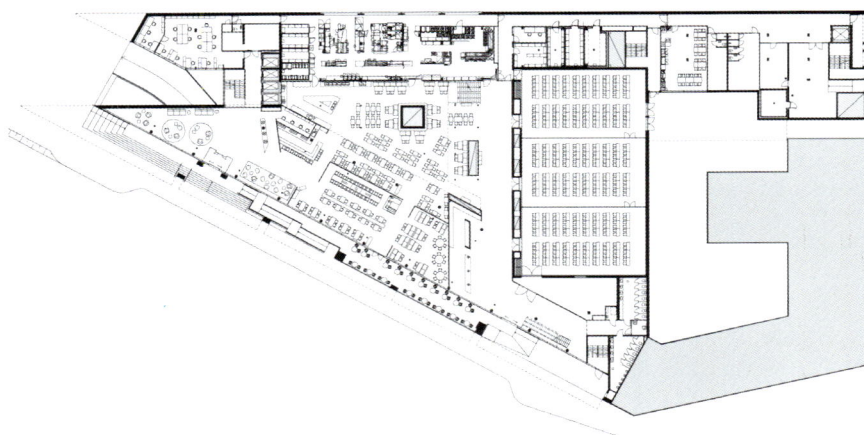

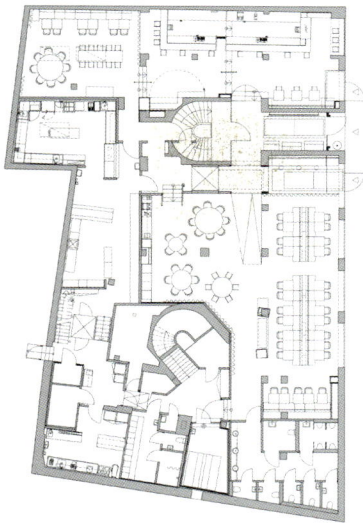

Story Hotel Restaurant

One of the main ideas has been that both the hotel and restaurant should have a timeless feel. New York pubs and bohemian hotels in Paris have been sources of inspiration. Much of the expression has also been picked up in the building's existing details. Beautifully worn walls have been saved and for example the old entry doors have been reused as bed frames. The especially made interior together with a well chosen mixture of vintage and new design gives a perfected result.

故事酒店餐厅

酒店和餐厅的设计都力求营造一种经典而永恒的感觉，其灵感来自于纽约的酒吧和巴黎的波西米亚酒店。翻新中也融合了建筑原有的细部设计：建筑优美的古旧墙壁得以保留，旧门板被用作了床的框架。室内设计融合了复古设计和新设计，达成了完美的效果。

Photo: Mikael Fjällström

Koncept Stockholm

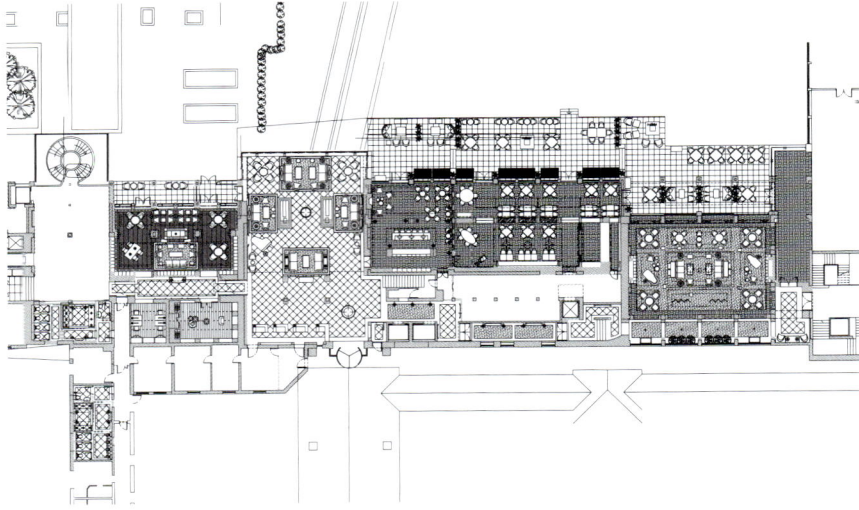

Bar

Ireland

Killarney

2008

Photo: HBA

HBA/Hirsch Bedner Associates

Crystal Bar

The Crystal Bar is clearly the hotel's 'jewel', featuring warm, elegant oak-plank flooring, engraved mirror tiles on the walls, and columns that reflect lights from the shimmering crystal chandeliers hung above the bar. The space simply glitters, spiked with eye-catching colour pairings of greys and vivid orange to create a fun and lively atmosphere. Contrasting to the glamour of the bar, the Brasserie has an airy and calm demeanour. Antiqued mirrors grace the panelled walls above putty-coloured, deep button-tufted banquettes. Relaxing guests can enjoy a leisurely conversation while chefs 'entertain' at open kitchen cook stations or mingle with other patrons at Cararra marble buffet counters.

Crystal酒吧

Crystal酒吧无疑是酒店的"瑰宝"，酒吧里面铺着温馨高雅的橡木地板，墙上铺着有刻画的镜面瓷砖，还有柱子反射着吊在吧内闪烁的水晶枝形吊灯发出的光。整个空间闪闪发光，惹眼的灰色跟艳橙色搭配更增强了这种效果，营造出写意而活泼的氛围。与酒吧的美艳形成对比的是具有轻快而宁静风格的啤酒屋。古董镜为嵌板墙壁增添了优雅色彩，下面是浅灰褐色带有深纽扣凹钉的宽阔长条软凳。

Thornton's Restaurant

Along with the clients, the designers create a layout that is functional and effortless and provide customers with a variety of chic, intimate spaces. Welcoming customers is a warm lounge and reception area. A partition wall of dark walnut and printed glass panels separates this lounge area from the main dining room. The glass is printed with a salmon-skin motif taken from a chef's photographs. The main dining room glows with warmth and is punctuated by sparkling accents. Comfortable and chic chairs in claret brushed velvet complement the deep palette. A portion of the ceiling is painted in a dark truffle hue, while the other consists of a curved Barrisol plane.

索顿餐厅

餐厅的布局集功能性和轻松感于一身，营造了时尚而亲切的氛围。进入餐厅，迎接顾客的是温馨的休息室和接待处。胡桃木板和印花玻璃将休息室与用餐区隔开，玻璃上的鲑鱼图案取自餐厅厨师拍摄的照片。温馨的用餐区装饰着间断的波纹。紫红色的天鹅绒座椅在餐厅深邃的主色调中显得舒适而别致。天花板的一部分漆成黑色，而其他部分则装饰着"巴立"的曲线型软膜板。

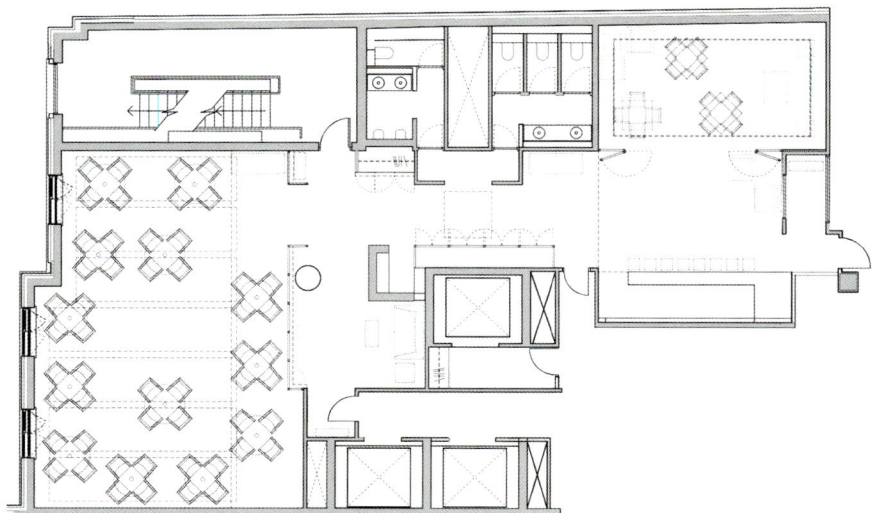

Photo: Kevin Thornton

100 Architect

Restaurant & Bar

Ireland

Belfast

2009

Photo: Tony Higgins

Project Orange

Restaurant & Bar in Fitzwilliam Hotel

The bar is deliberately moody and quite dark, tucked away behind the lobby and lined with intimate booths. A notable feature is the burnished brass coffers to the ceiling, a detail followed through in the brass front to the bar and the retro styled screens between booths. The restaurant likewise plays with ideas of openness and enclosure – the three distinct rooms being again lined with dining booths and divided by rotating oak mobiles – dynamic screens which diners can spin to open up sneaky views into the neighbouring booth, or close to provide total privacy. Each of the three rooms has a pair of oak refectory tables as the centre piece above which hang large, feature pendants made to the bespoke design.

菲兹威廉酒店餐厅和酒吧

酒吧的设计昏暗而阴沉,藏在酒店大堂后面,林立着一些私人隔间。最值得注意的特色是天花板上镶嵌的黄铜,从吧台的黄铜表面一直延伸到隔间之间的复古屏风。餐厅设计游弋于开放和封闭之间,三个独立房间与就餐隔间并列,中间是可旋转的橡木屏风,客人可以随意旋转开闭。三个房间中都有一对橡木长餐桌,上面垂下了特别定制的吊饰。

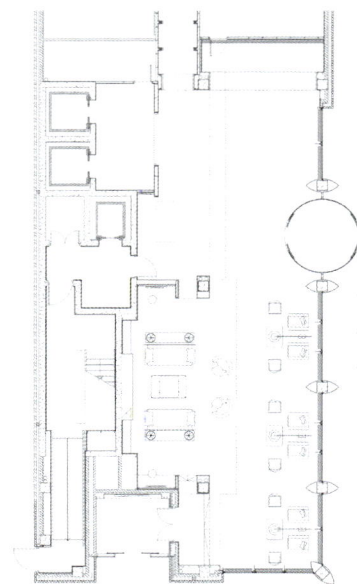

Café in Vincent Hotel

Located on Lord Street, the Vincent has the prime location in town and one of the finest in North West England. Southport is just a few hours by train from London. Six floors of understated luxury, including a boutique hotel with 60 guest studios, residences and suites, a destination café-deli, the V-Spa and gym, and the Galleria Suite for weddings and corporate events. After 7pm, the Café-Deli turns into Lord Street's top evening venue, with candles, dimmed lights, great music, a cool atmosphere, and an internationally-inspired menu. You can even dine alfresco when the weather's fine in the outdoor seating area overlooking buzzing Lord Street.

文森特酒店咖啡厅

文森特酒店坐落在勋爵街，那里是市区中的一个好位置。它也是英格兰西北部最好的酒店之一。南港离伦敦只有几个小时火车路程。晚上七点，咖啡美食厅里点亮了蜡烛、暗灯，美妙的音乐、绝妙的氛围，国际化的菜谱使之成为勋爵街上最美的一景。天气好的时候，客人还可以到户外座位区就餐，同时俯瞰繁忙的勋爵街。

Platillos Bar

Platillos is within the refurbishment of a historic square in Sheffield. The square has been developed to include many new bars, restaurants and a hotel. The square is one of the central features of a large urban renewal programme in Sheffield. Platillos is at the centre of this social scene and has become the place to be seen in. It was satisfying with Platillos that an innovative feel was established whilst having classic Spanish overtones. Another interesting recent project is The Akeman which combines a refurbished listed building with a contemporary new large extension. The new timber and slate clad building (both inside and out) crosses over to the old by a glass link block.

普拉提罗酒吧

普拉提罗酒吧坐落于英国谢菲尔德市古老的广场上，广场上还有其他很多新式酒吧和餐厅酒店。这个广场是谢菲尔德市城市重建计划的中心部分。普拉提罗酒吧这个项目给人以新颖的感觉,同时又带有西班牙风格特质，这一点是很令人感到高兴的。这个项目中新的木材和板材构成的新的建筑物横跨在原始建筑上方并在之间用玻璃连接块连接。

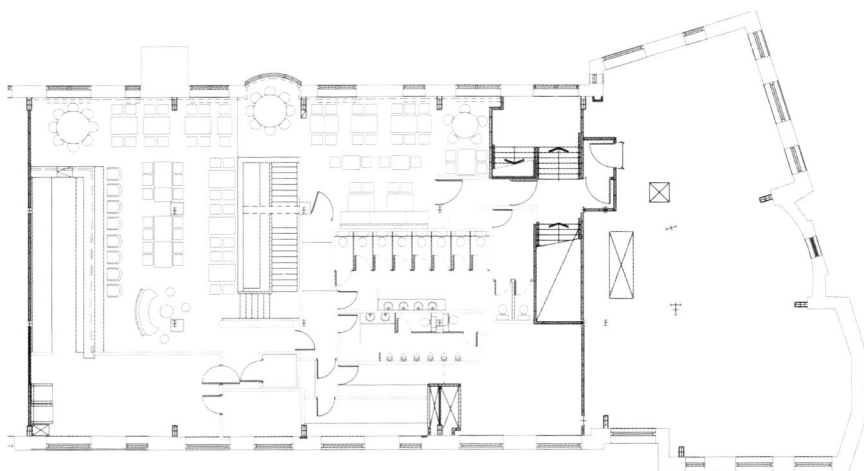

Mamasan at St Judes

Designers drew on references from Japan, China and Thailand. This is seen through the use of highly glossed black lacquered tables, black ash chairs and decorative bar front which carry a slashed motif reminiscent of traditional bamboo screens. Traditional Chinese paper lanterns add a decorative touch but they have been encased in black glass, which gives them added intrigue and diffuses the light. The main feature is a 5-metre-long photograph of a naked Geisha. She really is the spirit of Mamasan, elegant, sexy and playful. Along one wall individual booths are located. The backdrop for the booths is a collage of kitsch iconography of China.

圣朱迪斯妈妈桑酒吧

设计师借鉴了日本、中国和泰国的设计风格。酒吧内使用光亮的黑漆桌椅，入口处的装饰让人联想起传统的竹子屏风。天花板上，传统的中式纸灯外面罩着黑色的玻璃，营造出神秘而诡异的氛围。最醒目的是酒吧一面墙壁上一张5米长的裸女照片。它体现了酒吧的设计理念——优雅、性感、妙趣横生。酒吧的另一侧设有单独的座位，背景墙上是中国传统的拼贴肖像画。

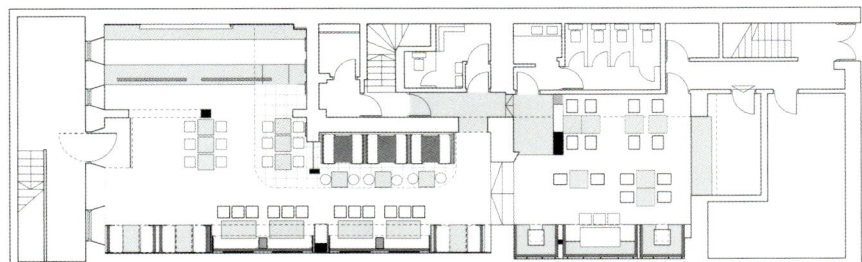

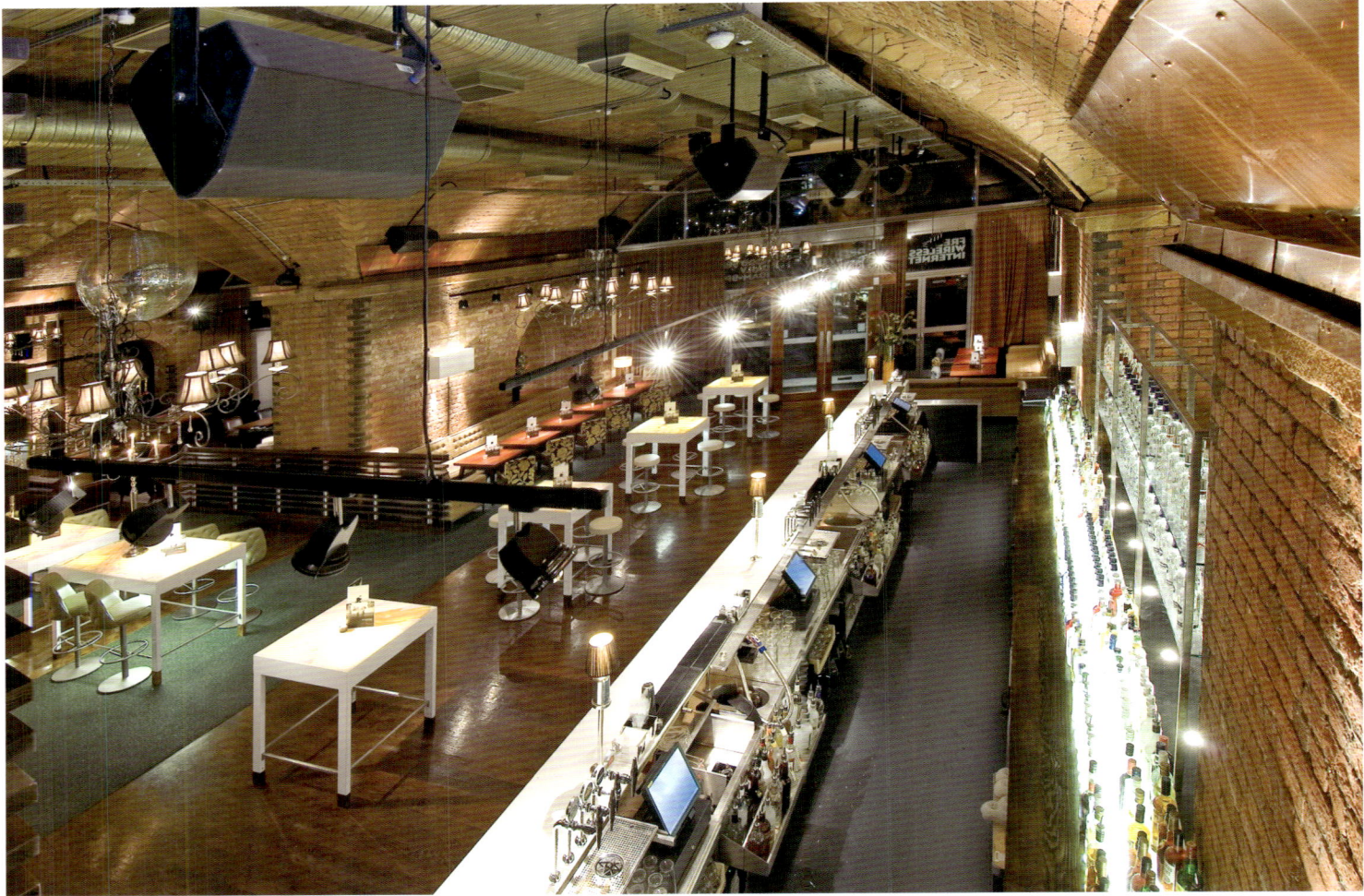

Deansgate Locks

Deansgate Locks has always been a flagship site for Revolution, having acquired the adjacent arch to the current site. Doubling its square footage, JTDA were charged with relaunching the site in July 2007. The dramatic original brickwork arches and double height volumes leant themselves to serious stylish interventions. Full height louvred walls define the space and a brilliant white bar runs the length of the main arch, faced directly opposite by a smaller sexy cocktail bar. The lower ground floor spaces form the main club rooms, working with the existing architecture.

迪安劳克酒吧

迪安劳克酒吧一直以来都是具有革命性的旗舰店，新颖的砌砖拱门和宽敞的空间赋予了酒吧现代的气息。高高的举架限定了空间的大小，白炽光的灯萦绕在拱门周围，拱门对面呈现在顾客眼前的是一间小型鸡尾酒酒吧。室内陈设散发出来的气息充满了整个空间，给人们一种永恒的、随意的、别致的惑觉。

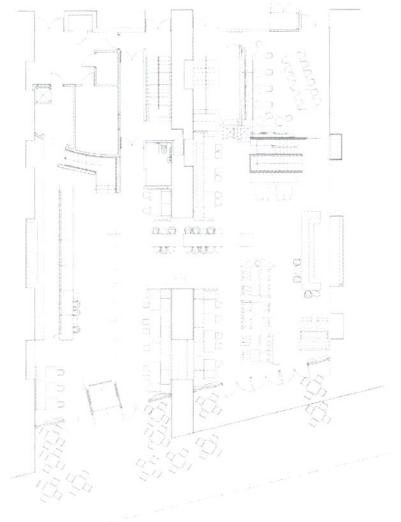

Digital Brighton

Digital made its first appearance in Newcastle Upon Tyne, and the second site fills a need on the south coast for a club that delivers on its promise of great music, great design in a great location. The site occupies six seafront arches with unrivalled views out to sea between the two piers. A striking new shopfront unifies the arches and shows off the extent of the club, allowing views from inside out across the beach and views in from the pedestrian walkway. Internally the club is made up of a main dance floor and stage, balcony bar and VIP areas, a small sexy bar in arch one and a second main dance space in the right hand arch.

布莱顿码头酒吧

显眼的店面赋予拱门一体感，使得酒吧的范围更显眼，人们不但可以看到从酒吧里向外延伸的海滩景色，也可以看到人行道上的景色。在酒吧的中心，有一个主舞台、包厢和贵宾席。同时在拱门下还有一个迷你吧，另一个主要舞会空间设在右边的拱门下。无框架玻璃的边缘在光的照射下，使得人们可以在上部的包厢区域看到所有空间最好的景色。

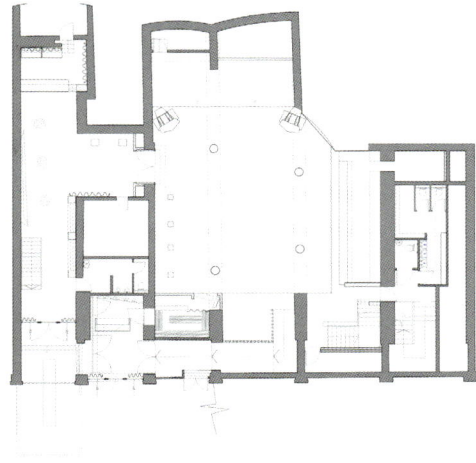

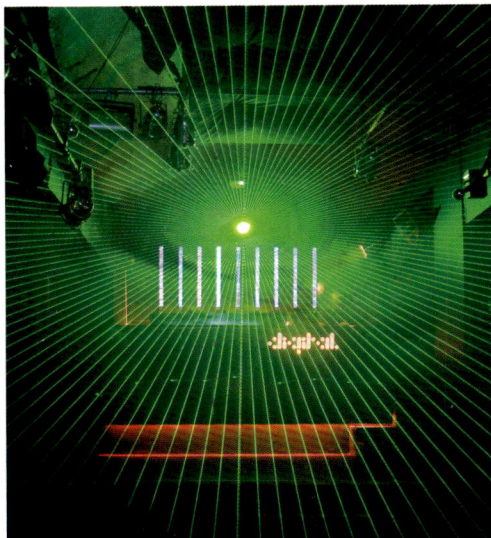

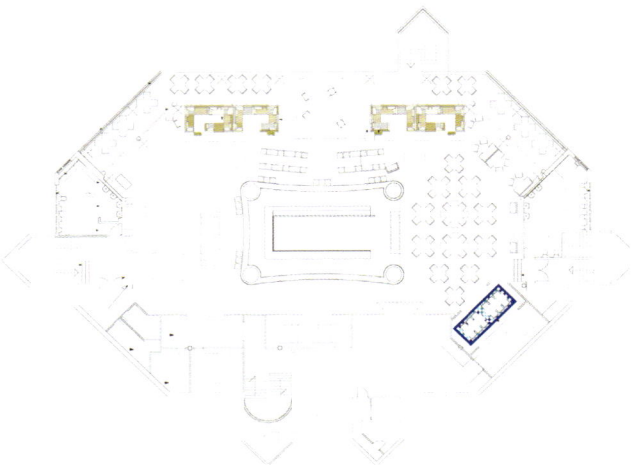

Aruba Bar & Restaurant

The venue, which spans a breathtaking 865 square metres, provides the largest single dining area on the south coast. Topped with polished zinc and overhung with wicker lanterns the bar is the centre-piece of the scheme and serves to separate the more formal dining area, a snug seating area and lounge area with day beds. The designers set a series of interventions including the 4.5-metre-high palm trees flanking the bar, sliding doors onto the terrace and outdoor showers, promoting a relaxed drinking and dining atmosphere. This combined with painted tongue and groove boarding and a mural of Poole Harbour chart on the ceiling creates a venue unique to the UK.

阿鲁巴餐厅酒吧

本项目占地865平方米，是南部海岸最大的餐厅。镀锌天花板和如柳条一样垂落下来的吊灯是酒吧中的突出亮点，它们将正式就餐区、休闲区和有坐卧两用沙发的休憩区划分开来。酒吧两侧种着4.5米高的棕榈树，阳台和户外沐浴间都装有滑动门，营造轻松的就餐氛围和体现异国情调。舌状喷绘和木质凹槽，以及天花板上刻画的普尔海港地图，这些元素在英国都是独一无二的。

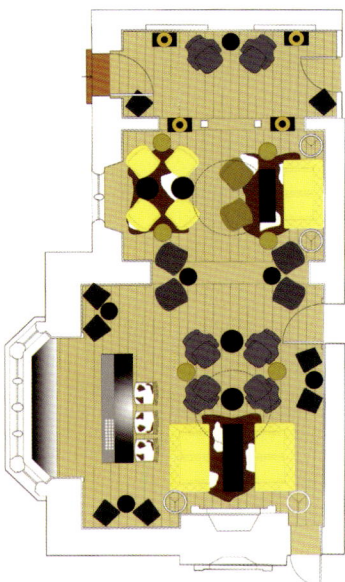

Bar in Eynsham Hall

Situated on the outskirts of the city of Oxford in extensive grounds, Eynsham Hall is a magnificent countryside mansion. Having suffered from various interim modifications while in use as a conference centre, the property is about to undergo an ambitious programme of works and once again will be transformed into a luxurious, country house hotel. The designers were invited to redesign the hotel bar in order to test design strategies and methods for modifying a historic building such as Eynsham. Originally the gunroom for the Mason family's collection of firearms, the new bar, now known as The Gunroom, fuses old with new, with a touch of art deco glamour. The bar counter is an island of decadent, black glass and polished chrome sitting in the gigantic, leaded bay window overlooking the grounds.

茵斯汉姆酒店酒吧

坐落在牛津城郊外的广阔地块上，茵斯汉姆酒店是乡村里一座很壮观的大楼。这座于1908年由Lady Mason设计的酒店重建后是一座非常棒的雅各布风格建筑。雅各布风格的建筑是20世纪之交风行的建筑。酒店的内部历经了多次改造，现在要在这里进行更大规模的改建：从会议中心再次改建成豪华的、乡村别墅型酒店。梅森家族收藏枪支的房间被用来做新的酒吧间。现在也被命名为军械库。

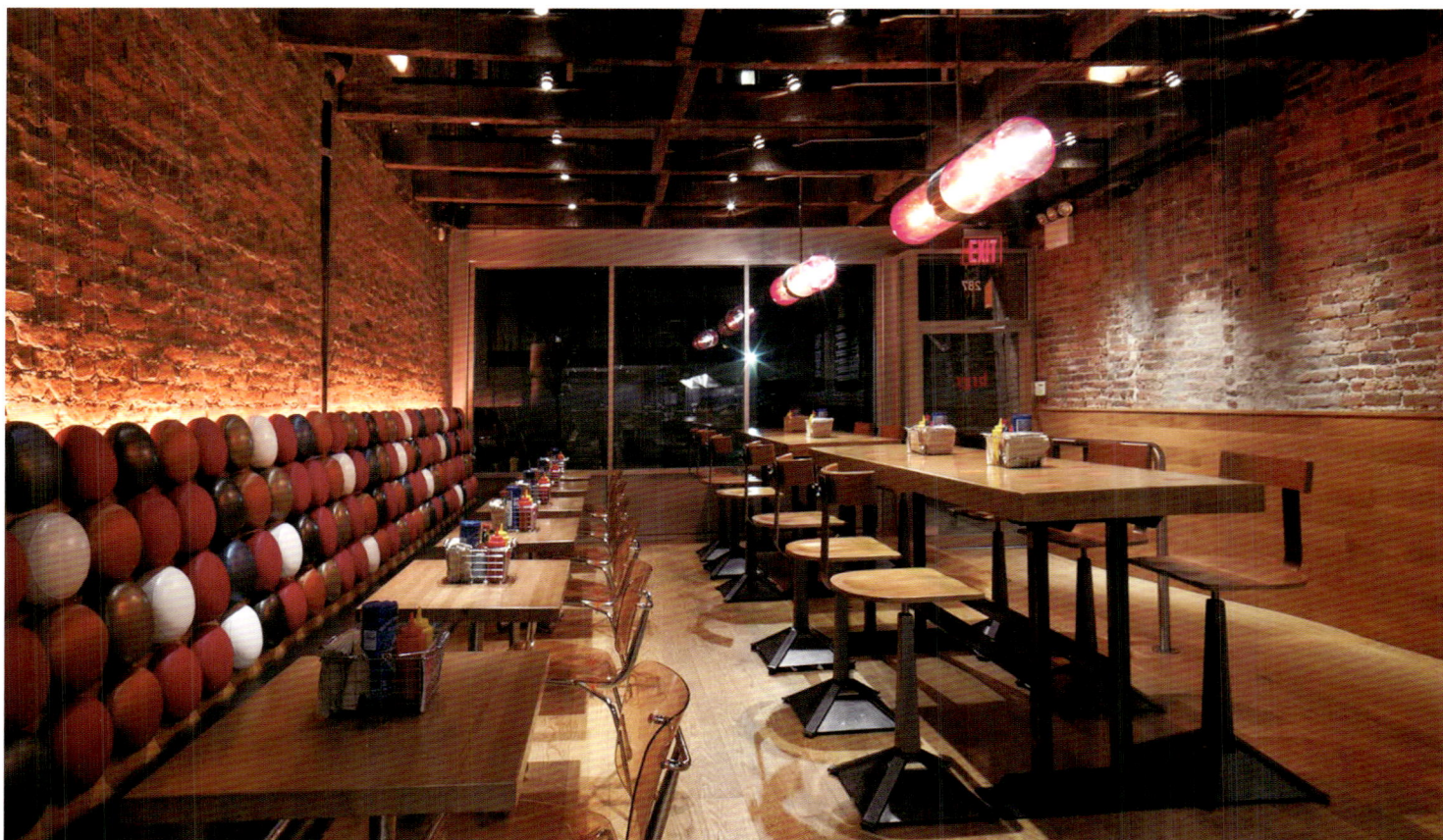

Brgr

Upon entering the restaurant, greeted by the custom designed logo, diners pick up menus and are directed to a counter where they place their orders. After ordering, diners choose to sit at either a table or a counter. Food is served in specially designed containers by wait staff in hip uniforms. The major interior elements create an environment that looks and feels like a chef's kitchen. The preparation and cooking area opens out into the front of house. A 14-foot-tall stainless steel hood suspended from the ceiling serves as a dramatic yet contextual sculptural element. Wire shelving above the dining area stores the restaurant's ingredients.

Brgr餐厅

进入餐厅就能看到特别设计的店标，顾客拿着菜单到柜台前点餐，然后选一张桌子或吧台坐下。食物会盛在特殊设计的容器中，由身着古怪制服的侍者送到顾客面前。餐厅内给人以厨房的感觉。开放式的备餐区和烹饪区向前面的就餐区敞开。天花板垂下4.3米高的不锈钢罩起衔接作用。就餐区上方设有钢丝网制成的搁架，摆放着餐厅的食材。

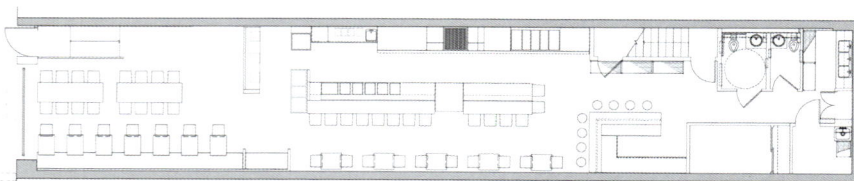

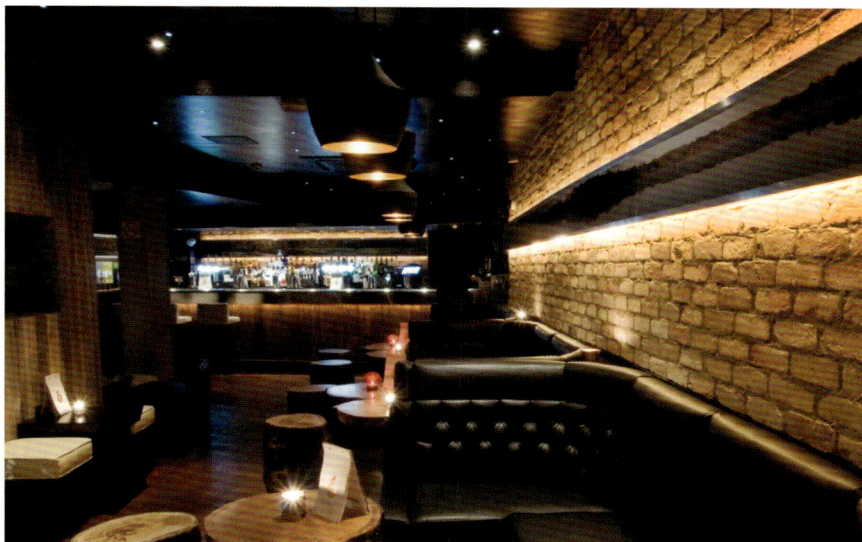

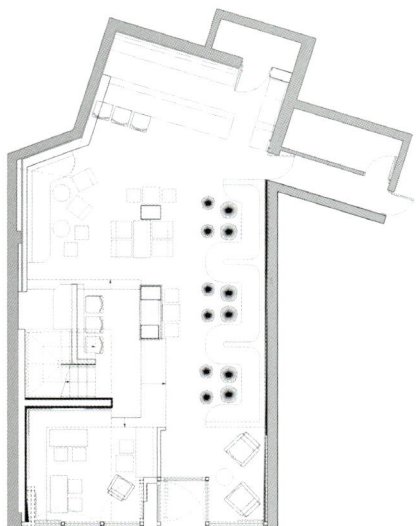

Wax Jambu

The project was to create a unique and non-pretentious environment based on principles of sustainability through longevity. The site is spread over three floors, on each of which a bar is designed. The fully glazed shop front allows views into the double height seating area, where full height voiles receive images projected onto them, which for a more relaxed day time feel, can be drawn back to reveal the eclectic mix of retro lamps, ornaments and artifacts on display. Reclaimed furniture and timbers used to reinforce the conceptual approach. Dramatic lighting effects, mixed with great food and beverage create an inviting environment to draw people in.

莲雾吧

该项目坚持以经久耐用为本，并打造出独特而又不张扬的空间环境。其共分三层，每层都设有一间酒吧。 店面通透的设计，使店内前厅视野清晰，高处的座位区均悬有投影布，可完美地呈现夜间演出影像，营造如昼日般的舒适感受。投影布还可自动收回，这时，复古灯饰、点缀饰品以及陈列的艺术品尽显眼前，令人赏心悦目。翻新的家具及横梁使设计与经久耐用的理念更近了一步。

Napa in Chiswick Moran Hotel

The restaurant goes by the name 'Napa', synonymous with great quality and laid back style. An aesthetic of cool marble, ebony laminates and crisp green leathers anticipate the fresh seasonal menu. A striking feature is a series of screens of polished stainless steel and rotating green Plexiglas ellipses affording glimpses into the resident's bar, a tucked away corner of moody smoked mirrors and cow hide upholstery.

克里斯莫兰酒店纳巴餐厅

餐厅名为纳巴（美国加利福尼亚州西部一城市，位于奥克兰以北，是纳巴山谷的中心，此山区是有名的葡萄园地区），象征着高品质和轻松的氛围。凉快的大理石板、仿乌木的层压板和亮绿色的皮革材料寓意着餐厅将提供四季的新鲜食品。一个更吸引眼球的地方是一面由多个椭圆形磨光不锈钢片及旋转的绿色有机玻璃片组合的屏风。透过屏风可以瞥见酒吧座位区—— 一个隐秘的座位区，布置了模糊的烟熏镜和牛皮座椅。

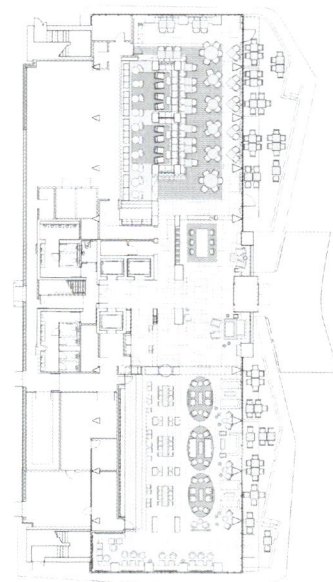

Olivo Restaurant

Simple and traditional materials here were widely used, such as solid recycled wood (as for the shop front, inner doors, bar counter, floor boards and pieces of furniture) and lime and natural pigments based plasters which entirely cover the walls, split into an upper area defined by a warm and deep shade of yellow, and an high ultramarine-blue dado, along the upper edge of which is a decorative stencilled band formed by recurrent roughly geometrical figures meant to remind the different shapes of Italian pasta.

橄榄餐厅

橄榄餐厅所选用的材料，简单传统，彰显淳朴典雅气息。店面、内门、吧台、地板、家具等皆采用普通的环保木质结构，延续简朴素雅风格；石膏墙壁运用柠檬色、黄色、青蓝色的巧妙搭配为整个空间增添出温馨之感。

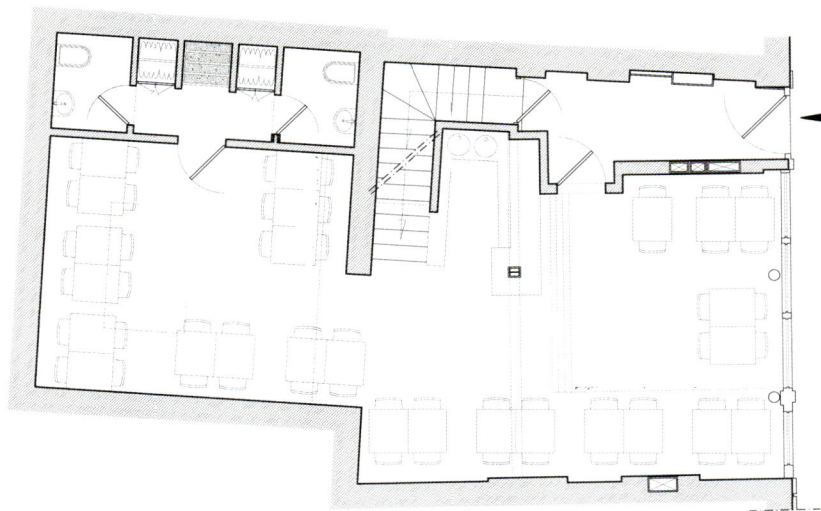

Obsidian Bar & Restaurant

Obsidian below the Aurora Hotel Manchester was conceived and built within six weeks from start to finish. JTDA worked closely with the development team to bring a slice of New York to Manchester. An all day operation the site had to cater for breakfast as well as late night dancing. Featured in the world's best bars Obsidian has gone from strength to strength in offering fine cuisine in the best of environments. 'Working with JTDA provided us with a creative and detailed concept that achieved all of our aspirations and adds a new dimension to the Manchester scene,' said Nik Basran, Authentic Food Company.

黑曜石酒吧餐厅

英国曼彻斯特极光酒店下面的黑曜石酒吧餐厅的设计构思和施工是在6周之内完成的。朱利安·泰勒设计公司与开发商密切合作把纽约酒吧风格带到了曼彻斯特。这里提供全天候服务，从早餐到深夜舞会。作为世界最好的酒吧之一，黑曜石酒吧餐厅在为顾客提供最好的食品的同时提供了最好的就餐气氛。

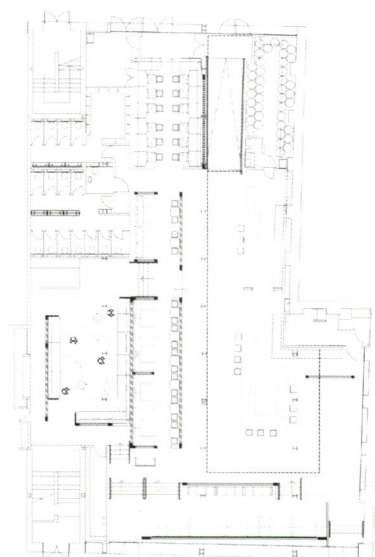

Olivomare Restaurant

Olivomare is a restaurant serving seafood. Apart from its name, such peculiarity has been highlighted through the use of more or less clear references to the sea world and environment. The most explicit among them undoubtedly are the wide wall entirely cladded with a large jigsaw-puzzle inspired by the works of the artist Maurits Escher, the linear sequence of luminescent 'tentacles' evoking a stray shoal of jellyfishes or sea anemones, the wide lozengy glazed partition reminiscent of fishers' nets and the wall cladding of the room at the rear, meant to evoke the sandy surface of the beach when moulded by the wind, all complemented by the intricate branch of a coral reef decorating the cloakroom's walls.

奥利弗梅尔餐厅

奥利弗梅尔餐厅主要供应海鲜佳品。除了它的店名之外，其或多或少的参照海洋世界进行装饰的就餐环境是又一亮点。奥利弗梅尔餐厅中最显眼的设计无疑是那面宽阔的墙壁，上面覆盖的拼图图案受到了毛里斯·埃舍尔作品的启发。发光"触角"的线性排列让人想起了离群的水母或海葵。菱形格子玻璃隔断则模仿了渔网。房间后部的墙壁有着沙滩的风情，而衣帽间墙壁上的珊瑚礁则进一步烘托了海洋的气息。

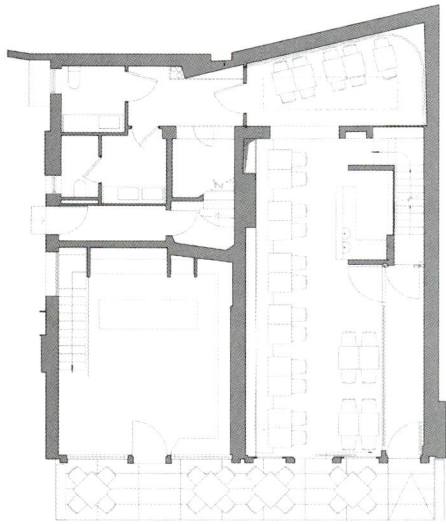

Photo: Archi. Pierluigi Piu

Archi. Pierluigi Piu

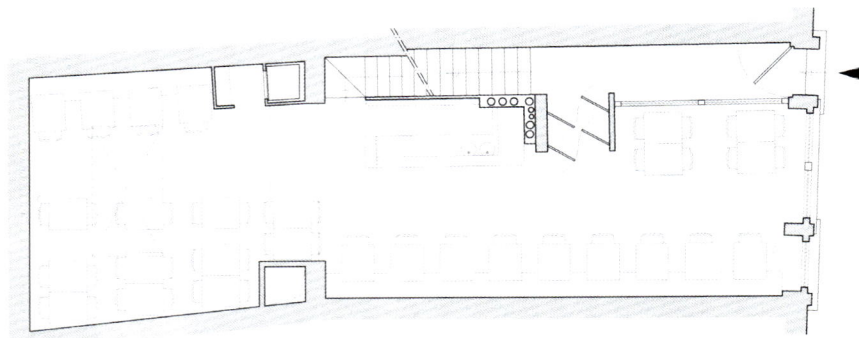

Oliveto Restaurant

The façade of the restaurant has been redesigned (to get rid of the excessively busy existing traditional wooden frame) with the aim of getting the maximum of transparency, both to view and daylight. It has been painted with a lime-green colour, which is very appealing to the trendy clients of Chelsea and Belgravia. Some made-to-measure stainless steel supports hold the shelves of the wash basins, which are covered with the same material and, in order to make the cleaning easier and give an impression of lightness.

奥利韦托餐厅

餐厅入口一改往日传统的木制框架结构，而代之以绿色透明图案设计，醒目时尚，对于前卫的切尔西和贝尔格拉维亚人士来说无疑是一个巨大诱惑。洗手间中特制的不锈钢洗手盆及支架设计，在灯光的烘托下分外夺目。

Vallorini

The interior concept was to be 'clean, white and crisp', with contemporary feature areas, moving the whole idea of an Italian café/deli away from both traditional sandwich bars and from a sentimental, rustic positioning. The flooring, tables and counter areas have a very clean look, with flooring in pale sandstone and the counter and tables in white technical stone (with chrome legs for the tables). The counter area is animated by large china bowls full of produce, giving the area a marketplace feel, with the produce all actually used and not just for display.

瓦洛里尼餐厅

"洁净、利落"是该项目的室内设计理念。结构严谨合理，突破传统的三明治酒吧设计模式。地板、桌子及吧台区依然遵循"洁净"的理念，光滑的表面在光线的作用下熠熠生辉。吧台区陈列的陶碗中盛放各种产品，风格独特。

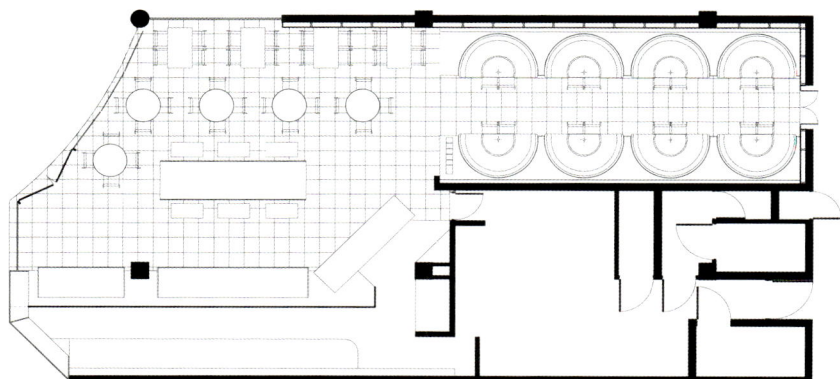

Revolution Vodka Bar

Revolution is a multiple unit Vodka bar brand. As both a drinking and dancing space it is served by three bars over the two floors. The spaces are articulated using triple height curtains with the main glazed façade controlled by half height white voiles to filter the light. All furniture is bespoke using fine hardwoods and inlaid veneers. This contrasts the raw industrial backdrop of blockwork and steel structure. The main bar is solid jarrah hardwood that seamlessly runs into the resin floor finishes. A jarrah canopy runs up behind the main bar and forms a dropped soffit in which the bar is defined.

革命伏特加酒吧

"革命"是一个多元化伏特加酒吧品牌。同时作为一种就餐喝酒和跳舞的空间，两层楼共设三个吧。室内光滑的正面结构铰接着三倍高度的窗帘和两倍高度的白色纱帘，控制正面的光线。 所有家具都是由优良硬木定制而成。酒吧间的地板是用结实的红柳桉木紧密铺成的，吧台后面的红柳桉木华盖形成拱腹的结构。

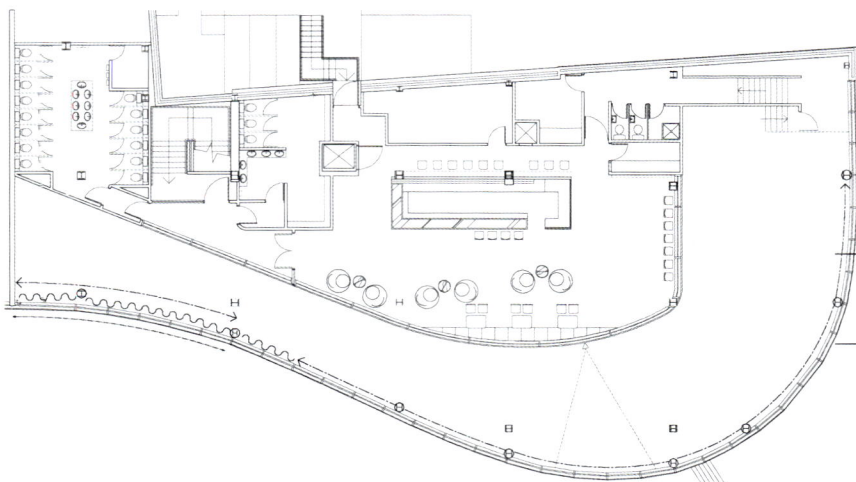

Grazia

Grazia is made up of two offers within one venue: Grazia Restaurant and Grazia Lounge. On the ground floor the restaurant offers the finest of dining experiences, while on the first floor the lounge offers the very best cocktails, late night lounging and dancing, all with fantastic views of the river. Designed with the feel of a LA villa, spaces are moulded by the sweeping curves of the structure and by the bespoke handmade chandeliers set into ceiling coffers throughout the venue. The spaces were designed as a series of rooms by the creation of a new grid formed from the arches that lead you from one space to another. Each area has its own bespoke furniture specifically designed to suit.

格拉茨餐厅

格拉茨餐厅为顾客提供两项服务，是由一个餐馆和一个休闲吧组成的。在一楼，餐厅提供了餐饮的最佳体验；而在二楼，休闲吧给人们提供了最好的鸡尾酒、深夜休闲吧以及舞蹈。同时，还可以欣赏河面上的奇幻景色。空间的设计由结构曲线所铸造，营造出一种别墅的感觉，同时手工树枝形的装饰灯安置在饰板上。

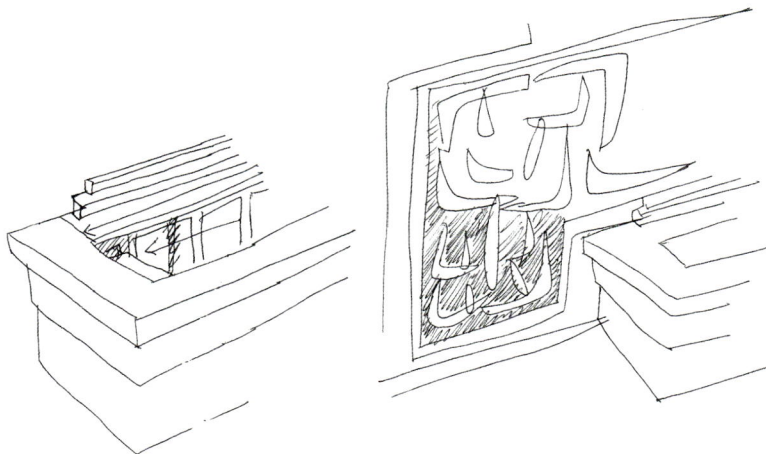

Pepe Nero Restaurant

An Italian restaurant situated within a new building, Docklands London. The design was inspired by the forms and colours of a traditional theatre auditorium. The main restaurant is a double height space with a sweeping staircase connecting to a balconied space above. Semi-circular banquette seating is formed echoing boxes at the theatre. The intention was to create a warm and dramatic environment that would create a sense of occasion, in contrast to the urban environment outside. The materials are red velvet with a bespoke oak cladding strip that is designed to go around corners. The interior style is simple and elegant, which is also rich modern.

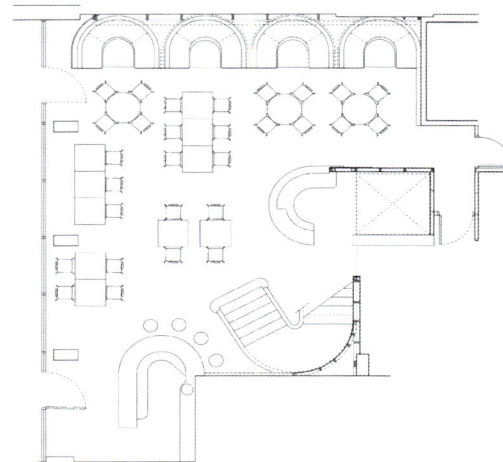

佩佩·尼罗餐厅

这是一家坐落在伦敦达克兰区一栋新建筑里的意大利餐厅，其设计受到传统剧院造型和色彩启发。主餐厅具有双层楼高，一个巨大的楼梯直通楼上的阳台。半圆形卡座与剧院里的回音箱相似。设计师试图打造一个温馨而戏剧化的环境，与外面的城市环境形成鲜明对比。设计材料选择了红色天鹅绒和定制橡木条。整个室内风格简洁优雅，极具现代感。

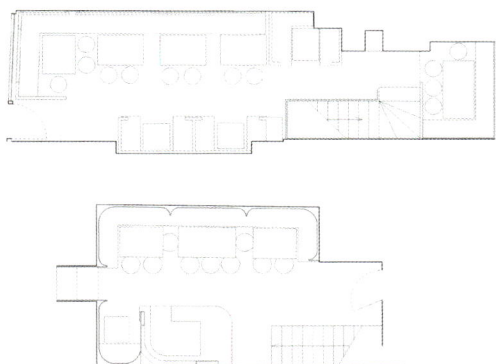

Rosa's Soho

The design is intended to bring to mind traditional British café interiors, but to also make reference to the Thai food served. The soft lighting and use of pink and red create a warm and inviting interior. High-back settles break up the ground floor, forming booths and partitions clad in modified oak ogee mouldings. The mouldings are manipulated to form coat hooks, lamps, a 'pie crust' edge to the tables, and cut away to reveal a waiter station and clock. Laser-cut brass plates, with edges echoing the oak profiles, decorate the walls, each featuring an image of a different animal from relating to dishes on the menu. The ceiling is made up from gloss pink panels in a brick pattern, set behind a deep frame.

罗莎餐厅

设计在试图模仿传统英式咖啡厅的室内设计的同时，又增添了泰国菜的元素。柔和的灯光与红粉色彩的运用打造了温馨而诱人的室内环境。高背椅在一楼形成了包间和屏风的作用，四周是弯曲的橡木模具，被用作衣架、灯架、桌沿、接待台和时钟。墙壁上的动物造型黄铜板与菜单上的菜品颇有联系。天花板由光泽粉色砖块图案面板组成，四周有深陷的框架。

Coda

Coda is situated on the upper level of the East portico of the Royal Albert Hall with unique views of the Prince Albert Memorial and Kensington Gore. The original space had been used as a café and the client wished to upgrade the service to incorporate fine dining in an appropriately upgraded interior within the heritage space. The layout was progressed using the inherent geometry and symmetry of the elliptical hall resulting in the relocation of the bar to the centre of the axis – creating a link to the central auditorium. The materials and fit out were devised to create a warm and elegant environment which related to its context.

终曲餐厅

餐厅所处位置独特，可以欣赏到艾伯特王子纪念堂及肯辛顿戈尔区的特有景致。最初是一间咖啡厅，客户决定提升室内环境及服务，将其打造成餐厅。设计师在椭圆型大厅的基础上，充分运用空间内连续的几何形状及完美的对称性，使其与中央大厅相连。材质以及设备的选择理念旨在打造温暖、典雅的环境。

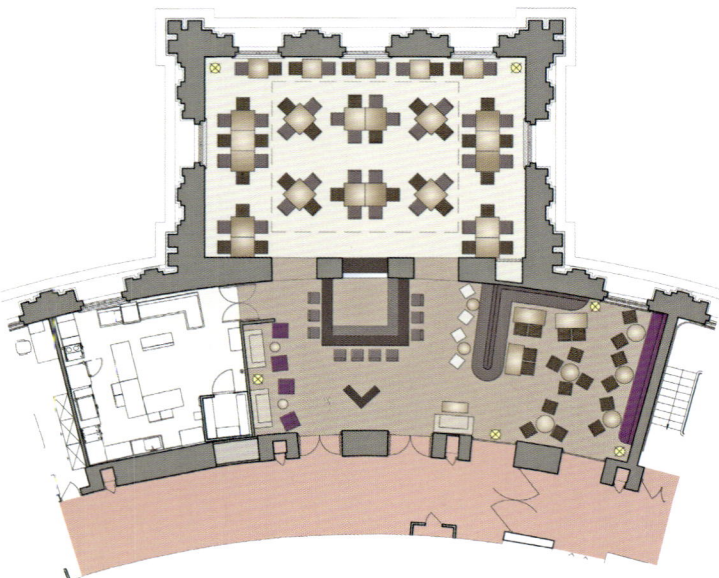

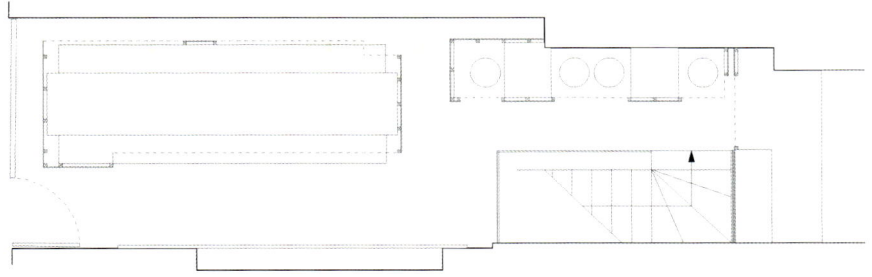

Noodles

The restaurant comprises a series of plywood booths, containing all the functions of the restaurant. Some wood elements are placed in a 'as found shop' in which all surfaces have been painted white. The glowing red interiors with arched entrances, illuminated arrows and signage are intended to acknowledge the restaurant's location in SOHO and reflect elements of the traditional (although disappearing) red light district. Diners sit within booths entered through individual arched doorways. A large booth, with sliced away sides is located towards the front of the shop occupied by a large communal table. The front of the booth is cut away to form the illuminated sign 'Noodles' which is visible through the shop window.

面条餐厅

餐厅由一系列的胶合板隔间组成，具备全套的餐厅功能。一些木制小玩意被摆放在全白的"发现式商店"里。拱门、闪光箭头和引导标示与热情洋溢的红色室内设计让人想起了正逐渐消失的红灯区。顾客们通过独立的拱门进入隔间就餐。朝向店门口的长隔间没有侧壁，可以作为公共餐桌。餐桌朝向大门口有一个发光的"面条"标识，从店外就能看到。

Photo: Gundry & Ducker

Gundryducker

Mint Leaf Restaurant

After previously completing the highly successful Mint Leaf Haymarket, it has long been the client's aim to take residency within the city of London and when the right site presented itself at 41 Lothbury the old headquarters of Nat-West bank the designers were brought back on board to head the team. The client wanted to create an exciting new bar and restaurant in the heart of the city in keeping with the design ethic of the previous Mint Leaf site while capturing a more daytime feel. Put simply, fine dinning and fine lounging.

薄荷叶餐厅

在薄荷叶赫马基特餐厅成功之后，委托人想在伦敦市开拓新市场，并最终选址在Lothbury41号，国民西敏寺银行总部原址。委托人想在市中心打造一个吸引人的新餐吧，保持薄荷叶的设计风格，并加入更多的日光元素，提供简洁、精致的餐饮和优良的休闲空间。

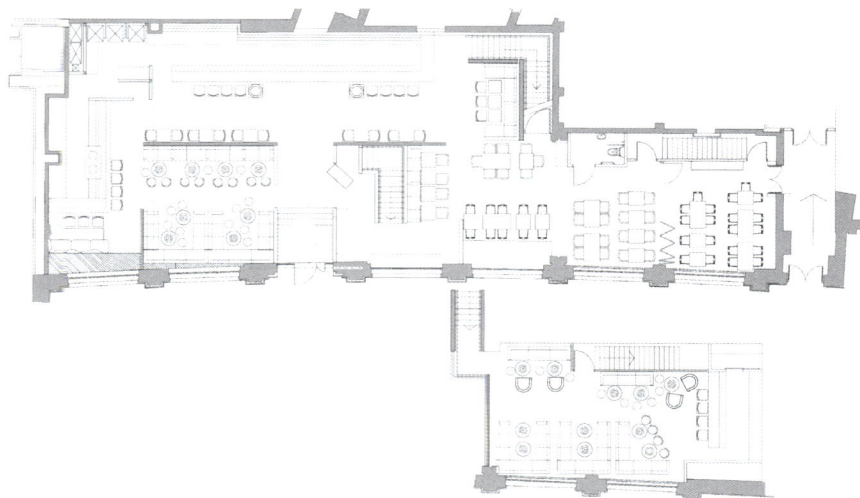

CDL Derby

JTDA were commissioned to design the public spaces at CDL Derby as part of Showcase Cinema's drive to providing a total entertainment experience that extends beyond the auditorium. The designers created designs that match the aspirations of their customers and provide them with high-class leisure facilities that better any available on the high street. The designs form a brand template both in the lobby areas and in the Directors Lounge and Studio One. The challenge within the lobby area was to create a more exciting and theatrical environment than other cinema offers and was achieved by the insertion of over sized shades, large feature frames and an illuminated tree. Soft seating and carpeted areas break up the expanse of tiled flooring and the high level velvet drapes running around the perimeter of the area add a dramatic backdrop.

德比CDL休闲吧

JTDA为电影院设计了公共空间——德比CDL休闲吧，以完善影院的全娱乐体验。设计师的设计满足了顾客的全部需求，为他们提供了商业街上最高品质的休闲设施。设计在休息大厅、导演休息室和一号工作室形成了一致的品牌模板。相比于其他电影院，这个休息大厅提供了更超凡的剧院环境，巨大的遮阳板、框架和发光树都让人感到与众不同。柔软的座椅和地毯与地砖区隔开，高档的天鹅绒窗帘打造了戏剧化的效果。

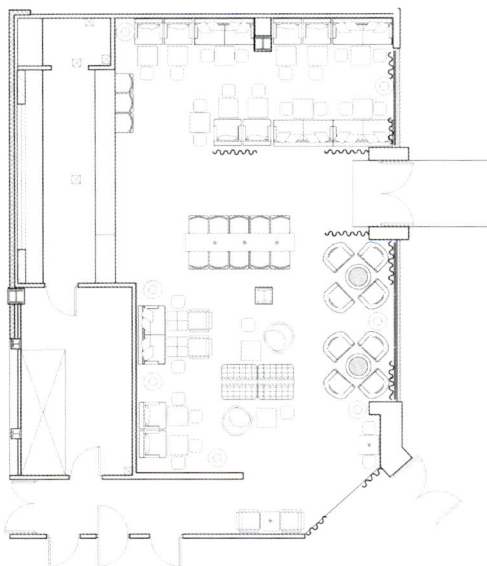

Photo: Julian Taylor Design Associates

Julian Taylor Design Associates

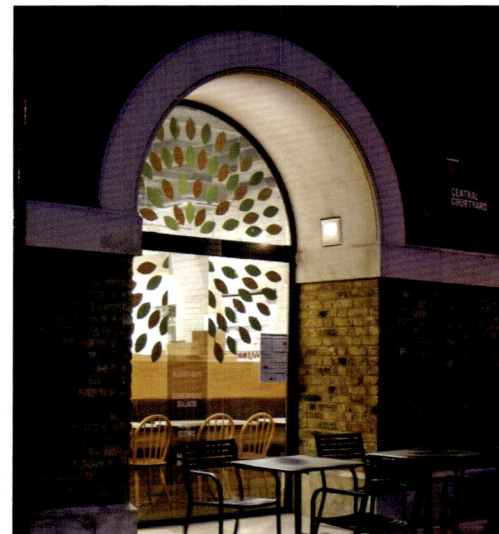

Pod Devonshire Square

This is an environmentally aware, 50-seater café in the heart of the City. Following its core business ethics, Pod on Devonshire Square will continue to provide nutritious and delicious food, freshly prepared from responsibly sourced seasonal ingredients and distributed in environmentally conscious packaging. The quality food and ethical business values are reflected in designers' simple but sophisticated design of the space. Sweet chestnut, coppiced in England, has been used for bespoke benches, seating and till units, lending a natural aesthetic. The chilled food encased within a traditional kitchen tile wall and exposed services give an industrial feel to the space and stainless steel in the customer area and the kitchen bring a contemporary edge to the restaurant.

德文郡广场豆荚咖啡厅

项目是市中心一家环保意识十足的咖啡厅，可接待50名客人。德文郡广场豆荚咖啡厅延续了品牌风格，提供美味又营养的食物，采用新鲜可靠的材料和环保食物包装。食物的质量和咖啡厅品牌价值体现在设计师简单而成熟的空间设计中。定制的长椅、座椅和家具都采用了英格兰典型的甜栗树木材，具有自然风格。冷藏食物装在传统厨房瓷砖墙的暗格上，开放式服务为空间营造出一种工业感，而餐饮区和厨房的不锈钢则为餐厅增添了现代气息。

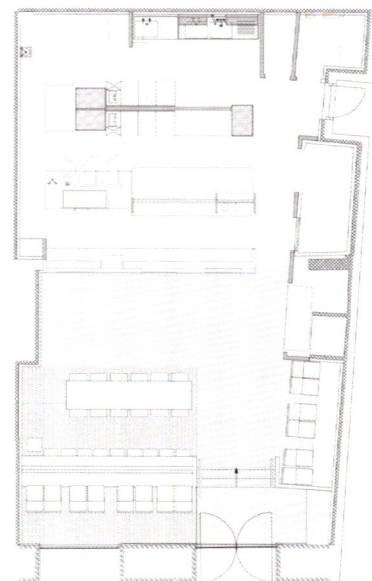

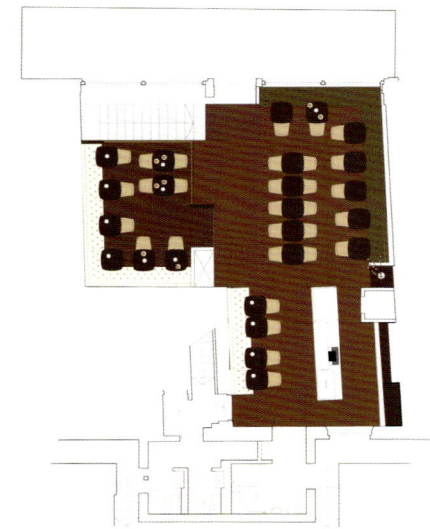

Raoul's

Raoul's is a new restaurant for Notting Hill where diners encounter a bold, new chic and enjoy an unusually modern, European sensibility. Boho meets rococo in this bijoux restaurant, located off the main Westbourne Park thoroughfare. Whilst the Raoul's name is already renown in Maida Vale as a 'slick brunch, lunch and dinner venue' the new Raoul's on Notting Hill's Talbot Road, promises all this and more, on a stretch of road traditionally bereft of glamorous eateries, because glamour is exactly what the architects have created here.

拉乌尔斯餐厅

拉乌尔斯是一家位于诺丁山的新餐厅，设计大胆而时尚，具有非同寻常的现代欧式感。这家精品餐厅位于西邦尔公园的大道上，融合了波西米亚和洛可可风格。拉乌尔斯品牌在梅达谷以"提供精致纯熟的早餐、午餐和晚餐"而闻名，诺丁山的新拉乌尔斯餐厅不仅会保持这一风格，而且还会进行改进，提供了一些不那么优雅的外带服务，因为建筑师已经将餐厅打造的极尽迷人了。

Photo: Gareth Gardner

Project Orange

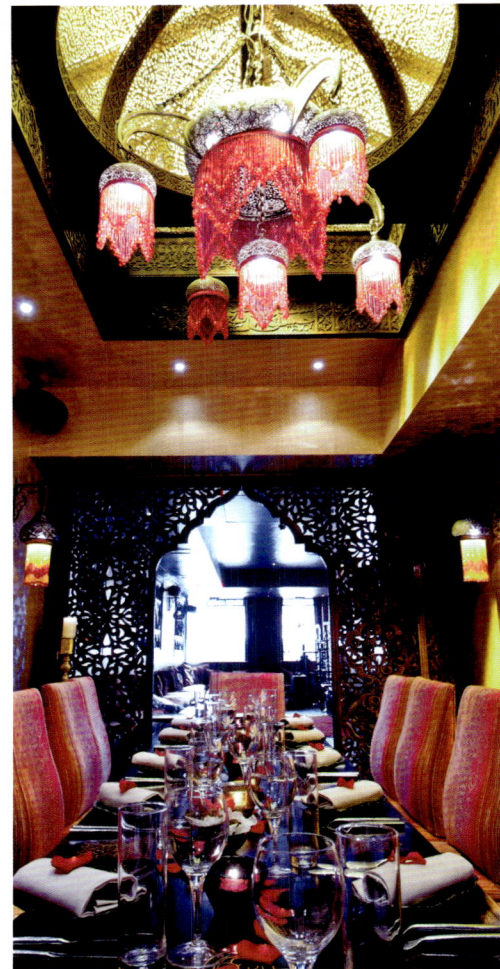

Pasha

Pasha, a Moroccan style restaurant was designed by B3 Designers and was approached by the owner, Tony Kitous to design, 'sensuous, authentic Morocco, complete with music, flowers, candles and belly dancers'. Pasha is based just off Hyde Park, London and was completed in June 2006. When designing the interior, the brief was to maximise the space, so the mezzanine was extended and the bar in the basement was repositioned to a better location, creating more space for seating and for customers to move around the restaurant freely.

帕莎餐厅

帕莎是一家摩洛哥风格餐厅，由B3设计公司设计。整体设计风格感性、具有摩洛哥风情，伴有音乐、鲜花、烛光和肚皮舞舞者。帕莎餐厅位于伦敦海德公园附近，于2006年6月完成。在室内设计过程中，设计师力求将空间最大化，扩建了阁楼，并将地下室的吧台重新设置，提供了更多的座位和活动空间。

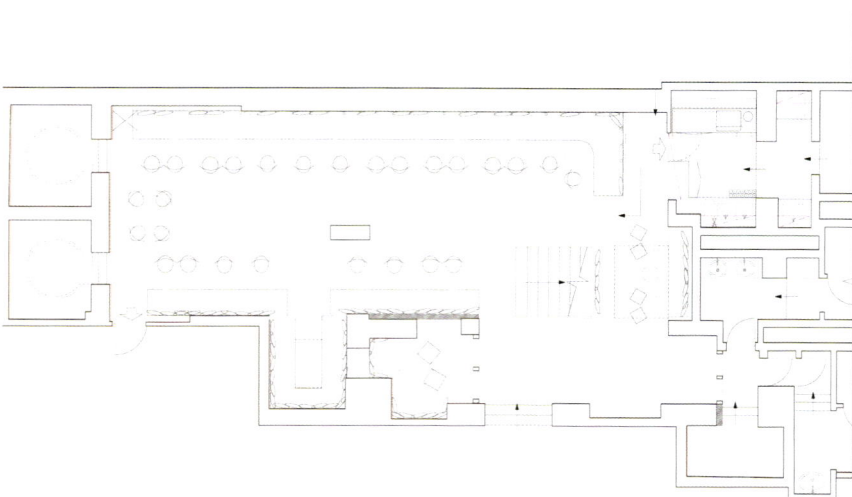

Trishna

The building where Trishna is located, dates back to the late 1800's that had some original features, such as wooden panelling which B3 Designers decided to preserve and incorporate within the new design scheme along with exposing the brick work where possible, being painted in a warm whitewash. Also helping with the interior design was the idea as to what head chef, Ravi Deulkar was going to be cooking, which was a subtle balance between modernity and authenticity, which B3 Designers translated into the design of the restaurant. The gold lighting, Alvar Aalto pendant that hangs lowly from the ceiling, completes the interior as it attracts customers with the warm glow that the lights reflect around the room.

斯纳餐厅

该建筑始建于19世纪后期，原有的木制镶板得以保留，覆以白色石灰的砖墙为室内营造出自然、亲切之感。此外，餐厅大厨拉维•达尔卡的烹饪手法崇尚新颖、真实的原则，设计师巧妙地将这一原则融入到餐厅空间的设计之中。悬垂于天花板之上的金色吊灯将柔和的灯光洒满整个空间，为客人营造宾至如归之感。地下室设有私人包间，一个大型橡木餐桌旁可供12位客人同时就餐，客人在就餐的同时可以饱览酒窖景致。

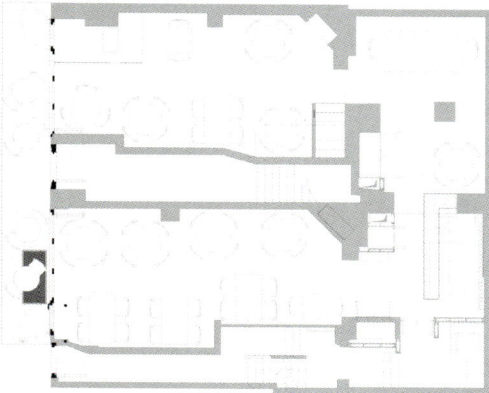

Nam

The interior design of Nam has an organic look, which combines the urban and traditional aspects of Vietnamese culture with colonial French heritage. There is large artwork on the wall, which have been directly painted onto the bare brick, which depicts scenes of Vietnamese village life. Also the shelves that hold displays are old wooden crates, combined together, giving a recycled look. The main features to this eatery are the handmade lampshades and the artwork along the walls, which have been created in very bright colours, attracting your eyes and adding a distinct character to the interior. Nam's exterior has large windows and has been painted grey, which shows off the bright colours of the interior design.

纳姆餐厅

餐馆的内部设计充分彰显了浓厚的带有法国殖民文化特色的越南传统文化与城市文化。裸露的砖墙上的图案流露出浓郁的越南乡村气息。陈旧的木质盛物柜拼接在一起，充分实现了资源的回收再利用。精致的手工灯罩和墙壁上的艺术图案，令空间五彩斑斓，吸引人们的靠近，同时与灰色的外观和大型玻璃窗形成鲜明的视觉对比。

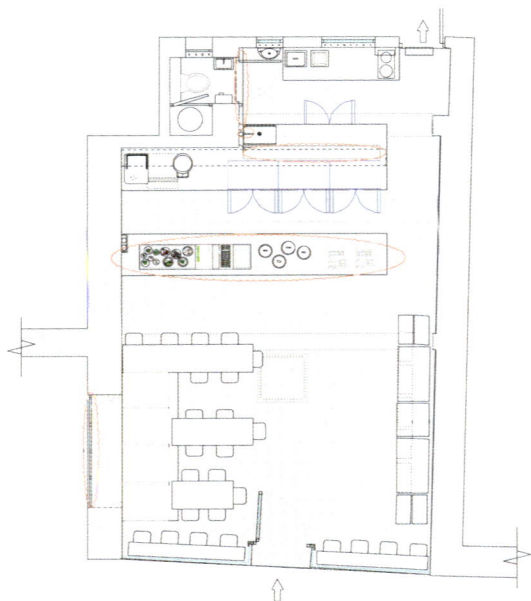

Mangiare

Mangiare is a 120 square metres upmarket Italian fast food outlet on London Wall in the City. Two internal structural columns were retained and a 5.5-metre-long Carrara marble clad communal table with purpose designed/made benches straddles these. Dark grey and bright yellow Formica and oak was used throughout to clad various internal elements. New servery counters with wany edged oak boards were inserted separating the public areas from the new kitchen with its imported Italian oven. The existing ceiling above the previous suspended ceiling tiles was left as the designers found it and painted dark grey to 'disappear'. Exposed mechanical services (ventilation, heating and cooling) were installed at high level and are carefully coordinated with the new lighting installations.

曼加尔快餐店

曼加尔是一家总面积120平方米的高级意大利快餐店，位于伦敦墙附近。设计保留了两根结构廊柱，5.5米长的卡拉拉大理石长桌搭配着特别定制的长椅。深灰色和亮黄的胶木和橡木几乎覆盖了全部的室内元素。新备餐间和边缘参差不齐的橡木板分隔了公共区域与带有意大利烤箱的厨房。设计师在原有的天花板上漆上了深灰色，隐藏了瓷砖。裸露的机械设施（如通风、供暖、制冷系统）被设在较高的位置，与新的灯光设施精心搭配在一起。

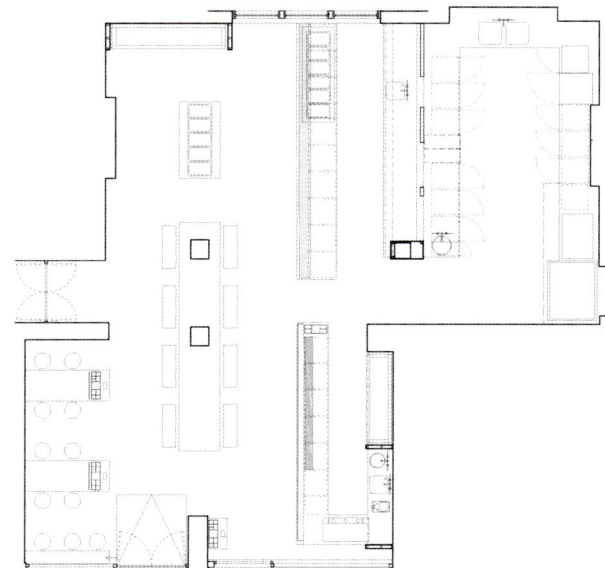

Cheltenham

Formally a Grade 2 listed church, Revolution Cheltenham comprises of a two-floor development over a ground floor and mezzanine level. The design concentrates on enmphasising the original features of this individual space and called for an original creative direction to allow a successful solution. The design therefore allows stylish contemporary additions to compliment the classic features which fuses the history and prestige of this fabulous building into a popular, intriguing venue. The ground floor is dominated by a 12-metre-long marble bar with seating both sides to create a central focus point in keeping with the architecture. Booth seating has been designed to allow groups privacy while raised areas create varying views and different spaces allowing the user to explore.

切尔滕纳姆酒吧

切尔滕纳姆酒吧原址是一个二级教堂,由两层主楼、一层基座和一个阁楼组成。设计三要强调独立空间的原创性,追求成功的解决方案。风格化现代元素与经典历史设计融合在一起,形成了一个备受欢迎的迷人空间。一楼是12米长的大理石吧台,两侧都设有座位,保证了建筑的中心对称。包间坐席为团体提供了私密的空间,高低起伏的区域为酒吧提供了不同的视角和探索空间。

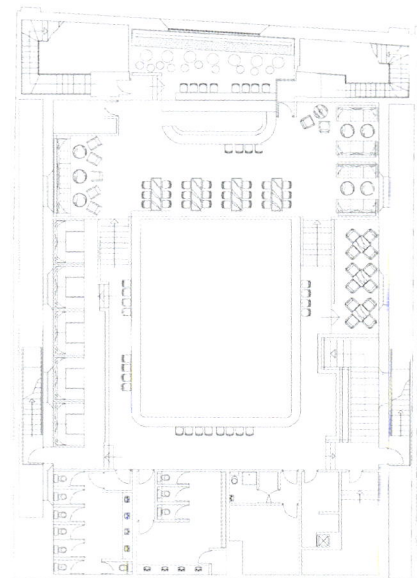

Tokyo

Tokyo is the signature site of Stereo Corp, a multiple unit operator, renowned for quality, music, service and drinks offer. The site is laid out over three floors, and the basement is an exclusive VIP bar for 40 persons, accessed only by a swipe card entry system. As the individual rooms are so small, each item was planned and down to the smallest detail. The velvet curtaining of the VIP area promotes a soft, welcoming high class feel, small mirrors with individual candles sit in front of the curtains. In the centre of the space sit hand cut log tables to add to the quirky one off nature of the site.

东京餐厅

东京餐厅是立体声集团的标志性场地，是一家多元化经营者，以品质、音乐、服务和饮品而闻名。餐厅共分为3层，地下室是可容纳40人的VIP酒吧，持卡才能进入。因为每个包间的空间有限，包房内的每个细节都经过精心设计。VIP区域的天鹅绒窗帘营造出柔和、热情的高档氛围，窗帘前是小镜子和蜡烛。空间的中心是手工制作的餐桌，为餐厅增添了奇异的氛围。

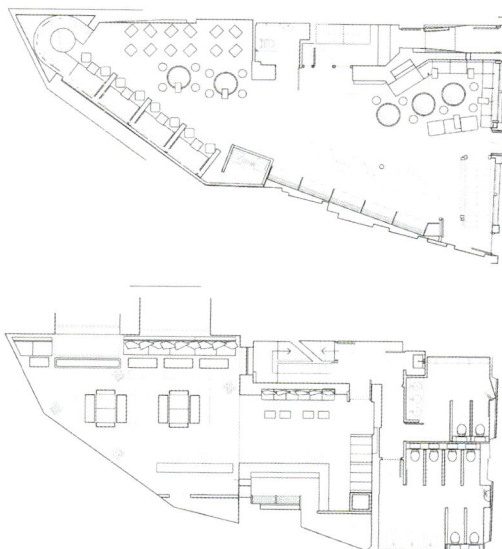

29

The restaurant is situated on the first floor and upon entering the user is initially presented with the oyster bar. Classic furniture and exposed brickwork compliment the space which uses aspects of the established design in the previously opened One Up bar with a tin panelled clad bar front and beveled mirror back bar. The floor finish throughout is a rich Jarrah adding to the elegance and warmth of the site. The main dinning space is characterised by black chandeliers inset into gold coffers with an eclectic mix of furniture adding to the sophisticated dining experience. An open kitchen and wine display add intrigue into the space keeping the user visually stimulated and with the offer concentrating on great food and beverages 29 has fast grown a reputation for style and exclusiveness attracting many celebrity guests.

29餐厅

餐厅位于二楼，顾客通过门口的牡蛎吧进入。经典家具和裸露的砖结构完善了整个空间，设计运用了原来吧台的正面锡板包面和背面斜面镜设计。地板采用红桉木，为餐厅增添了优雅和暖意。就餐空间以嵌入金色花格镶板的黑色吊灯和折中主义家具为特点，提供精致的餐饮服务。开放式厨房和酒架为空间增添了情趣。29餐厅所提供的美食和情调为它赢得了众多名人的青睐。

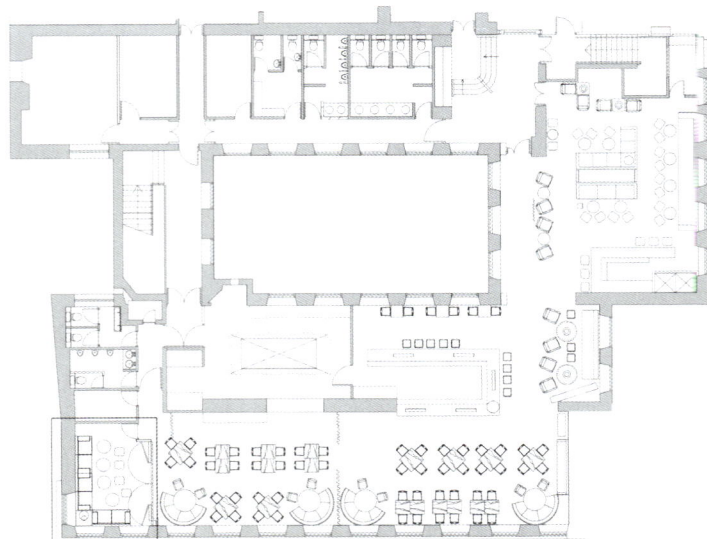

Tinderbox

Tinderbox is a 200-square-metre espresso coffee bar on two levels situated within the N1 Shopping Centre in Islington and replaces the old Tinderbox on Upper Street. At ground floor level is a takeaway kiosk while at first floor is the main coffee shop. The building works involved the complete fit-out of a 3.6- metre-high 'shell & core' space that was originally intended as a crèche for the N1 Centre. Linking the kiosk to the main level is a staircase that was transformed from a dark and claustrophobic escape stair by removing walls, cladding the walls with stair ed timber strips and new lighting.

打火匣咖啡吧

打火匣是一家总面积200平方米的咖啡吧，共分两层，坐落在伊丝灵顿N1购物中心，替代了旧打火匣。一楼是外卖亭，二楼是主咖啡厅。项目的设计工作包含全套的3.6米高的空间，其原来是N1购物中心的托儿所。外卖亭和主咖啡厅之间由楼梯相连。设计师移除了昏暗、幽闭的防火楼梯的墙壁，在上面增添了条纹和新灯饰。

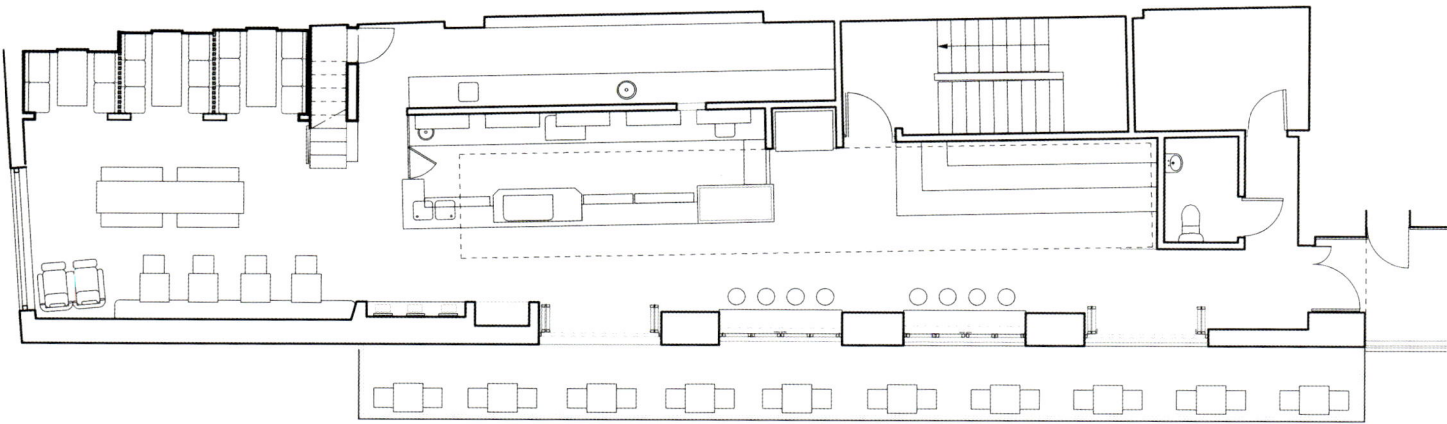

Cafe
UK
London
2009
Photo: Richard Dean
Jonathan Clark Architects

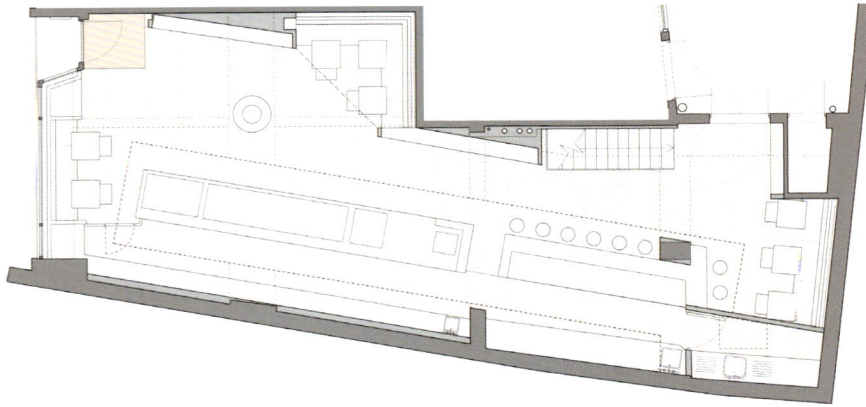

William Curley

This is a new shop and dessert bar for renowned chocolatiers William & Suzue Curley. They already have a small shop in Richmond which has received numerous accolades and reviews in the national press. This project involved the complete strip-out and subsequent fit-out of a 90-square-metre ground floor retail space and 60-square-metre basement. At ground floor level is a 7-metre-long chilled cabinet for displaying fine patisserie, chocolates and ice cream in a temperature controlled environment. Adjacent to this is a dessert bar with seating for 7. There are also three separate areas of banquette seating to give another 16 covers.

威廉·科里甜点店

这是一家为知名巧克力制造商威廉和苏祖·科里而设计的甜品店。此前，他们在里德满的小店在国内媒体上赢得了广泛好评。这个项目包括90平方米零售空间和60平方米地下室的翻新和装修工程。一楼有一个7米长的冷藏柜，用来储藏法式糕点、巧克力和冰激凌。其旁边是一个可容纳7人的甜点吧。另外的3个独立区域提供了16人的卡座位置。

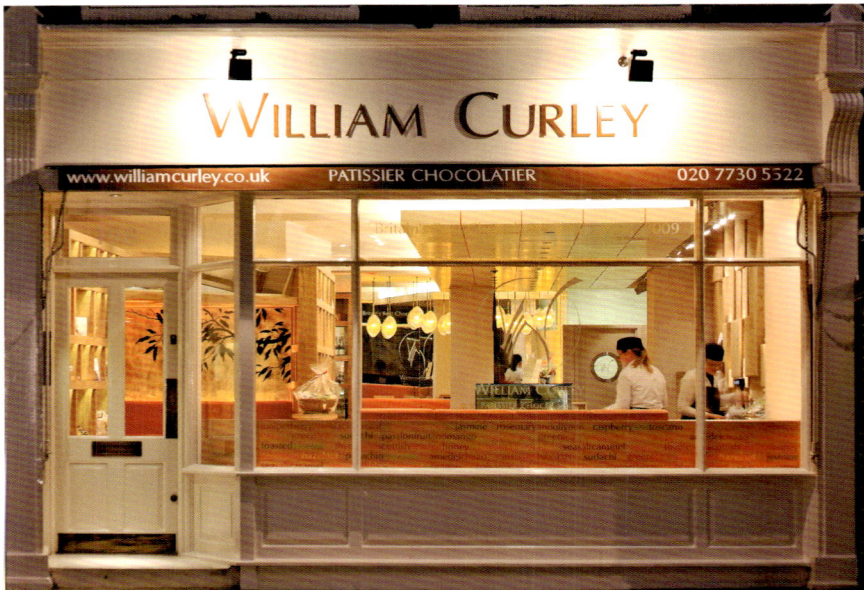

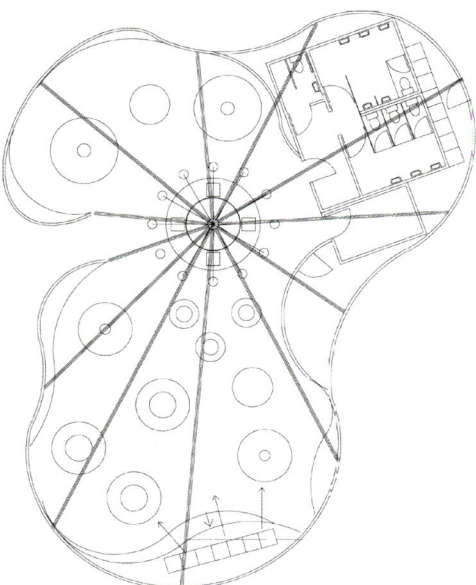

Bailissimo Travelling Bar

The focus of the design proposition has been to help in opening up the drinking occasionality, and communicating the concept of 'infectious playfulness' for Baileys. The designer's approach has been to create a series of interlinking experiences of differing scale – each designed to bring to life a particular drink mix. This is achieved through creating a series of 'poured' forms flowing from the ceiling, lit and dynamic, each terminating in a series of different configurations, encouraging the visitor to sample in a context that supports the drinking occasion. For example one poured form terminates in a lounging configuration, encouraging people to kick back – one form becomes a vertical drinks table encouraging groups to congregate. Others encourage conviviality by seating people together around a notional 'hearth'.

百利斯莫旅行酒吧

设计定位在帮助酒吧推广酒文化和百利酒的感染力。设计师打造了一系列不同规模的联动体验，每个都能为生活带来独特的酒类混合体验。天花板上"倾泻而下"的造型结构、灯光和动感的氛围都吸引着顾客，鼓励他们进行不同的饮酒体验。例如，一个造型垂直酒桌的结构就模仿了休息室，鼓励团体聚会。而另一个结构则模仿了壁炉，让人们围坐着进行欢宴。

327

Whitechapel Dining Room

The approach to the design of the Dining Rooms has been to try and forge a synthesis between the history and character of the original Arts and Crafts building, and the contemporary cutting edge of the exhibited works. In contrast to the expansive gallery spaces, these public areas are intimate and cosy characterised by their timber panelling, pendant lighting and leather upholstery; with reclaimed library units contrasted with modern detailing, fixtures and fittings. The effect is to create a timeless space that seems at once modern and traditional, and where materials will change and improve with age and use. The gentrification of east London gallops onward.

白教堂餐厅

餐厅的设计试图融合艺术工艺建筑的历史个性和现代先锋作品。与普通画廊不同，木质面板、吊灯和皮质家具让餐厅的公共区域私密而舒适。图书馆旧家具与现代化细部、装置和家具形成了鲜明对比。这是一个既现代又传统的经典空间，材料的运用会随着时代和用途的变迁而变迁。

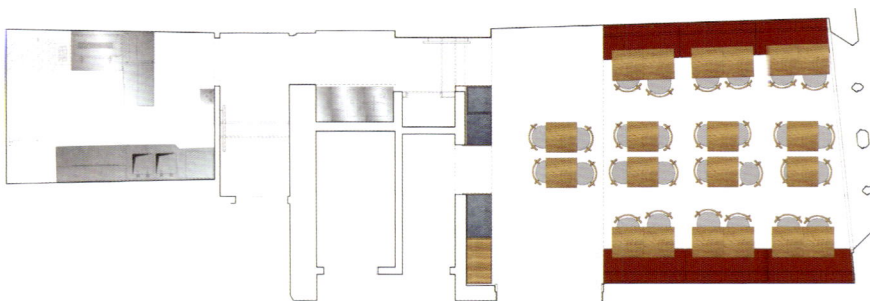

Jamie's Italian Restaurant

The restaurant exterior has a great L-shaped space (approx 100-square-metre in total) on Westfield's Southern Terrace, which offers al fresco dining beneath a canopy. The restaurant's interior is very long (55-metre from the main door to the door by the antipasti area towards the rear of the restaurant), with a particularly narrow central area, which wasn't ideal for seating. This gave rise to Blacksheep's ideas for the optimum space-plan and for zoned areas, from the main restaurant space 'Piazza' to a central 'Market Place' area, without seating, which made a virtue of the narrow central space and where customers can see views of the kitchens and fresh pasta making or can linger over a bigger retail area than ever before, a route to either the toilets or else to the restaurant's rear dining space, The Back Room, a special, more intimate area with lower ceilings and a slightly more 'bling' treatment.

杰米意大利餐厅

餐厅的外部是L形造型（总面积约为100平方米），位于西田区的南平台，在一个华盖之下提供露天餐饮体验。餐厅的室内空间非常狭长（从大门到露天区的门长55米），中心狭窄的区域不适合就坐。设计师借此进行了与众不同的空间规划和区域设计，中心"市场区"没有座椅，顾客可以在狭窄的空间里看到厨房和意大利面的制作过程，在零售区买商品，由此走向洗手间或餐厅后部的其他就餐区。餐厅后部是一个独特的私密包间，天花板较低，但是装修略显华贵。

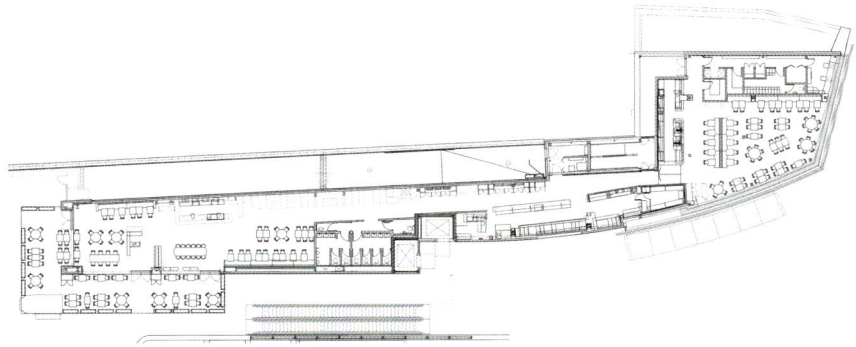

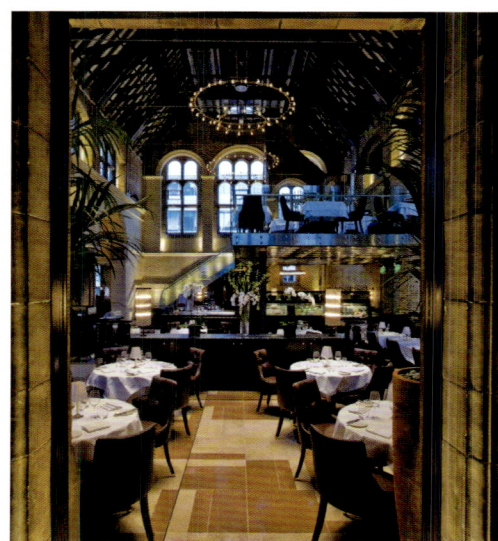

Galvin La Chapelle

Galvin La Chapelle would serve classic Galvin Bistrot food, whilst the all day café du Luxe and aperitivo bar would reflect a slightly more contemporary feel housing the iconic Great Eastern Bar and serve great food from their wood fire oven. There was minimal natural back of house space and so the biggest challenge was to place in a substantial kitchen and allocated space in high visibility areas; other key challenges consisted of how to heat, cool and ventilate without the mechanical design taking over the overall space aesthetically, linking the café de luxe and la Chapelle whilst having their own entrance also needed to appear seamless so the whole site appeared as a cohesive flowing space. Lighting the hall also created many challenges to bring out the natural beauty of the hall and achieve understated elegance.

加尔文·拉夏贝尔餐厅

加尔文·拉夏贝尔餐厅提供加尔文招牌美食，全天候高级咖啡和餐前小食吧反映了典型东部吧的现代感，柴火烤箱制作的美食令人流连忘返。设计最大的挑战是在可见区域打造一个设备齐全的厨房和配菜空间。其他的挑战包括如何在保证空间美观的前提下，进行供暖、制冷和通风等机械设计。餐厅和咖啡厅之间的连接需要十分紧密，使其成为一个整体。

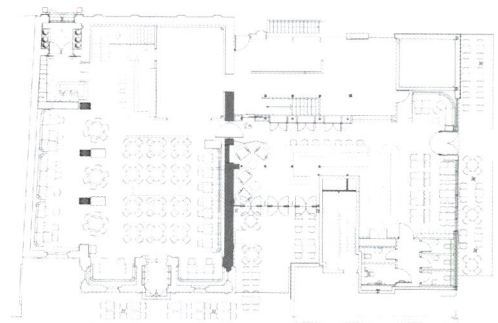

Carbon Bar

The Carbon Bar is a welcome addition to the Cumberland Hotel featuring an interior inspired by industrial architecture and brings low-key, yet chic Shoreditch style to the West End. The interior of Carbon is a fusion of concrete, brick, steel, mesh and leather; contrasting against the inviting, outsised Chesterfields, beveled mirrors and sketches of 21st century industrial living that cover the walls. Elegantly architectural, Carbon has been designed to maximise space, privacy and the ability to be seen all at once. The venue contains a large 14-metre bar, as well as a mezzanine champagne bar which hangs suspended over the lower ground floor. To access the mezzanine, guests climb the stairs adjacent to a two-storey champagne wall, filled with some of Taittinger's most expensive and rare bottles.

碳元素酒吧

碳元素酒吧位于坎伯兰酒店中,具有典型的工业特征,为伦敦西区带来了低调而时尚的东伦敦新兴文艺区风格。其室内设计融合了混凝土、砖石、钢铁、网格和皮革,与舒适的长椅、斜面镜和21世纪工业生活墙纸形成了鲜明对比。碳元素的设计力求将空间和私密感最大化,由一个14米长的大吧台和一个阁楼香槟吧组成。客人需要爬上两层楼高的香槟墙(收藏着泰廷爵香槟最珍贵的酒瓶)到达阁楼。

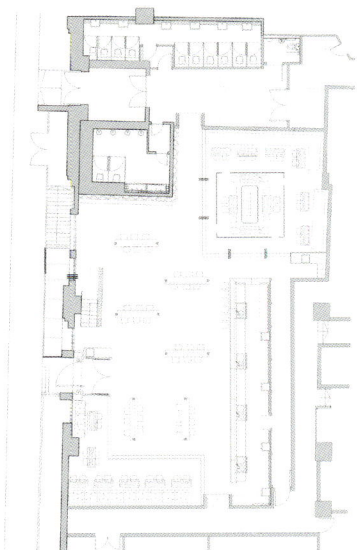

Label

Label was conceived as an all day and all night operation, functioning as a bar and restaurant on the ground floor whilst transforming into a full on night club on the lower ground floor at night, in which mirrors surround the dance floor as a place to see and be seen. Comfy and cool it offers quality service, drinks and food yet is refreshingly attitude-free. The all day offer dictated that the ground floor maintain its light and airy feel, the light surfaces and materials are designed to reflect light and open up the space. Within that space are a series of staggered louvred screens that capture views through the bar and the frame vistas. The natural materials contrast the clean white lines of the ceiling and the walls giving it a very contemporary feel.

标签餐吧

标签餐吧全天候营业，一楼是酒吧和餐厅，晚上下层则变身为夜店。舞池周围的镜子让周遭的一切都清晰可见。标签餐吧提供高品质的服务、饮品和食物，是个舒适放松的空间。一楼的全日制餐吧具有清新轻快的氛围，清新的装饰反映了空间的特点，让空间看起来更开阔。错列的百叶屏风后面是酒吧和窗外美丽的风景。自然材料与天花板和墙壁上干净的白色线条共同打造了舒适的感觉。

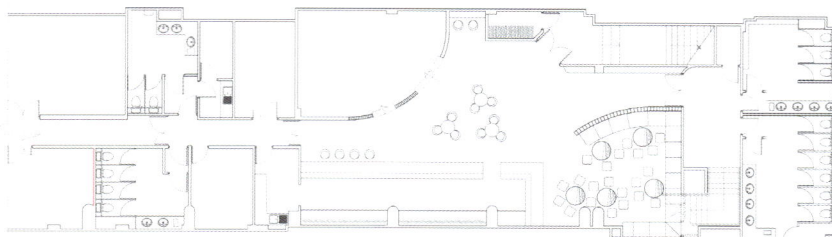

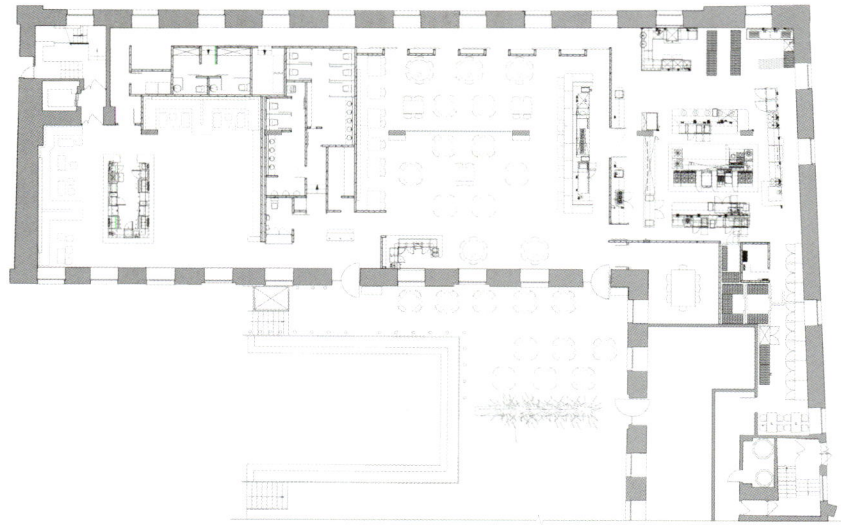

Anise

Anise Bar is attached to Cinnamon Kitchen, which is based in Devonshire Square, London that serves an extensive choice of cocktails. Like Cinnamon Kitchen, Anise's interior design was inspired by the Indian culture of geometrical and floral patterns, which is incorporated amongst the design, such as the floor pattern and the silver-plated perforated handmade lamps that are hung in the window alcoves. Surrounding most of the interior space in Anise is deep banquette seats along with mirrors that are on a slight angle that reflects light from the candles, which are spread around the interior, giving a romantic atmosphere from the flickering light. Hanging from the centre of the bar are bottles and glasses that are on display, making it an attractive feature and visible throughout the bar. Also the bar glows as it is lit up, making it stand out, attracting people to the feature to buy more drinks.

茴香酒吧

茴香酒吧附属于肉桂小厨，位于伦敦德文郡广场，供应种类繁多的鸡尾酒。与肉桂小厨一样，茴香酒吧的室内设计采用了印度几何和花朵图案，这些图案体现在地板和窗格的手工艺灯具上。室内空间的四周是卡座和镜子，微倾的镜子可以反射遍布室内的烛光，营造了闪烁的浪漫氛围。吧台正上方吊着玻璃展示架摆放着酒瓶，是个吸引人的特征。吧台仿佛被点亮了一般发着光，十分显眼，吸引着人们来购买更多的酒。

Babel

Babel Bar needed an update from its existing form, so Faucet Inn contacted B3 Designers to modernise the bar, which is based on the Northcote Road in Clapham, London. The result of the interior is a mixture of styles from different eras such as the reclaimed nineteenth century bar, 50s style furniture and the numerous types of lamp shades and mirrors. The general colour of the bar interior is neutral pastels but the lighting brings in the various colours due to the different lampshades. The idea behind the cluster of lamps hanging from the ceiling is to entice passers-by from the high street into the bar, especially at night as they stand out due to the range of colours eliminating off the lampshades around the room. The exterior of Babel has luminous seating, which reflects the range of colours coming from the interior's lighting.

巴别酒吧

位于伦敦克拉珀姆区的巴别酒吧的装修需要升级，B3设计公司受龙头旅馆的委托对其进行现代化装修。酒吧的室内风格融合了各个时代的特征，如19世纪的吧台、20世纪50年代风格的家具和不计其数的灯饰和镜子。酒吧的主体色调是中性的粉蜡色，但是灯光为空间带来了丰富的色彩。天花板上灯饰吸引着商业街上的路人进入酒吧；特别是在夜晚，室内色彩丰富的灯罩特别明显。巴别酒吧的室外部分采用了照明座椅，与室内的灯光相得益彰。

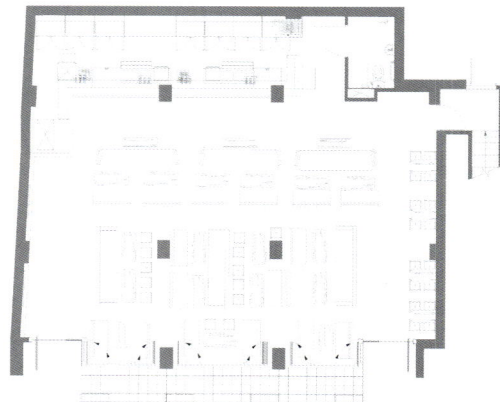

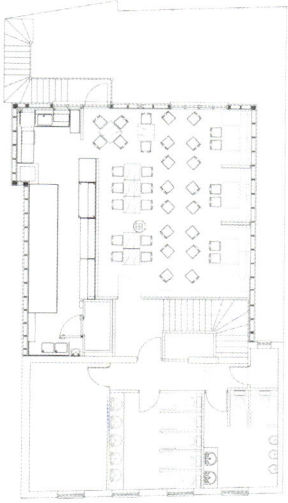

Qube

Qube is a two-floor bar and grille; on the ground floor a 12-metre bar is designed to dispense cocktails from all over the world. The site is designed to be feminine friendly and to offer the highest class of product and service throughout. Sexy, eclectic furniture and a casual chic atmosphere combine to make this a highly sophisticated destination venue. The grille on the first floor serves the finest local produce from an open kitchen; this unique offer allows customers to pick their own produce and see it cooked to their own exacting requirements. The low key casual nature of the design creates a relaxed informal atmosphere with the finest food.

Q 吧

Q吧是一间拥有两层楼的酒吧和餐厅，在一楼设计12米长的吧台是用来摆放来自世界各地的鸡尾酒。这里洋溢着柔和的气息，为客人们提供最优质的产品和服务。性感的、中性的家具风格以及幽静气氛贯穿了整个酒吧内部。二楼的餐厅提供当地最优质的美食，开放式的厨房设计可以让顾客自己挑选想吃的食物并可以看到厨师做菜的全过程。

335

Photo: Innova:Designers Studio

Diego & Pedro Serrano (Innova::Designers Studio)

Restaurant L'Ancora

The small size of the room is offset by the recovery of the total height and location of the toilets. The kitchen, fully seen from any point of the local, along with the use of a natural finished palette, as phyllite stone, bamboo platform and maple furniture, represents the traditional part, while most technological materials as LED lighting, methacrylates, vinyl on walls, projections and environmental graphics, the most avant-garde, with a brazenness that content, aim to merge in the set, rather than stress itself. This fusion becomes a metaphor in the bathrooms, where the cutlery from the old restaurant becomes hangers 'improvised'.

拉安科娜餐馆

设计师通过调整空间的高度和洗手间的位置，改善了餐厅空间的狭小。从餐厅的任何角度都可以看到厨房。室内有千枚岩、竹板、枫木家具等传统的天然材料，墙面上装饰的LED灯，树脂和乙烯构成夸张前卫的图案，十分引人注目，并与整个空间融为一体。洗手间也体现了整体融合，日餐厅用具转变成了这里的简易衣架。

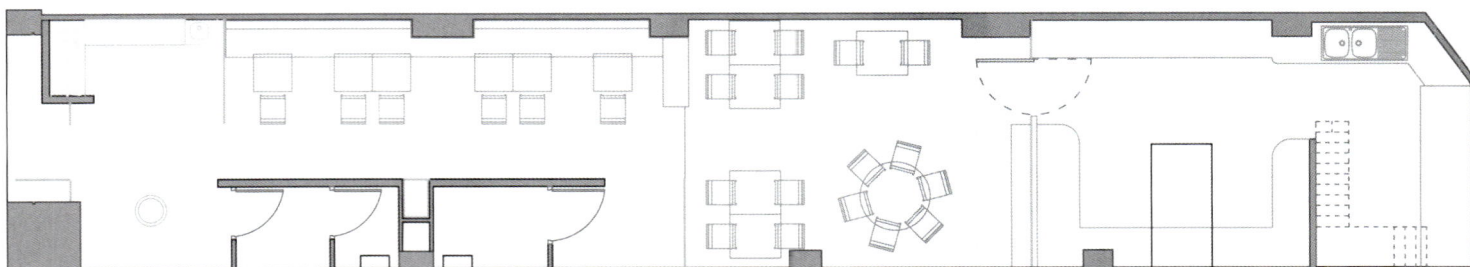

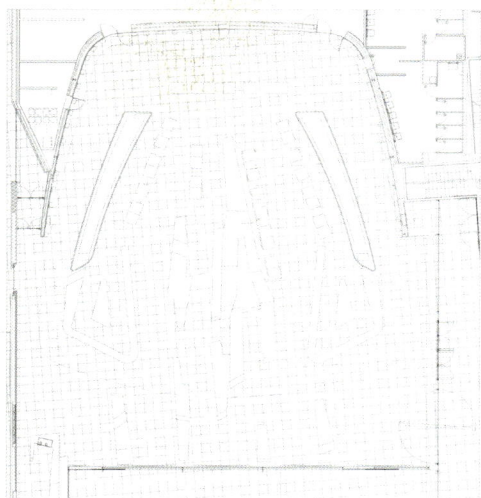

Reina Sofia Museum Restaurant

When commissioned to design the new interiors of the Arola's Madrid Restaurant at the new Reina Sofia Museum, Vidal y Asociados arquitectos-VAa team faced the challenge of developing a space conceived by the brilliant architect Jean Nouvel, Team B-720, and Alberto Medem. VAa also had to achieve the demands of one of the new masters of the culinary universe – Sergi Arola – whose sincere, driven by passion and creative cuisine was the inspiration for the whole concept. The space, located on the ground floor of the museum borders a very busy street and the museum's courtyard, making the perfect meeting point and the excellence of entertainment.

索菲娅王妃博物馆餐馆

VAa建筑事务所对马德里的索菲亚王妃博物馆餐馆进行新的设计，VAa建筑事务所面临着一个巨大挑战，因为原有建筑是由著名建筑师吉恩工作小组设计的，VAa设计小组将人的感觉进行类推、暗示、幻想和回忆展示出来，并逐渐呈现出它的真实性。家具的设计非常特别，而且具备了不同的功能。设计师严谨地选择了装饰材料、织品和涂料，使设计和整个环境融合为一体。

Photo: Ignacio Álvarez-Monteserín and José Gad Peralta Iglesias

Restaurant

Spain

Madrid

2008

Photo: Javier Peñas

James& Mau Architecture - Mauricio Galeano and Jaime Gaztelu

Estado Puro

The designers reinterpreted the typical Spanish image in the same way Paco Roncero reinvents traditional Spanish flavours and unite tradition with innovation while avoiding kitsch or fashion botox. In order to maximise on the extraordinary location of the restaurant (front of two major monuments of Madrid: Plaza de Neptuno and Prado Museum), the architecture had to enclose those monuments by accentuating, amplifying and directing the space towards them. The 'skin' covering wall and ceiling helps to generate a sense of continuity and a directional tension towards Neptuno. 'La Peineta': barrettes or Spanish comb was chosen to create the skin: combining with humorous Spanish folklore and sophistication. The objective was to allow both Spanish and tourists to identify themselves to the Spanish culture in a modern, sophisticated and funny way.

纯粹状态餐厅

设计师试图打造如柏高•罗恩瑟罗一样经典又创新的西班牙餐厅，避免俗媚或是跟风设计。为了让餐厅在特殊位置（朝向马德里的两大古迹：喷泉广场和普拉多博物馆）上最大化，建筑的设计必须包容这两个景点，强调并扩大朝向它们的空间。护墙和天花板的设计具有连续性，并且具有指向广场的张力。设计师选择了"La Peineta"——一种西班牙发夹或是梳子作为主要图案，极具西班牙风情。无论是西班牙本地人还是游客都能在餐厅里体会到独特的西班牙文化。

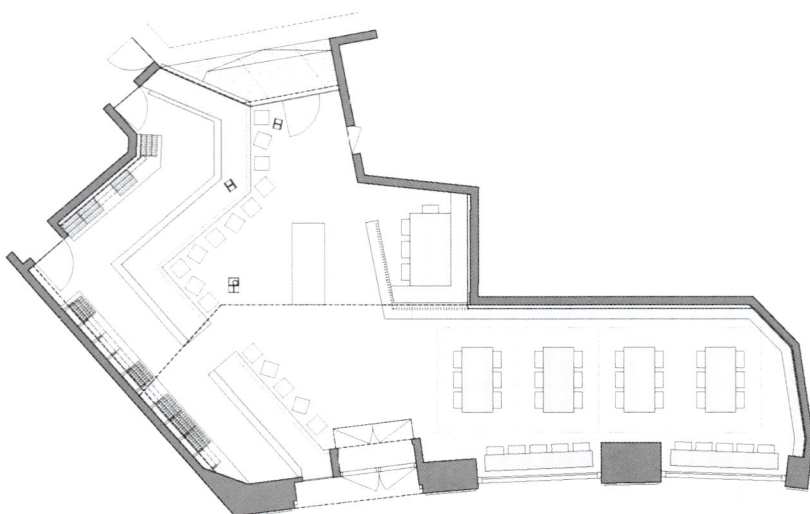

Bar Lobo

Through the front door, one walks into an open space of Mediterranean fiesta and brightness. The display of the openings is set in black iron, which has been made to fit the old style, and therefore maintain the unity of the façade. To the left, a series of tables alongside the wide windows, displayed beside a line of built in benches, one irregular and one more lineal. Both follow the line of the façade and both give movement and direction to the bar. To the right, the bar, in black Zimbabwe granite with its display of woks, and the access to kitchens and stairs, are more rectilinear and slightly more formal.

罗伯酒吧

人们通过正门进入一个开阔的地中海风情空间，明亮而欢快。开口处的框架采用了黑铁，与古老的风格十分相称，从而保持了与外墙的统一度。左侧，一系列的餐桌沿着宽大的窗户摆放，内侧是一排嵌入式长椅，一个呈不规则形状，一个呈直线型。二者沿着墙线而建，直通吧台。右侧是津巴布韦花岗岩制成的吧台，通往厨房和楼梯的通道更加笔直，也略显正式。

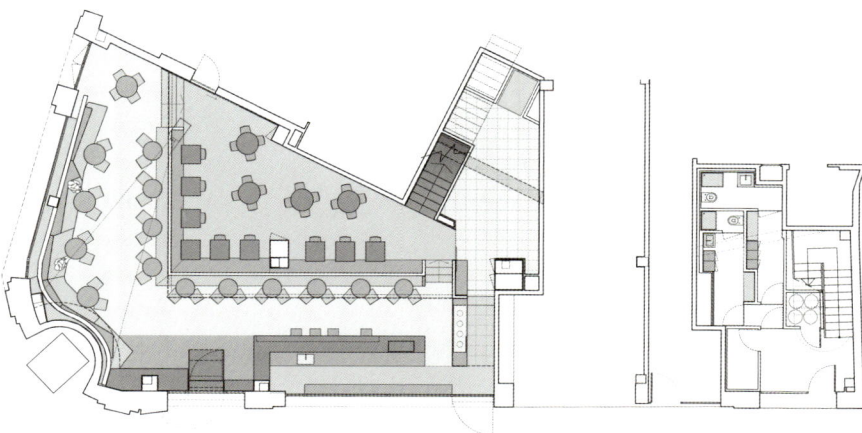

Bar

Spain

Barcelona

2006

Photo: Eugeni Pons

Francesc Rifé

339

Restaurant

Spain

Sevilla

2009

Photo: Fernando Alda

Francesc Rifé Studio

Restaurant Gastromium

It's a space of 242 square metres placed on the ground floor of a residential building. The restaurant has two façades; one of them is used by the clients and the other for private use. In the main façade there are vast sliding doors made of black metal sheet integrating the main entrance, creating a space of transition between the exterior and the interior. This transition effect is created by the black sheet of the sliding door in contrast with the luminosity of the white tone of the interior of the restaurant. The vertical light in the main entrance represents the sunlight of the South. This light element continues horizontally for the perimeter of the restaurant. The back façade integrates air systems. Also, it's a huge black metal sheet, contrasting with the residential building.

加斯特米姆餐厅

餐厅位于一座住宅楼的一楼，总面积为242平方米。餐厅有两面外墙，一面供顾客使用　一面是员工通道。主外墙的入口处有巨大的黑钢玻璃滑门，为室内外提供了一个转折点，黑色的骨门与餐厅内纯白的色调形成了鲜明对比。入口处的垂直灯光代表着南方的阳光，这一灯光元素在餐厅的四周得以延续。后墙的设计融合了通风系统，也有一扇巨大的钢板，与住宅楼区分开。

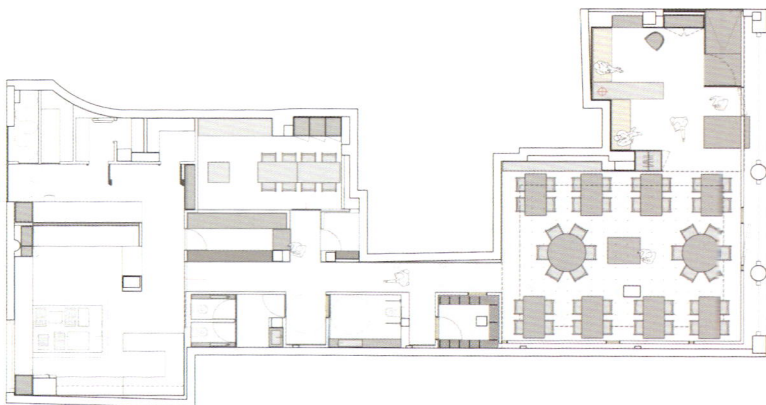

Tomate

The new Tomate is the first restaurant of the Tragaluz Group in Madrid, designed by Sandra Tarruella. The project entails the rehabilitation of an existing restaurant, with a collage of materials and textures that contribute to a fresh, informal, dynamic and natural image. The original space, an L-shaped floor plan, had two very distinct areas. The 'street' zone, more luminous, was the main dining room, and the remaining area, much darker, was used as a bar zone at night. Both sections were very diverse in aesthetics and use. One of the main ideas of the new project was to unify these two spaces by means of a concrete shelf.

番茄餐厅

新番茄餐厅是Tragaluz集团在马德里成立的首家餐厅。该项目旨在于原有建筑的基础上，通过材料和材质的完美搭配，创建一个温馨、自然、活力四射的新型餐厅。原有空间呈L形布局，分为两个不同的区域。光亮的"街区"部分是就餐区，而色彩较暗的区域则专为午夜酒吧而设。二者因功能不同而风格各异。该项目意在于两个区域之间创建一个盛物架，将二者有机衔接在一起。

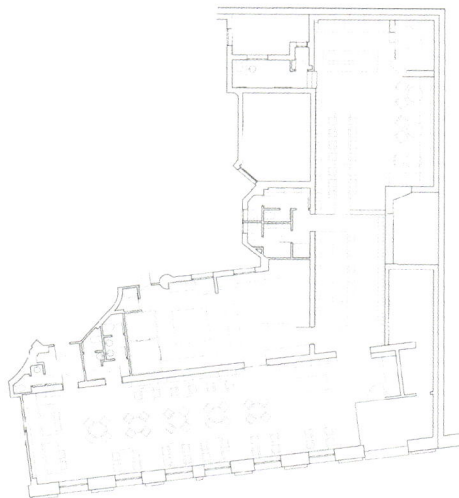

Restaurant

Spain

Teruel

2010

Photo: Stone Design Studio

Stone Design Studio

Lapiaz

Within the building the designers have created a self-service restaurant and a chocolatería (a café serving hot chocolate as a speciality) where they have tried to make the customers the main protagonist, allowing them to enjoy the warm, welcoming space full of references to the surrounding mountains. This building has a big presence and a large personality whose glass facade could provide the view of the ski runs and the surrounding forest. Using natural wood, to contrast with the building's pillars, and the whole range of greens which the designers find in this natural setting, they create a restaurant for all types of clients adapted to the needs of a skier.

Lapiaz餐厅

空间中包括自助餐厅和热巧克力专售区两个部分。餐厅的设计从客人的角度出发，力图为其营造一个温馨、热情、舒适的就餐空间。大型玻璃外立面能够将窗外森林、滑雪道完美呈现给就餐的客人。餐厅以木质家具为主题，色彩鲜艳，试图满足各种滑雪者的不同需求。

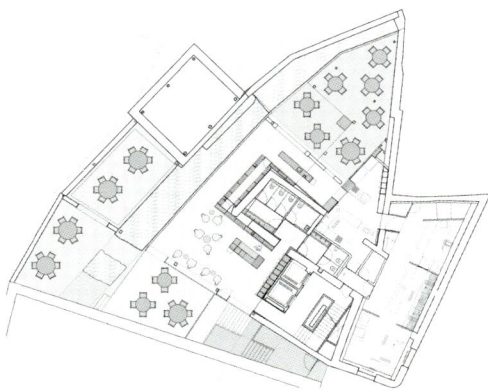

Arrop

With a simple, restrained and elegant design, the Arrop project is strongly linked to the persona and style of chef Ricard Camarena. The restaurant forms part of the Hotel Palacio Marqués de Caro in Valencia, which is itself still under construction, although the restaurant has been designed to work independently from the hotel. The project, which required a highly technical solution, needed to preserve three important elements: the old city wall from the 16th century, the aljibe (a rainwater collection tank) and a 17th century Gothic arch. The challenge for the desienger was to integrate these three elements into the modern design which characterises the hotel as well as the restaurant.

阿洛普餐厅

阿洛普项目简洁、内敛而优雅,具有其主厨里卡德•卡玛里纳的个人性格特征。餐厅是巴伦西亚帕拉西奥侯爵酒店的一部分,酒店还没有完工,餐厅目前开始独立运营。项目的工艺要求很高,需要保留3个重要的元素:16世纪的旧城墙、雨水收集箱和17世纪的哥特式拱门。设计师所面临的挑战是将这3个元素融入酒店和餐厅的现代感设计之中。

Restaurant

Spain

Valencia

2010

Photo: Fernando Alda

Francesc Rifé Studio

Restaurant

Spain

Pamplona

2008

Photo: José Manuel Cutillas

Vaillo&Irigaray + Galar

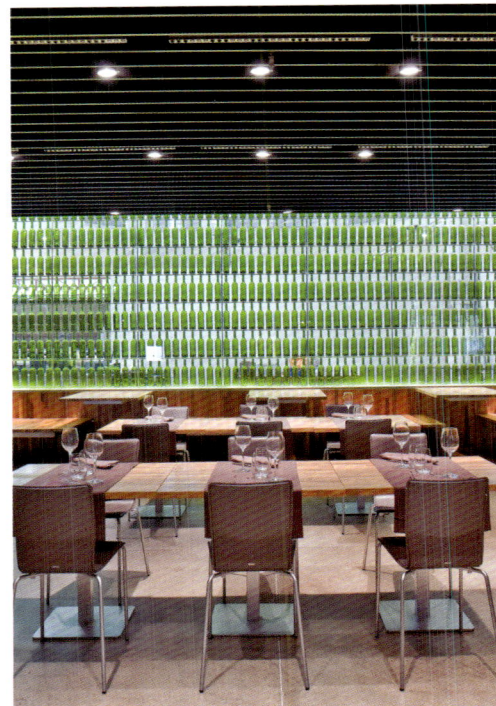

Restaurante El Merca'o

The design essentially aims to create an atmosphere: warm, affectionate, sober, timeless, moody, elegant, where the light can accentuate the day and night, the party and privacy. The restaurant occupies two floors: the first connected with the market and the second located in semi-basement: the ladder organises access from the street as a central landing. It also offers two spaces, two different modes of being: a daily, open, flexible, linked to the circulation of the street and the market itself, organising the circulation and access through the bar, and a deeper, austere, linking its atmosphere hold with the concept of underground place and providing a receptacle calm, serene and calmer.

Merca'o餐厅

设计旨在营造一种温暖、理性、经典、情绪化而又优雅的氛围,灯光可以渲染白天和黑夜、派对和私人空间。餐厅分为两层,一层与市场相连,一层位于半地下室,两层之间由中心楼梯相连。这是两个截然不同的空间:与市场和街面相连的空间开放而灵活;地下的酒吧更深沉、简洁,具有地下的氛围,相对沉稳而幽静。

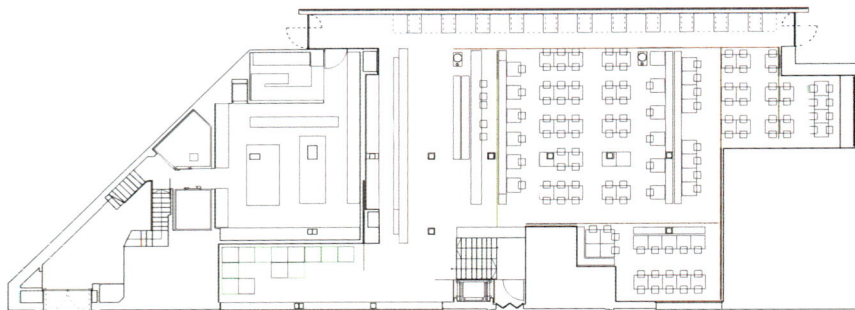

Lolita, Infrastructure for Events and Meals

Roadside restaurants are a rare species within the increasingly prestigious restaurant world. Such places superpose their condition as an infrastructure adapted to the commercial, informational and social flow of the road network on mythical scenarios taken from road movies and literature. In recent years, their structures have evolved in order to offer services for large-format events without this having involved anything more than a change in scale. The project rose to the challenge of changing this trend by building a structure capable of managing a programme subject to constant reorganisation, with the presence of a heterogeneous public and the expectation of diverse uses, a flexible space capable of setting itself up as a scenario for almost any type of activity. The aim was to transform a roadside restaurant into a versatile infrastructure for events and meals.

洛丽塔餐厅

路边餐厅在日渐奢华餐厅世界越来越少见。路边餐厅似乎只存在于公路电影和文学作品中,为路上的商业、通讯和社交人群提供餐饮服务。近年来,它们的规模和结构有所演变,以便为大型活动提供餐饮服务。项目的挑战就是赶上这种趋势,打造一个能够为不同客户群体举办各类型活动的灵活空间。其目的是将路边餐厅转变为一个多功能活动、餐饮中心。

Restaurant

Spain

Zaragoza

2010

Photo: Miguel de Guzmán

María Langarita and Víctor Navarro

Lounge

Spain

Girona

2006

Photo: Lluis Ros

Eduardo Gutiérrez, Jordi Fernández / ON-A

LOU_5 Sentidos Lounge Bar

This is a lounge bar with a special premise: the singularity and the exclusive treatment of the client. It is a unique space with a unique architectonical solution that generates a high quantity of different perceptions: visual, chromatic, audio, sensitive ones, etc. The idea arises from the ground plan of the bar as well as from the analysis of the different types of customers of the lounge bar. There is a bar zone, as well as some reserved spaces (however, not completely closed) for exclusive groups, an entrance zone of a more general use, and other, more quiet zones of variable configuration characterised by the central position of the privets. This way any client can find their preferred space, the one that most adjusts to their personality.

卢5摇滚休闲吧

这家休闲吧对待客人有一种独特的方式，是一个具有特殊建筑结构的空间，提供高品质的视觉、色彩、听觉和感官体验。其设计理念来自于休闲吧的空间布局和对不同顾客群体的研究。吧台区和一些预留的空间为特殊团体使用，入口区的功能相对广泛，其他相对安静的区域以一棵水蜡树为中心。这样一来，各种客人均可以找到适合自己的最佳场所。

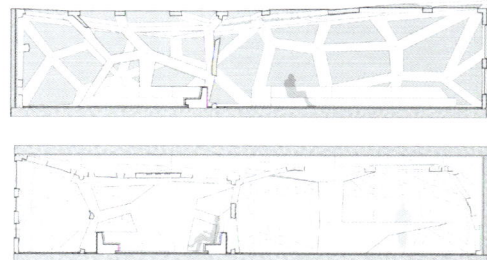

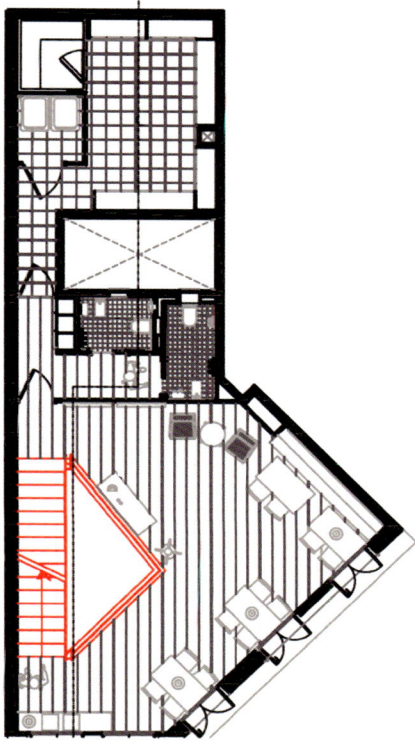

Federal Caffederal Café

The client's brief for this existing two-floor building and terrace, located on a corner street in the neighbourhood of San Antonio, in Barcelona, was a cosy and informal café/bistro/lunch spot restaurant. While the facing of the façade has been restored with minimal intervention, to remain in keeping with the neighbouring buildings, radical interventions have been made to the interior space, which has been meticulously renovated. Upon entering, the eye is guided to the upper floor by the double height and the concrete steps made in situ, and then up again to the garden-terrace. On the ground floor the sight is extended to the small patio, which is an element of separation between the public and service space, suggesting the idea of an open space. The bar is originally set against the wall. It has been custom designed, like the dessert table, strategically placed in the centre of the sliding window and the axis of symmetry of the courtyard. Both the bar and the furniture are in plywood, coated in plastic laminate of different colours.

联邦咖啡厅

委托人的目标是将这个位于巴塞罗那圣安东尼区街角的建筑打造成一个舒适的餐厅，兼具咖啡馆、酒馆和午餐吧的功能。建筑的外墙只做了极小的改动，以保持和周边建筑的一致性。与此相比，室内的改动相当大。一进门，视线便被楼梯吸引到通往楼上和花园平台。一楼的视野一直延伸到分隔公共和私密空间的天井，显得更加开敞。吧台倚靠着墙壁，经过特别定制，与甜点桌类似，巧妙地设置在滑动窗的中央和庭院中轴线上。吧台和桌椅家具都是胶合板制成的，表面是不同色彩的层压塑料。

Café

Spain

Barcelona

2010

Photo: Christian Schallert

Barbara Appolloni

Lounge

Spain

Navarra

2008

Photo: Jose Manuel Cutillas

Vaillo+Irigaray (Antonio Vaillo i Daniel and Juan L. Irigaray Huarte)

Lounge Ms

The new space is conceived as a continuation of the existing fence, wrapping it all, but hidden: expressing their new identity, not as a constructed element, but as a re-forested. A new plant species grown in the surrounding area. The new scalar similarity between elements of 'plant' makes the proposal a new understanding of the link with the existing connection. This species builds a base of recycled plastic tubes of different colours similar to the natural, which like 'reeds' is organising a braided flexible and deformable organic capability to adapt to any situation and geometry.

Ms休闲吧

新空间是原有栅栏的延伸，将栅栏完全包围了起来：显示了他们的新形象——不是作为建筑元素，而是一片再造森林，一种新物种。这种类似于植物的新元素在新设计和原有设计之间打造了一种全新的联系。这些由回收塑料管制成的植物贴近自然，宛如"芦苇"，可以被编织起来，灵活地变形，以适应各种场景和造型需求。

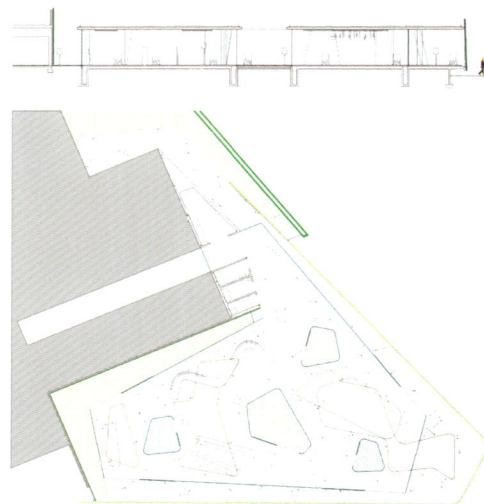

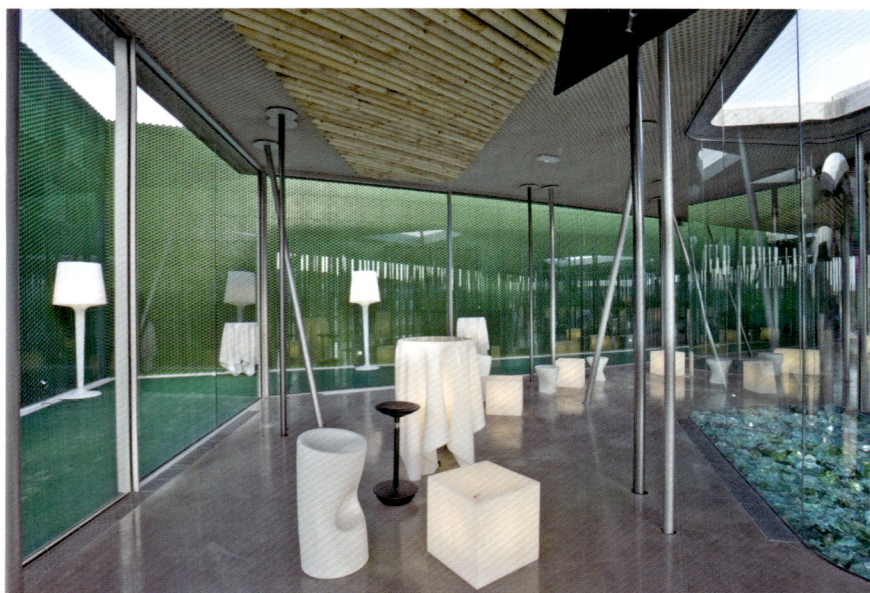

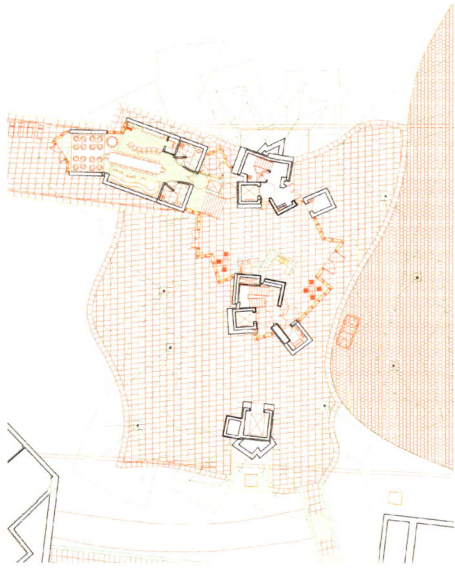

Restaurant in Marqués de Riscal

The world of wine and the world of gastronomy meet at The Marques de Riscal Hotel. With the most exclusive crockery at their finger tips, guests can enjoy the creative and traditional Basque-Riojan cuisine of Francis Paniego. The hotel also offers a more informal Wine Bar and a rooftop lounge with breath taking panoramic views. Its cosy fire-place will be perfect for guests who want to sample the thousands of wines from around the world found in the hotel's cellars or enjoy the reading of one of the 1000 books of the library, all of them with topics related to the luxury world. Guests can also experience the entire process of wine production, from the most traditional to the newest and technologically most advanced processes.

利斯卡侯爵酒店餐厅

美酒与美食的世界在利斯卡侯爵酒店相遇了。客人可以在餐厅内使用独特的餐具享受主厨弗朗西斯•帕尼艾哥制作的美食。酒店还提供一个非正式酒吧和一个屋顶休闲吧,可以俯瞰城市的美景。舒适的壁炉周围是客人品尝酒窖内上千种酒类和阅读图书馆里藏书(全部与奢华的世界有关)的好去处。客人可以体验运用传统和最新工艺的整个红酒制作过程。

Restaurant

Spain

Alava

2006

Photo: The Luxury Collection

Frank O. Gehry

Restaurant & Bar

Spain

Madrid

2007

Photo: Héctor Ruiz-Velázquez

Héctor Ruiz-Velázquez

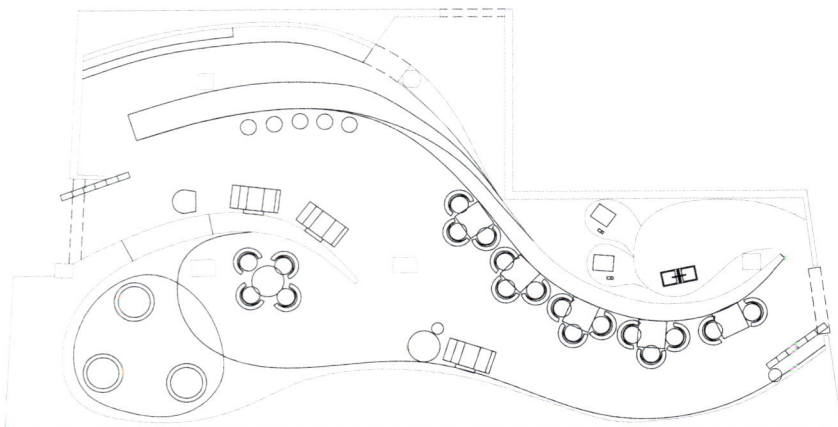

Bar-Restaurant Glenfiddich

This bar is a theme bar which is designed for the Glenfiddich Distillery. The bar-restaurant Glenfiddich is an allegory to senses and pleasure. As if frozen in time, the space was created to transport guests to that moment of sensation, freedom and pleasure which they experience when we pour ourselves a glass of whisky. This bar restaurant, with its open and fluid aesthetics incites the visitor to reinvent shapes and postures from the past. This public space allows them to freely enjoy its organic shapes as in the process of maturing whisky, with its wooden finishing, adding richness in textures, smells and colours, creating an atmosphere that is alive and latent. Structurally, the space is created by the tension between the diametrically opposite entrances which create an axel or passage way in which one either circulates or remain. This canal, formed by pieces of wood, seems to provoke the laws of gravity and logic, constantly changing the spatial perception as the visitor moves throughout.

格兰菲迪酒吧餐厅

项目是一间专门为格兰菲迪酒厂设计的主题酒吧。格兰菲迪酒吧餐厅是个极尽感官和愉悦的场所，将客人带到了喝过一杯威士忌的自由、愉悦的状态。酒吧餐厅开放和流畅的美学让客人想起了过去的日子。空间采用木质家具、增添了质感、气味和色彩，打造了一个活跃而隐秘的氛围，人们可以在这里尽情地享受威士忌。在结构上，空间利用了两扇正对的门，打造了一个中轴走廊，供人停留或是走动。这个通道由木板装饰而成，似乎利用了重力和逻辑法则，在客人移动时不断转换视角。

Restaurant

Spain

Madrid

2008

Photo: Elvira Blanco

Elvira Blanco & Jose Maria Aguilar

Restaurant in Hospes Madrid

Originally an affluent apartment house with wrought iron balconies designed in 1883 by architect José María de Aguilar, the handsome red brick Hospes Madrid is an icon of Bourbon Restoration period architecture. The Senzone restaurant offers a formal dining room with an oak coffered ceiling and stucco work – in contrast to the luminous interior patio where guests can relax on comfortable sofas and listen to whispering water as they drink at the hotel bar. Hospes Madrid's imaginative array of modern additions accentuate an overall sense of permanent privilege and eternal elegance.

马德里霍斯佩斯酒店餐厅

这原本是一座公寓建筑，设计于1883年，有着锻铁阳台，漂亮的红砖，这座"马德里霍斯佩斯"是波旁皇族复辟时期是一座代表性建筑。森众餐厅提供了一个正式的用餐环境，里面有橡木格子天花板和拉毛粉饰装饰品——这跟明亮的室内平台形成对比，客人可以坐在平台上舒适的沙发上放松自己，一边在酒店的酒吧里喝点儿什么，一边听着潺潺的流水声。

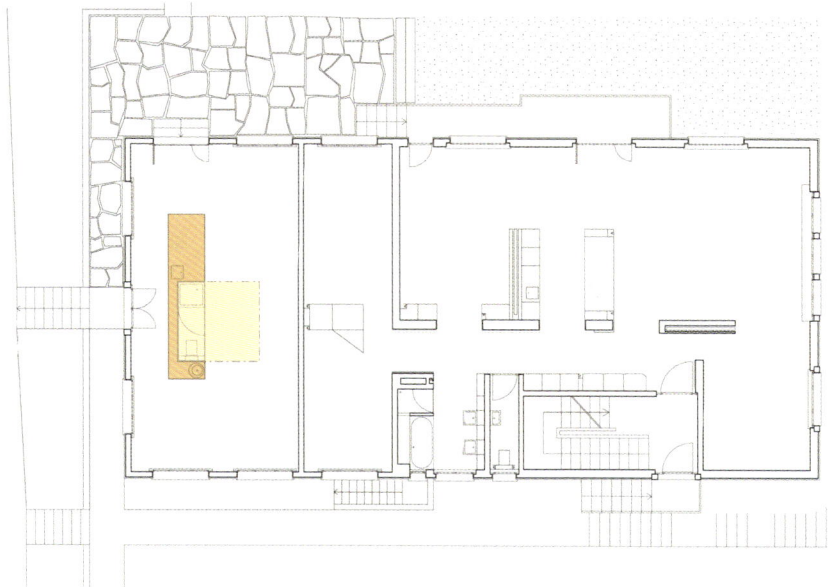

Restaurant

Spain

Barcelona

2007

Photo: Robert Justamante Antolin

FFWD (Laia Guardiola & David Benito)

Plató Restaurant

The 220-square-metre restaurant's main space is conceived as an area that can be easily adapted to different uses and situations. The dining room becomes a set where the main characters are the fellow guests. Table arrangements and lighting can be modified to create the right atmosphere depending on different needs. A non-ending platform is built to define the dining room occupation area. Four three-dimensional structures divide the space transversely, lodging the technical installations of the room. Next to the street you will find the bar area, a space in constant change, always in motion.

柏拉图餐厅

总面积为220平方米的餐厅的主要空间能够适应不同的场合需求。就餐区的灯光和桌椅设置可以根据不同的气氛进行调整。就餐区边缘是一个无边际平台。四个3D结构对空间进行了横向划分，里面设置着餐厅所需的电机设施。临街的一侧是吧台，气氛不断变换，充满了动感。

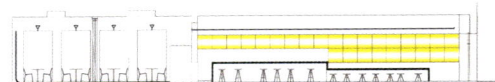

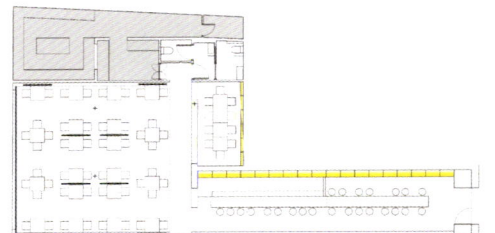

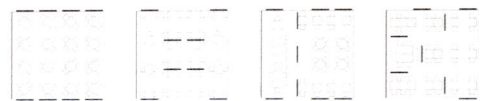

Restaurant & Bar in Murmuri Barcelona

Restaurant Murmuri is a spacious and luminous space where it is worth noticing the theatrical velvet sofa that the same interior designer has created. With a selection of original and delicious dishes the restaurant is nowadays the best Asiatic gastronomic establishment in town. But there is even more. The bar, Marfil by Murmuri, brings back a 1920's classic that will be open all day and till 2 o'clock in the morning every day with a refreshing and light range of choices to be enjoyed both inside and on the privileged terrace. This will be the new meeting point for Barcelona's most fashionable people.

巴塞罗那默默里酒店餐厅和酒吧

默默里餐厅的空间宽敞而明亮，里面夸张的天鹅绒沙发也出自同一个室内设计师之手。独创的美食让餐厅跻身于城内最佳的亚洲餐厅之列。默默里的象牙酒吧将人们带回到了20世纪20年代的经典环境，全天候开放，客人们可以在酒吧内和露台上尽情享乐，是巴塞罗那潮流人士的新据点。

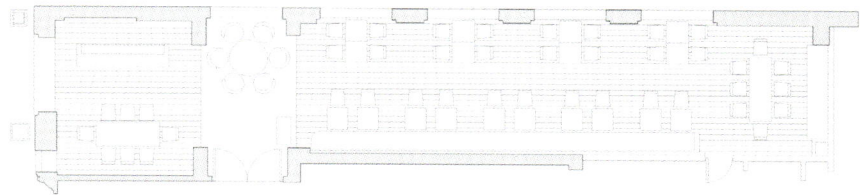

Photo: Miguel de Guzmán

María Langarita and Víctor Navarro

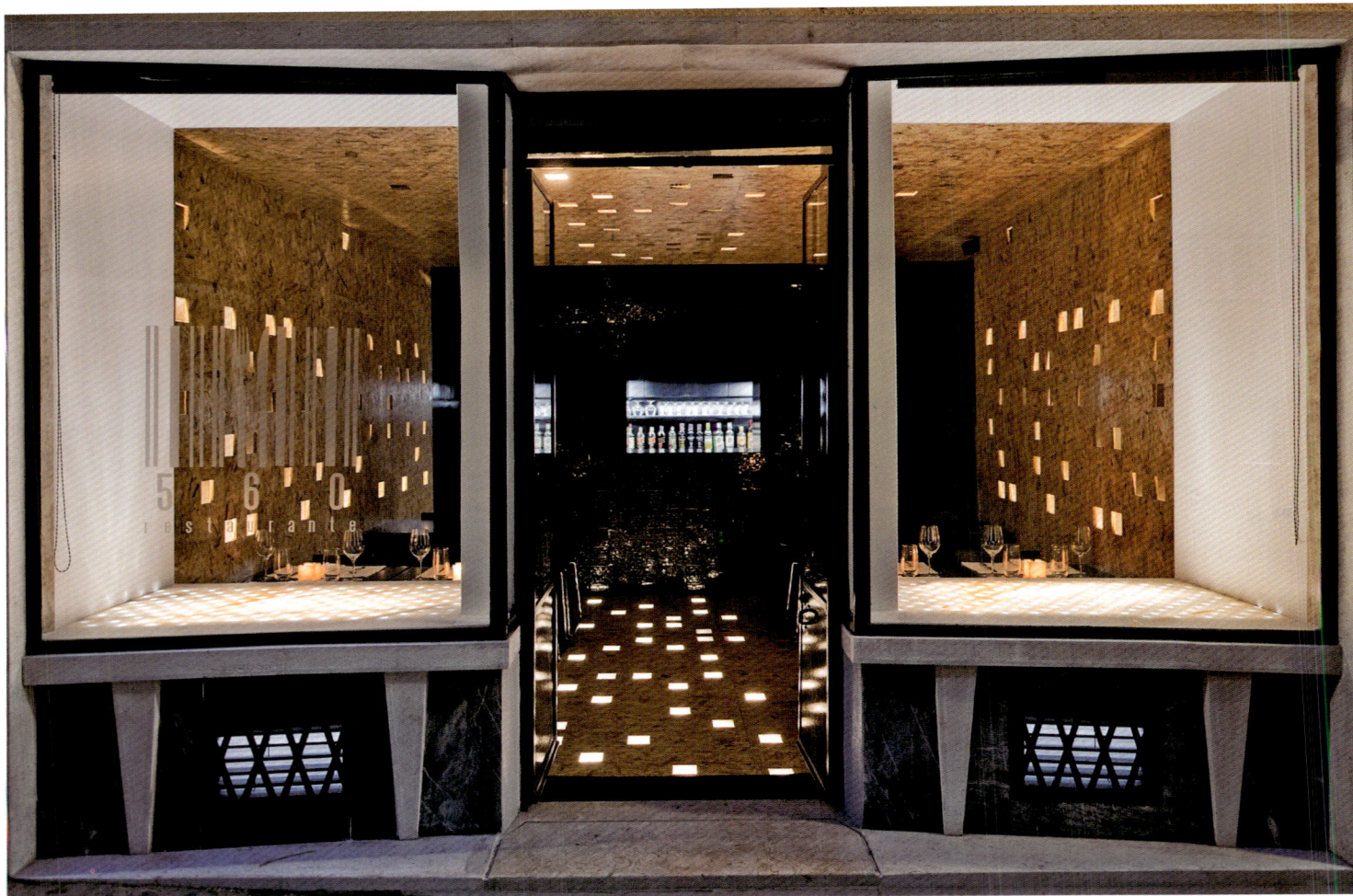

Restaurant

Portugal

Lisbon

2008

Photo: Fernando Guerra / FG + SG

João Tiago Aguiar – Acarquitectos

Restaurant 560

The restaurant is placed on the ground-floor of an old building in Bairro Alto, Lisbon's city centre. The space itself is divided into two separate parts – one corresponds to the two dining-rooms and another one to all the services inherit to its well function. The entrance is made by two separate doors and each one of it sided by two huge windows. The separation between the eating area and the services area is made by a Portuguese black tile wall, which marks the change in materials and draws the line to where the kitchen, bar and toilets are. This wall doesn't touch the perimeter in both ends in order to be read as a volume and at the same time creating two entrances – to the bar on the left and to the toilets on the right.

560餐厅

560餐厅坐落在里斯本市中心一幢老建筑的一层，分为两个独立的部分，各自拥有单独的入口。其一包括两个就餐室，另外一个则为服务区。两部分之间通过黑色瓷砖墙壁间隔，突显材质变化的同时，更指明了厨房、酒吧及卫生间的方向（墙壁上设有两个入口，左侧通往酒吧，右侧通往卫生间）。另外，墙壁两端并未与空间相连，自成一体。

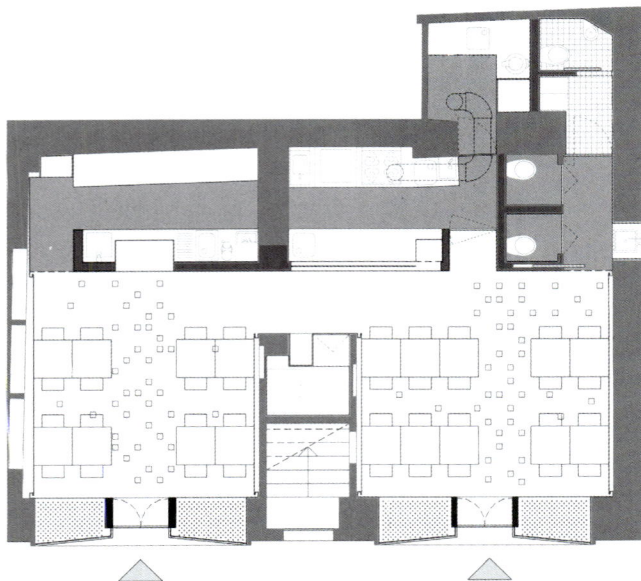

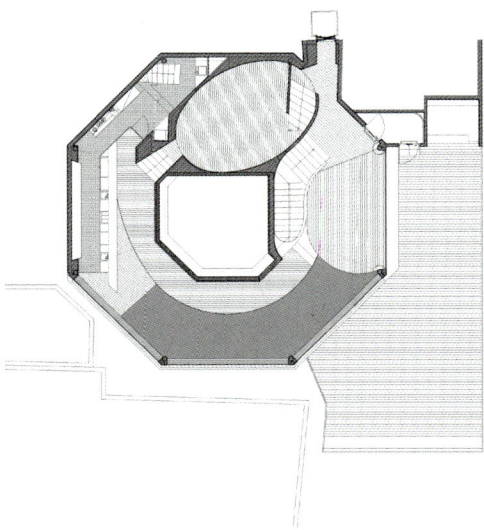

Bar

Portugal

Lisbon

2008

Photo: Nuno Sousa Dias, Tiago Silva Dias

Silva Dias - Arquitectos

Silk Club

Silk Club occupies the rooftop of a building located in one of the more dynamic and charismatic neighbourhoods, for its cultural heritage, of Lisbon historical centre – called Chiado. The uniqueness of the space, distributed by two floors and a terrace, with a view over Lisbon, required a deep rehabilitation for the transformation of a degraded space into a sophisticated and modern private night club. The designers took advantage of the intricate and difficult configuration of the building, and the areas were moulded in order to create a space system, flowing in its articulation, allowing a dynamic view between the different functional areas, making possible the creation of differentiated environments taking advantage of each spatial formal specificities.

丝绸俱乐部酒吧

丝绸俱乐部位于里斯本市中心迷人的奇亚多街区，在一座建筑的顶层。俱乐部分为两层，还有一个平台，可以俯瞰里斯本的美景。设计工作是翻新原有的空间，将其打造成一间精致现代的私人俱乐部。设计师利用了建筑错杂的造型特点，将各个区域模块化，打造了一个空间系统，既在不同功能区之间建立了动态联系，又保证了每个空间的特异性。

355

Ginkgo

The name Ginkgo – a species of trees with at least 270 million years old – came as reminder of the longevity of the good 'design' – in this case, by nature. Assuming a minimalist attitude the Ginkgo bar is a unique project that promises to renovate the image of the architecture in the south of Portugal. It's a lounge place with a fresh environment and a light and pleasant atmosphere, sublimed by the great selection of Jazz and Bossanova music. Formally very clear, pure and objective, with simple geometric shapes, the image of the building proposed by the architect Tiago Rosado assumes a strong character. Projecting itself to the terrace the building is delimitated by a frame that is assumed as a frontier between the bar and the surrounding area, conferring a protection and comfort sensation.

银杏吧

银杏吧的名字"银杏"是一种具有至少2.7亿年历史的古老树木，暗示着好设计，即自然设计的持久性。银杏吧是一个独特的项目，是葡萄牙南部建筑的创新。这个休闲空间具有新鲜的环境、轻快和愉悦的气氛，飘荡着爵士乐和波萨诺瓦舞曲。建筑简洁纯粹而目的明确，其简单的几何造型让人印象深刻。建筑的外框直达露台，将酒吧与周边区域分隔开来，传达了一种安全感和舒适感。

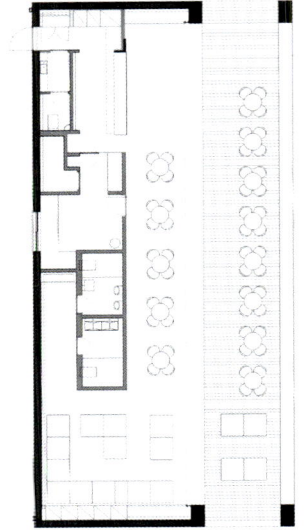

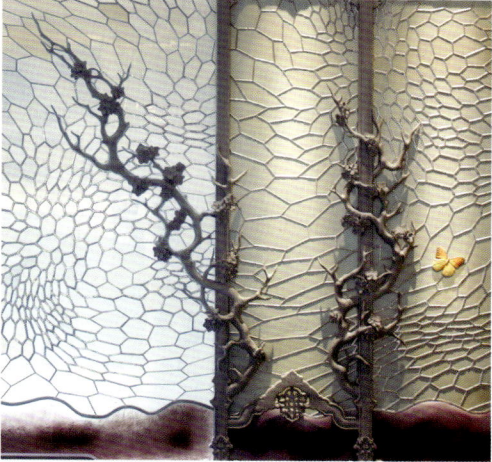

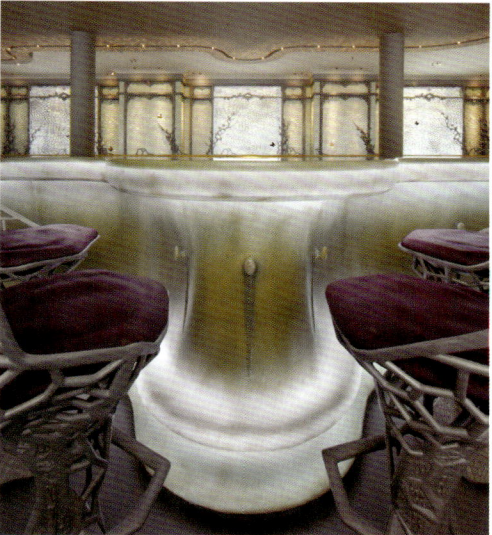

Le Ladurée Bar

The hieratic chairs seem to form the coral cladding of a majestic sunken ship. At the same time, it would also be true to say that this vast, organic, alveolar structure has come together like a chrysalis hiding the gestation of an imperceptible/invisible world, between two infinities. Let's look a little closer at these chairs. The seat, a purple cluster, forms the primordial chaos, the foundation, of this expanded aluminium structure. The following structure, where each cell of this subtle architectonic is arranged around convection points replying to the framework of a cosmos is reinvented from the base to the summit, from the infinitely small to the infinitely large.

拉杜雷酒吧

教堂风格的座椅宛如沉船上的珊瑚。与此同时，这个巨大的有机蜂窝结构又像是隐藏在未知世界的茧蛹，处在两个极端之中。仔细看这些座椅，紫色的坐垫是纷繁的源头，而底座则采用了延伸型铝结构。其他的每个网眼结构都精妙绝伦，围绕着最初的原始框架而展开，不断将基点提升到最高点，将绝对小的元素拓展到无限大。

Bar

France

Paris

2008

Photo: Thierry Malty

Roxane Rodriguez

Restaurant

France

Paris

2009

Photo: Jérôme Spriet and Patrick Gries

Matali Crasset

La Cantine de la Ménagerie de Verre

When the client asked the designer to give some thought to a reception area for the Ménagerie de Verre, the designer tried to retain its breathing, its evanescence, the reason why one likes it, it is an ethereal space where time has no hold and this is why the intervention is discreet, as people are sometimes invited by scenography. It is therefore a place to relax, to wait, to take refreshment before or after a show, when a company is working and is in residence. The place asked for serenity whence the designer chose birch plywood, a very smooth and discreet material. Around a cellular structure in wood, islands of tables, low armchairs and pouffes are arranged, in this way offering two types of comfort. The armchair and the table take inspiration from the trestle, one of the most basic languages, elementary in furniture. It becomes a device that can be removed at any time to free up the spaces.

玻璃动物园餐厅

当委托人向设计师请教如何设计玻璃动物园的接待处时，设计师试图保留动物园的气氛 体现出人们热爱它的原因。这是一个飘渺的空间，没有时间的概念，因此任何设计都要小心翼翼。人们在这个空间里休息和等候，也许是在两场表演之间的间隙里休息，也许是等待正在工作或在家的同伴。空间需要宁静，因此设计师选用了桦木胶合板——一种光滑而谨慎的材料作为主要装饰材料。桌台、低扶手椅和厚垫椅围绕着一个木制蜂窝结构而展开，异常舒适。扶手椅和桌子的设计灵感来自于栈桥，运用了最基本的家具设计理念。桌椅可以随时被移开，让整个空间变得开放。

Restaurant

France

Lyon

2009

Photo: Renaud Callebaut

Studio Patrick Norguet

Restaurant in Sofitel Bellcourt Lyon

The design of the common areas of the hotel – the lobby, the Le Melhor bar, the Silk brasserie, the Les Trois Dômes gourmet restaurant, the gym, the spa and the adjoining garden with its bar – is the result of an extremely detailed contextual analysis by the designer. The aim is to give the hotel a true sense of identity and to avoid the merely decorative. Mission accomplished with the integration of both local and global features the designer placed centre stage and his communication of specific sensory experiences. The sustainable and convivial sensory aspects of Norguet's new interiors also feature harmonious contrasts in the décor, such as the coffee and silvery colours in the Trois Dôme restaurant and the red and black in the Le Melhor bar, the whole theme punctuated by understated yet original furnishings with attitude.

里昂索菲特酒店餐厅

酒店公共区域——大堂、摇滚酒吧、丝绸酒馆、三穹顶餐厅、健身房、水疗中心和花园酒吧的设计都经过了设计师的精心策划。其目的是增强酒店的辨识度的同时又避免纯粹的过度装饰。通过将本地和国际化元素融合在一起和加入自己的特殊感官体验，项目完成的异常出色。装饰的和谐对比也是项目室内设计值得注意的地方，如三穹顶餐厅中咖啡色与银色的对比和摇滚酒吧中红色与黑色的对比，整个主题朴素低调而又不失个性。

Restaurant

France

Paris

2003

Photo: Luc Boegly

Christian Biecher

Fauchon

For this place which evokes Paris and its multiple facets, the architect Christian Biecher was inspired by the codes of the Fauchon brand: the historic pale pink, which is the most important colour to Fauchon, with its sweet and appetising tone; a black and white graphic, and also a grape motif. He decided to bring them together in an environment of very strong light that is nevertheless comfortable, perfectly adapted and conceived down to the last detail. The interior fit-out of the 300-square-metre restaurant rests on the play of reflections, on the transformation of the atmosphere according to the time of day and the changes of light.

馥颂餐厅

为了设计这代表巴黎精品时尚及巴黎多样化面貌的场所，建筑师克里斯提安·毕谢的灵感取自馥颂品牌最具象征性的符码：粉雾桃红色、黑白文字设计，或甚至还有葡萄图案。建筑师将这些元素融入一个灯光极为强烈却拥有舒适氛围的空间中，连最微小的细节都巧妙融入了这些设计符码。室内空间设计主要在营造镜面反射和材质反光效果，以及跟随时辰变化的气氛和多彩多姿的灯光照明。

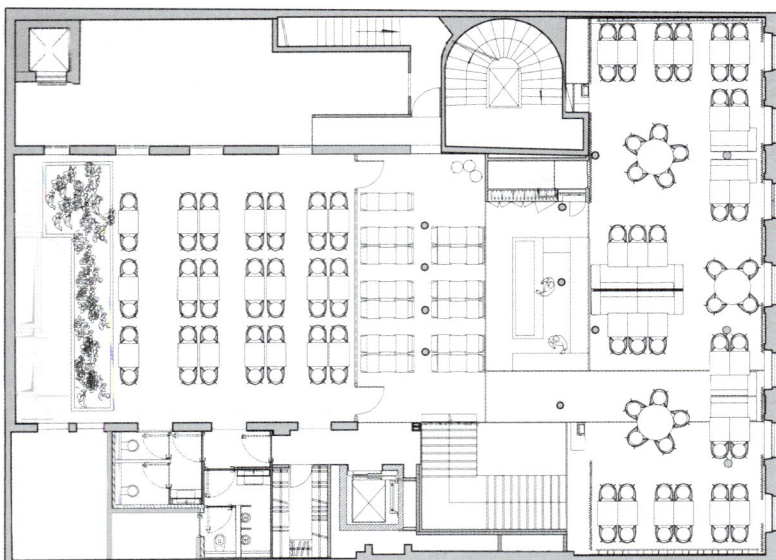

Le Boudoir

The design of the Boudoir is based on two colours, red and black; two wall materials, MDF 'curtains' and black Perspex; two types of lighting, candles and digitally controlled lights; two veiled references, the ceramic tiles with a monograph design on the floor and the beaten iron of the baobabs – all those to bring out two ambiances: day and night. Serious attention has been paid to the lighting, allowing the Boudoir to metamorphose throughout the day. Through sophisticated programming, the lighting creates specific ambiances for the very chic lunchtime service, the aperitif and lounge restaurant in the evening, and finally the nightclub version.

卜朵儿餐厅

卜朵儿餐厅的设计构思透过两个色彩(红与黑)、两种墙面材质(有褶裥的布帘与黑色有机玻璃)、两款灯光(烛光与资讯化照明设备)以及两个巧思(截印在地上的陶土与锻铁面包树)来表现，这一切都为了营造出两种时空气氛：日与夜。精心设计的照明系统使餐厅依着时辰化身转变：午餐的商业时尚氛围、晚餐前的轻松光线、晚间沙发酒吧餐厅的柔和灯光，还有夜店式疯狂的五光十色。

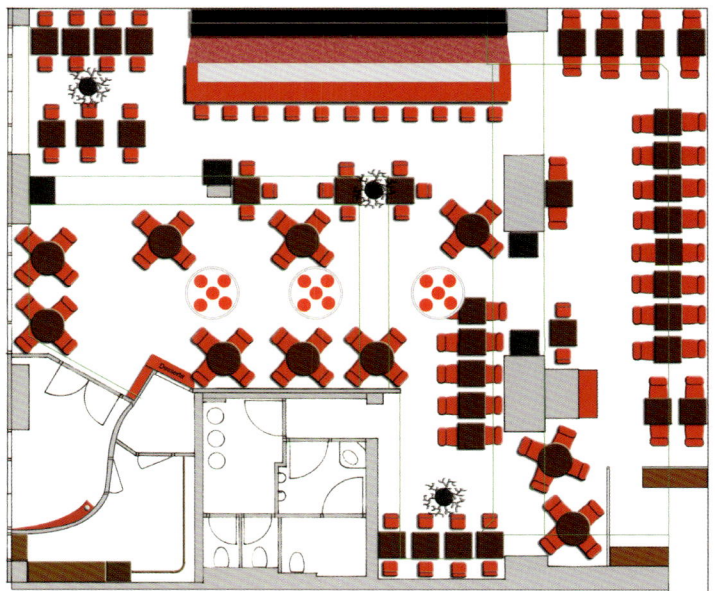

Restaurant

France

Lyon

2008

Photo: Eric Saillet

Donatelle Piana & Philippe Batifoulier

Restaurant

France

Paris

2003

Photo: Wijane Noree

Ralston & Bau

Sur un Arbre Perche

Sur un Arbre Perché is a story that is told while eating. Its inspiration naturally comes from the fable of Jean de la Fontaine. Paris is an astonishing and marvellous city, and as such is teeming with movement and stimuli. The designers wanted to create an environment that would make its visitors feel really cared for. An authentic and natural space that would provide a peaceful pause in the midst of the commotion of the city. The idea of the fable came up as something natural, simple and obvious. The perched cabins are made from authentic old Swedish barns to give a form to childhood dreams, without compromising on the materials and the way they are perceived.

栖息树上餐厅

"栖息树上"是个要坐在餐桌上边吃边讲的故事……其设计灵感显然来自法国作家拉封丹的寓言"乌鸦与狐狸"。餐厅的设计师们希望创造出一个能够真正关照到客人的环境,为他们构想一个独特、自然的空间,在热闹喧嚣的城市中营造出一个可让人放松享受的天地。厅内所有架高的棚子都是由真正的旧谷仓建材而成。

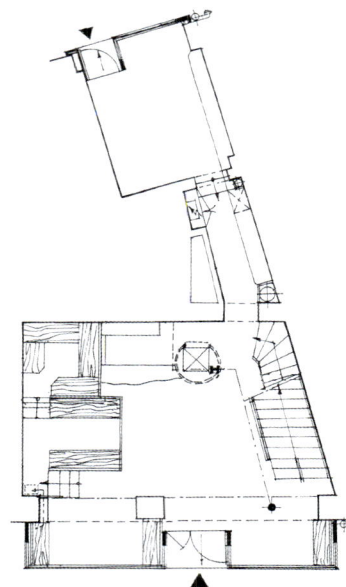

La Tassée

The architectural concept of La Tassée is based on the idea of restoring pride in the historical referents of this old Lyonnaise institution that has been in the same family for three generations. The timeless colour scheme of browns, taupe and sand was thus retained, and given new harmony through sober and elegant lines. The walls alternate between wooden panelling inset with architectural lighting, and a pony skin effect. The resin floor gives a contemporary look to the setting and is echoed in the drop ceiling, into which are set chandeliers that have been customised with silver wine-tasting cups, a symbol of the restaurant.

天时餐厅

天时餐厅的建筑设计构想专注在发扬这间目前已由第三代子孙掌厨的古老里昂餐馆的历史背景。因此，设计师采用了棕色、深黑色和沙土色，配合高雅简洁的线条，呈现和谐的整体。墙上交替呈现镶嵌建筑式灯光的木制墙饰和小马毛皮的材质效果。树脂地面赋于整体空间一种极具现代感的气质，并与天花板横梁的突出结构相呼应。

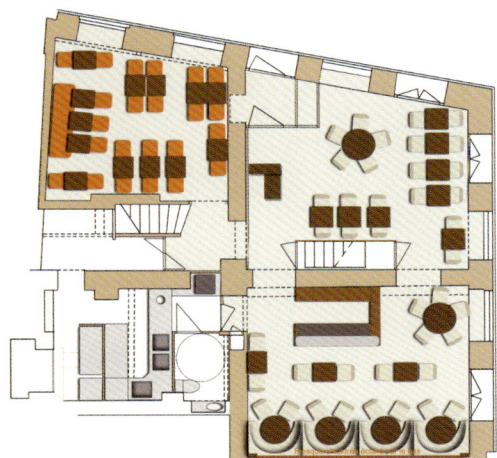

Restaurant

France

Lyon

2008

Photo: Alain Rico

Donatelle Piana & Philippe Batifoulier

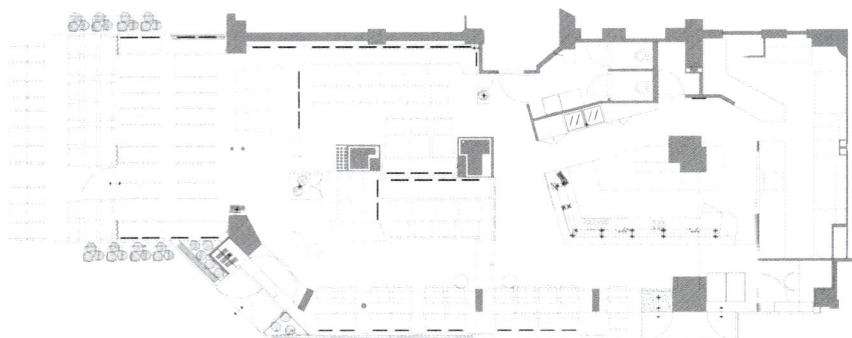

Restaurant

France

Paris

2006

Photo: Jacques Gavard

Olivier Gossart

Le Bosquet

The name 'Le Bosquet', so evocative of nature, served as the main theme from which to design an elegant and homogenous whole using fine materials such as leather and wood with the same background colour of 'earth', between beige and taupe. This warm environment provides a setting for plants and trees such as olives, maples, Christmas trees in the window and a bar made from a collection of small birch trunks. An adjustable lighting scheme allows one to change the ambiance. Long ceiling lights in carbon fibre in the shape of bottles are visible from the outside and define the brasserie visually. Projected on the wall are artworks or artists' films.

波斯科餐厅

餐厅取名波斯科，令人联想到大自然的意象，于是"树林"便成为餐厅构思主线，以设计出优雅和谐的整体空间。设计师利用一些高贵的材质，如皮料或木材等，处理成一个具有"土地"色感的空间背景，介于淡土褐色与暗褐色之间。这个具有温馨色调的环境衬托出厅里植物树木的美感，就像那些被奉在橱窗里的迷你橄榄树和小枫树，还有那运用一块块桦树树干堆叠成的吧台。

Restaurant

France

Paris

2006

Photo: Matali Crasset

Matali Crasset

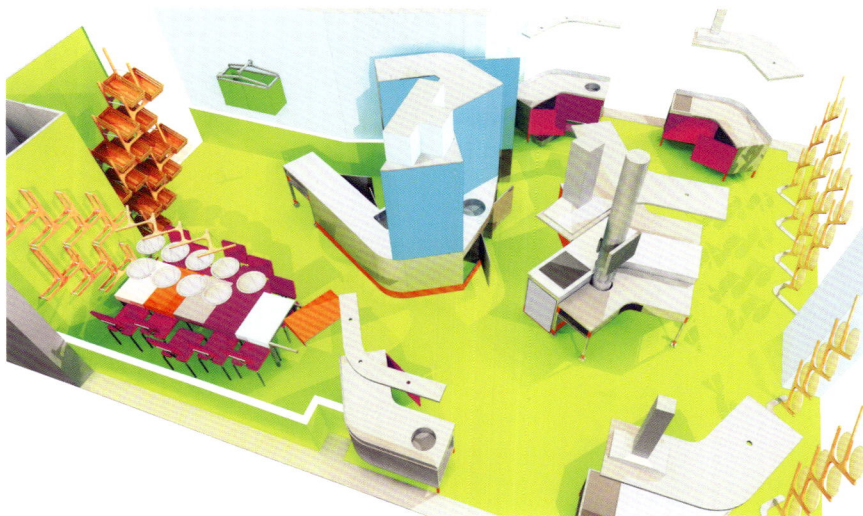

La Cuisine Fraîch' Attitude

Matali Drasset Design is characterised by its unique design and framework for the kitchen design. Here is a flow of space, streamlined style. The project was proposed and demonstrated with a new topic, which is modular in the kitchen. The construction of this multi-functional kitchen is from the characteristics of the cylinder to attract people's vision. This compact space lacks the inclusion of exquisite design. All the details of the design of the mosaic are the master. For the colour, he still loves pink and blue; the work area using orange, reflects the clear warm and romantic atmosphere. Materials are stainless steel and blue glass-based resin.

酷赛餐厅

设计的特色是其独特的框架设计以及厨房设计。这个项目提出并且展现了一个新的议题，那就是模块化的厨房。厨房拥有多功能的特性，以中心的圆柱向外发散，吸引着人们的视线。所有细节的设计、工艺的镶嵌都是精心制作的。对于设计中色彩的选择，依然选用他酷爱的粉红色和蓝色，工作区采用橙色，体现了温馨浪漫的清晰氛围。材料是以不锈钢和蓝色的树脂玻璃为主。

Bar

France

Saint-Tropez

2008

Photo: Olivier Martin-Gambier

Christophe Pillet

Bar du Port

After two months of intensive work, Le Bar du Port in its 2008 version has been entirely remodelled based on a subtle play of mirrors, shadow and light. The latter two elements, which are so presented in Saint-Tropez, were treated as key to the new decoration. The floor is in slate-coloured ceramic tiles, which contrast with the white lacquered bar stools. Glass, a ceiling of smoked mirrors that doubles the height of the bar and gives it a freer space, walls in the white, roughened 'Saint Hubert' stone and white furniture also create a dazzling ambiance in the Bar du Port, which is subtly transformed over the course of the day by three LED screens diffusing changing mood colours.

海港酒吧

经过两个月密集施工后，海港酒吧2008年新版已完全改造出来，整个空间巧妙运用了镜子的反射效果、光与影的对比；在圣特罗佩，光与影两者都是此城处处可见的美景特色，因而也成为此次室内设计的重要关键元素。厅内地面是深灰色的缸瓷石板，与放置其上的白漆吧台高凳形成对比。天花板运用玻璃和烟镜，让人感觉整个餐厅的高度增倍，使空间感觉更自由舒畅。

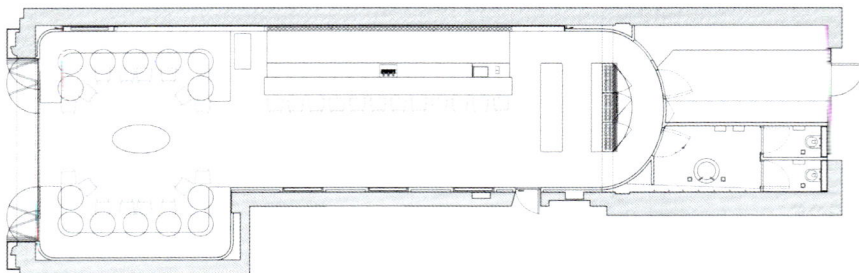

Guilo Guilo

Echoing the famous novel 'In Praise of Shadows' by Jun'ichiro Tanizaki, the designer conceived a space drawn with light and shadow, where simplicity and discretion are of the utmost importance. In the middle of the main room, bathed in clear light that illuminates the dishes, is the chef's station, where the preparation of the cuisine takes place. An immense, metallic, rectangular block overhangs and dominates this space, serving technical requirements while also centring the view and the feeling of presence towards this 'kitchen'. Guilo Guilo immerses the diner in a complete experience: everything in its own time.

枝鲁枝鲁餐厅

设计师受日本文学大师谷崎润一郎的名著《阴翳礼赞》的影响，建构出一个呈现阴影与光线对比的空间，秉持单纯简约、含蓄审慎的设计原则。一个巨大长方形金属体从天花板高高向下延伸，主导整个空间，如此设计不只是烹饪技术层面的需要，也引导客人视线投向这个"厨房"空间。枝鲁枝鲁邀请客人参与一段完整的美食之旅，先体验视觉震撼，再满足味蕾享受。

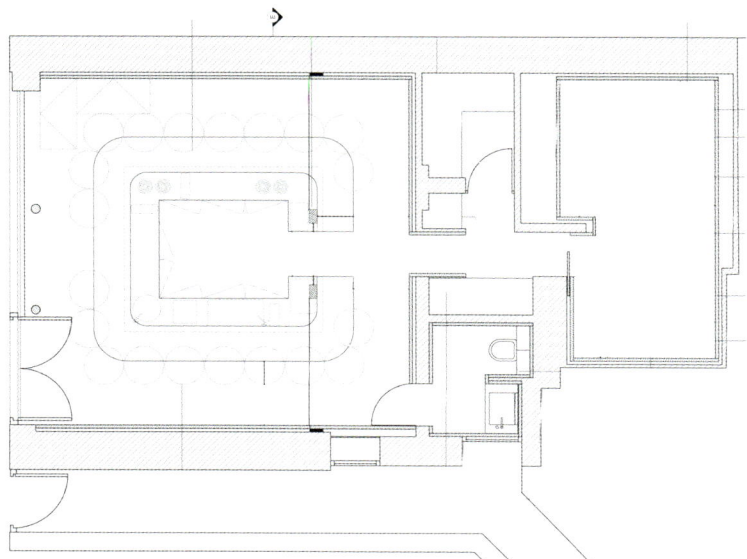

Restaurant

France

Paris

2008

Photo: Olivier Martin-Gambier

Christophe Pillet

367

Tokyo Eat

A large, translucent box hangs from the ceiling. The sun's rays enter through its south-facing window, cross the space and cast long shadows when they encounter the diners, thus playing a part in the animation of the space. This box contains water closets from all over the world. These translucent toilets play on the relationship between seeing and being seen, inherent in the design of fashionable restaurants. This collection of world toilets actually forms part of the museum's collection, exciting curiosity for the art itself, and includes a German WC, a Japanese WC with its self-cleaning bowl, a large American WC, an Indian WC, an Italian urinal and a French child's WC...

东京食餐厅

餐厅宛如天花板上悬挂着的巨大透明盒子。阳光透过朝南的窗户照射进来，营造出光影效果，使空间更显活跃。餐厅里收藏着来自世界各地的洗手间，起到了装饰作用。形形色色的洗手间也是博物馆馆藏的一部分：德国洗手间、日式自动清洗水槽、巨大的美式洗手间、意式小便池和法式儿童洗手间……这些洗手间也是吸引着游客一大因素。

L'Auberge de L'ill

The glass-roofed entrance porch, clad in horizontal wooden laths like the traditional local tobacco drying sheds, draws attention to the attractive main door. A pathway of ceramic tiles and brushed carpet leads visually towards a wrought-iron sculpture of a fish. Serious thought has also been given to the lighting, whose intensity and colour evolve as the meal progresses. In the second half of the evening, the intensity slowly resides towards a very warm atmosphere, while, in counterpoint, LEDs change from amber to light blue.

伊尔河客栈餐厅

客栈重新装修后，吸引了络绎不绝的宾客。餐厅入口处用玻璃架构出来的玄关空间外围装了水平向排列的木条，就像当地的老式烟草干燥屋的外观，使客栈的大门更为醒目。整个客栈的灯光设置也作了深入的改装，跟随用餐进程，灯光的强弱与色彩也跟着改变，到了晚餐的后半时段，灯光渐趋柔和，营造出温馨的氛围，而相对地，LED灯也渐渐从琥珀色转为淡蓝色。

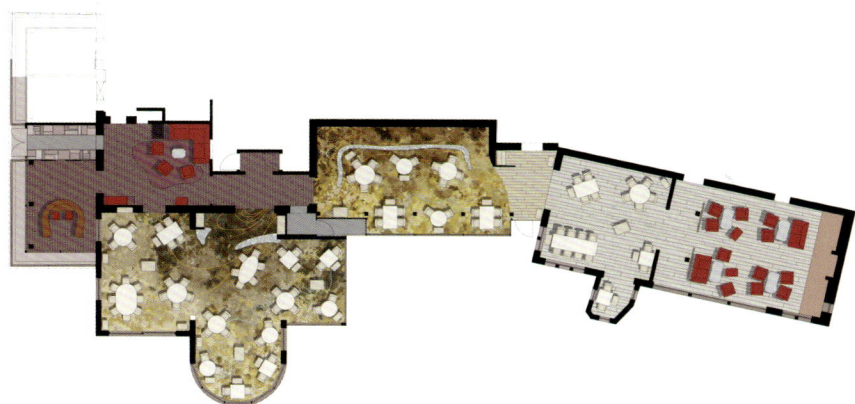

Restaurant

France

Illhaeusern

2007

Photo: Eric Laignel

Patrick Jouin

Restaurant

France

St. Barthelemy

2007

Photo: Jean-Philippe Pitter

Penny Morrison

Restaurant in Hotel St Barth Isle de France

Located on one of St. Barthelemy's most stunning beaches, Baie des Flamands. The intimate Hotel St Barth Isle de France beautifully combines the charm of the West Indies with the sophisticated style of the south of France. The Restaurant 'La Casa de L'Isle' overlooking Flamands bay offers a wide variety of dishes including fusion food, salads and fresh grilled fish. The restaurant overlooks the ocean and has additional seating known as 'La Cabane de L'Isle de France', directly on the sand for relaxed lunches.

法国圣巴斯岛酒店餐厅

这家酒店坐落在圣巴泰勒米岛最令人叹为观止的海滩上。圣巴泰勒米酒店内的装修亲切温馨。完美地结合了西印度群岛的魅力和法国南部精雕细琢的设计风格。小岛之盒餐厅俯瞰弗拉蒙斯海湾的美景，提供各种各样的美食，包括：融合菜式、沙拉和新鲜的烤鱼。餐厅俯瞰大海，并且拥有直接位于沙滩上的额外坐位，客人可以在那里享受休闲午餐。

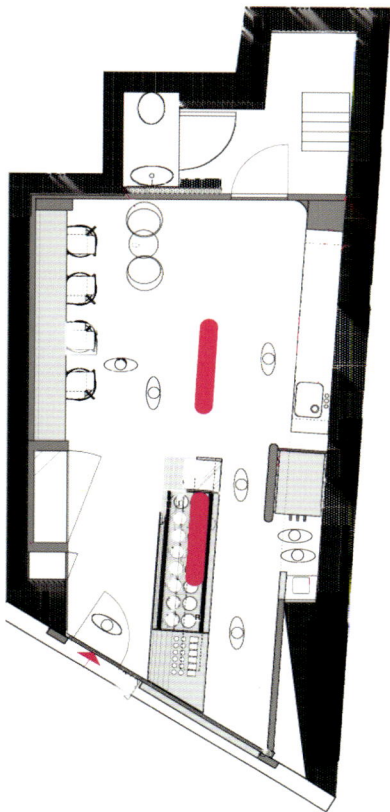

My Berry Ice Cream Bar

A space led by the white, with curved and elegant shapes, which is the illustration of the restoration concept: '0% fat yoghourt'. This space was designed with the same movement once needed to serve a cup (cutting) of ice-cold yoghourt. To say it in another way, the gesture of the maitre glacier was extrapolated to synthesise it in an architectural space. The outside façade clearly reveals this idea of Claudio Colucci's concept and already suggests the tasting. The bar counter is not more than the logical continuity of this white, elegant, dynamic and curvy material. It is a place with no free effects, a variation of refreshing colours just like the products which will come to complete the yoghourt.

吾之莓冰淇淋馆

这是一个以白色为主调、以圆弧线条及优雅形体来设计的空间，符合"零脂肪优格"的新餐饮概念的形象。设计师的灵感来自挖一杯优格冰淇淋的动作，也就是说，设计师将撷取冰淇淋的动作转换成设计此建筑空间的构想线条。店面的外观设计已明显地揭露设计师的这个构思，而且呈现出产品的诱人美味。冰淇淋柜及吧台理所当然地延续了这白色的质感及优雅又有活力的曲线。

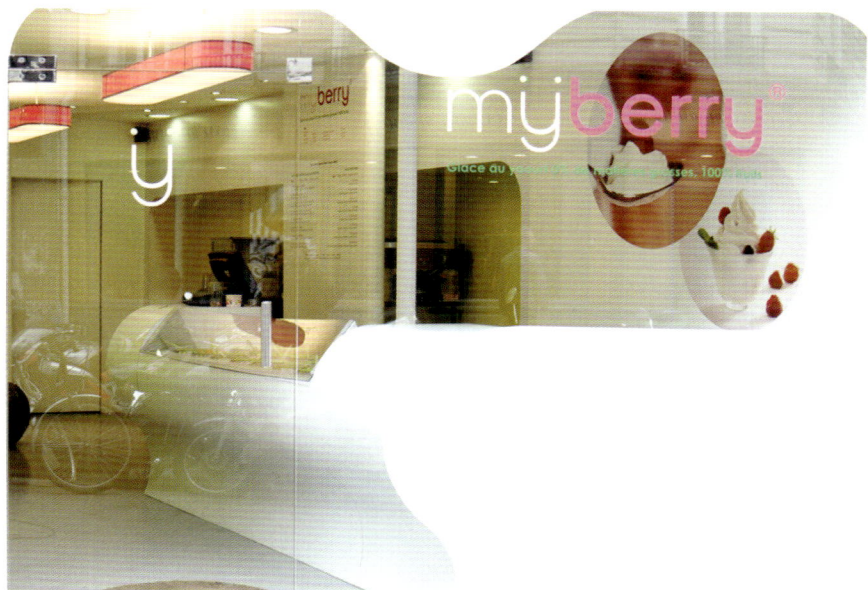

Bar

France

Paris

2008

Photo: Gilles Toledano

Claudio Colucci

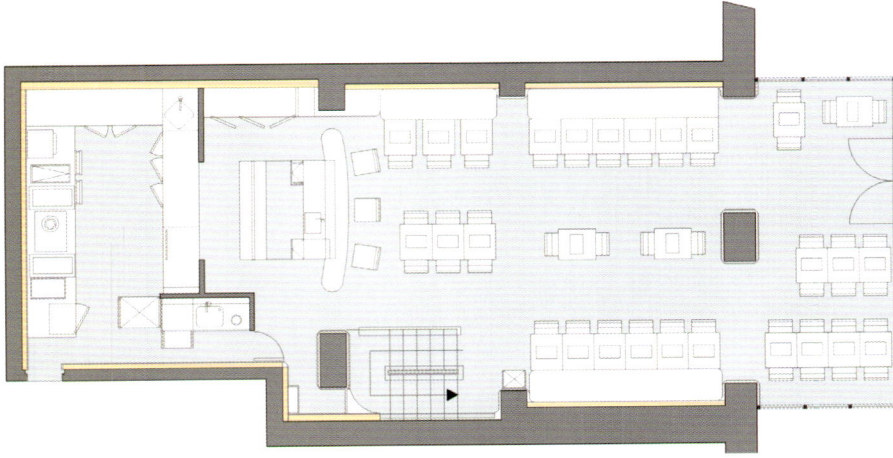

Restaurant

France

Paris

2006

Photo: Osmose

Olivier Gossart

Osmose

Thought of as a cocoon entirely moulded in plaster and covered with white and matt lacquer, the restaurant has a black parquet floor that contrasts and sets off the furniture created by Olivier Gossart. The bar in the form of a brilliant white pebble is in 'osmosis' with the shell of the restaurant. Behind this, a curtain of water between the dining room and the kitchen allows one to make out the silhouettes of the kitchen staff. The banquettes and chairs are lacquered white and covered in white fabric. The tables are in lacquered dark grey glass and initialled 'Osmose'. Lights are set into the plaster ceilings, with LEDs highlighting the banquettes and create changing coloured ambiances.

欧思摩兹餐厅

设计师将餐厅构思成一个蚕茧般的空间，整体室内用纤维灰浆模制并覆盖一层白色雾面漆，而黑色镶木地板正好与其产生对比。吧台设计成鹅卵石形状，与整个餐厅的蚕茧构思有异曲同工之妙，不过吧台采用的是白色亮面材质。吧台后面设置了一幕水帘，介于用餐区和厨房之间，客人可透过水帘瞥见厨师们的身影。

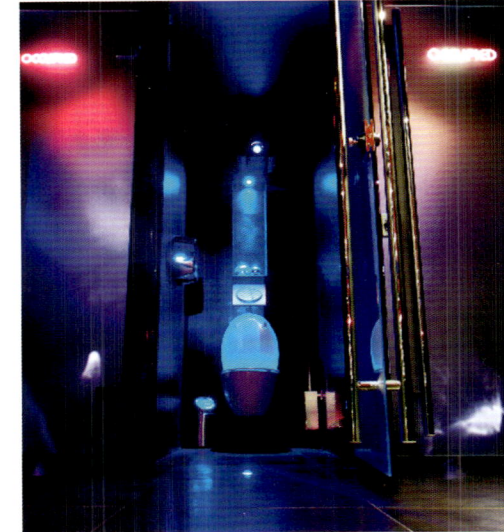

Restaurant in Hotel Daniel

The Lounge and the Restaurant Daniel are the perfect location for intimate conversation against a colourful and elegant backdrop. The walls are decorated with gilt glass panels portraying exquisite Chinese gardens, created by the artist Gérard Coltat, who is based in Castagniers in the south of France. The dark wooden tables are inlaid with mother-of-pearl. A silver-backed display case houses a collection of multi-coloured glasses brought back by the owners from exotic locations. Throughout the hotel, the specialist painting of the walls was carried out by the artist Anne Laure Thuret, who was also responsible for applying the silver leaf eglomisé to the back of the display cases in the Lounge.

丹尼尔酒店餐厅

丹尼尔餐厅是进行私密会话的绝佳场所，拥有多彩多姿而优雅的环境。墙壁上装饰的镀金玻璃板上描绘着中式园林，由法国南部卡斯塔格尼尔的艺术家杰拉德·克尔塔特创作。黑木餐桌嵌入了珍珠母。银色的展示柜里摆放着主人从世界各地带回来的彩色玻璃杯。酒店墙壁上独特的绘画作品全部出自艺术家安妮·劳雷·杜莱特之手，她还负责在展示柜上绘制了银色叶子。

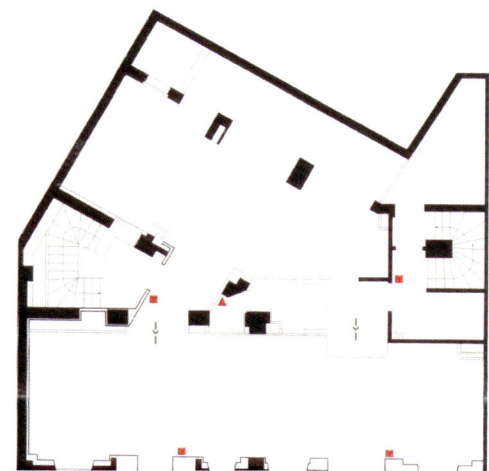

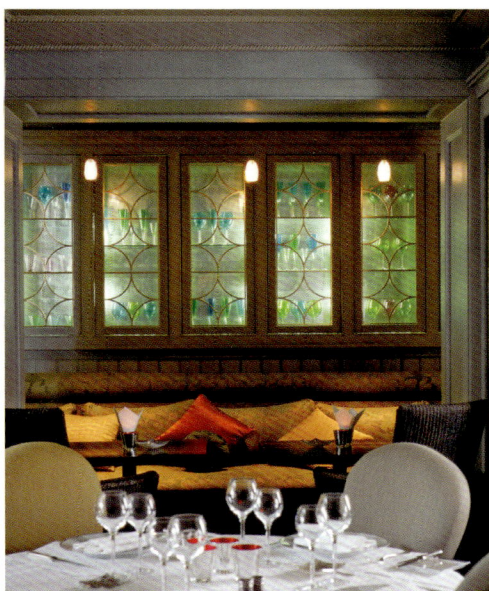

Restaurant

France

Paris

2009

Photo: David Wormsley

Tarfa Salam

Restaurant

France

Paris

2008

Photo: Yvan Moreau

Valérie Serin

Nabulione

The restaurant Nabulione has an extraordinary setting facing the D.me des Invalides, at the heart of one of the most confidential neighbourhoods of the French capital. The guest of honour is Napoléon, who is called Nabulione in Corsican. The story recounted here is like a private visit to the Invalides, in particular its Dome and Salon d'Honneur. The place gives the impression of having always been there, and only lightly touched by the designer. The presence of Napoléon is invoked by porphyry, the marble of Emperors. All the architectural and decorative codes suggest the Dome, and the Grand Salon d'Honneur, with its chimneypiece and panelling.

拿破里昂餐厅

拿破里昂餐厅有着格外特别的地理位置：面对着巴黎荣军院的金色圆顶，位处法国首都机密性最高的地区中心。餐厅的荣誉贵宾就是法国历史英雄拿破仑，其家乡柯西嘉方言则发音成拿破里昂。宾客来到此餐厅就犹如私下参访荣军院，尤其就像参观了圆穹和院中的荣誉厅。项目有种屹立不倒的气势，设计师只做了极小的改动。斑岩——大理石之王彰显了拿破仑的气势。所有的建筑设计元素都与金色圆顶和荣誉厅相似，尤其是壁炉架和镶板设计。

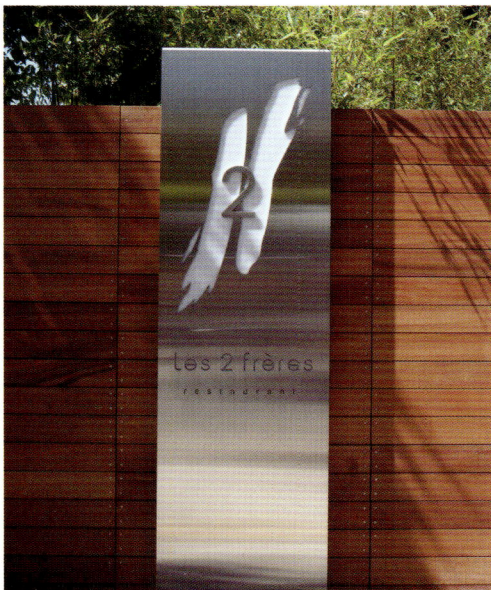

Les 2 Frères

The exteriors of the restaurant Les 2 Frères are defined by high boarding in exotic wood and by bamboo hedges lit up in an indirect play of light. Jean-Christophe Sabarthes is the architect: "My project leans on the properties of simple, geometric forms, and the idea is to create a spare but welcoming decorative scheme. For that I have used common elements in the interior and exterior decoration and 'real' materials such as glass, wood, metal and concrete." The main entrance of the restaurant is a portico in stainless steel clasped on either side by windows, responding to the serving hatch that is found opposite in a perfect symmetrical alignment.

兄弟餐厅

兄弟餐厅的外围是运用带异国风情的进口木材所制之高墙和一丛丛竹篱来界定空间，并搭配了间接照明系统所营造出的灯光效果。负责此设计工程的建筑师诠释道："我对此方案的构想着重在呈现单纯的几何形体，意在创造简洁优雅却不失温馨的装修空间。为此，在室内及室外，我都使用相同的建材，运用玻璃、木材、金属和混凝土这些真材实料来构建。"

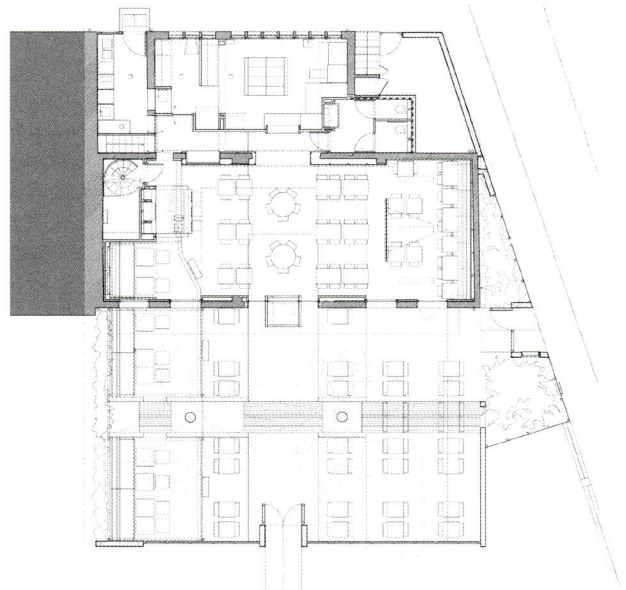

Restaurant

France

Aix-en-Provence

2006

Photo: Les 2 Frères, Agence Caméléon

Xavier Luvison & Jean-Christophe Sabarthès

Restaurant

France

Paris

2007

Photo: Arnaud Rinuccini

Atelier FB

376

Hanawa

The premises (spread over three floors of 1,500 square metres) is composed of nine spaces with distinct themes, each zone being fitted out specifically in line with the cuisine on offer. A vast ground floor welcomes visitors and houses a tea salon with a serene ambiance in counterpoint to the delicate patisseries. At the three below-ground counters with their futuristic design, chefs prepare a gastronomic Western-style cuisine cooked on heated plates in front of the clients – Teppanyaki. On the first floor, three rooms are arranged around a Japanese Tsubo-niwa garden, where lanterns in lava stone and miniature vegetation are in harmony with the traditional Japanese cuisine.

汉纳瓦餐厅

餐厅有三层楼，共1500平方米，由9个不同主题的空间组成，每个区域都依照其所提供的餐点而做出特殊的空间设计。迎接宾客的是一个广大的地面层空间，这里并设有一间气氛宁静的品茶沙龙，搭配着精致多样的糕点。地下楼层的装修具有未来主义设计风格，厅内有三座铁板烧台，师傅当着客人面，在铁板上烹饪着西式的铁板烧美食。

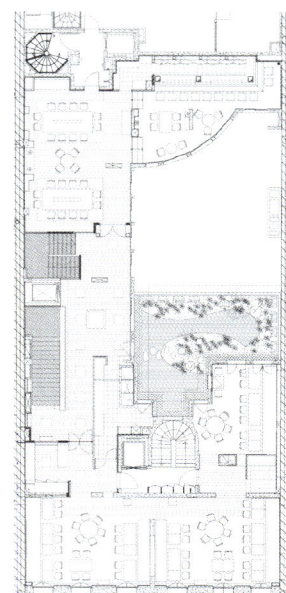

Le Pré Catelan

Le Pré Catelan first opened at the beginning of the 20th century and was a very popular place. A century later, the new design required an important aesthetic and intellectual thought. Pierre-Yves Rochon's aim was to give the place back its grandeur. Starting from the existing construction, the designer wanted to draw up a modern and striking concept that would emphasise the elegance of the original setting, in harmony with the surrounding nature. This idea manifests itself through the furniture as well as the materials, the colours and the lighting. The predominant colours are green, black, white and silver. Black and white are used alternately to emphasise the differences between each area.

佩卡特兰餐厅

佩卡特兰餐厅早在20世纪初期就已开幕，而且一直是很受欢迎、高朋满座的餐厅。此次全新的室内设计需要具有非常敏锐的美学与智慧思考。建筑师的目标在于使这栋建筑重现其宏伟堂皇的气质。因此，设计方案的主导原则是从现存的架构着手，进而勾画出一个非常现代且令人惊叹的设计概念，以此来强调原有建构的宏伟。

Restaurant

France

Paris

2007

Photo: Catherine Jaillon, Claude Weber

Pierre-Yves Rochon

Restaurant

France

Paris

2007

Photo: Toustem

Matali Crasset

Toustem

Matali Crasset thought it a good idea to claim the space through the floor: an orange resin has been poured over it, seeming to splash in its wake over furniture, doors, plinths and even the plates. The restaurant's character is far from being constrained and the concept can express itself at will. In the cellar room, a false ceiling of stretched fabric covers the raw stone and breaks with the rustic and almost austere character of the place. Finally, Hélène Darroze had fun by sliding very personal touches into the decoration: an impressive bright pink light covering at the entrance, a dark wood counter and high stools that are perfect for enjoying tapas and cocktails, and a massive spit.

图斯坦餐厅

设计师认为本方案必须从地面的处理来掌握整体的空间，因此她采用了橘红色的树脂来作为铺地的材料，鲜艳的色彩于是一路泼洒到桌椅家具、厅门、踢脚板，甚至蔓延到餐盘上来。这个餐厅的装潢布置并非就此一成不变，它的设计基础概念能够依主人的兴致所致，而延伸出其他多种变化。

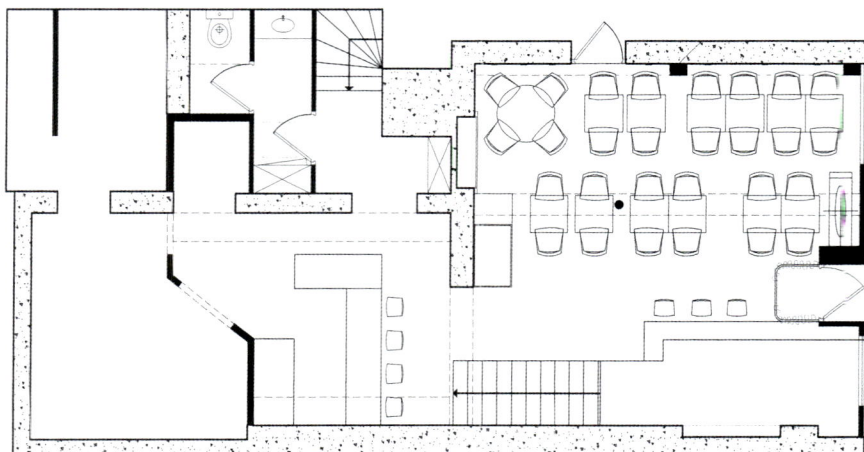

L'epi Dupin

The main project was to move the bar to the back of the restaurant in order to group together the practical areas and to free up the reception space on the façade side. This allowed the owners to offer a different, less formal kind of welcome by creating a communal table as a prolongation of the bar. The continuity of 'communal table-bar', with the masses in good proportion to each other, also allowed them to give the place its own character. The materials were chosen with simplicity and rigour in mind, setting up a contrast between the warmth of solid wood and leather, fine and natural materials, and the black synthetic resin that forms the bar, the communal table and the bases for the wooden table.

艾皮杜庞餐厅

设计方案主要的重点是先将吧台移至餐馆最里面的位置，把工作空间合并在一起，也因此空出靠店面外边的位置以作为接待客人的空间。这样的格局变动能够提供一种截然不同的接客方式，比较轻松，避免刻板，并留出空间摆置一张大主宾桌，延伸了吧台空间。"主宾桌–吧台"这条连续动线构成一个看起来高朋满座的整体空间，也因而成为此餐馆的特色。

Restaurant

France

Paris

2008

Photo: Jean-Charles Valienne

Frédérique Gormand & Christophe Vendel

Restaurant

France

Paris

2006

Photo: Bureau Betak

Alexandre de Betak

Black Calavados

This little theatre has been ideally conceived and put in the spotlight (or rather the shadows) by Alexandre de Betak, the artistic director who is the darling of the couture shows (Dior, Viktor & Rolf, Victoria's Secret, etc.). The simple volumes of the restaurant are on a human scale. Betak here uses all the nuances of black, also working with the light in a very individual fashion. The furniture, also designed for the place, offers smooth, brilliant, matt or satin surfaces. Resin, black stainless steel, smoked mirrors answer each other in a space where everything has been done to deflect stares, or to see and be seen.

黑色卡尔瓦多斯餐厅

这舞台般的餐饮空间是由亚力山大–德–贝塔克精心设计，他是高级服装走秀界(迪奥、维果罗夫、维多利亚的秘密等)最抢手的艺术总监，此次他负责整个内部的规划布置，直至最细微部分的设计。餐厅简洁的空间场域设计符合客人的舒适需求。贝塔克巧妙运用黑色各种色调，也以十分独特的方式设计灯光。为餐厅量身定做的家具呈现光滑、亮丽、雾面消光或上光的表面质感。

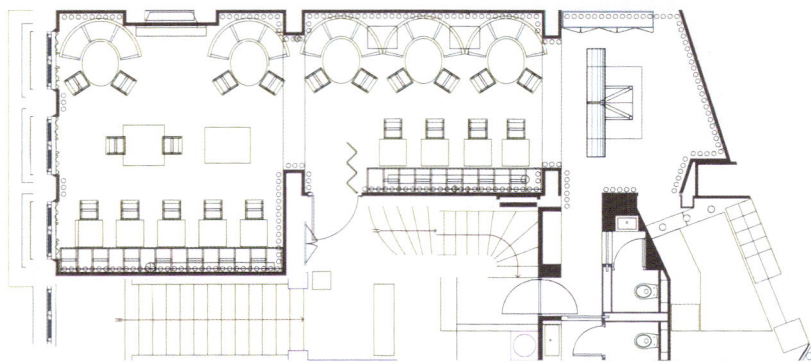

Restaurant in Gabriel Hotel

Located in the busy heart of the city, Hotel Gabriel is a metropolitan oasis in the truest sense of the term. The design reveals a firm dedication to guest rejuvenation and relaxation entirely in keeping with the theme. Guest room interiors celebrate Zen minimalist charm in white and pastel, while the bar lounge features a more surrealist decor in shades of white. Linear furniture with cruise-liner portholes and curves creates a harmonious balance best expressed in the guest room centrepiece, a vertical 'cabinet' comprised of a mirror-hidden television, a folding desk and a mini bar. Signature ballerinas throughout the property add a whimsical touch to the dreamy 1930s' feel.

盖博瑞尔酒店餐厅

盖博瑞尔酒店坐落于繁华市中心，其设计以"为客人营造舒适放松的环境"为主要理念，堪称都市中的绿洲。客房内彰显禅宗的简约风格，采用白色或单色调装饰，而酒吧则以超现实主义风格为主。此外，客房内的特色元素还包括带有舷窗的家具———一个由电视柜、折叠桌及小茶几构成的组合柜。随处可见的芭蕾舞演员画片更是增添了20世纪30年代的梦幻气息。

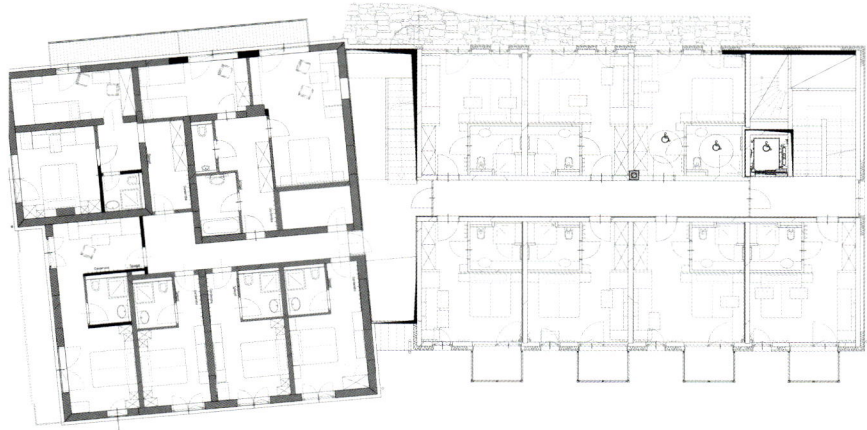

Restaurant

France

Paris

2007

Photo: Silent Factory, Philippe Ruault

Philippe Boisselier

Le Saut du Loup

The restaurant Le Saut du Loup is accessible by two entrances, one via the Musée des Arts Décoratifs and the other through the Carrousel garden. In garden design, a 'saut de loup' is an invisible limit, made up of a ditch and a wall leaving the perspective intact. Used as protection, it allows for perfect continuity between two landscapes. The whole of Philippe Boisselier's project is linked to this idea of a 'saut de loup'. On two open levels between the museum and the garden, the project plays with the contrast between black and white to construct the space.

防狼堑壕餐厅

"防狼堑壕"位于巴黎市中心，可从瑞弗利路装饰艺术展览馆入口或卡鲁榭花园方向进入餐厅。在园林艺术领域中，"防狼堑壕"指的是一道看不见的界限，由壕沟与堑墙组成，留下完整无碍的透视远景。整个方案即是依循这著名的"防狼堑壕"的建构原则。这个两层楼餐厅位于装饰艺术馆和花园之间，方案设计以黑白对比来建造整体空间。

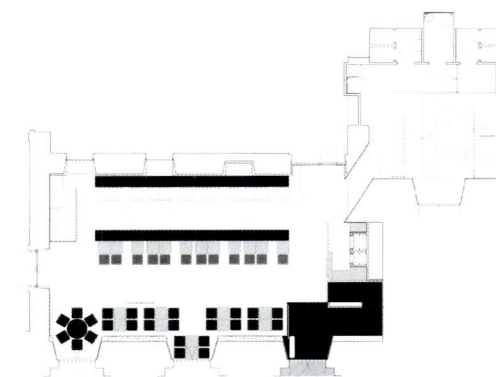

Le Restaurant 1947

In Courchevel, in the hotel Le Cheval Blanc created by Sybille de Margerie, the restaurant Le 1947 makes a daring statement, in harmony with the surrounding nature. A refined world, it rejoices in a play of shadows and light with the mountains as its backdrop. Details range from stone slabs in an Opus Romanus pattern, a fireplace, bronze doors patinated gun-metal grey, the quality of the skirting and ceilings in gingerbread-coloured walnut to the door lintels in alabaster. Bronze and alabaster wall lights and lamps with silk shades diffuse a subtle light.

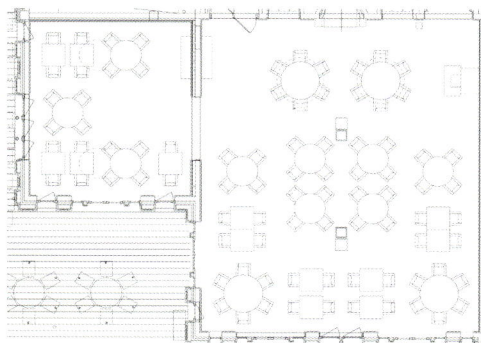

1947餐厅

1947餐厅位于高雪维尔城，在席比勒•德•马格丽设计的白马酒店内，餐厅的整个构思十分大胆，却又与四周自然环境协调呼应。这个设计极其精致的餐厅当中富含着无数的巧妙光影效果，而且拥有珍贵的山岳景致。室内设计呈现出丰富的细部巧思，例如：具有特殊纹路的压花石材、壁炉、黑灰色光泽的铜门、高质量的建筑底座、蜜糖蛋糕色的胡桃木天花板、以雪花石膏精制出来的门侧。

Restaurant

France

Courchevel

2006

Photo: Marc Bérenguer

Sybille de Margerie

Restaurant

France

Paris

2009

Photo: Philippe Dureuil

Idoine Agency

Tiger Wok Restaurant

Tigerwok Restaurant is a happy blend of flavours from East and West in relaxed Zen ambiance. An informal system of self service allows the customer free rein to invent a multitude of possible combinations of flavours himself. A surprising mix which invites you to try out recipe ideas from around the world. In order to give a greater boost to its success, the Tigerwok team appealed to the Idoine design agency to optimise its concept for subsequent openings. The gamble paid off, since a brand new Tigerwok has just opened at Ivry, at the foot of a Pathé cinema complex, where the graphic and architectural codes now express this concept better than ever.

老虎锅餐厅

老虎锅餐厅在一个充满禅意的放松环境中供应东西方美食。一个非正式的自助服务系统上顾客可以自主选择自己喜爱的口味，品尝世界各地的美食。为了扩大餐厅的影响力，老虎锅营销团队委托Idoine设计公司来为品牌进行全套的设计。设计的成果在伊夫里大街的新老虎锅餐厅中展现出来，新的平面和建筑设计都更贴切地传达了其品牌理念。

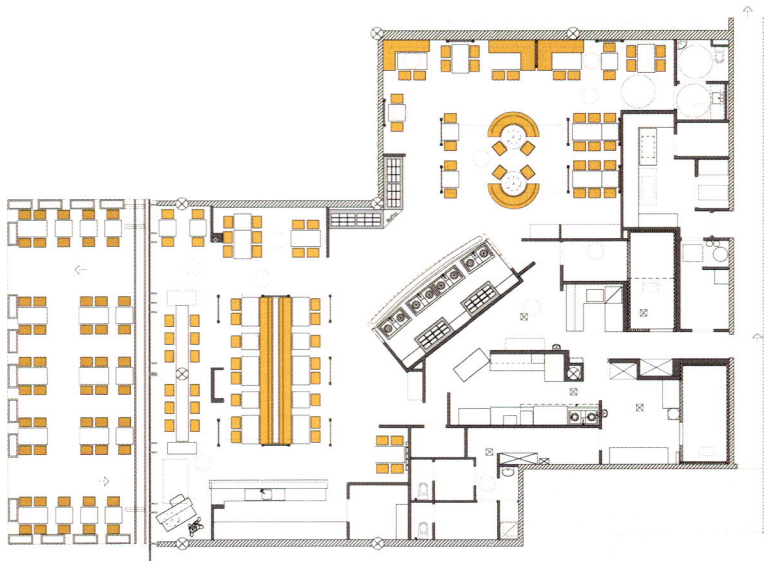

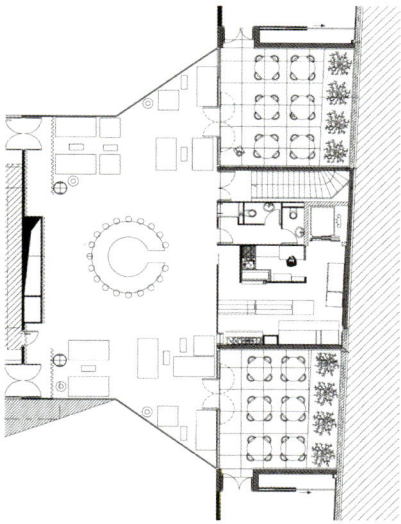

Bar

France

Lyon

2008

Photo: Cyrilledruart

Cyrilledruart

Lounge Bar

The lounge spreads across the building and is based on a symmetrical layout. In its centre stands the black cone-shaped bar, entirely made of Dupont Corian. Not only a technical challenge to build, it becomes a sort of sculpture and a user-friendly furniture, as everyone faces each other. Around the bar, the lunch or diner areas, made of wooden low tables and grey sofas. From some seats, visitors have a direct view over the simulation areas, which participate to the show. The lounge should be apprehended as a separate world. A world of dreams and mystery, it is an intimate, refined and sophisticated space that also offers exceptional cuisine.

休闲酒吧

休闲酒吧在建筑中呈对称结构展开，其中心是一个全部由杜邦可立耐材料制成的黑色锥形吧台。吧台具有雕塑效果，也让使用者感到舒适——圆形结构人们可以面对面而坐。吧台四周是由木制矮桌和灰色沙发所组成的就餐区，从一些位子可以直接看到表演区的表演。休闲酒吧是一个与世隔绝的桃花源，充满了梦想和神秘气息，在一个私密而精致的空间里为顾客提供非同寻常的美食。

Restaurant

France

Suresnes

2009

Photo: Dragon Rouge

Dragon Rouge- Georges Olivereau

Villa Plancha

The Casino Group, one of the leading retail distribution groups in France, is also active in the catering industry. In order to diversify its offering in this area, Casino wanted to create a new catering concept with restaurants located on the outskirts of urban areas. 'Villa Plancha' is the name given to this new concept, a name created by Dragon Rouge to express a restaurant that immerses its patrons in a world of evasion, allowing them to discover the delights of the à la plancha style of cooking. Dragon Rouge also created the Villa Plancha logo: aniseed green against a bright pink background that includes the shadow of an olive branch.

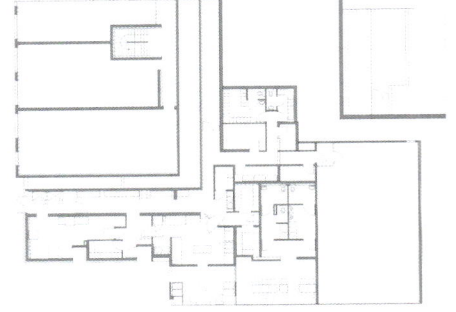

铁板别墅餐厅

作为法国顶尖零售集团之一，赌城集团在餐饮业也十分活跃。为了发展餐饮业，赌城集团决定再打造一家新式概念餐厅。餐厅被命名为"铁板别墅"，其寓意在于让客人在铁板美食中找到一个世外桃源。设计师还为餐厅设计了新标识：亮粉背景上写着茴香绿的字母，背景上还装饰着橄榄枝阴影。

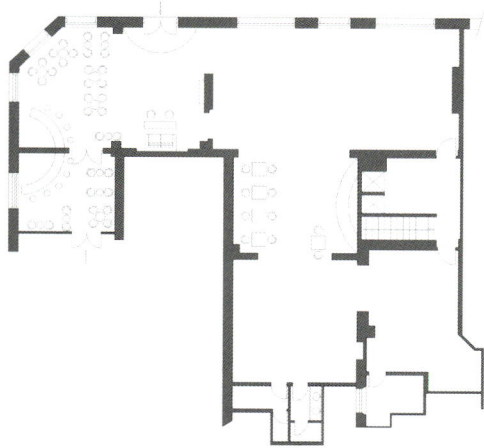

Les Enfants Terribles Paris

The project of 'Les Enfants Terribles' is a restaurant in Paris, which means in English 'terrible children'. Two years after opening the restaurant 'Les Enfants Terribles' in Megève, Jocelyne and Jean-Louis Sibuet have created a restaurant bearing the same name just off the Champs-Elysées. It is a new address in Paris to boast traditional quality of French cuisine. This new venue reflects the desire for the creators of the Fermes de Marie to pursue the development of their group and its concept. It is the Sibuet son, Nicolas, who supervised the metamorphosis of the restaurant that formerly belonged to Johnny Hallyday (a famous French singer) under the name of 'Rue Balzac', which now has a capacity seating of over one hundred.

可怕的孩子餐厅巴黎店

"可怕的孩子"是一家位于巴黎的餐厅。由于"可怕的孩子"在梅杰夫、乔斯林和让路易的餐厅大获成功,餐厅主人希布特决定在香榭丽舍大街上开设一家同品牌餐厅,专门供应法国传统美食。餐厅延续了"玛丽的农庄"的设计理念。希布特的儿子尼古拉斯负责整个餐厅的设计改造工作。餐厅原属于法国知名歌手约翰尼·哈里代,能容纳100多人。

Photo: Frédéric Ducout

Group Sibuet / Nicolas Sibuet, Hervé Thibault, Jocelyne Sibuet

387

Restaurant

France

Paris

2009

Photo: Ralston Bau, Vincent Baillais

Ralston & Bau

Sous les Cerisiers

The latest restaurant design by Ralston & Bau is an intimate space in Paris that is dedicated to a balanced fusion of Japanese and French gastronomy. 'Sous les Cerisiers' (under the cherry trees) is an invitation to taste the delicious cooking of Sakura Franck. Both cultures have influenced the theatrical interior concept. The classic opera and geisha cultures from France and Japan can be seen in the scenographic layers, costumes and shadows cast throughout the interior. A contrast of dark and light spaces divides the room: the bright area with the bar as a centre point is used for cooking courses during the daytime, and the dark part, following a perspective angle, including a VIP space to enjoy a gastronomic menu at night. Moveable and translucent walls separate the seating areas for privacy.

樱花树下餐厅

Ralston & Bau公司所设计的"樱花树下"餐厅完美地结合了日式和法式美食，其寓意为邀请客人到樱花树就餐。餐厅的设计理念受到日法两和文化的影响：法国的歌剧文化和日本的艺妓文化在室内设计的布景、装饰和光影效果中都有体现。空间分为明暗两个区域：明亮的区域以吧台为中心，提供日间餐饮；昏暗的区域包含贵宾区，可以在晚上观看艺妓表演。可移动的半透明墙壁保证了座位间的私密性。

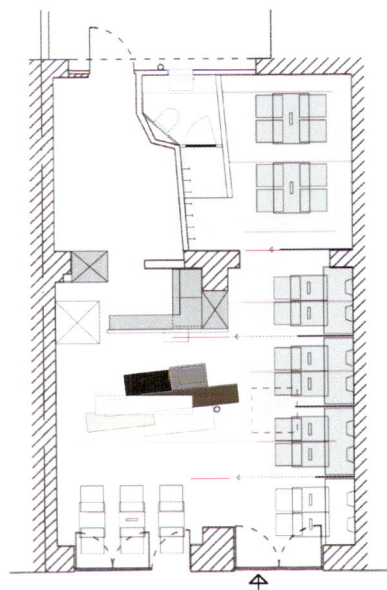

Restaurant

France

Paris

2008

Photo: André Morin

Kengo Kuma & Associates

Jugetsudo

The Saint-Germain-des-Pres area is definitely trendy and welcomes the very first Japanese tea house in Paris. Bright and natural, harmonious atmosphere, the Jugetsudo is a Zen place where the tea ceremony is a lifestyle. Created by Maruyama Nori, the tea house Jugetsudo perpetuates the Japanese tradition and the cult of beauty of nature. The tea is drunk here with reverence and mobilises the five senses. The designer of the tea house wanted to create a space like a bamboo thicket. In the thicket floats a different kind of air and light from those of our daily lives. At the centre of this unique space, the designers placed a solid, jointless board of Japanese cypress. Cypress was a special tree in that it was believed to smoothen the things put on it, so people could feel the nature of Japan on that board.

寿月堂

寿月堂位于巴黎圣日耳曼区，是巴黎第一家日本茶室。明亮而自然和谐的氛围让寿月堂充满了禅意，在这里，饮茶是一种生活方式。寿月堂茶室融合了日本文化和对自然美的崇拜。人们怀着敬意进行饮茶活动，并以此调动五感。设计师试图将茶室打造成一片竹林，里面是与人们日常生活截然不同的气氛。空间的中心是一块无接缝日本扁柏木板，柏木具有能够柔化物品的功能，人们可以透过它感受到日本文化的本质。

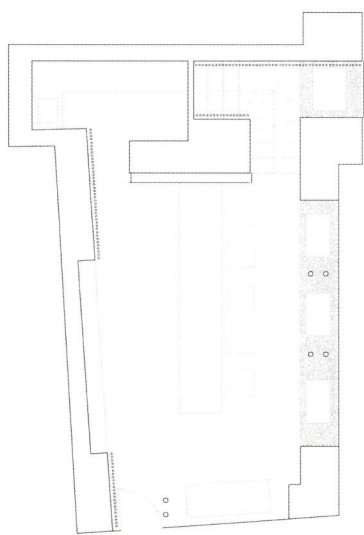

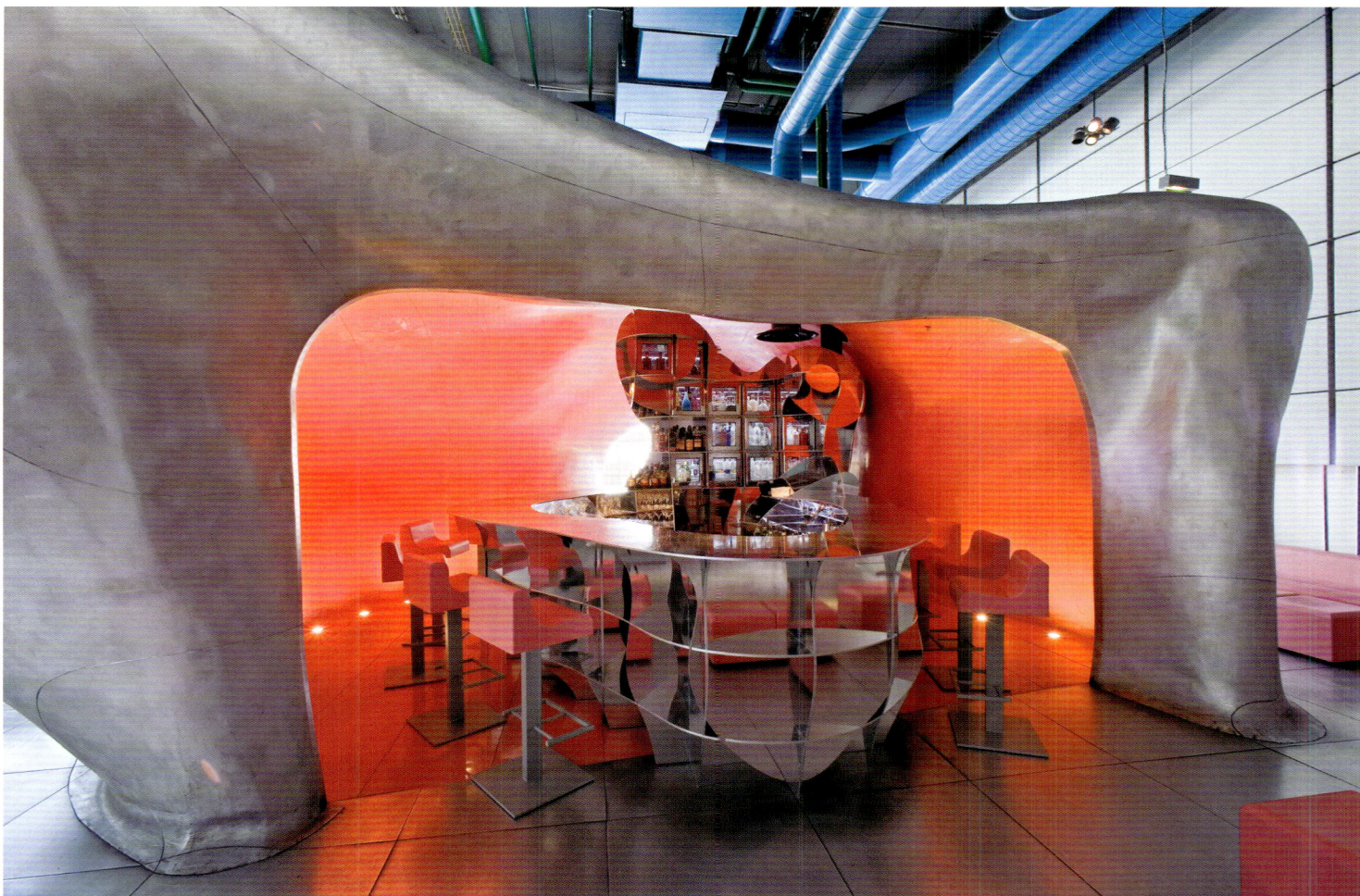

Pink Bar

The bar is a new project, built inside the volume previously housing the cloakroom and restrooms. This new 'Pink' space, as second generation project, is conceived as a resultant element based on the same grid of the original restaurant project. The designers designed a bar for champagne piper-heidsieck in conjunction with piscine. The striking piece is crafted from Suryln, which is a material that is often used in molded luxury goods and morphs into a satellite champagne bar which draws its inspiration from the brand's iconic and extravagant red colour.

粉红吧

粉红吧的原址是衣帽间和化妆间。新建的粉红空间在原有餐厅的网格规划上进行了改造。设计师特别设计了一个白雪香槟吧台，旁边是饮酒台。这个大胆的造型结构由Suryln制成。Sury n经常运用奢侈品建模，由它制成的香槟吧的灵感来自于酒吧的标志性夸张大红色。

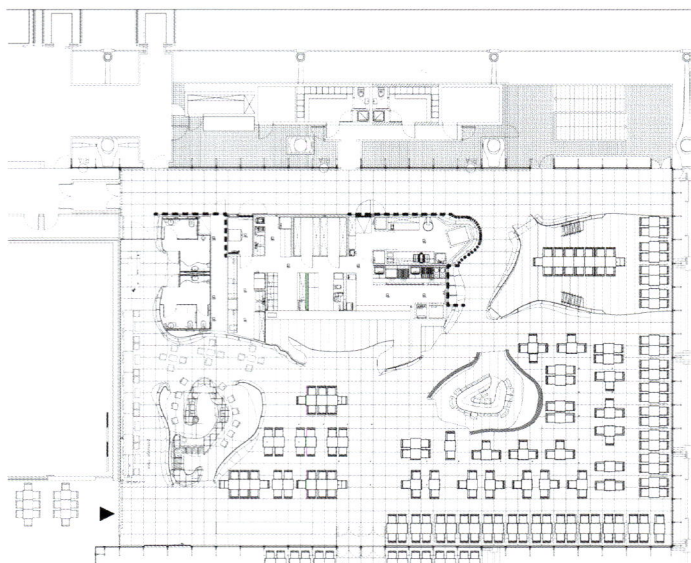

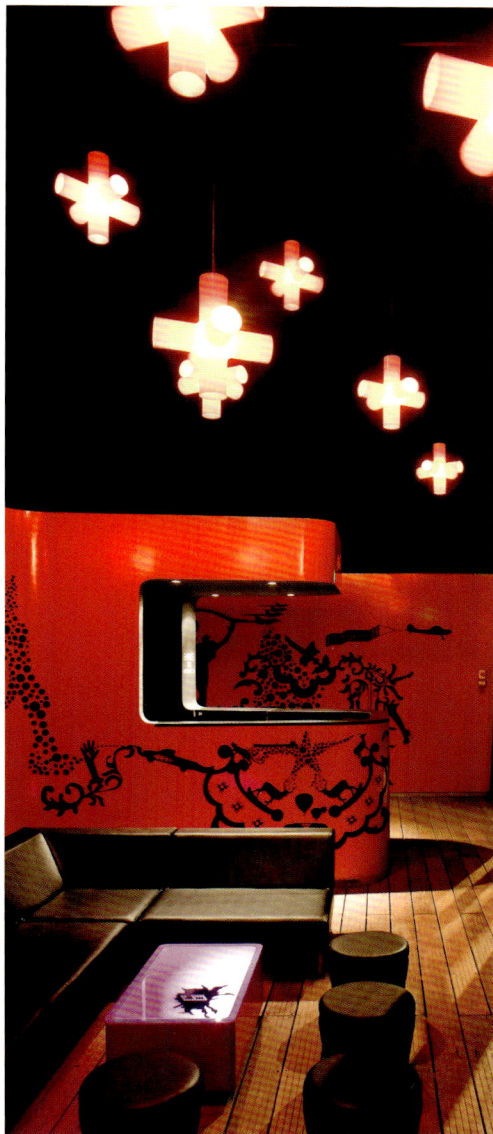

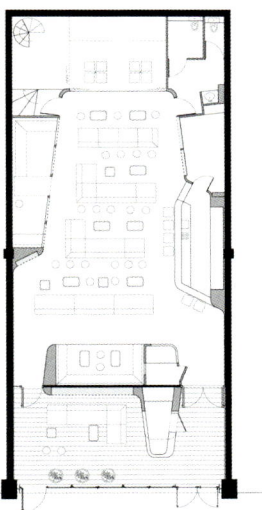

La Suite 21 Club

The design's inspiration comes from the lace patterns for the computer-cut stickers and all the graphic design work. The old time provincial's cabaret mixed with contemporary pop culture. The red lacquered and perforated ribbon wrap up all the club's functions (bar, VIP lounge, DJ booth, lap dance rooms). This ribbon draws a large lounge and an indoor terrace where the furniture is set up. The sofas allow various and comfortable positions the see the show and chat with friends. The graphic ribbon is like a large fresco with different levels of meaning, from the lace pattern to small provocative stories and hidden jokes. It's a really good and hot space for people to relax and happy together with friends.

21号套房俱乐部

设计的灵感来自于数码贴画和其他平面设计作品的蕾丝图案，融合了怀旧乡村酒吧和现代流行艺术风格。红色亮漆镂空缎带遍布了整个俱乐部空间（酒吧、VIP休息室、DJ间、舞厅）。缎带勾勒出大型休闲吧和室内平台的轮廓。人们可以坐在各式各样舒适的沙发上观看表演或是聊天。缎带图案就像一个巨型的壁画，具有多层寓意，如蕾丝图案、煽动性小故事和隐晦的笑话。这是一个能让人和友人一起愉悦放松的热力空间。

Photo: Stéphane Chalmeau

Joran Briand and Arthur de Chatelperron

OCTO

Inside the restaurant, the menu board presents the food in a clear way echoing the graphic identity displayed on the façade: lively, fresh and dynamic colours which arouse desire and stimulate the appetite. In addition, the interior architecture of the first establishment gets round the limitations of the small space and plays on the consistency of the codes: the colours and shapes used in the façade are echoed here: square seats in acid colours (yellows, green, orange, red) share the space with simple and light furniture. Tables and surfaces for eating standing subdivide the eating area in a regular and harmonious way without impeding circulation or blocking the way. The choice of materials notably the floor parquet reinforces the natural and warm aspect of the concept.

OCTO餐厅

餐厅内的菜单看板以一种清晰的方式展示着店内所提供的食物，与餐厅外墙的图案装饰遥相呼应：生动、鲜活而具有动感的色调能够刺激人的食欲。室内设计成功解决了小空间的限制，也具有一致性，外墙的色彩和造型在这里都有所体现：方形座椅采用了黄、绿、橙、红颜色，空间设计简洁，采用了轻质家具。桌子和站立就餐台自然而然地划分出不同的就餐区，不会阻碍人们走动。在材料的选择上，镶木地板增添了空间自然、温馨的效果。

Restaurant

France

Marseille

2009

Photo: Philippe Dureuil

Idoine Agency

Restaurant in Hotel Benkirai

The international group Charm & More requested Patrick Jouin to make the Benkiraï a Garden of Eden, refined but not precious, luxurious but not flashy. The restaurant opens onto a long terrace and adapts to suit the time of day: sun beds disappear, to make way for candlelit dinners around the swimming pool. The indoor air-conditioned bar doubles up as an outdoor bar with exotic, organic parasols. The stainless steel masts dance together supporting Batyline fabric lattice parasols, like a multitude of gigantic toadstools creating shady areas. When night falls, they are lit up, inviting you to relax underneath.

彭吉哈伊酒店餐厅

酒店被设计成一座伊甸园，考究却不造作，奢华却不虚饰。餐厅朝向一个长长的平台开门，而且能够适应一天中的不同时间段：太阳落山后能够在泳池周围享受烛光晚餐。室内装有空调的酒吧延伸到户外，变成一个户外酒吧，有异国情调的绿色遮阳伞。不锈钢杆子支撑着巴迪林织物的格子遮阳伞，仿佛许多巨大的蘑菇，创造出阴凉的区域。

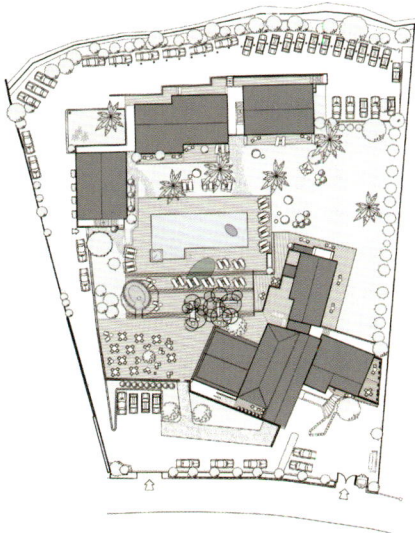

Restaurant

France

Saint-Tropez

2006

Photo: Eric Laignel

Patrick Jouin

Bar

France

Paris

2008

Photo: Romee de Goriainoff

Coco Black

Experimental Cocktail Club

This Neo-Baroque lounge in Paris's second arrondissement was inspired by the owner's wish to create a small intimate space that would suggest the intimacy of an elegant 'Salon'. Inside, one realises the history attached to the space. Majestic hand hewn ceiling beams and rough hand chiseled stone walls unveil the centuries old patina and character of the building. The designer Coco Black, chose to embrace the natural finishes of the rooms and all architectural surfaces were allowed to remain as they had been for hundreds of years. The chandeliers are constructed from acrylic plastic sheets cut with a CNC laser in a profile design that mimics the candle and crystal components of a traditional glass chandelier.

"体验" 鸡尾酒吧

酒吧位于巴黎第二大区，客户要求在新巴洛克风格的空间内营造一种犹如"沙龙"般的私密环境。手工制作的天花梁柱尽显壮丽，手工凿制的石墙质感十足，突出了建筑一直保留的古典特色。丙烯塑料打造的枝形吊灯模仿蜡烛以及古代玻璃吊灯的外观，巧妙的设计格外吸引眼球。

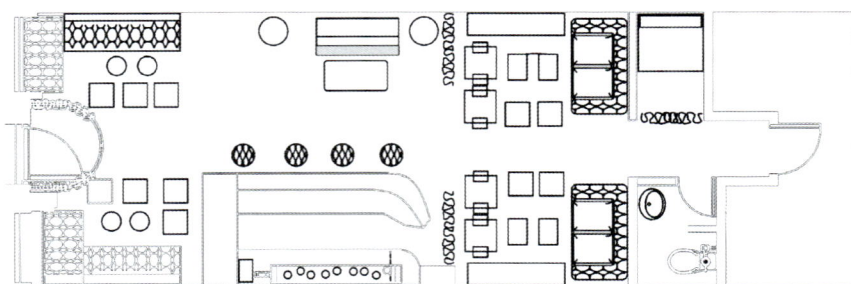

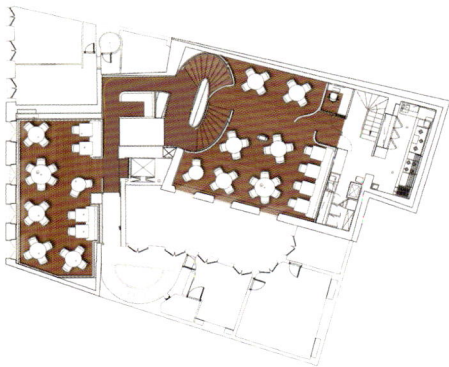

Restaurant

France

Paris

2008

Photo: Eric Laignel

Jouin Manku

Oth Sombath Restaurant

The inspiration runs through the bar and three dining rooms, each offering a unique atmosphere to enjoy Oth's sumptuous flavours – one in scintillating gold, one in bold orange, and one in soft creams – the colour palette calling to mind in turn the temple treasures, bold colours of the women's dresses and rice paddies. Saffron-hued upholstered walls pop in the restaurant's multilevel interior. Textures also play reference in the restaurant – that of the curving wall of the bar looks like the hairstyles on Buddha sculptures while the warm wooden floor on the rez-de-chaussée echoes the dark wooden artisans' creations in Thailand. The arch of a dragon's back resonates with the sweeping curves of the stairway connecting the three dining rooms.

奥斯餐厅

设计师的独特灵感在酒吧和三间就餐室内完美展现，赋予各自不同的空间氛围。就餐室风格各异，一间饰以闪亮的金色，一间以橙色为主，另外一间则完全采用柔和的奶油色。织物同样被用作主要设计元素，装饰在酒吧的墙壁上使其看起来如同大佛雕塑的头饰。楼梯蜿蜒而又壮观，将就餐区连通。

Restaurant

The Netherlands

Haarlem

2008

Photo: Tjep

Frank Tjepkema, Janneke Hooymans, Tina Stieger, Leonie Janssen, Marloes Pronk, Bertrand Gravier, Camille Cortet

Pluk

Pluk is a new take-away shop offering fresh juices, yogurt shakes and special salads. The essence of the formula is to enjoy good food. The emphasis on health is implicit. The interior design tries to reflect this juxtaposition of healthiness and fun. The counters contain fruits and vegetables in three colour groups. They are actually fake, but because of the way designers integrated a special gradient effect the whole counter becomes simply delicious, and it took ten months to develop the exact right colour, gradient and fruit/vegetable combination to get this result: people who enter Pluk are overwhelmed and just can't resist ordering one of the wonderful drinks.

普拉克外卖店

普拉克外卖店主要提供新鲜果汁、酸奶昔和特制沙拉。"享用优质食物，体验健康生活"的设计理念在室内空间尽情地体现。柜台共有三种颜色，盛放着仿制的水果和蔬菜，看上去十分美味诱人，香甜可口。设计师花了几个月的时间来寻找水果和蔬菜颜色的正确组合，才达到如此完美的效果。进入普拉克的人都会情不自禁地为自己点一份饮品。

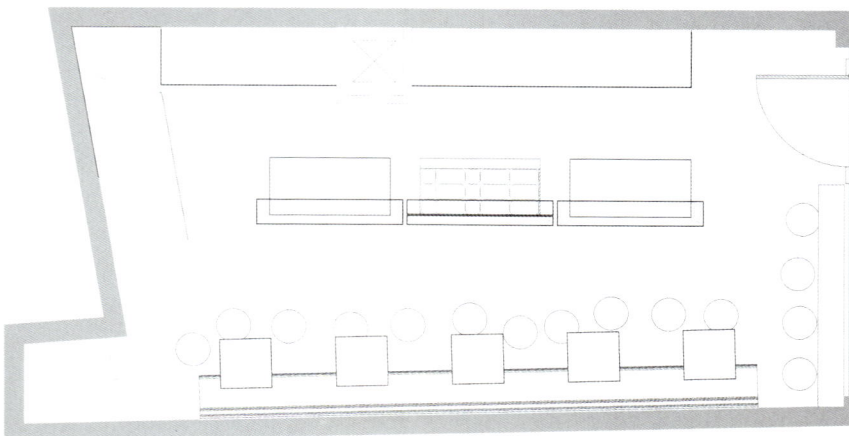

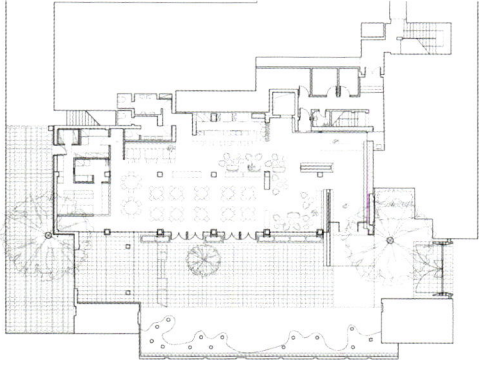

Restaurant in Kruisheren Hotel

An outstanding feature is the newly-installed mezzanine where guests are served breakfast wh le taking in views of Maastricht through the chancel windows. It's also open every day for light lunches and dinner, when the chef presents a variety of culinary surprises. The monastery, surrounded by cloister corridors once home to the 'Order of Crutched Friars', now hosts most of the guestrooms; these feature both original architectural elements and daring modern interventions. Split-level Kruisheren Restaurant located in the old monastery offers panoramic views of the church and the city of Maastricht, serving buffet breakfast and light lunches with a selection of wines from the Wine Bar.

克鲁舍伦酒店餐厅

酒店的一大特色是新建的一二楼之间的中间层,酒店住客在这里享用早餐,同时又能透过窗户欣赏马斯特里赫特市的美丽风景。中间层每天白天也开放,提供清淡的午餐和晚餐,主厨会为大家带来各种意想不到的美味。克鲁舍伦多层餐厅位于原修道院内,能俯瞰教堂和马斯特里赫特市的风景。餐厅提供自助早餐和清淡的午餐,酒窖里的各种美酒可供选择。

Henk Vos

Photo: Designhotels

2006

Maastricht

The Netherlands

Restaurant

Restaurant

The Netherlands

Amsterdamm

2006

Photo: Roel Heine

Mood Makers, Roel Heine

Mashua

The design is a contrast between the contemporary and classical styles, which are blended surprisingly well. These contrasts are clear to see in every detail of the design. Eye-catchers are the large black crystal chandeliers, which are flanked by simple steel cylinder lamps. Designers chose materials according to the fusion concept as well. On one hand, there are warm materials like dark wooden floor, leather couches and chairs, and on the other hand the steel of the lamps and the back wall of the bar. The wall at the back of the restaurant is decorated with colourful patterned wallpaper, which makes the space seem wider than it actually is.

玛莎餐厅

餐厅的设计体现了现代与古典的对比与融合。现代与古典风格的对比在每个微小的细节都凸现出来。巨大的黑色水晶吊灯旁边点缀着简洁的钢筒灯在空间十分醒目。设计师在材料的选择上同样依照融合统一的原则。使用黑色实木地板、皮革沙发和座椅等古典元素，而钢筒灯和餐厅的墙壁又展示了现代风格。餐厅后墙使用彩色花式的墙纸装饰，使空间看上去更加宽敞明亮。

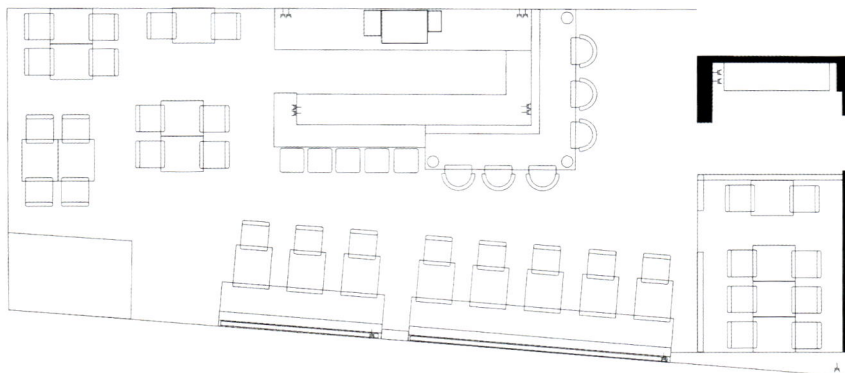

Praq Amersfoort

The space is characterised by a monumental farm style roof composed of huge massive wooden beams. Within this space the designers tried to create a playful world by placing furniture for example, a table becomes a window, a bus or a kitchen. The space is warm and interesting with a palette of white and orange; the floor is white and brown, supporting the active atmosphere; the six-metre-high construction in the centre looks like an abstract cloud evoking something of a colourful game while contrasting nicely with the handcrafted architecture. One space is reserved for children and their parents while the other is reserved for adults.

普拉克餐厅

餐厅的屋顶由许多巨大的横梁搭建而成，体现出乡村特色。设计师利用家具的摆放，创造出一个妙趣横生的世界，如桌子被制成窗户、巴士、厨房等造型。餐厅以白色和橙色为主，温馨而和谐；棕白相间的地板衬托出活泼的气氛；高达6米的中央屋顶就像一朵云彩，精美的手工家具将餐厅变成了游戏场。餐厅中一部分空间专为孩子和他们的父母打造，另一部分则是成年人的领地。

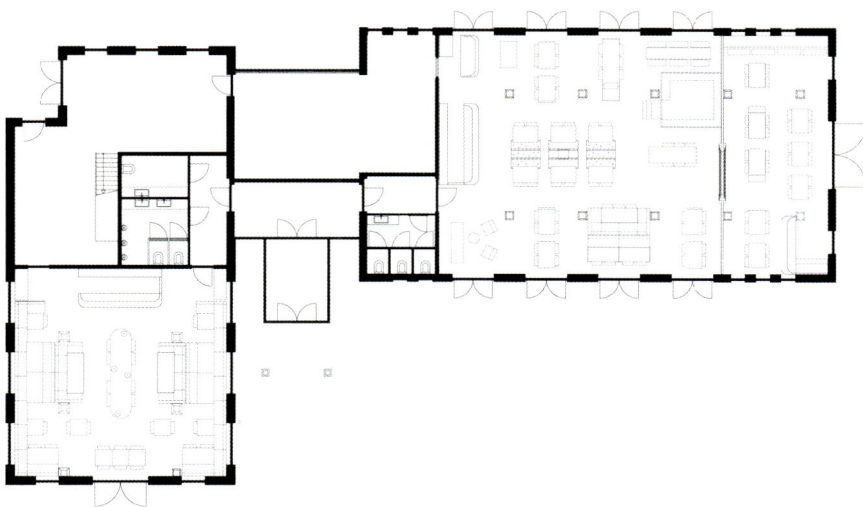

Restaurant

The Netherlands

Amersfoort

2008

Photo: Tjep

Frank Tjepkema, Janneke Hooymans, Tina Stieger, Leonie Janssen, Bertrand Gravier, Camille Cortet

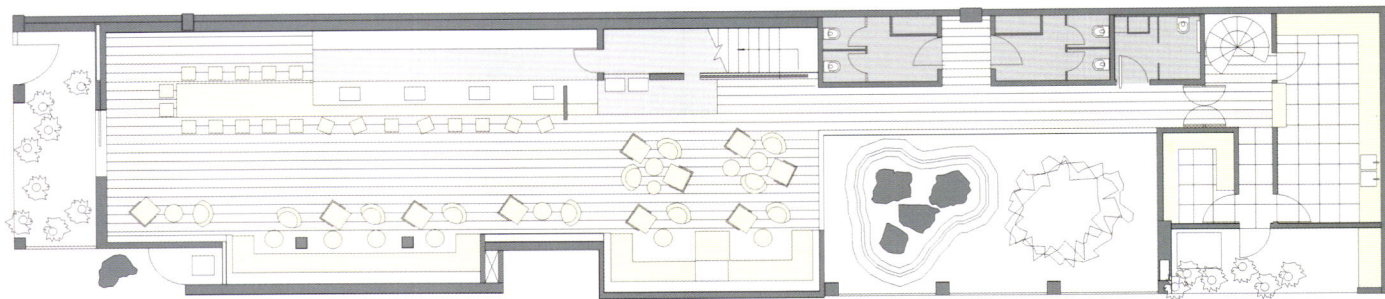

Escape Bar

The designer's idea was to create a bar, despite of its contemporary environment as any other in the world, where people could get together, meet friends, have some drinks, 'tapas' and chat. The Escape Bar atmosphere is both sophisticated and trendy. The lighting colours change according to the moment, the people, and the DJ sound creating different scenes for each moment. Several trees that are encircled by glass walls and ceiling with spidery pattern emphasise natural atmosphere in the bar. The long wine cabinet covers one whole wall which becomes a special interior scene. The long counter is the meeting place and the mezzanine for private conversations.

"逃" 酒吧

设计师的想法是创建一个酒吧，不管世界上其他地方的形势如何，人们都可以在这个酒吧里聚会、会见朋友、喝点东西、吃些小吃并且聊聊天。酒吧里的气氛时尚而底蕴十足。灯光的颜色在不同的时刻，在不同的人和DJ播放的不同音乐中变化，每一刻都是不同的风景。酒吧玻璃幕墙内种植的几棵树和天花板上蜘蛛网状的图案强调了这座酒吧自然的氛围。长长的橱柜延续了整面墙壁，形成了独特的室内风景。长吧台是会面的场所，而阁楼更适合私人会话。

De Kuyper Royal Distillery

De Kuyper Royal Distillers, housed in a listed building in Schiedam, wanted a new arrangement for the reception areas in order to appeal to contemporary tastes – the world of chic cocktail bars frequented by a young, successful, dynamic, international, trend-setting and sophisticated audience. The visit must be an instructive experience, where people can get acquainted with the Distillery's history and sample the latest products through sight and taste. The new design is based on the reception programme, a route that leads through different ambiences. Contemporary additions stand out through their form, adding a new world that does not detract from the dignity of the monument.

皇室鸡尾酒吧

皇室鸡尾酒吧项目是要将老接待区进行翻修，从而迎合当代人的品位。设计的宗旨就是要打造一个别致的鸡尾酒吧，能为那些年轻有为、活力四射、来自世界各地、喜欢流行时尚和体验古典高雅的人带来欢乐。新设计基于传统接待程序的理念，一条主干道可以体验不同的氛围，通过形式、材料和色调上的改变，使新的设计富有当代风格，还不失纪念碑的尊严。

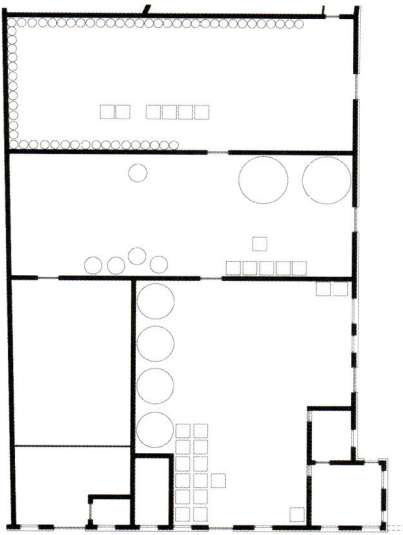

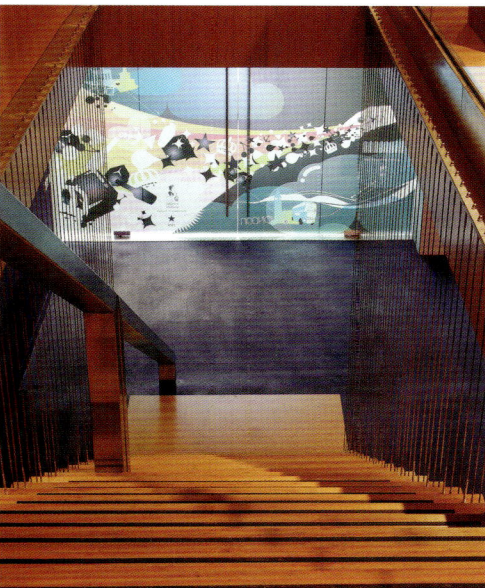

Bar

The Netherlands

Schiedam

2007

Photo: Rob't Hart Fotografie

TVS Interiors

Canteen

The Netherlands

Haarlemmermeer

2009

Photo: Ewout Huibers, Richard Powers

Concrete Architectural Associates; Rob Wagemans, Erikjan Ver-meulen, Jeroen Vester, Erik van Dillen, Matthijs Hombergen, Sofie Ruytenberg

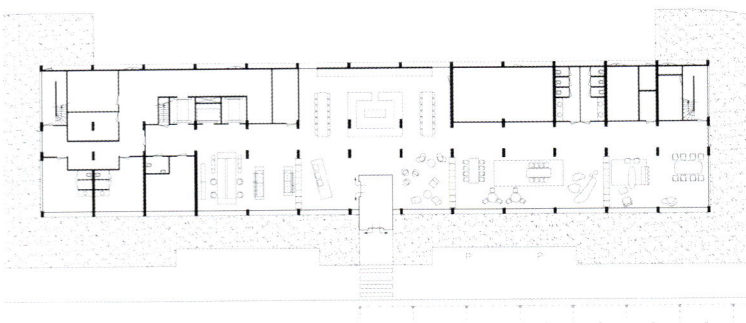

CanteenM in Citizen M Hotel Schiphol

The Food & Beverage area is mainly self-service. An 11-metres-long cabinet contains 7 refrigerated glass cases with drinks, sandwiches, salads, sushi and other kinds of snacks and meals. Visitors take the food and drinks of their choice and order coffee and pay at the two cash desks in the red service bar. Products can be consumed at one of the large red bar-tables or in one of the living rooms. The front of the red service bar contains a glass case with magazines, newspaper and merchandise. At night parts of the cabinet are closed and extra barstools and different lighting create an intimate atmosphere and change the room to a bar-area.

市民M酒店自助餐厅

餐饮区基本采用自助方式。11米长的橱柜内装着7个冷柜，里面摆放着饮品、三明治、沙拉、寿司以及其他点心和主食。客人可以自主选择食品和饮品，点一杯咖啡，然后到红色的服务台付账。红色服务台前面的玻璃橱内摆放着杂志、报纸和商品。晚上，橱柜的部分会关闭，高脚凳和不同的灯光效果将餐厅转换为酒吧的氛围。

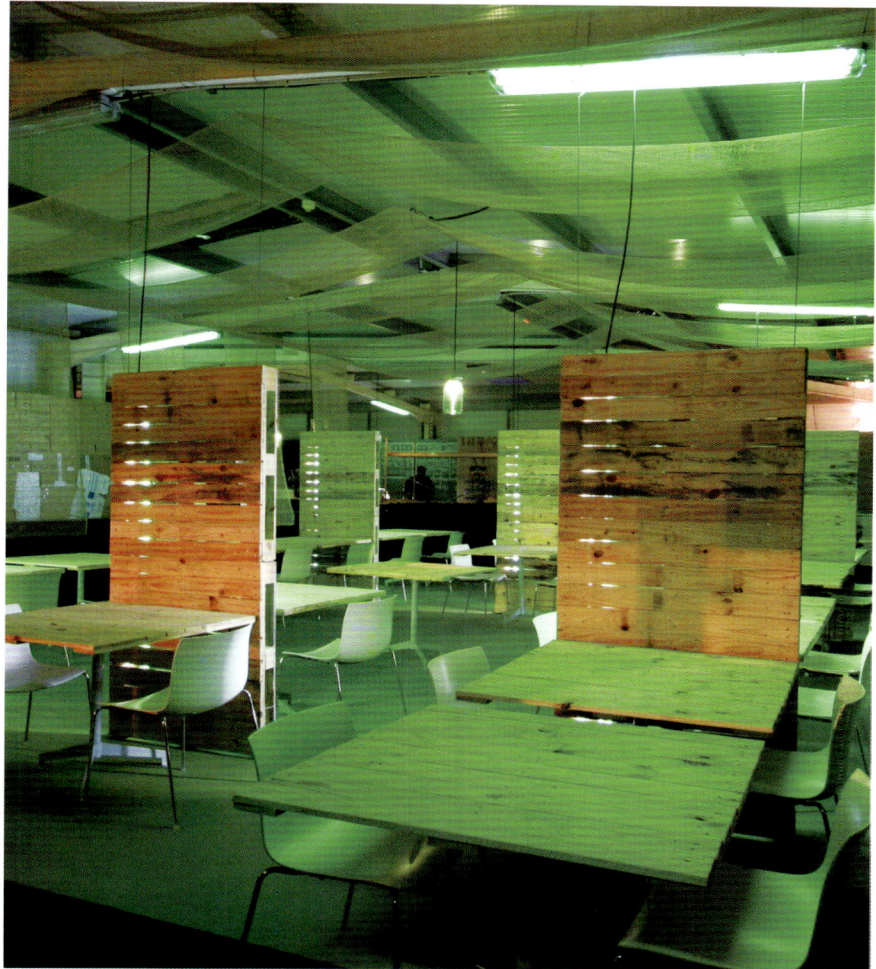

Restaurant Sostenible Cassia

The workshop has been designed to give way, through an artistic language, a subject of rabid news: recycling. The proposal is to reuse all the waste material generated by the assembly of the fair, so that it is the raw material used in the decoration and assembly of the cafeteria and restaurant. This project reused materials with the creative direction of designers, in addition to the participation of several members of the Artists Gallery EC and logistical support of Big Grup. What deserves special mention as a contribution to the Restaurant is that, the designers developed a special letter on the occasion of this workshop, using a large amount of the natural resources.

卡西亚餐厅

设计以艺术的方式表达了如今最热门的"循环利用"主题。设计师使用回收的废料,并将它们作为食堂和餐厅装修的原材料。该项目由创意设计师、EC艺术画廊的艺术家和大集团的后勤人员合作完成,实现了物品的循环使用和可持续性。卡西亚餐厅的设计中最值得一提的是设计师将可循环使用进行到底,利用了大量环保健康的自然资源。

Photo: innova::designers studio

Diego & Pedro Serrano (innova::designers studio)

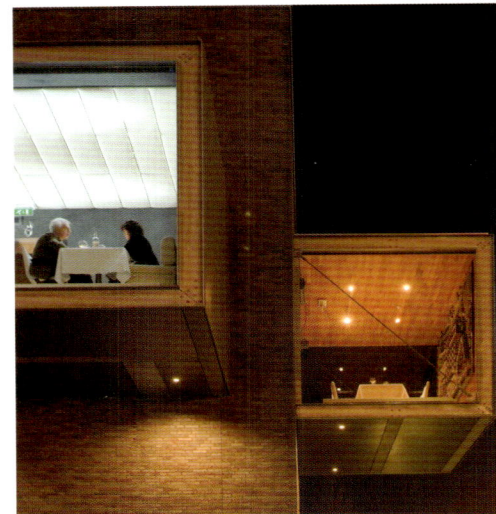

Restaurant Divinatio

Fish restaurant Divinatio is located at the historical harbour of Utrecht, The Netherlands. The objective was to realise a restaurant with a maritime feel for the upper segment of the market. The building is designed to be functional. The contrasts – rough on the outside and smooth on the inside – have been translated in the design. The interior has to appeal to a broad and contemporary group, without discouraging the more traditional public. When choosing the materials, Liong Lie (the architect) took both the visual as tangible effects into account. Leather, blue glass and Corian have been used in the design. The open kitchen is framed in white Corian and has a dark back wall which creates the illusion of cooks standing in a theatre. The walls have been covered with an acoustic clay plaster that has a mother of pearl effect which shimmers in the sunlight. At night the curiosity of the passerby is raised by the big illuminated glass bay window.

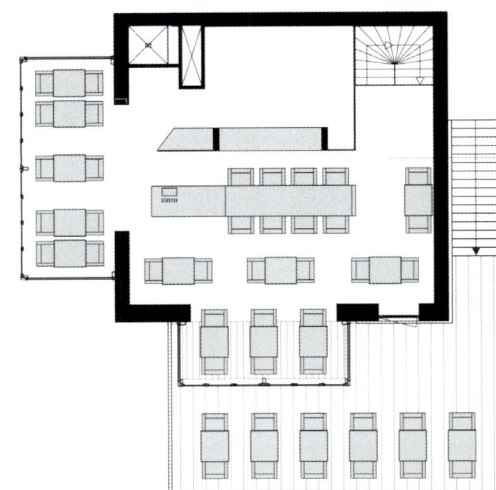

迪维娜迪奥餐厅

迪维娜迪奥海鲜餐厅位于荷兰乌特勒支港口，旨在打造一个高档滨海风情空间。设计的内外形成了鲜明对比——外部粗犷，内部细腻。室内设计既能吸引现代群体，又不排斥传统大众。在材料的选择上，建筑师认真考虑了视觉效果，运用了皮革、蓝色玻璃和人造大理石。在白色人造大理石框架中的开放式厨房有着深色背景，给人感觉好像在剧院中烹饪一样。墙壁上的涂料具有珠光效果，在阳光下会闪闪烁烁。晚上，点亮的大型飘窗吸引着街上路人的注意力。

Restaurant

The Netherlands

Utrecht

2008

Photo: Christiaan de Bruijne

123DV

De Bijenkorf Kitchen

The starting point of developing a new Bijenkorf restaurant is that the restaurant has to be part of 'de Bijenkorf' and its philosophy. The restaurant as part of the department store, the restaurant as a kitchen' of de Bijenkorf. The idea for the Bijenkorf restaurant is an open kitchen consisting of different cuisines from all over the world. The best ingredients from various countries make each dish into something very special. Because it is a design of a restaurant in a department store, the main focus lies on the 'goods' – the cooking material and food. The designers show the guest what ingredients the various dishes are composed of and what the dishes are based on.

蜂箱小厨

蜂箱小厨的设计必须符合"蜂箱"百货的风格和经营理念，因为它是百货公司的一部分，也可以说是百货公司的厨房。蜂箱小厨拥有一个开放式厨房，提供世界各地的美食，各地美食的融合让它变得特别。因为餐厅处于百货公司之内，那么设计重点就应该放在"商品"——烹饪材料和食物上。顾客可以清晰地了解各种菜肴的原材料和制作过程。

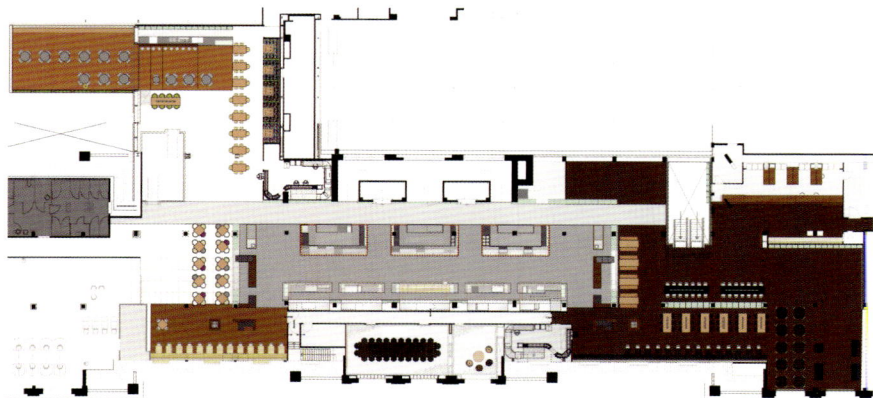

Restaurant

The Netherlands

Amsterdam

2008

Photo: Ewout Huibers (www.ewout.tv)

Concrete Architectural Associates

Restaurant

The Netherlands

Laren, Noord-Holland

2009

Photo: Ewout Huibers

Concrete Architectural Associates

Loetje, Laren NH

The overall design concept is to place an urban interior in a more traditional and rural surrounding. Starting point was to create a 'grand café' with a bar-like atmosphere. The intention was to create two worlds within the experience, while the appearance should reflect one world. The designers found the solution in materialisation, one type of flooring and the ceiling of light shades, which connect the bar with the restaurant area. The bar and restaurant differ in furniture and have a variable table setting, round tables in the bar and rectangular in the restaurant. The bar is the central point in the restaurant. Additionally the designers connected the restaurant with the kitchen. Not in a constructive way but by means of a 14-metre-long photoprint, suggesting craftsmanship, quality and tradition of Loetje.

罗埃蒂耶餐厅

项目的基本理念是在相对传统的乡村环境中打造一个都市室内空间，出发点是打造一个带酒吧的"大咖啡馆"，其目的是在一个相同的外观下营造两种截然不同的体验。设计师在材料上找到了解决方案，酒吧和餐厅采用了同样的地板和灯光装饰。酒吧和餐厅的家具不同，前者是圆桌，后者是方桌。吧台是整个餐厅的中心。设计师通过14米长的影印画把餐厅和厨房连接了起来，体现了罗埃蒂耶的工艺、品质和传统。

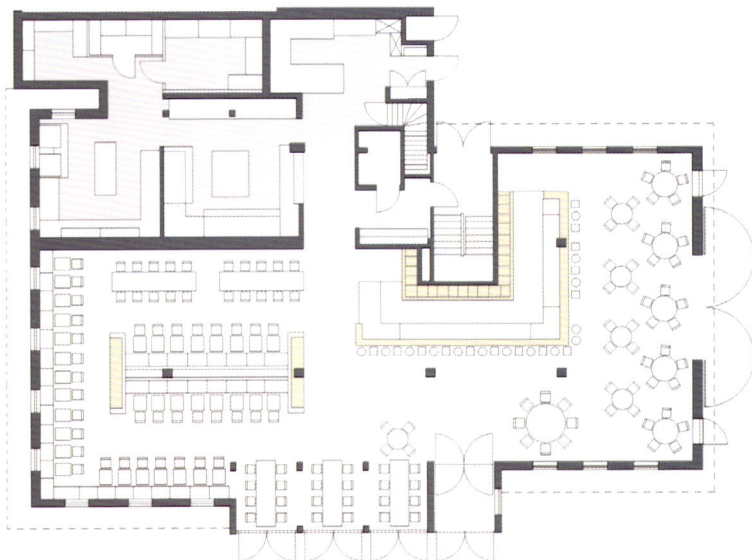

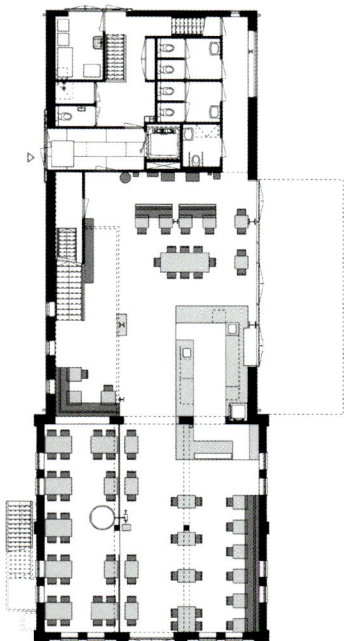

Restaurant and Cooking Studio Zijlstroom

'Eating is knowing' became the motto for the restaurant and basis for the brief. On the terrain an orchard is created as well as a kitchen garden and greenhouse, where vegetables and herbs are grown. These are directly used in the kitchen of the restaurant and the cooking studio. The architecture supports the slow food concept. The flour mill has been restored to accommodate café, restaurant, cooking studio and meeting room. To improve light and space on the ground floor the upper floor was partially removed. No further attempts were taken to modernise the building largely. Specific details such as the hoist winch and flour funnel from the factory are maintained, the latest in a new feature: a barrel with visible rainwater. With its large canopy, hovering over the terrace, it occurs as a no-nonsense farm barn.

基思卢姆餐厅兼烹饪教室

餐厅的口号是"吃即是学",这也是餐厅设计的基础。餐厅外面有个果园,种植着蔬菜、香料,可以直接拿到餐厅的厨房和烹饪教室使用。建筑设计也支持着慢食理念。原来的面粉工厂被划分为四个区域:咖啡厅、餐厅、烹饪教室和会客室。为了保证一楼的空间和照明,二楼的部分设施被拆除了。建筑师并没有做更多的改动,工厂里的起重机和面粉漏斗被保留了下来,用来储存雨水。露台上巨大华盖让餐厅看起来像一个正式的农家谷仓。

Restaurant

The Netherlands

Leiderdorp

2007

Photo: Rene de Wit

Kingma Roorda Architects

Bar

The Netherlands

Amsterdam

2009

Photo: Ewout Huibers

Concrete Architectural Assosiates

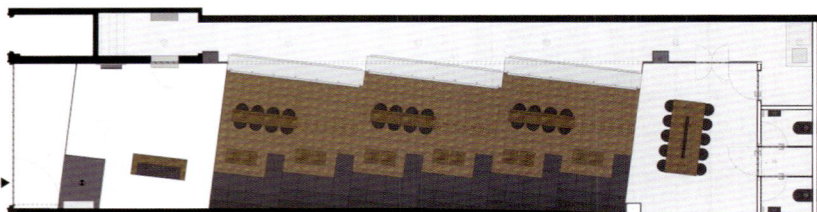

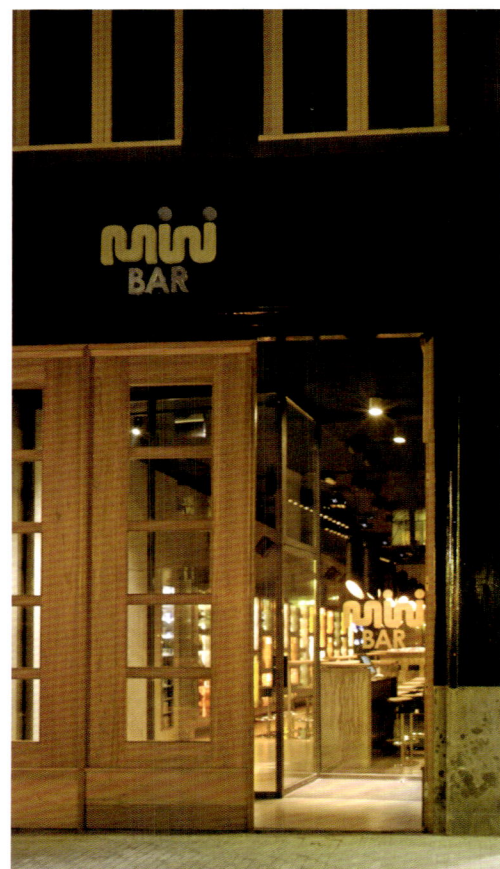

Minibar

The founding concept for the Minibar was created by three Dutch friends, who were looking for a different way of having a drink in a relaxed environment. No more queuing, trying to catch the bartender's eye, or being ignored at the bar, but a comprehensive selection of drinks taken at your own pace, from your own personal minibar. The designers translated this idea into an experience. At the Minibar you'll be welcomed by a receptionist, checked in and given a key to your own personal minibar. There are beer bars, champagne bars and regular bars filled with combinations of drinks, allowing you to go straight, mix up a storm or try an exotic beer or two. Every bar has a list of the drinks and the prices, and if you're a little hungry there is a delicious selection of nuts, for something more, Sushi and Curry can be ordered and delivered.

迷你吧

迷你吧的概念由三位荷兰友人提出，他们试图找到一个能够喝一杯的放松环境：不用排队、不用费心找酒保、也不会在吧台被忽略，可以在私人迷你吧尽情挑选自己想喝的酒。设计师实现了这个概念。在迷你吧，接待员会带给你一把钥匙，带你走入自己的私人酒吧。酒吧里有啤酒吧、香槟吧和组合酒吧，可以满足客人的各种需求。每个私人酒吧都有自己的饮品价格单，如果客人有就餐需求，酒吧还提供坚果、寿司和咖喱等小吃。

Café-Restaurant OPEN

The railway bridge on the Westerdokseiland in Amsterdam is one of the last surviving pivot bridges in the Netherlands. The City of Amsterdam was therefore keen to preserve this showpiece on the river and decided to give the monumental bridge a public function. Because of the location and the panoramic views it was decided to use it as a café-restaurant. The City Council organised a design competition in 2005 and the entry by de Architekten Cie. was chosen from the 14 submissions. OPEN Café-restaurant is a pure, transparent, glass volume that fits precisely onto the existing bridge. It is composed of a floor, a roof and a glazed façade that is formed entirely of pivotal windows, all of which can be opened.

开放咖啡餐厅

阿姆斯特丹的Westerdokseiland铁路桥是荷兰仅存的枢纽桥梁之一，因此阿姆斯特丹市政府十分注重其保护工程，想要给它增加一些公共实用功能，并最终决定建造一个咖啡餐厅。2005年，城市委员会组织了一次设计竞赛，de Architekten Cie.的设计从14项入围作品中脱颖而出。开放咖啡餐厅是一个纯粹、透明的玻璃空间，与桥梁契合得刚好，由楼板、屋顶和玻璃外墙组成，外墙上的窗户可以打开。

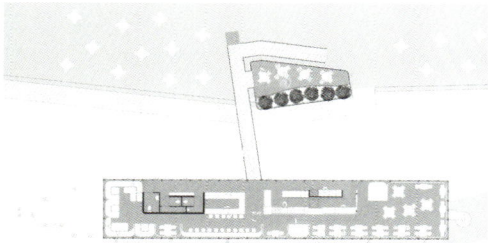

Photo: Rob Hoekstra

de Architekten Cie

Bar

The Netherlands

Tilburg

2006

Photo: Maurice Mentjens

Maurice Mentjens

NWE Vorst

The black room was originally the hunting room, as it can be seen from the ceiling on which animals are depicted in the plasterwork. In this room, the designer was keen to pay homage to well-known animals from literature and theatre. They are depicted in high gloss black on the black matt walls and ceiling, and seem to float like stars in a pitch-black sky. The white room next door features a large white sound-proofing panel. The bars in both rooms are reminiscent of an altar, with a silver 'triptych' as a bottle rack in the black salon (silver representing the moon) and a golden triptych in the white salon (gold representing the sun).

纽维·沃斯特酒吧

餐厅内的黑色房间以前是狩猎室，石膏天花板上还留有动物图案的浮雕。设计师敬畏文学或戏剧作品中著名的动物形象，因此将亮黑色图案画在黑色亚光墙壁和天花板上，像漆黑夜空中的点点繁星。白色房间中最醒目的是巨大的白色隔音板。酒吧的两个房间都让人联想起教堂的圣坛。黑色房间里的银色酒柜象征月亮；而白色房间里的金色酒柜则象征太阳。

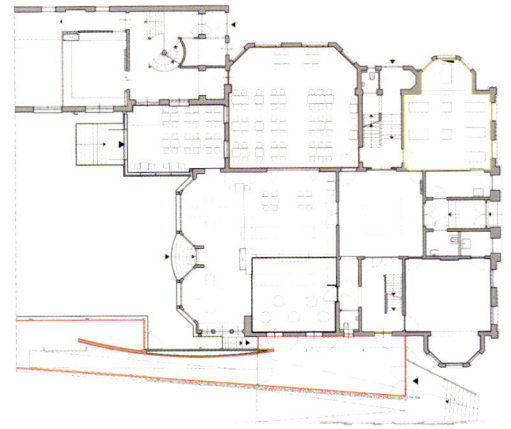

Pavilion

Wood lives, wood loves. The pavilion represents modern rural architecture: simple, practical and eloquent. Black and white are the traditional colours. The pavilion shows how minimal means can lead to large results. The building leaves a poetic impression of purity. The no-nonsense manner in which this 'simple barn' has a large panoramic window shows a pleasant openness to conventional styles.

木之亭餐厅

该项目充分体现了当代田园风格建筑的特点，崇尚简单、实用、美观的设计理念。经典的黑白色搭配，打造了浪漫诗意的个性空间，令客人身心愉悦。特殊的"仓形"设计令游客轻松自然，将美景尽收眼底。

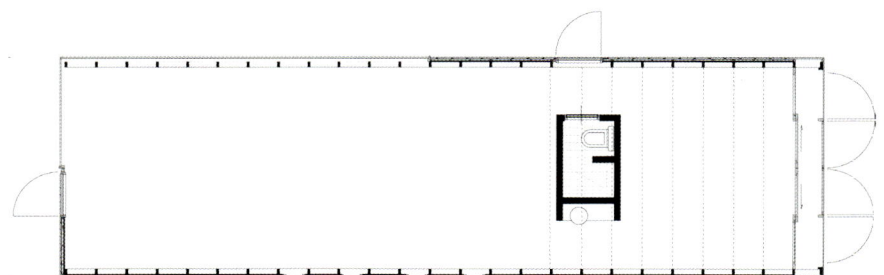

Restaurant

The Netherlands

Oeken

2008

Photo: Faro Architects

Faro Architects

Restaurant

The Netherlands

Amsterdam

2010

Photo: Ewout Huibers

Concrete Architectural Associates; Rob Wagemans, Ulrike Lehner, Marc Brummelhuis, Sofie Ruytenberg (graphic design), Femke Zumbrink (Graphic Design), Erik van Dillen

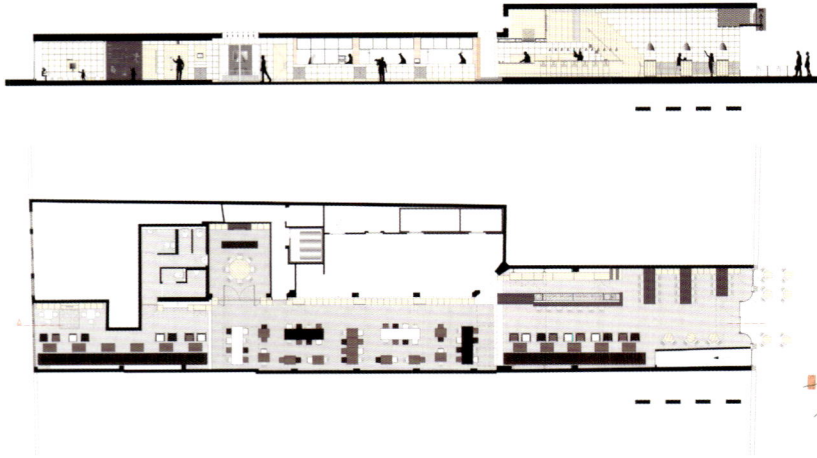

Mazzo

Mazzo is the Italian sister of Brasserie Witteveen – both in hospitality concept and in the interior there is an obvious connection. As with Witteveen the name and location of Mazzo share a history in the city of Amsterdam. Mazzo was a famous and notorious disco in the monumental building on the Rozengracht. It's a typical Amsterdam building: narrow and very deep spaces fused together with different floor and ceiling levels, which automatically provides a natural positioning for the restaurant. The diversity of the fused spaces and the natural restaurant layout need a connecting element: a huge wooden cupboard across the whole restaurant linking all the spaces to each other and organising them at the same time.

玛索餐厅

玛索餐厅在经营理念和室内设计上都与威特芬酒馆类似，也同样拥有浓厚的历史氛围。玛索曾经是玫瑰运河一座历史建筑里的迪斯科舞厅。这是一座典型的阿姆斯特丹建筑：空间狭长，混杂着不同的楼层和天花板高度，为餐厅提供了得天独厚的条件。巨大的木橱柜横跨整个餐厅，将各个空间组织、联系到一起。

Brasserie Witteveen

Making Witteveen into a present-day brasserie was the starting point of the design. Typical of a brasserie is that it is a place for both eating out and having a drink, though more particularly it is a place where everyone can relax and feel at home. In spite of the diverse world of bars and restaurants, many Dutch people are more at ease going to a pub. In Dutch pub interiors the Persian rugs on the tables play an important role and give guests a familiar feeling. The new design for Witteveen will recall this familiar feeling, but without reminding of an old-fashioned dark Dutch interior. In Witteveen the Persian rug can be found on the floor, not as a classic rug, knotted and soft, but as a mosaic floor. This way the Persian pattern stays subtly recognisable and the characteristics of the cement tiles; each tile is handmade and with unique colour shades creates a lively effect on the floor.

威特芬酒馆

威特芬酒馆是一个可以就餐并喝一杯的地方，人们在这里会得到在家一般的放松。尽管现在有林林总总的酒吧和餐厅，许多荷兰人还是更喜欢酒馆。在荷兰式酒馆中，桌上的波斯毛毯起到了重要的作用，让顾客有一种熟悉感。威特芬酒馆的设计保留了这种熟悉感，但是摒弃了荷兰旧式昏暗的室内设计。波斯毛毯图案在镶嵌地板上得以显示，这样一来，波斯图案依旧可见，其独特的色调也在地面上打造了活泼生动的效果。

Restaurant

The Netherlands

Amsterdam

2009

Photo: Ewout Huibers

Rob Wagemans, Ulrike Lehner, Joyce Kelder, Erik van Dillen, Sofie Ruytenberg (graphic design)

Café

Belgium

Brussels

2009

Photo: Serge Brison and Bernard Boccara

B612 Associates

La Cafétéria Viaduc

The interior spaces of the day-care itself is totally oriented so to maximise the comfort of the children while apart from their parents. Here the material used is wood rather than steel or aluminium. The purpose is to use the quality and warm light colour of the wood for the evocation of human contact. The project is designed so to largely bring in pleasant natural light while the external sun-screen avoid dazzling light and overheating. The transparency of the screen and its variations induce a rich relation with the outside, the spatial forms generated by the unfolding of the wooden panels are playful. Their continuous, generous and dynamic movement contribute to the sensation of dilatation and intensity of space and light while allowing to use the created in between spaces for the integration of artificial lighting.

高架桥自助餐厅

项目的内部空间完全是为了让离开家长的孩子们的舒适而设计，运用了木材而不是钢铁和铝材。其目的是以木材质感而温暖的轻快色调来促进人际交流。一方面，室内有令人愉悦的自然光；另一方面，外置遮阳板也避免了刺眼的阳光和高温。遮阳板的透明度和可调节特性丰富了室内外的联系；木板的折叠角度很有趣，它们增添了空间和光线的膨胀感和张力，也为人工照明提供了空间。

Bar

Belgium

Brussels

2008

Photo: Frank Gielen

Puresang Team

Grey Goose

Grey Goose Vodka asked Puresang to design a full operating bar in their headquarters in Brussels. The design became a very exciting play of blue mirrors, cut in triangles and mounted in 3D against the wall to create more depth and tactility in the rather small space. The client asked for a high-roller drinking environment that could also be used as a meeting room. That's why the high gloss bar runs over in the meeting table. The atmospherique light structure is custom made for this project and reflects the sparkles of ice. The wall at the end of the bar gives the room a dramatic feel without being too 'product placing'. The bottles were used as a screen that feels as an ice wall.

灰雁酒吧

灰雁伏特加公司委托Puresang在他们布鲁塞尔的总部设计一间酒吧。设计巧妙地运用了三角形的蓝色镜子，它们呈立体形态装饰在墙壁上，增加了小空间的深度和感知度。委托人要求酒吧具有高档氛围，同时又要能作为会客室使用，因此酒吧里还有个会议桌。极具气氛的灯光设施是为项目特别定制的，映在冰块上，闪闪发光。酒吧后部的墙面上摆满了酒瓶，宛如一面冰墙。

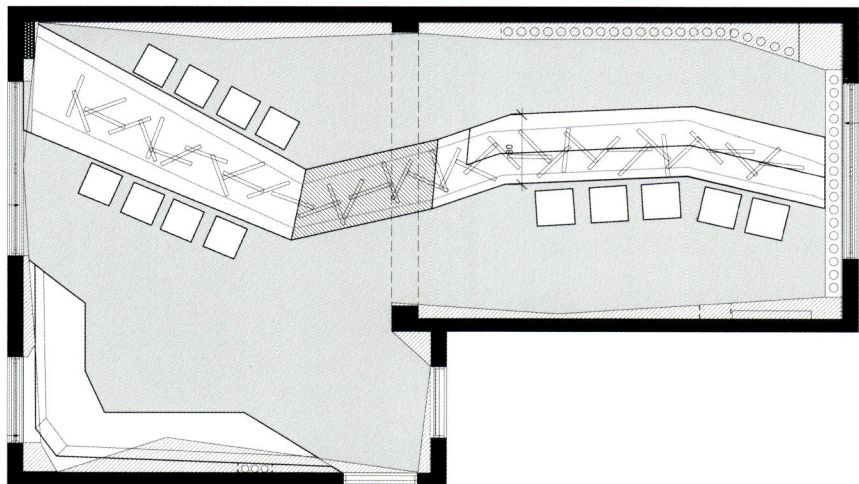

Restaurant

Belgium

Brussels

2009

Photo: Serge Brison and Bernard Boccara

SAQ Architects

KWINT Brussels, BE

Located in the centre of Brussels, the restaurant KWINT serves as a central meeting spot for the newly opened 'Brussels Square Conference Centre'. The length of the space is emphasised by both the padded side-wall and the sculpture hovering over the dining tables. This 30-metre-long sculpture is a creation of Arne Quinze and erupts from the bar at the end of the room almost like a living and articulated organism. Floating above, the installation gives the seated customers a sense of protection and intensifies the notion of gathering. The upholstered wall not only functions as a perfect acoustic absorbent for the ambient conversations, it houses all the essential technical elements and infrastructure: the heating and ventilation system, the electrical installations, but also the annex rooms such as the kitchen, the toilets, and the lounge. The wall was garnished with an irregular pattern of dots as if formed by the shadow of the moving sculpture.

布鲁塞尔科文特餐厅

科文特餐厅位于布鲁塞尔中心区，是布鲁塞尔广场会议中心的主要会面场所之一。加垫侧墙板和餐桌上方悬挂的雕塑品共同拉伸了空间的长度。长30米的雕塑是阿纳·奎兹的作品，从餐厅后部的吧台喷薄而出，仿佛鲜活的生命体一般。这个悬浮在上空的装置让顾客有着一种安全感和聚集感。装软垫的墙壁不仅仅是一个隔音装置，还具有基础设施功能：其内部具有供暖和通风系统、电气设施、以及厨房、洗手间、休息室等附加结构。墙上的不规则斑点宛如雕塑的影子。

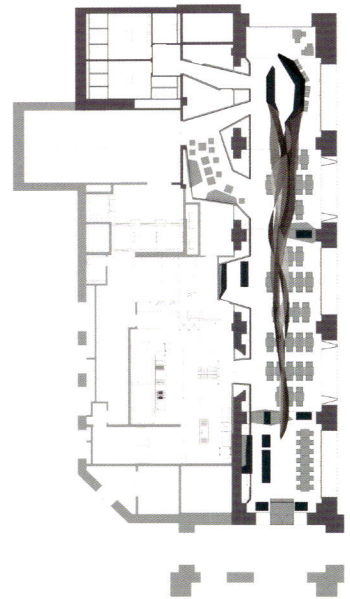

The Dominican Hotel

Tucked behind Brussels' famous theatre and opera house, The Dominican is a new hotel that offers a strong sense of history mixed with forward-thinking, eclectic design in the European Union's capital city. The hotel's sweeping archways are a clear reference to a Dominican abbey that stood on this site in the 15th century. Its original facade has been integrated into the new construction by the Belgian Firm Lens Ass Architects. Guests enter the lofty, high-ceilinged public spaces and their breath is taken away by the attention to detail. A stroll through the Monastery Corridor evokes an almost medieval feeling of elegance with original Belgian stone flooring. The Grand Lounge, considered the heart of the hotel, calls to mind the spirit of old European decadence in soaring windows and metalwork at the same time as attracting a definitively style-conscious modern clientele with its cutting-edge design.

多米尼加酒店

坐落于布鲁塞尔著名的剧院和歌剧院的后身，多米尼加是一个崭新的酒店，营造了很强的历史氛围，同前瞻性的思想和欧洲联盟不拘一格的设计融合在一起。酒店的宽大的拱门可以让人们很自然的联想到十五世纪位于这里的多米尼加修道院。在比利时水晶艾斯设计公司的精心策划下独具匠心，它原来的门面已经被嵌入到新的建筑之中。客人们来到举架高大宽敞的公共空间，这里的细节设计令人们叹为观止，留有原始比利时石阶的地板向人们彰显着中世纪的典雅风格。

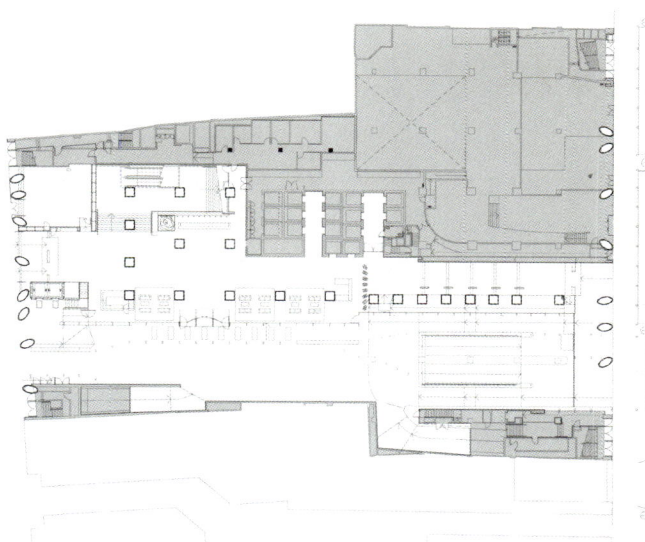

Restaurant

Belgium

Brussels

2007

Photo: Lens Ass Architects

Lens Ass Architects

Lánchíd 19

Named after Budapest's famed 'Chain Bridge' spanning the Danube and situated near both it and the Buda Royal Castle, the Lánchíd 19 has become a contemporary architectural landmark that attracts a new kind of cosmopolitan crowd. The design of the restaurant features a natural air. Outside the giant French window is the scenery of the courtyard, reflected in the interior by the potted plant. The tables and the chairs are all designed to be delicate and artistic, making the restaurant simple and clever and offering guests a relaxing and comfortable dinning environment. The lighting design of the restaurant contributes to the overall style.

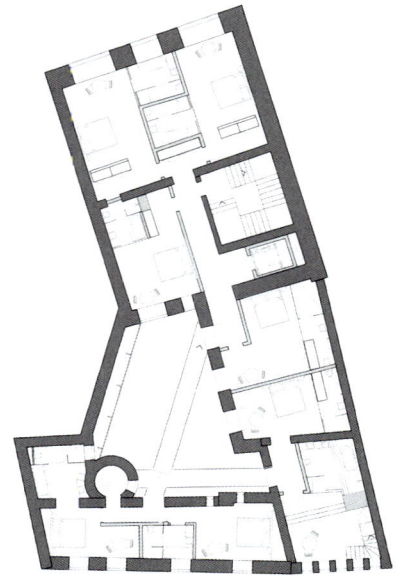

链桥19号设计酒店餐厅

链桥19号设计酒店毗邻横跨多瑙河的著名"锁链桥"与皇家城堡，现已成为这一地区的当代建筑里程碑，吸引着来自世界各地的人群。餐厅设计充满自然气息，开敞的落地窗外就是庭院景观，餐厅里摆放的盆栽和外面的植物相互呼应。桌椅都是纤巧而有艺术感的，餐厅设计简约而灵动，为顾客创造了自然舒适的就餐环境。餐厅的灯光设计也延续了这一风格。

Restaurant in Museum of Modern Art- Grand Duc Jean

The restaurant's structure contains two long tables made of Douglas fir tree, which seem to be levitating between the two rows of Thonet chairs. Its clear coloured tiles roof seems to be the meeting point between a blue morning sky and the beige Bourgogne stone floor. Its shadow protects the eaters from the strong sun, only allowing a few pointillist rays to go through its even cracks. Padded by the thick tiles which absorb the sounds and create a muffled atmosphere, it has become a comfortable gathering spot; a soothing stop in a museum trail.

当代艺术博物馆餐厅

餐厅内两张长长的绿松木餐桌如同悬浮在两排椅子之间，格外吸引眼球。上方的彩色琉璃瓦屋顶恰似淡蓝的晨空和米色地面的交汇点，保护着食客们免受烈日灼晒的同时，还可以吸收噪音，营造一个安静的氛围。这里不仅仅是聚会的舒适场所，同时也是那些在博物馆中游览的客人停歇的驿站。

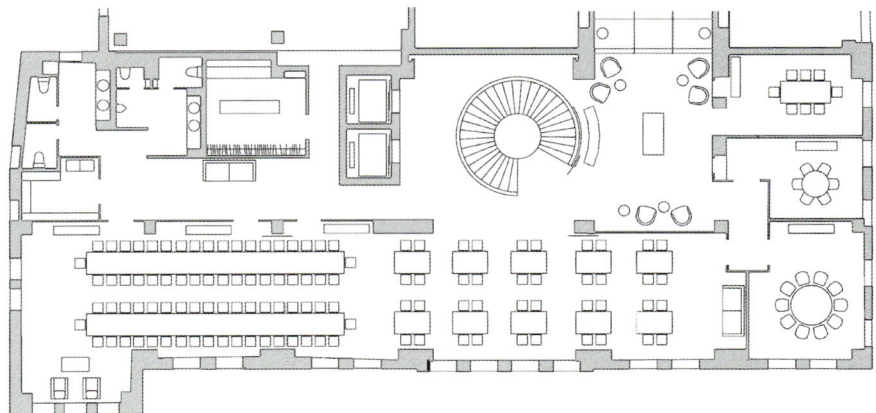

Restaurant

Luxembourg

Parc Dräi Eechelen

2007

Photo: Ronan & Erwan Bouroullec

Ronan & Erwan Bouroullec

Café

Germany

Berlin

2006

Photo: Joden Mozer Design

Joden Mozer Design

Cylinder Café

Cylinder updates that idea, in the same way that the new Beetle is a refreshing version of the old Beetle. The design is an analogy between cars and buildings. The restaurant is 'parked' in a glass building at the base of a museum, 'a rounded object in a box'. Inside, the walls are curved and molded like the forms inside and outside a car. The kitchen façade is like the dashboard of the car. The restaurant seats are inspired by automobile seats. Walls, ceilings and floors blend into each other. Lamps are molded into the ceilings seamless, smooth and aerodynamic. Everything for the restaurant was custom designed and manufactured by Jordan Mozer and Associates.

汽缸咖啡馆

像新甲壳虫是旧甲壳虫的升级版一样，汽缸理念也让餐厅设计有所升级。设计以汽车比喻了建筑。圆形的餐厅位于博物馆一楼的玻璃幕墙内。餐厅的墙上装饰着汽车的零部件。厨房的外观像是汽车仪表盘。餐厅的座位就像汽车座椅。墙壁、天花板和地板融为一体。设计师利用空气动力学原理，直接在天花板上做出灯的造型。全部设计都是由设计师为餐厅量身打造的。

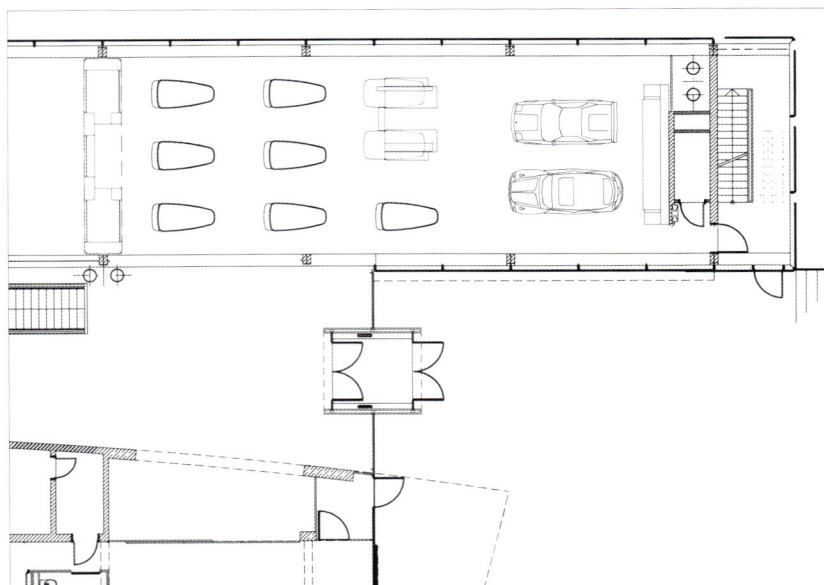

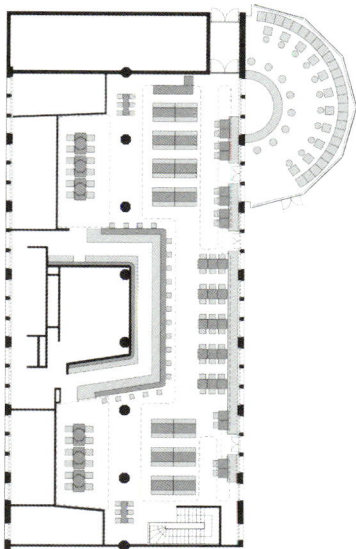

Mash Stuttgart

The Mash brewery, a popular party destination, was seeking to reposition itself through comprehensive refurbishment and the introduction of a mixed concept. The new Mash is designed to cater for events as a club, restaurant and all-day bar. The large bar, which is stunningly illuminated by more than 2,500 plastic batons, takes centre stage in the space. It divides the room into three zones in which two niches enclosed behind silk curtains display a more intimate character. The individual areas are connected by continuous mauve-coloured flooring as well as a superimposed ceiling of white printed tiles with a surrealist collage.

斯图加特麦芽浆餐厅

麦芽浆啤酒厂是一个潮流景点，力求在翻修中融合各种理念。新麦芽浆餐厅可以提供俱乐部、餐厅和全日制酒吧服务。酒吧里最抢眼的是天花板中心2,500多根塑料棒。它们将空间划分成3个区域，丝绸帘子后面两间小屋提供了更加私密的空间。淡紫色的地板和叠加式白色瓷砖天花板将这些独立区域联系到了一起。

Restaurant & Bar

Germany

Stuttgart

2008

Photo: Zooey Braun

Ippolito Fleitz Group - Identity Architects

Photo: Design Hotels™

Klaus Peter Lange, Regine Schwethelm & Sybille von Heyden

Gastwerk Hotel

Seasoned hotelier Kai Hollmann turned Hamburg's 19th-century municipal gasworks ('Gaswerk') into Europe's first loft-style hotelby as easy as dropping a 't' into its name, instantly transforming the complex into a 'guest works'. Sheer curtains on either side point the way to the bar and restaurant or to the seven conference rooms. Pure functionality is still expressed by exposed pipes and naked brick walls, but interior designers Regine Schwethelm and Sibylle von Heyden found that the brick actually provided a warm backdrop. They set about contrasting the building's rough edges with soothing textures, slick and simple MDF furniture, oversized armchairs and touches of Asia – an imported style prevalent in this seafaring city.

房客之家旅馆餐厅

卡伊·霍夫曼是一名旅馆经营者，他将汉堡这家19世纪的市政所属的煤气厂变为欧洲第一家阁楼式的旅馆。两边垂下的帘幕指示出通往酒吧、餐厅或者7间会议室的路。通过外露的管道和裸露的砖墙也体现出纯粹的功能性，但是室内设计师雷吉娜·施维瑟姆和西比尔·凡·海登发现，其实砖能够提供一种温暖的背景。于是他们着手将建筑粗糙的边缘跟光滑的质地、简单光亮的家具、大号的扶手椅和亚洲风情形成鲜明对比；尤其值得一提的是，亚洲风情是这座水上城市里的一个重要流行风。

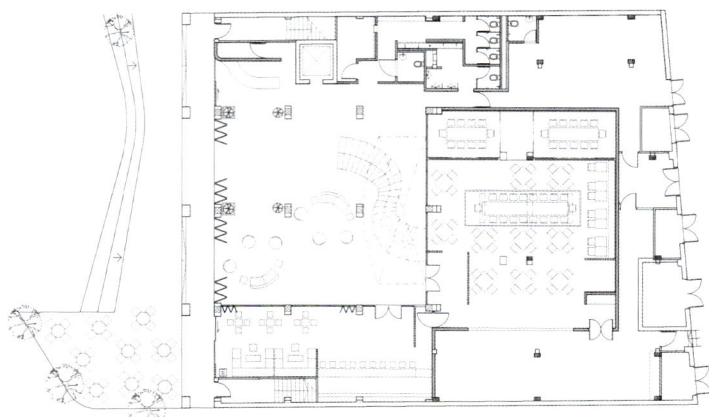

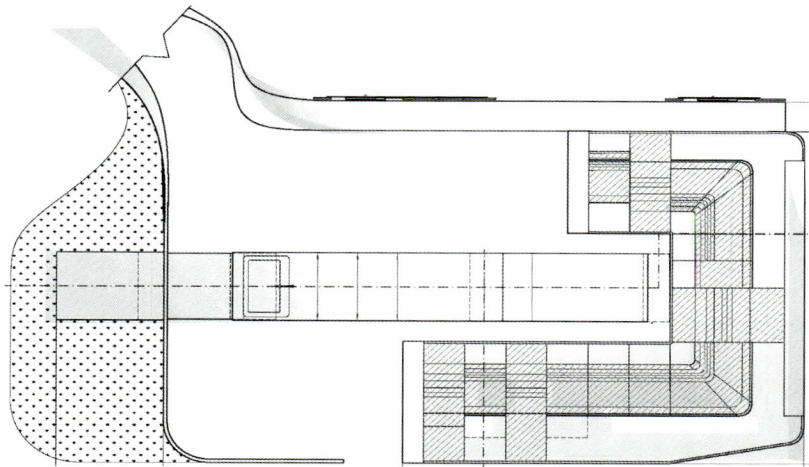

Bar Lounge 808

Like many modern, stylish cocktail lounges, bar lounge 808 was intended to accommodate professionals in the fashion, art and entertainment. The designers used a lot of beautiful, refined woods and filigree metal supports to create an impression of lightness and levity. It achieved an element of subtle elegance that is very comfortable and light. Bar Lounge 808 occupies a wedge-shaped, corner site, with floor-to-ceiling windows forming a 40 metres (131 feet) glass façade. In fact, the interior proved to be too 'open' and overexposed, so diaphanous and sheer curtains were chosen to line the windows.

808酒吧

同许多时尚流行的酒吧一样，808酒吧客户的定位就是时尚人士和艺人。当然大量美丽精致的木头和金银丝的装饰的采用，营造了一个变幻莫测的氛围。同时还夹杂了少许的典雅别致，明亮舒适。灯光的设计并不是问题，808酒吧作为角落，断面成楔形，从地板到天花板镶嵌了一个40米高的落地窗。外立面的设计过于开放，感光度过高，因此精致的珠帘装饰在窗上。

Bar

Germany

Berlin

2006

Photo: Fritz Busam

Plajer & Franz Studio

Restaurant

Germany

Munich

2007

Photo: Fnp Architects

Fnp Architects

CAW

It is embodied in a simple and clean modern style. The red wall in front of the store is striking and attractive. The five colours of interior space are elegant white, bright red, green and yellow and natural grey, which show fresh and clean quality of the whole design. Interior design is also very simple. The sunshine through the large rectangular windows on the side walls floods the room. Through the window, people can watch the scenery. Wood desktop without ornaments retain the original colour, giving a feeling of nature. Contrast with the dark surrounding environment, the restaurant makes a strong impression on people.

CAW餐厅

该设计体现的是一种简单明快的现代的设计风格。餐厅内由四种颜色装饰，典雅的白色、亮丽的红色、自然的黄绿色和朴素的灰色。整体设计给人的感觉是清新自然的。室内设计风格简约，阳光从侧面墙壁上的一扇大型长方形窗户射进屋内。透过这面窗，人们可以观赏外边的景致。木质的桌面没有加入太多修饰，保留着木头原有的颜色，给人一种亲近大自然的感觉。

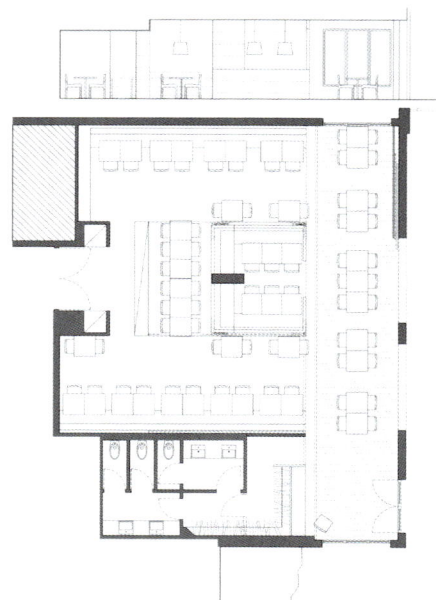

Bar Foufou

The FouFou's L-shaped barroom was arranged around a large bar counter. Mirrors affixed to the wall behind the bar expand the space further still. The entrance area is heralded by a large, fabric-covered ceiling light that serves as a room opener. The bar itself is upholstered in a metallic greenish-gold diamond pattern, which is continued on the ceiling in the rear area of the barroom. The walls are executed in a complimentary shade of classic green. This forms an effective backdrop for the white, original case tree windows. The bar can be opened up to the outside during the summer, enabling customers to be served outside.

否否酒吧

否否酒吧的L形酒吧间围绕巨大的吧台展开。吧台后方墙壁上的镜子在视觉上扩大了空间。入口处天花板上灯饰吸引着人走进酒吧。皮革软垫质感的金属绿色吧台上压出了菱格图案，并且一直延续到天花板和酒吧间后部的设计之中。墙壁也采取了少许的绿色元素，为白色的原木窗户提供了有力的背景。夏天，顾客还可以到酒吧外面接受服务。

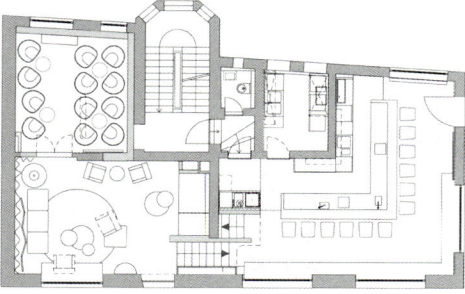

Bar

Germany

Stuttgart

2009

Photo: Zooey Braun

Ippolito Fleitz Group - Identity Architects

Restaurant

Germany

Munich

2007

Photo: Gabriele Allendorf

KSP Engel und Zimmermann Architects, Munich
Lighting Design: Gabriele Allendorf – light identity, Munich

Roche Diagnostics Cafeteria

Cafeteria analogous to casino, ceiling panels in the cafeteria also activate the ceiling pattern. But in this case planar luminaries take turns with pendant luminaries. The down swinging light bulbs regenerate a new level in the room and produce a comfortable relaxed atmosphere. Through the vertical adaptation of the stripes on a wall of a stairway a decorative element will be created, which signalise about function of a vertical circulation. Through different art of placement of this stripes and/or possible colour difference the identification of each single upper floor will be possible.

罗氏制药自助餐厅

自助餐厅与度擦汗功能类似，顶棚嵌板能够让天花板的图案更具活力。在这个项目中，曲面灯和吊灯交替作用。下垂的灯泡为空间打造了一个新层次，营造出舒适休闲的氛围。楼梯侧面墙壁上的条形灯既是一种装饰元素，也暗示着垂直路径。自助餐厅通过条形灯位置、色彩的不同，标示了不同的楼层。

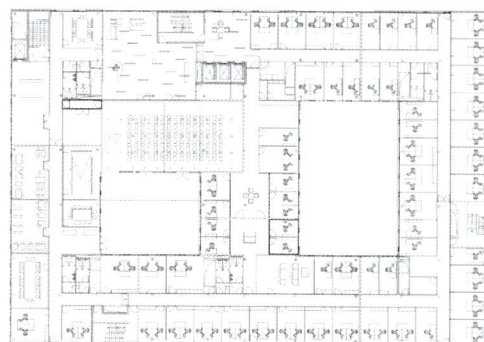

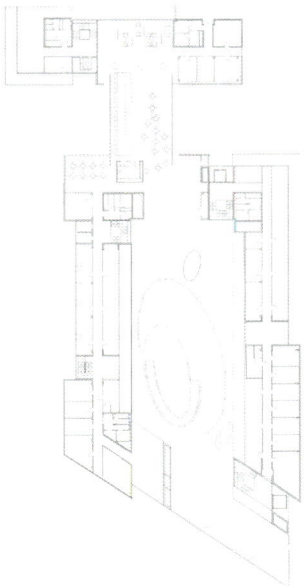

European Patent Office Prestigious Areas

The European Patent Office (EPO) sees architecture, art and light as a single entity. In the imposing foyer, a huge spiral staircase forms the architectural focus: Like a DNA helix from which all evolutionary life originates, it spirals through the upper floors to end visually in the sky. The ceilings are illuminated with coved lighting of fluorescent lamps right there where verticals blend the horizontals. This accents the rhythmical dimensioning of the room while giving orientation simultaneously and revealing the flanking rooms for visitor's view. Passing the joining gastronomic area the formal discretion is continued and new lighting qualities are enhanced. Especially manufactured minimalist light objects illuminate the lounge.

EPO物业信托投资公司餐厅

欧洲专利局秉承"建筑、艺术、灯光互为一体"的理念。大厅内，宽大的螺旋状楼梯构成了主要特色——如同DNA螺旋结构一般，所有的一切都从这里起源，蜿蜒向上，似乎消失在天际线处。天花板上装饰着荧光灯，突出空间的节奏感。休息室内，专门定制的简约风格灯饰照亮了整个空间。

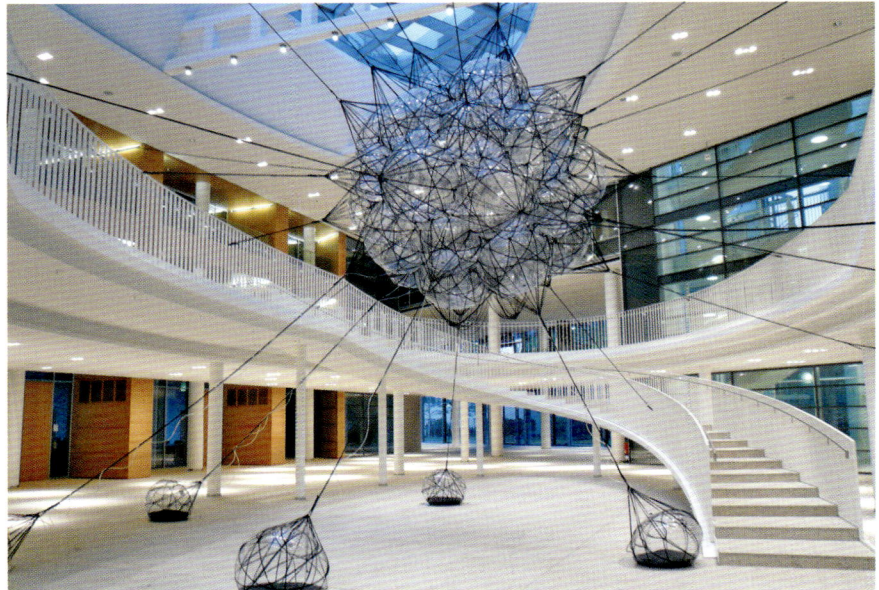

Restaurant

Germany

Munich

2007

Photo: Gabriele Allendorf

SIAT, Munich

Wald- und Schlosshotel Friedrichsruhe

An onyx bar is the centrepiece of the bistro. Thanks to its height, this bar is very comfortable for the guest. Here he can sit and eat in his bathrobe, take a quick snack at a coffee table or the intimacy of one of the niches, as in a classical bar. A large-scale wine fridge filled with carefully selected wines forms the background of the bar. Ordering is easy as the staff is close at hand to provide meals cooked in the restaurant. The floor made of solid oak combined with French lime stone and teak-furniture surrounded by a cosy collection of natural fabrics in sunny colours - create an atmosphere of luxurious relaxation.

瓦尔德酒店酒吧

小酒馆的中央是一个缟玛瑙材料的吧台。它的高度适中，让客人感觉很舒服。客人可以穿着睡衣在这里吃东西、喝酒，在咖啡台处或在一个私密的角落里吃快餐，就像在典型的酒吧里那样。大型的葡萄酒冰箱内放着精选的葡萄酒，成为了酒吧的背景（墙）。在这里点菜是很方便的，服务员就在周围，等待给客人上餐厅的菜肴。地面铺着硬橡木和法国的石灰石，柚木家具被环绕在颜色明快、舒适的天然织物中，创造了一个豪华的休闲氛围。

Photo: Niki Szilagyi Interior Architecture

Niki Szilagyi Interior Architecture

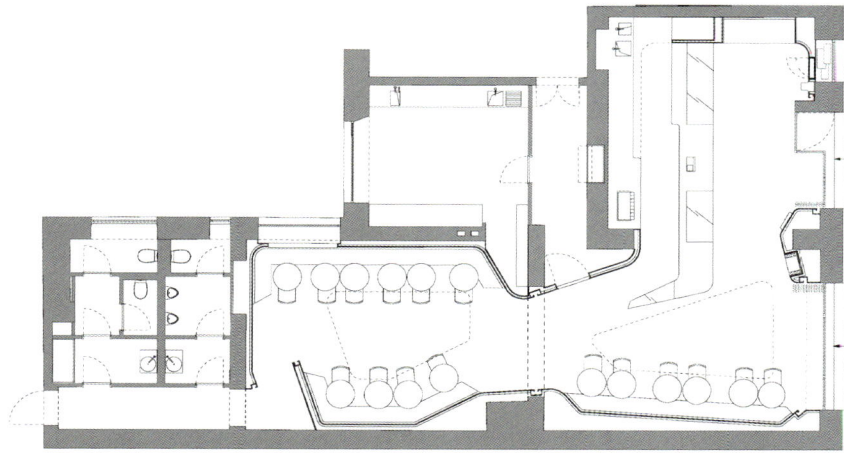

Giacomo

The outstanding concepts in the food sector follow the same rules that prevail in the retail industry. Beside singularity and strong brand recognition, authenticity and emotional aspects are the most crucial factors. The Giacomo gourmet fast food concept is not contradictory in itself. Rather it draws on the best of both worlds. Giacomo meets the yearning desire for quality and emotional authenticity in times where monotony abounds. In a world where time has become the greatest luxury and where work and leisure time tend to blend more and more, it is essential to provide the customer with fast and proper service. Therefore it is not inadequate to sell fast food, because for Giacomo fast food does not equate to eating quickly but to be attended to quickly and exclusively .

吉亚科莫快餐店

餐厅的设计与零售业设计有异曲同工之妙。除了个性和强烈的品牌辨识度之外，可靠性和情感因素也至关重要。吉亚科莫快餐店的设计完美地融合了二者。吉亚科莫满足了消费者对质量和情感两方面的需求。在这个时间已是奢侈品、工作与休闲也日渐混合的世界，为消费者提供快速而适当的服务至关重要。吉亚科莫所提供的快餐并不是快速食用的概念，而是快速而独家的服务。

Restaurant

Germany

Berlin

2009

Photo: Plajer & Franz Studio

Plajer & Franz Studio

Restaurant

Germany

Hamburg

2007

Photo: Studio Uwe Gärtner

Eins Architects

Nat. Fine Bio Food

The purpose of the newly-founded chain, Nat. Fine Bio Food, is to make fast food healthy, to offer delicious organic food in a timely manner – and this in a contemporary environment. nat. brings together Organic and Lifestyle, and with this follows the increasingly-important LOHAS-Trend (Lifestyle of Health and Sustainability). The claim, 'nature comes to the city' guided the design concept. The ceiling allows association with cloud formations; the columns remind one of tree trunks; backlit walls dissolve the spatial boundaries through oversized, blown-up plant and herb-panoramas. The main contradiction of a 'natural' restaurant in an artificial urban environment becomes the core of the design.

奈特健康生态食品餐厅

奈特健康生态食品连锁餐厅提供及时而健康美味的有机食品。奈特将有机食品和生活方式带到了一起，符合乐活族（健康、可持续的生活方式）潮流。"自然进驻城市"的口号是贯穿设计始终的重点。天花板宛如云的形状，圆柱仿佛树干，背景墙上是巨大而繁茂的植物和香草。设计的核心是自然餐厅与人造城市环境所形成的冲突和对比。

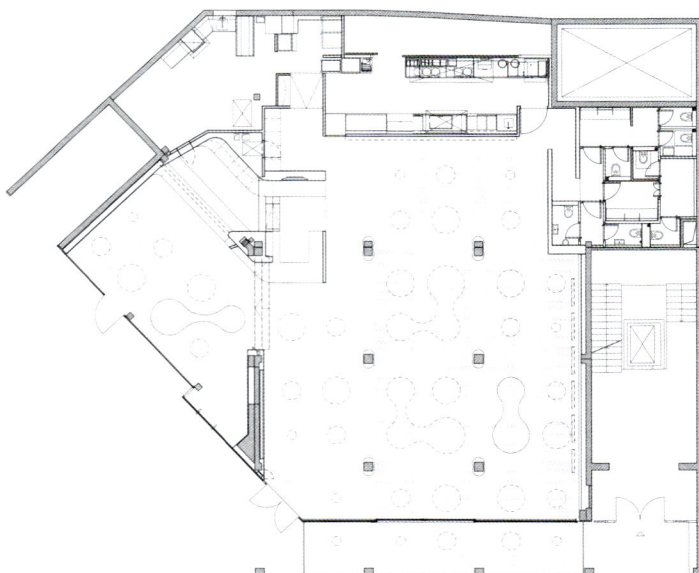

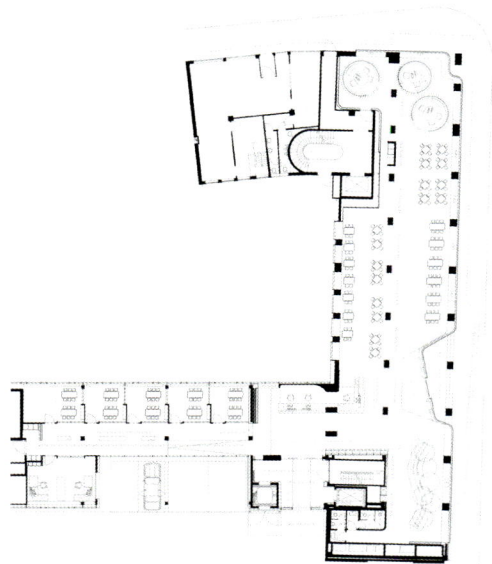

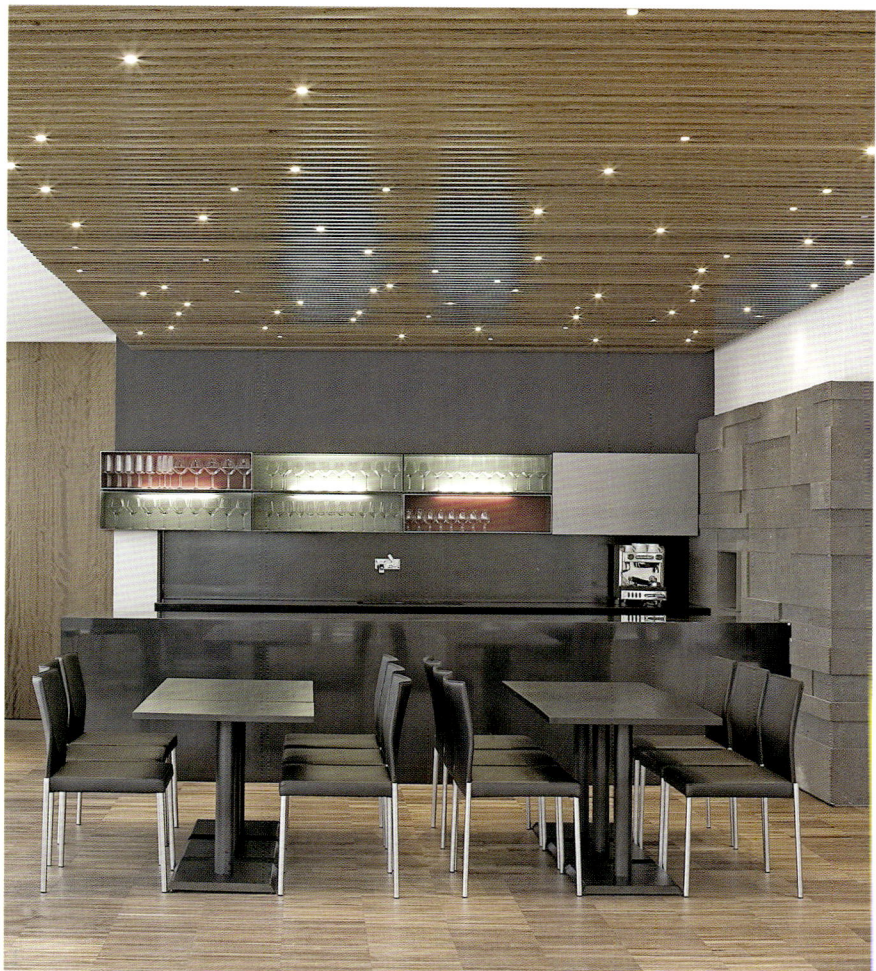

Restaurant

Germany

Trier

2007

Photo: Ingbert and Jutta Schilz

Ingbert and Jutta Schilz

Restaurant in Becker's Hotel

Amidst the vineyards and rolling hills of Trier, one of the oldest cities in Europe, dating back to Roman times, Becker's Hotel & Restaurant grounds itself both in nature and in a family tradition that goes back five generations. Yet Becker's self-professed heart is its kitchen. Done in black and white, the restaurant it serves suggests a sophistication befitting a black-tie event or a white wedding, while the wine bar's deeper, darker tones evoke a cosmopolitan night out. Here, too, are the ubiquitous basalt stones, but the simple rectangular shelving for the wine glasses adds a repetitive, almost artistic visual element to the room, perhaps citing the reserved dignity of a Donald Judd sculpture.

贝克之家酒店餐厅

贝克之家酒店坐落在始于罗马时期的欧洲古城——特里尔，四周环绕着美丽的葡萄园和起伏的小山丘。厨房可以说是整个酒店的中心。黑白色打造的餐厅极为精致，适于举办各种正式晚宴及婚礼。深色的酒吧营造了节日夜晚的狂欢氛围，随处可见的玄武岩以及简约的长方形酒柜增添了艺术气息。

Restaurant

Germany

München

2010

Photo: Zooey Braun

Ippolito Fleitz Group - Identity Architects

Wienerwald – Interior Concept for Restaurants

The new interior design underscores the realignment of the brand, while translating the chain's traditional strengths of high quality, comfort and German cuisine into a contemporary design idiom. Materials and colours reflect the principles of freshness and naturalness, which find their expression in materials such as wood, leather and textiles, as well as in the dominant green tones that complement the fresh white. Gold is used as an accent colour, conjuring up associations of quality and the crisp, gold-coloured skin of the main product, the Wienerwald grilled chicken. The space has been organised to ensure good visitor guidance, crucial in a self-service restaurant, as well as respecting the need for a differentiated selection of seating. Upon entering the restaurant, the guest is guided towards a frontally positioned counter, which presents itself as a clearly structured, monolithic unit.

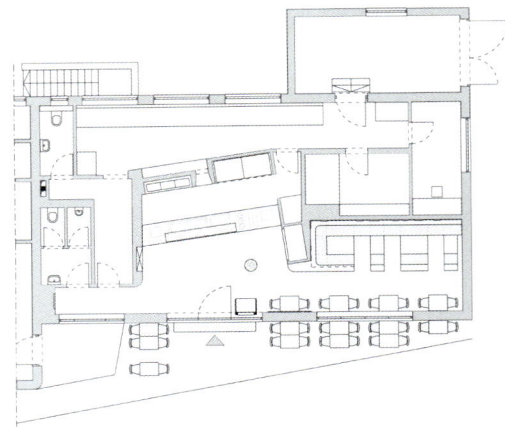

维也纳森林餐厅

新的室内设计强调了该品牌的重组，以一种现代设计方式诠释了品牌一贯的高品质、舒适度和其所提供的德国美食。材料和色彩的选择反映了餐厅清新自然的原则，如木材、皮革和织物，以及点缀着白色的整体绿色色调。金色反映了餐厅的品质和它的招牌菜——维也纳森林烤鸡。空间的设置便于顾客走动，也充分考虑了座椅的安排。当顾客走进餐厅，便会被引领到一个朝前设置、设计简洁的前台。

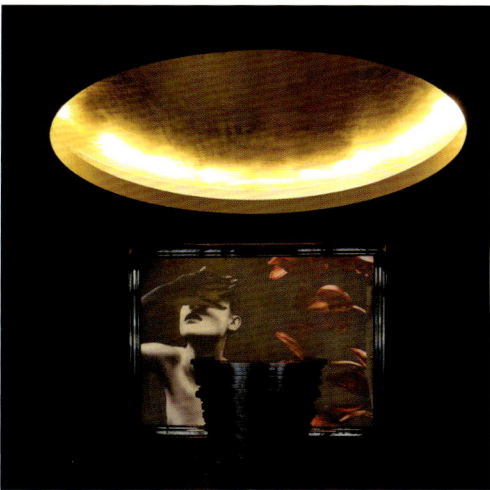

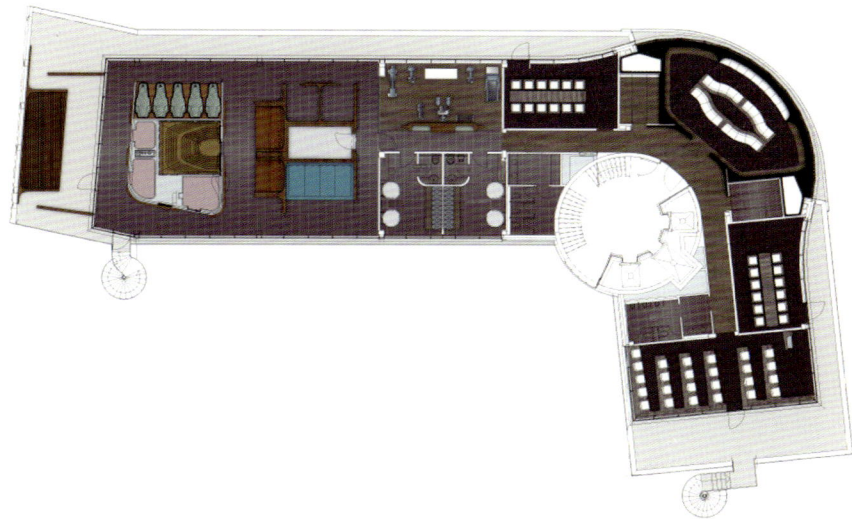

Restaurant in Roomers Hotel

Formerly a quaint office building, Roomers Hotel has now morphed into the exact opposite: a fresh, cosy space full of swirling dark colours interpreted in new ways. The Roomers restaurant is everything but a typical ordinary hotel restaurant. Niches for two promise secluded moments, group tables are ideal for celebrating in style with friends. 130 seats, some of them open air. Along with that comes smooth music from a DJ, which can be turned up for occasional events – and all that without additional essential technique.

房客酒店餐厅

这座建筑原本是普通的办公楼，现在却变得完全不同：现在这里是一个生机勃勃的空间，一个舒适宜人的酒店，在这里，以全新的手法诠释的深沉的颜色仿佛漩涡般把你卷入其中。房客餐厅并不是一家普通的酒店餐厅。分组餐桌是友人聚会的绝佳场所。餐厅共有130个座位，DJ所播放的优雅的音乐可以根据活动气氛进行任意的调节，无需过多的变动。

Restaurant

Germany

Frankfurt

2009

Photo: Roomers by Designhotels
Photographer Top level: Ernst Stratmann

Oana Rosen in cooperation with Mickey Rosen and Alex Ursenau

Restaurant

Germany

Berlin

2010

Photo: Klm-Architekten

Klm-Architekten

Restaurant Tipica

The centre of this Mexican Gourmet Taqueria is the silver-gold coated bar with an elegantly shaped, seamless surface of white mineral material. Centrally the bar locates along a glass wall to the visible show-kitchen. The white wall, canvas ceiling and leather cushion shine like the typical Mexican house facades in sand-coloured surfaces. Six large screen lights reflecting golden light, hanging above the tables illuminated coral lamp shades of foamy ceramic. It creates a soft atmospheric light that radiates steamed into the room. Silky fleece wallpaper adorn the wall. The glass cube on the wall with typical Mexican spices and hot chillies are the background for the culinary experience.

蒂皮卡餐厅

这家墨西哥美食餐厅的中心是金属吧台，吧台表面装饰着白色的矿物材料。沿着吧台的玻璃墙壁后面是可见式厨房。白色的墙壁、帆布天花板和皮垫像典型墨西哥住宅的沙色外墙一样。餐桌上方，6个巨大的灯饰照射出金色的灯光，营造出一种柔和的感觉。墙壁上装饰着丝滑的羊绒墙纸。墙壁上玻璃格子里的墨西哥调料和红辣椒为这次美食提供了完美的背景。

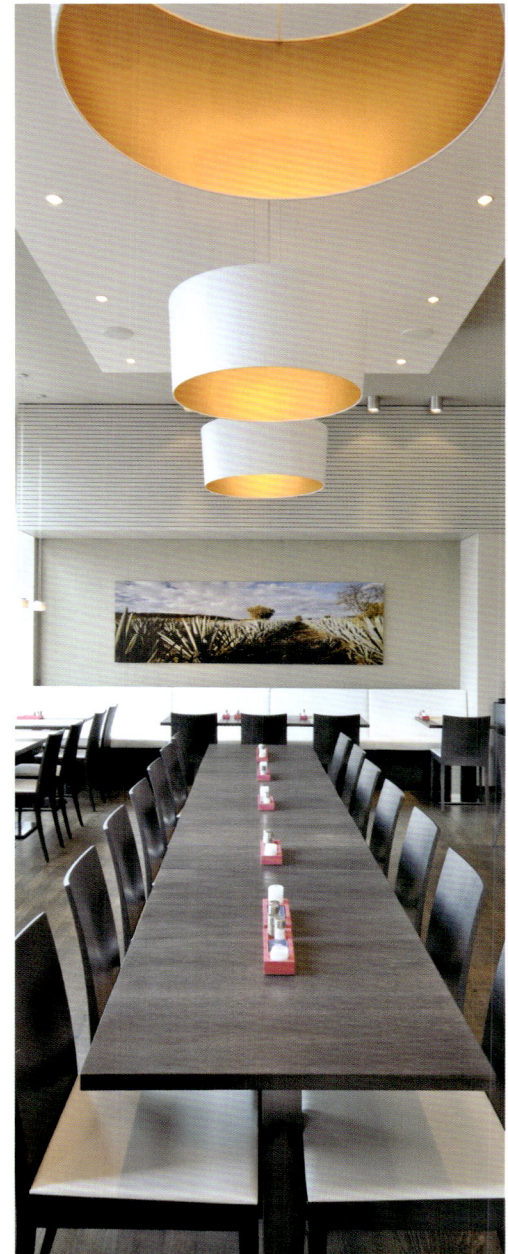

VLET Restaurant

The graceful arches of the barrel-vaulted ceiling, steel beam supports and textured brick and plaster walls are enhanced by newer design elements such as the rustic timber plank floors, rich leather chairs and driftwood 'sculptures' lit from below to define their shadows against the irregular wall surface. The heavy, rough-hewn natural finishes of these structural design features are offset by the smooth, shiny modern materials of the stainless steel dining table bases and barstool frames, clear crystal wine glasses and sparkling translucent pendant lights. The visually exciting contrast is further emphasised by the bold flash of fresh lime green behind the bar and on its counter surface.

韦丽特餐厅

优雅的拱形屋顶、古老的钢梁支柱以及质感十足的砖石墙壁在木板地面、皮质椅子以及底部照明的木质雕塑的陪衬下更加韵味十足。粗糙自然的装饰特色与光滑的材质相得益彰，不锈钢餐桌和凳子、晶莹透明的玻璃酒杯以及闪闪发光的灯饰格外引人注目。此外，吧台后面及表面一抹淡雅清新的黄绿色调将原有的对比特色进一步深化，带来了活泼生动的气息。

Restaurant

Germany

Hamburg

2008

Photo: JOI-Design GmbH

JOI-Design GmbH

Restaurant

Germany

Hamburg

2008

Photo: Andreas Brücklmair

3meta Design Devision

Restaurant Mangold

The restaurant is located in a 1870s' building which is part of the a hotel. Basically the design had to be integrated in the old existing brick architecture. So the choice of material is very classic. The most important thing was the haptic of whatever the guest touches. Green velvet, natural tanned leather, brass and oyster coloured massive oak with a somewhat rough look and feeling. Partitions with an industrial design, antacid coated and glazed seem to have always been there, and isn't a designed modern statement. A big bar which shelters the breakfast buffet in the morning changes into a front cooking desk in the evening, where the guests can order his fish directly and talks with the cook over the way of preparation. The restaurant is equipped with generous and very comfortable benches. The seating, consistently with armrests, doesn't simplify the choice of where to sit.

曼戈尔德餐厅

餐厅在一座19世纪70年代的历史建筑里，是酒店的一部分。因为整体设计必须配合古老的砖结构建筑，所以必须选择经典的材料，最重要的是客人的触感。绿色天鹅绒、天然棕色皮革、黄铜和橡木有种略微粗犷的感觉。具有工业感的隔断设计毫无违和感，恰到好处。吧台白天提供自助早餐，晚上则成了烹饪前台，客人可以当场点菜，并且与厨师聊天。餐厅拥有舒适的长椅和带扶手的座椅。

Restaurant Calla

The pure cream-and-white colour palette creates a sublime sense of serenity, while the textured stucco sculptures gracing the walls add a nuanced contrast. The shape of the waves can also be found through the gold-plated banana tree leaves scattered at the base of these wall insets. Their metallic surface subtly casts sparkles of light, which, along with the halogen wall-washers, are reminiscent of the glints of sunlight bouncing off the nearby water. The restaurant itself can be divided into two parts by a well-disguised curved wall, allowing for additional flexibility with private parties. With thoughtfully planned spaces and clean sculptural finishes, Restaurant Calla exemplifies modern sophistication.

卡拉餐厅

餐厅内，米白色调奠定了高雅恬淡的基调，墙壁上的雕塑装饰则增添了另一种韵味。香蕉树叶图案凌乱地排列在墙壁的底脚处，镀金表层闪闪发光，让人不禁想到波光粼粼的水面。一面蜿蜒的墙壁将整个空间一分为二，为那些寻求安静就餐环境的顾客们提供了舒适的场所。

Restaurant

Germany

Hamburg

2008

Photo: JOI-Design GmbH

JOI-Design GmbH

Hotel Ritter -Wilden Ritter

Beside the traditional restaurant for the sophisticated but down-to-earth German dishes, there are several other locations: a new lobby bar, the restored knight cellar, classic and cosy in its vault, close to the new wine cellar, and as a new highlight in the restaurant scene, the new gourmet restaurant 'Wilden Ritter' (wild knight). The most impressive material, which was chosen beside the apple wood veneer with its high gear wood grain, is the wallpaper. A wallpaper made according to the input of the interior designer was created only for this project.

骑士酒店——狂野骑士餐厅

酒店内除了老练而实际的传统餐厅之外,还有一些其他的好去处。大堂酒吧、舒适的骑士酒窖和新美食餐厅"狂野骑士"。除了纹理清晰的苹果木薄木板之外,最引人注目的装饰材料便是壁纸,餐厅的壁纸是特别为这个项目定制的。

Le Parc in Le Meridien Parkhotel

The restaurant 'Le parc' with its winter garden has changed into a modern restaurant. Elegant colours like mauve and beige are interchanging with dark wooden parquet and brown carpet. The big surfaces like wall and ceiling are kept in crème to the point of light brown tones and stand in contrast to the elegant wooden fixtures in 'oak'. The highlights are the illuminated boxes, which were inserted into the walls, where custom-made decoration items are placed.

艾美公园酒店公园餐厅

公园餐厅和它的冬日花园被改造成了一个现代餐厅。优雅的淡紫和米黄与黑色镶木地板和棕色的地毯相互映衬。墙壁和天花板等表面采用了奶油色系，浅棕色与优雅的橡木色形成了对比。嵌入墙壁的发光盒子里面放置着特别定制的装饰品，是设计的重点。

Restaurant

Germany

Frankfurt

2009

Photo: JOI-Design GmbH

JOI-Design GmbH

Restaurant

Germany

Unterschleissheim

2010

Photo: JOI-Design GmbH

JOI-Design GmbH

Redox in the Dolce Hotel

The name 'Redox' refers to the chemical process that naturally occurs in antioxidant foods and, as such, directly corresponds to a tenet of Dolce's redefined corporate philosophy of 'nourishment', or nurturing the spirit, mind and body of its guests. JOI-Design interpreted this concept through the gourmet restaurant's contemporary focal point, a backlit 'wine wall' which showcases the bountiful- and antioxidant-rich- varietals vinified in the region. Gracing either side of the access gallery between the wine displays, 'sculptures' of traditional male and female 'trachts', the embroidered national costume of Germany, are a light-hearted homage to the locale's provenance. Evidence of this intimate, 40-seat dining area's Bavarian inspiration can be found through the roughly textured antler chandeliers that provide a distinct counterpoint to the smoothness of the crisp white tablecloths and rich dark stained oak; together they evoke the spirit of a fine, time-honoured German dining hall.

道尔斯酒店氧化还原餐厅

"氧化还原"指的是防老化食物中自然进行的化学进程，也指道尔斯酒店的经营信条"滋养品"，即让客人的身心得到滋养。设计师将这一理念通过餐厅的"酒墙"体现出来，里面摆设着大量富含抗氧化剂的本地葡萄酒。入口走廊两侧的酒架之间分别摆设着德国男女传统刺绣服饰"雕塑"，对本地文化进行小小的致敬。餐厅的巴伐利亚风情体现在纹理粗糙的鹿角吊灯上，它与干净整洁的白色桌布和暗色斑点橡木遥相呼应，共同创造了精致、经典的德国餐厅氛围。

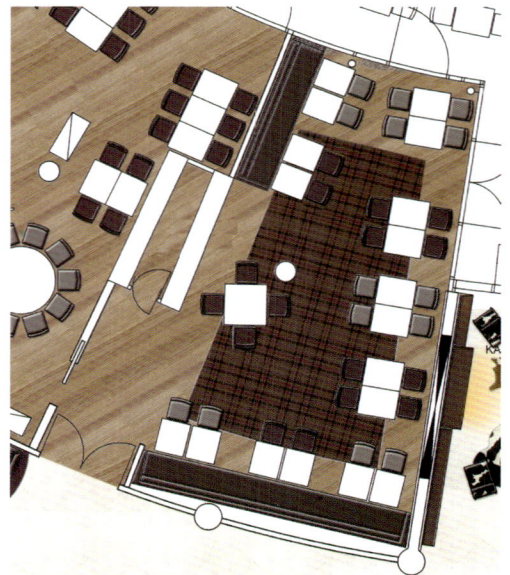

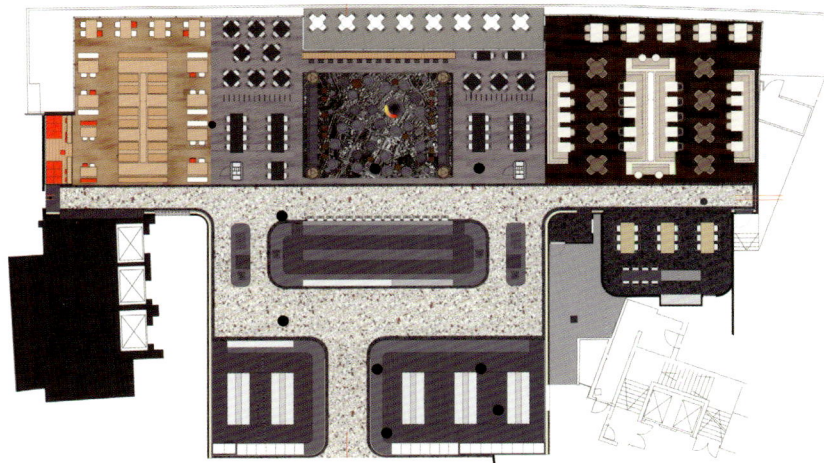

Restaurant

Germany

Stuttgart

2010

Photo: Ewout Huibers

Rob Wagemans, Melanie Knüwer, Charlotte Key, Sofie Ruytenberg (graphic design), Rik van Dillen

Karls Kitchen

The guest's initial experience is on coming in, walking through the kitchens, barely separated from the preparation of fresh products by a large glass wall. To the left you can see exclusive sandwiches, fresh salads and homemade desserts being prepared. On the right you can take a sneaky peek into the warm kitchens, where traditional local classics, Asian wok meals and European specialities are prepared. The materials used reference the classic kitchen of grandma's time; white tiles, black natural stone in combination with a terrazzo floor. Opposite the kitchens is the bar, a free-standing element between the free-flow restaurant and the seating area. The bar's function is twofold: on the kitchen side, cakes and pastries made in-store, fresh juices and beverages are offered. The bar adjacent to the lounge area tempts one to sit at the bar and enjoy a nice glass of wine, a glass of local beer or one of their coffee specialities.

卡尔斯厨房

一走进卡尔斯厨房就能看见玻璃墙后面的新鲜食材，左侧是三文治、新鲜沙拉和自制甜点；右侧是热厨，供应当地传统美食、亚洲美食和欧式菜品。厨房材料的使用再现了家庭氛围：白瓷砖、黑色天然石材和水磨石地板。厨房对面是位于餐厅和休息区之间吧台，具有两种功能：一是在厨房的那一侧提供蛋糕、甜点、鲜果汁和啤酒；一是在休息区的一侧让人们坐到吧台边品尝美酒或咖啡。

Photo: Christian Richters, Andrea Flak, Gunter Gluecklich

David Chipperfield Architects

Restaurant in Empire Riverside Hotel & 'Brauhaus'

This four-star hotel and a retail and office building in Hamburg, St. Pauli is part of a 110,000-square-metre housing and office development on the former Bavaria brewery grounds. The extraordinary nature and history of the Bavaria grounds – overlooking the river Elbe – inspired the research for this project. The relatively high ceiling and the glass curtain wall made the restaurant seem extremely spacious. The regular load-bearing columns add a noble sense to the space. Modern tables and chairs are placed by the windows, enabling spectacular views when dining. In such an environment, you would experience the luxury of modern life. The chandeliers are a perfect decoration during the day and at night.

河边帝国酒店餐厅

位于巴戈利亚酿酒厂旧址的110000平方米的住宅及办公项目开放区囊括了一幢集四星级酒店、零售店及办公空间于一体的综合建筑。较大的层高和玻璃幕墙让餐厅显得异常开阔，整齐的柱子除了承重的作用也增加了室内空间的大气。富有现代感的座椅摆放在窗边，就餐时有开阔的视野，尽享现代奢华生活。精心设计的吊灯无论在白天还是晚上都是精美的装饰。

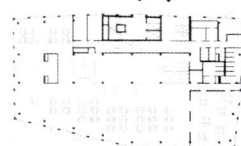

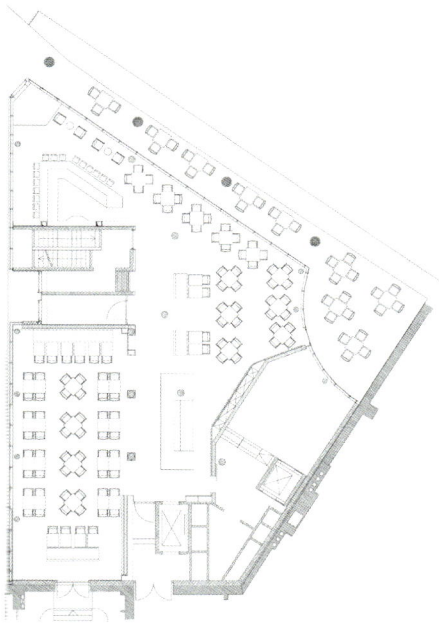

Restaurant Garamond

Garamond, the restaurant of the Hotel Adrema shimmers in bronze, black glass, and lacquered surfaces. The hotel's insignia, the barcode, originating from the register machine Adrema, was playfully interpreted and transformed into an ornamental band for the new interior theme. Variations of bands and lines were employed in the furnishing. A luminous ornament is engraved in a monolith-like form, made of Hi-macs, crowning the buffet. A honey gold hue shines through this undulating band creating elegance in its vicinity. Contrasting piping highlighting the chair's shape augments a similar curly ornament embroidered on the backs of the leather chairs, specially designed for Garamond. Likewise, the glossy, black light band complements the illuminated fluting in the ceiling above.

加拉蒙餐厅

加拉蒙是艾德玛酒店的餐厅，主要装饰是铜、黑色玻璃和亮漆墙面。艾德酒店的标志和条码，都被诙谐地转变成餐厅设计主题里的装饰花纹。室内装饰着变化多端的丝带和线条。巨大的人造理石台面上装饰着发光的带状花纹，提升了自助区的品位。起伏的花纹上蜂蜜般的金黄色闪闪发光，营造出优雅的意境。皮质椅背上绣着类似的波纹，与椅背的饰边相互映衬，图案是为这家餐厅特别设计的。闪亮的黑色光带与天花板上的灯槽互为补充。

Restaurant & Bar

Germany

Berlin

2005

Photo: David Hiepler, Fritz Brunier

Gisbert Pöppler

Café

Poland

Lodz

2009

Photo: Ula Tarasiewicz, Jakub Stepien

Wunderteam.PL

Café, Bookstore and Entrance Zone in Muzeum Sztuki in Lodz

The Museum of Art in Lódz has an avant-garde tradition. It contains 19th and 20th century art, at the same time being contained in a 19th century middle-class palace. In order to modernise the ground floor of the building and adapt it for new functions, it was necessary to pay respect both to the avantgarde tradition, as well as the building's historic function. The previous functional division of the rooms at the Museum's entrance failed to create an impression of accessibility. The Museum also lacked a common space as a venue for social life. The objective was to create a place with its own personality, encouraging people to stay in the museum after an exhibition, discuss their views and share their opinions.

艺术博物馆咖啡厅、书店和大厅

罗兹艺术博物馆具有先锋意识，收藏19、20世纪的艺术作品，位于一座19世纪中产阶级建筑之中。为了让建筑的一楼更具现代感并增加新功能，其设计既要具备先锋意识，又要最终建筑的历史传统。博物馆大厅原有的空间设置缺乏通透性，急需一个能够社交的公共空间。设计的目标是打造一个具有独特个性的空间，让人们在展览之后停留一段时间，进行作品讨论和意见交换。

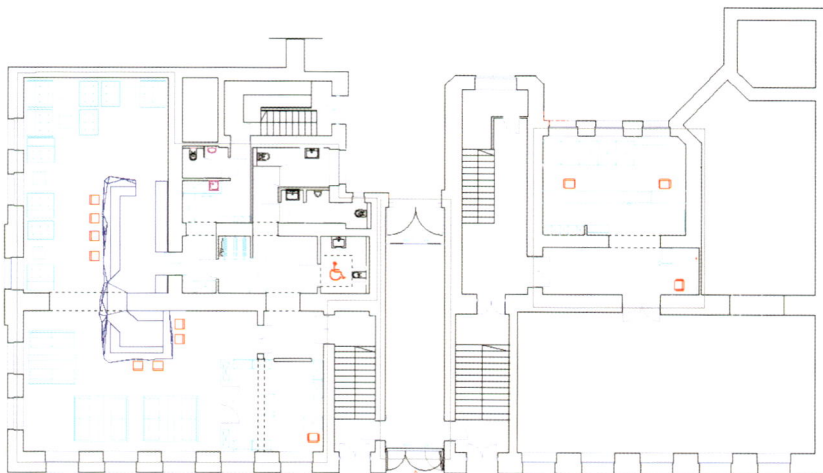

Andel's Lodz

In 2009, the former cotton mill was transformed into a sophisticated hotel by Jestico + Whiles. It includes 278 designer style rooms and suites, 3,100 square metres of conference space, ballroom for 800 people and fine-dining restaurants and bars with seats for more than 450 people, swimming pool and wellness centre. The bar's design seems to be simple, because the building itself has a distinct character. Exposed beams, columns and brick walls are part of the interior decoration. The interior structure and small vintage windows look mysterious and pure. White tables and chairs in dark colour enhance this feeling.

罗兹安德尔酒吧

2009年，建筑事务所将纺织厂改造成为一座精品酒店。酒店拥有278间设计客房和套房、3,100平方米的会议室、可容纳800人的舞厅以及450个坐席的精品餐厅和酒吧、一个游泳池和一个健身中心。酒吧的设计看似并不复杂，但是因为建筑本身的结构显得极富特色，裸露在外面的梁柱砖墙等结构成为了室内的装饰之一，室内结构和复古小窗让室内呈现出神秘而纯净的特色，白色的餐桌和深色木质的餐椅加强了这一特色。

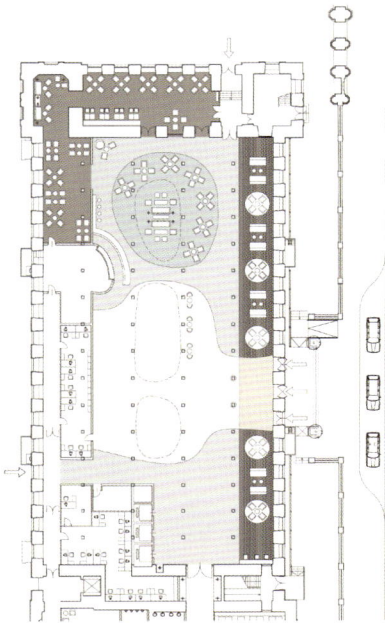

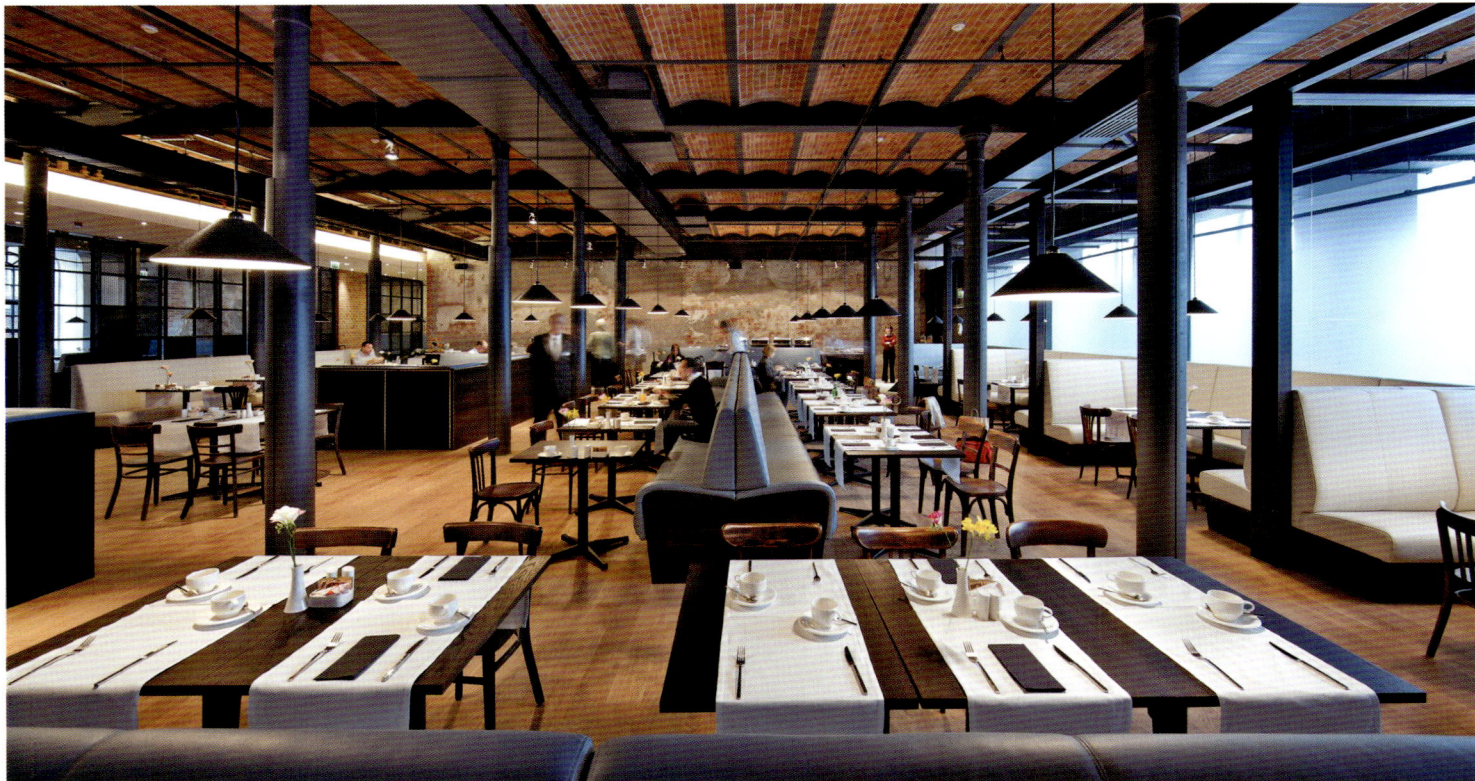

Bar

Poland

Lodz

2009

Photo: Ales Jungmann

Jestico + Whiles

Restaurant

Switzerland

Basel

2008

Photo: HHF Architects

HHF Architects

Mensa Kirschgarten

The project is an integration of a cafeteria into an existing open lobby of an important school building located in the city centre of Basel. Ribbons made of wood were put above the existing floor, wall and ceiling. In between the concrete structure these ribbons reach out to the courtyard, creating a sunscreen and a terrace where people can sit and eat. This gesture links the interior space to the outside and enables a direct access from that courtyard. A glass façade works as thermal barrier. With that the outer appearance of the building with its slim columns could be kept as before.

门萨幼儿园餐厅

设计师在位于巴塞尔中心的一座学校的开放式大厅内打造了一间餐厅。木质条带装饰在地面、墙壁及天花板表面，并一直延伸到庭院内，形成了遮阳屏及楼梯状的座椅。巧妙的设计将室内外空间连通，玻璃外观如同天然隔热屏障，反射强烈的光线。建筑外部细长的柱子则得以保留。

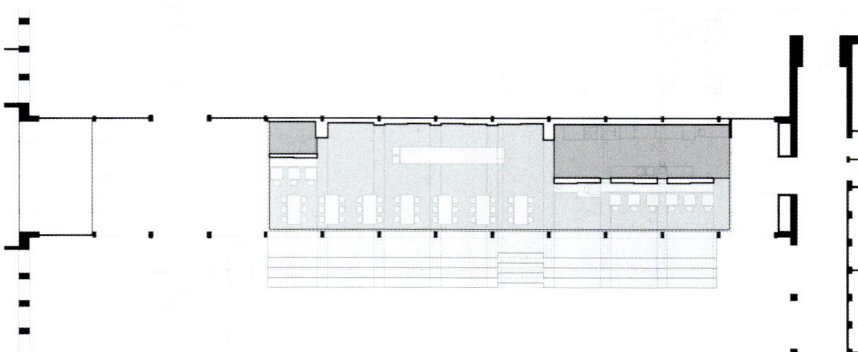

Confiserie Bachmann, Basel

With this renovation, the local, but in Switzerland well-known brand has received a bright, elegant, and contemporary expression that adopts the characteristic qualities of the previous chocolate shop despite the visible changes. The renovation optimises its location by reorienting itself anew to not only the street, but also around its corner to the covered passageway with outdoor seating. An inviting coffee bar flooded with light, as well as the arrangement of the glass sales cabinet, the counter, the furnishings, and the bright colours, create a new openness to the street and the covered passageway. Although the space is small, the renovation creates a spacious, yet intimate atmosphere with the use of reflection and colour.

巴赫曼糕点店

通过翻新，瑞士知名品牌糕点店巴赫曼拥有了明快、优雅、现代的新外观，又保持了原有的品质和特性。翻新工程不仅重新调整了店面的街道朝向，而且在转角处设计了带顶走廊，便于露天休息。极具吸引力的咖啡吧光线十足，玻璃展柜、柜台、家具的设置和鲜明的色彩让店面焕然一新。尽管空间并不大，翻新工程令糕点店看起来宽敞而又拥有私密的气氛。

Photo: Tom Bisig, Basel

HHF Architects

Restaurant

Austria

Vienna

2009

Photo: Lea Titz

Denis Kosutic

Orlando di Castello Restaurant

The idea of uniting the worlds of Queen Elizabeth, the rapper 50 Cent and a girl from Tyrol all in one room and of forming from these associations, which are in contradiction to each other, a new kind of harmonic composition are the dominant factors of the outline. Symbols, such as delicate, stylised little flowers and hard metallic nuts appear in the room in countless versions, thereby making for strong contrasts. The exciting sense of space appears through the examination of the new proportions: baseboards transform themselves into wall claddings, floor lamps become ceiling lights, panels of fabric divide into small note-like cloths, benches explode into small, kidney-shaped segments. The result of these alienations, often ironic, is a surreal atmosphere full of surprises.

奥兰多城堡餐厅

餐厅设计融汇了伊丽莎白女王、说唱歌手50美分和提洛尔女孩的风格，这些风格看似互相矛盾，却又在这里和谐地组合在一起。精致的碎花图案和硬金属狂热符号在同一个空间里形成了鲜明的对比。令人兴奋的空间感体现在新的空间比例上：护壁板变身为墙体，落地灯成了吸顶灯，纤维板被切割成小碎片，长椅变成了小的肾形座椅。

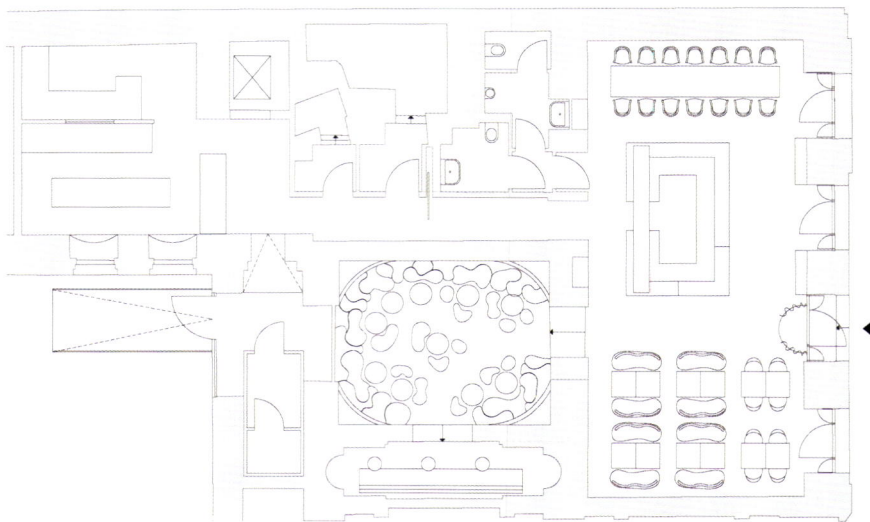

Café Restaurant Esterhazy

The Esterhazy Palace in Eisenstadt is a Baroque castle with supra-historical, and cultural and historical significance. The ensemble consists of the castle park, orangery, stables, domain administration, Jewish Town, and a farm. As an important measure of the reconstruction, ceiling-high four holes were broken through the six-foot-thick sandstone walls of the northern façade and opened the neoclassical, permanently installed shutters, turning to the restaurant's overlooking relationship with the castle. Inside, the past decades have been eliminated and an entirely new infrastructure has been created. All fixtures are made of walnut.

埃斯特哈齐餐厅

位于艾森斯塔特的埃斯特哈齐宫殿是一座巴洛克风格的城堡，具有超历史的、文化的和历史的重要性。整体包括城堡公园、橘园、马厩、领域管理、犹太城和一座农场。这次重建的一个重要手段，就是在北面的6英尺厚的砂岩墙上，在天花板的高度上开了四个洞，并且开了永久安装的新古典主义风格的百叶窗，将餐厅跟城堡的关系显现出来。餐厅内部，过去几十年的痕迹全部抹去，换以全新的装修。所有的装置都是用胡桃木制成的。

Klaus-Jurgen Bauer Architect

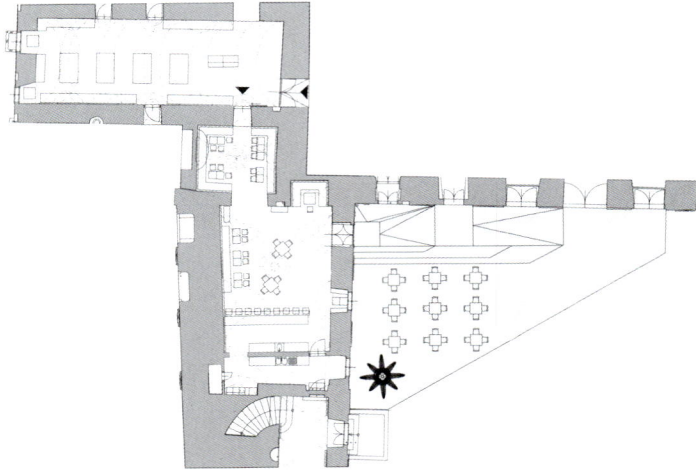

Café

Austria

Vienna

2009

Photo: Rudolf Hemetsberger

Klaus-Jürgen Bauer

Café Maskaron

The three rooms formed an architecturally organic unity in a subtle way, to which fine interventions were necessary in the historic masonry. Outside, a wooden terrace was set in the courtyard of the castle. Through the design of this extraordinary interior design is now a place created that seemingly exists beyond time and space: one of the windows was extended to the door, on which is a black carpet. Black leather lined with benches in back wall and corners; the square tables are also black. The white chairs of synthetic material with the thin steel legs are floating light as paper. On the front wall provides a bar with refined backlit glass for elegant glamour. Two extravagant chandeliers hang from the vault like the cloud, which is like an artificial sky by painting.

假面咖啡厅

通过对古老石结构建筑的精心改造，咖啡厅的3个空间形成了一个有机的建筑整体。外面的木板平台设在城堡的庭院之中。室内设计似乎存在于时间与空间之外，其中的一扇窗户伸展到了门外，地面上铺设着黑色地毯。黑色皮革长椅沿着后墙和转角摆放，方桌也是黑的。合成材料制成的白色座椅的细铁椅腿显得异常轻盈。前墙有一个吧台，背光玻璃板散发出优雅的光芒。拱顶上两盏奢华的吊灯宛如云朵，让天花板宛如一个人造天空。

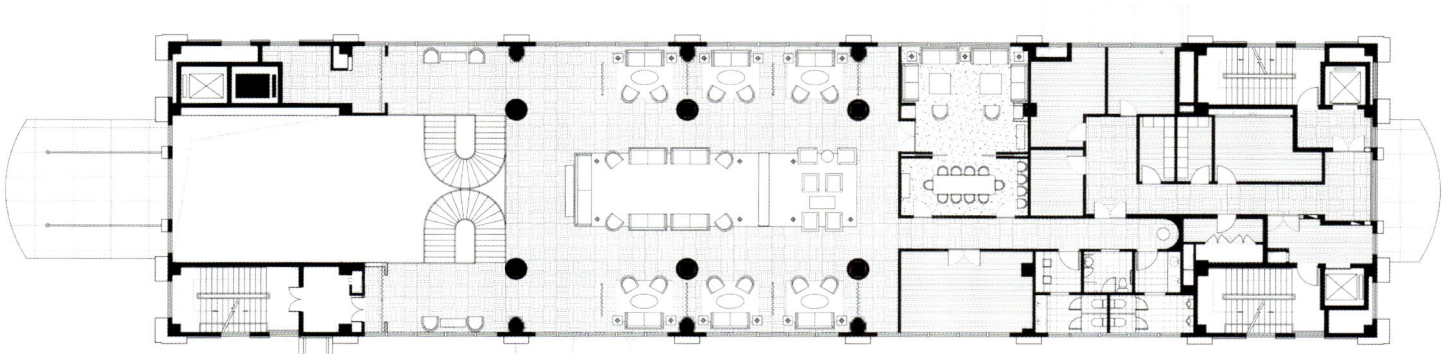

Restaurant & Bar

Austria

Eisenstadt

2006

Photo: Curt Themessel & Ioan Nemtoi

Curt Themessel & Ioan Nemtoi

Restaurant & Bar in The Levante Parliament

Just behind Austria's parliament building in central Vienna, The Levante Parliament is located around an elegant 400-square-metre courtyard. Here, separated from the bustling city, guests find a relaxing lounge ambience; an oasis of tranquillity detached from summer heat and winter's frost. The seventy-five-seat restaurant, the bar and the Mediterranean garden were designed by the international glass designer and serve international cuisine. The glass designer Ioan Nemtoi was the inspiration for The Levante's Restaurant. He designed the four fire frames and the unique glass bar. They are the eye-catcher and the optical centre of the restaurant.

莱万特国会酒店餐厅&酒吧

在维也纳的中心，奥地利国会大厦后身，就是这家莱万特国会酒店，坐落在一个占地400平方米的雅致的庭院里。门图伊餐厅酒吧，这间能提供75个座位的餐厅、酒吧和地中海花园都是国际知名的玻璃艺术设计师门图伊设计的，这里供应各种国际美食。玻璃艺术设计师约安•门图伊是莱万特餐厅的灵感之源。他设计了这里独特的玻璃酒吧，非常吸引眼球，是整个餐厅的视觉中心。

Restaurant

Austria

Vienna

2005

Photo: Ronacher Architects

Ronacher Architects

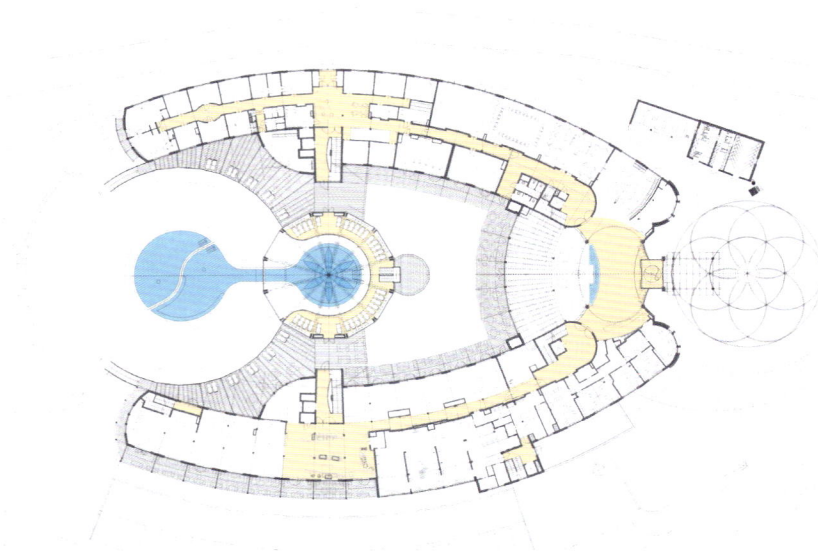

Restaurant in The Larimar Hotel

This hotel is named after the Caribbean precious stone Larimar, whose colour alternates between light blue and turquoise. This stone is reputed to provide a positive effect with physical and emotional healing processes. The basic idea behind the design was the development of an oval as the main structure which should provide a maximum of warmth and security in both the internal and external area. The four elements of air, water, fire and earth were reflected in the entire design of the interior. For this reason, four types of room were designed according to each of the elements. In addition, the wellness and restaurant areas also draw their inspiration from the four elements.

兰瑞玛宾馆餐厅

该酒店是根据加勒比海宝石兰瑞玛而命名的，这种宝石的颜色可以在浅蓝色和青绿色之间变换。而且此宝石由于对身体健康和精神愉悦有很好的作用而闻名于世。设计的基本理念就是要打造一个椭圆形的建筑，这样就可以在内部空间和外部空间的设计上提供一个宽敞的、温暖的以及安全的环境。整个项目的设计理念是以水、火、土、风为基本元素。据此，根据不同元素设计了不同风格的房间。另外，健康中心和餐饮中心也在四种设计元素中汲取灵感。

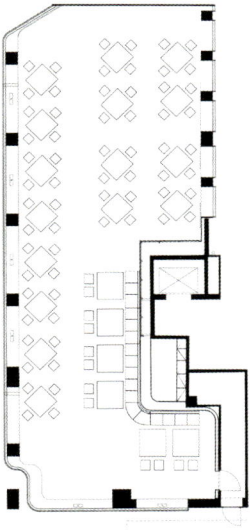

Restaurant in Balance Holiday Hotel

In the heart of the famous 'Europa Sport Region Zell am SeeKaprun' affording views of snowcapped Alpine peaks just a few steps from the lake and the skiing area, Balance Holiday Hotel is the destination for wellness and recuperation. The candle light in the restaurant enhanced the elegant and noble atmosphere of the space. The regular tables and the armchairs are perfectly matched to produce an unexpected effect. The central glass wine cabinet and the dining tables placed around it are the focal point of the space. The arrangement in the interior is unique. Besides, the decoration on the wall and the pieces of artwork are delicate too.

巴兰斯假日酒店餐厅

巴兰斯酒店位于著名的采而湖欧洲体育中心，与湖区和滑雪场仅有几步之遥，可以欣赏阿尔卑斯山的壮美雪景。餐厅里的烛光烘托了餐厅富贵典雅的风格。中规中矩的方桌和扶手椅在设计师的巧妙搭配下呈现出新的风格。中央的玻璃酒柜以及围绕酒柜摆放的餐桌成为空间的中心。室内布局独具一格。墙壁的装饰及艺术品的摆放同样考究。

Restaurant

Austria

Zell am See

2008

Photo: Niki Szilagyi

Niki Szilagyi, Evi Märklstetter

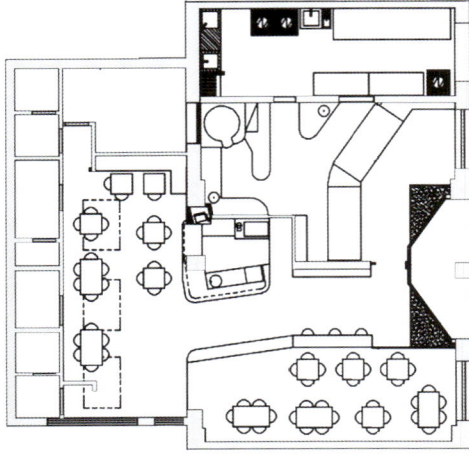

Restaurant

Italy

Montecatini

2007

Photo: Sandro Bonaccorsi and Riccardo Cioli

Arkimisti®

Tocqueville Restaurant

The project provides a revival of an old restaurant used for catering, increasing the initial surface and work processes. In the old catering division, which has been renewed too, there are collocated a modern restaurant and a wine bar. The new space has been created in a uniform way, taking also advantage of the height of the old structure discovering the old trusses. Great attention has been reserved to the distributional plant, creating a homogeneous space for the various new work processes required. Minimalism, suffused lights and pastel coloured walls complete the project for an informal, but at the same time, high detailed meeting place.

塔克威利餐厅

本项目是将一个古旧的餐厅项目旧貌换新颜。对室内空间进行重新设计，工作流程具体细致划分。已经翻新的古旧餐厅包含一个具有当代气息的餐厅和时尚前卫的酒吧。餐厅是在原来项目的基础上进行翻新改造，运用了和谐统一的方式，利用了原项目旧体制框架的高度。餐厅周围分布的植物全部经过精心的修饰，创造了和谐统一的幽雅环境，适用于各种工作需求。极简主义灯光和浅色墙壁让项目更加平易近人。

Restaurant

Italy

Reggio Emilia

2008

Photo: Diego Parolini

Studio M2r Atelier D'architettura (Luca Medici, Luca Monti, Lorenzo Rapisarda)

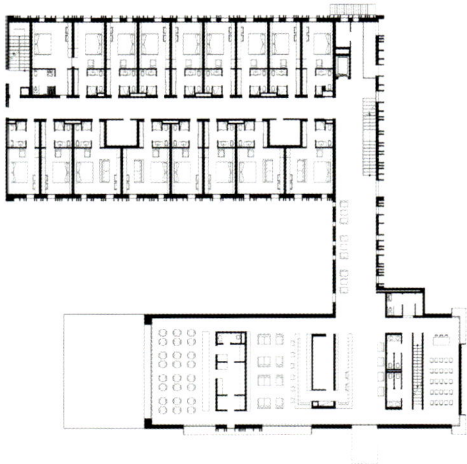

Restaurant in Casalgrande Hotel

The building area is surrounded by a lovely rural area with small country houses, groups of trees and big vineyards. This rural landscape creates strong suggestions that influenced the whole project. The idea of the project combines sustainable environmental strategies in a strong architectural language achieved through the use of very contemporary materials, the reinforced coloured concrete and the cor-ten steel. The one-floor building contains all the public spaces, namely the reception near the entrance, the big hall, the bar area and the coffee area overlooking the countryside, a cosy meeting area and the service spaces. Special attention was paid on studying the decorative features, which are characterised by basic lines and made of wengè wood.

卡萨尔·格兰德大酒店餐厅

酒店的四周环绕着精致的山间小屋、郁郁葱葱的树林和宽阔的葡萄园，浓郁的田园特色赋予其独特的魅力。一层的结构用于设置所有的公共空间，包括入口处的接待台、大厅、酒吧、咖啡厅、会议区以及其他服务区。设计师在装饰上花费了大量心思，运用简洁的线条和鸡翅木材料创造了非凡的效果。

Bar

Italy

Venice

2008

Photo: Alvin Grassi

Alvin Grassi

Bar at Palazzo Barbarigo sul Canal Grande

The hotel is located next to the Canal Grande, not far from Rialto Bridge, and just few minutes walk from Piazzale Roma and the railway station. Hotel Palazzo Barbarigo sul Canal Grande is an Art Deco wonderland that amalgamates both the emotion and playfulness of the Venetian style. The Art Deco soars in the beautiful hotel glass bar back-painted black, a space recalling this unique city's Golden Twenties. The dark colour tone makes the bar look elegant and mysterious.

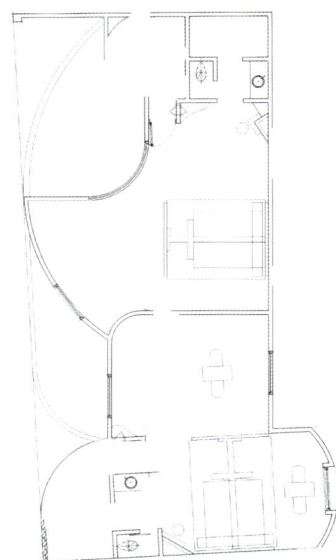

大运河巴尔巴里戈宫酒店酒吧

酒店紧邻大运河，在里奥尔桥附近，距离罗马广场和火车站只有几分钟的路程。大运河巴尔巴里戈宫酒店是装饰艺术的仙境，同时又巧妙地融入了威尼斯风格。装饰艺术在酒吧的黑色色调中得到了完美的体现，这里让人想起了城市的黄金时代——20世纪20年代。深沉的色彩使整个酒吧显得华贵而神秘。

Espera Coffee Bar and Wine Bar in Hotel EOS

The former Continental Hotel has been completely renewed with new structures, new name, new facades covered with 'leccese' stone. The Espera coffee bar and wine bar is a dynamic and youthful setting, furnished with exclusive details and decorative elements that make the atmosphere warm and relaxing, in perfect equilibrium between the tradition of the materials and sophisticated design with sleek and modern lines. The wooden furniture and the wooden floor in the bar offer guests the home feeling. To echo the overall style of the interior design, the designers selected the furniture that has a strong sense of design.

EOS酒店咖啡厅和酒吧

原来的五洲酒店经翻新之后更名为"EOS酒店"，外观采用拉察地产石材打造一新，开创了酒店设计的新理念。Espera咖啡厅和酒吧以动感十足的背景为主，独特的细节和装饰元素带来了温暖休闲的气息，在传统材质和现代设计中打造平衡感。酒吧的木质家具和木地板给顾客家一样的感受，设计师精心挑选了充满设计感的家具，以适应室内设计的整体氛围。

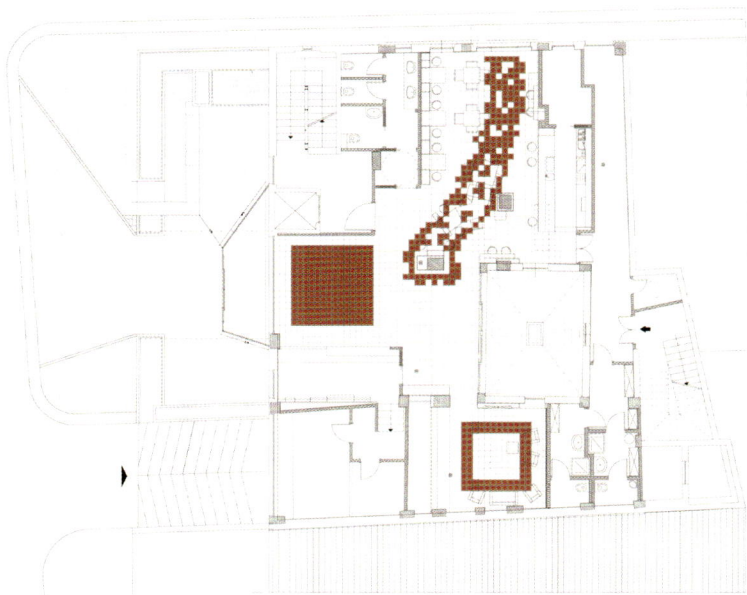

Bar

Italy

Lecce

2009

Photo: Marino Mannarini

Scacchetti

Café

Italy

Rome

2005

Photo: Arch. Luigi Filetici / Barilariarchitetti

Run/dom_Barilariarchitetti

Des 1 Teel Café

The project was an opportunity, for the designers' studio, to identify what they considered to be the 'levels of separation' through which an interior-design project is no longer an 'ornamental' issue but a 'substantial' one: This is the first 'change of state' between the two: the introduction of the 'structural aspect'. The hanging-part projection of the structure is approximately 3.5 metres. The second 'change of state': This basic code is the concept of 'comfortable' and so they 'wrapped/hugged' the whole space with detachable 'layers' enclosing an 'empty centre' where people would stay. The third 'change of state' was the 'space issue': Because of the small dimensions of the bar.

德斯餐厅

本项目为设计师提供了一次思考的机会，使他们重新定义了室内设计项目。它带给人的不仅仅是"视觉"的享受，更是一项"关系重大"的严肃问题。项目设计理念独特创新：第一，首次体现了"结构内容"。第二，设计注重整个项目的"舒适度"。第三，设计注重"空间结构"问题。

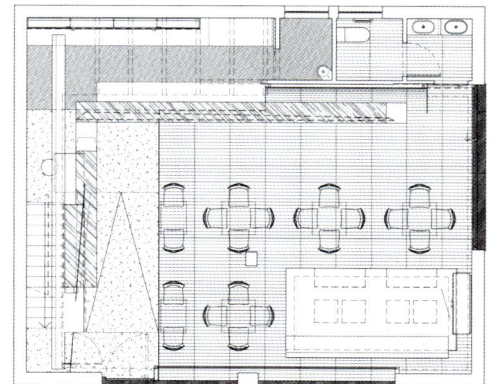

Restaurant

Italy

Milan

2006

Photo: Mussapidesign

Arch. Massimo Mussapi

Frantoi Celletti & Cultivar Restaurant

The renovation included new internal wall divisions to adapt the areas to the new needs and to place in the spotlight the shed covering and steel structure that sustains the mezzanine floor. The open-to-view kitchen could not be ignored. It is more than 20 metres long and offers clients a view and guarantees hygiene, while also stimulating the staff to consider themselves protagonists. Interior design features are rugged and sophisticated, fresh and inviting at the same time; the dominant colour is white, plus dark green, yellow, and the colour of terracotta. In pleasant contrast with contemporary architecture, antique objects and equipment used in the past to produce oil are positioned in strategic places.

弗朗伊托餐厅

项目的翻新需要分隔墙壁，调整空间结构来满足新的设计需要。在天花板和支撑阁楼的钢筋支架上安装聚光灯。开放式厨房不容小觑。它长达20多米，是一道室内风景，在保证卫生的同时，也让厨师们意识到自己才是餐厅的真正主角。室内设计的特色是粗糙而不失精细，清新而又媚惑动人。色调以白色为主，配以深绿色、黄色和土红色。古董物品和曾用于榨油的设施与现代建筑形成了愉悦的对比。

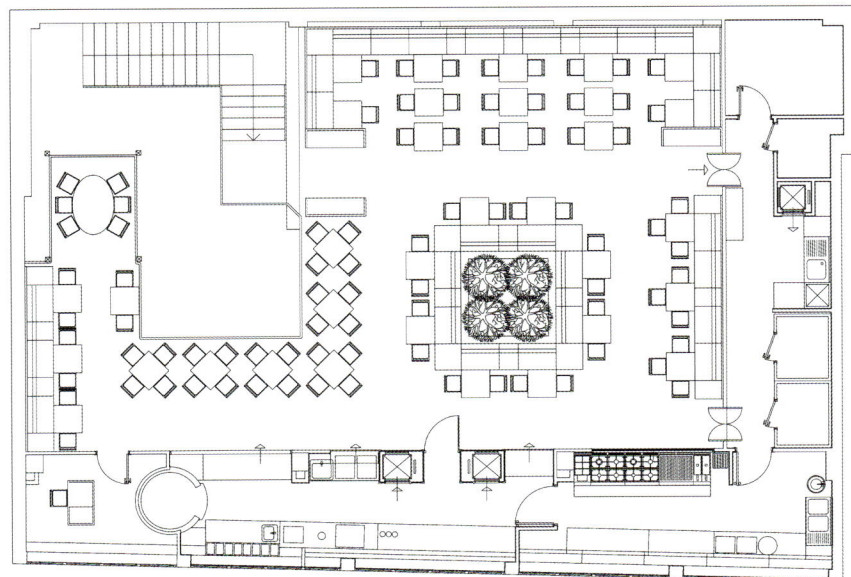

Restaurant

Italy

Verona

2007

Photo: Ciro Frank & Schiappa

Andrea Aloisi & Enrica Mosciaro

Restaurant in Hotel Mod05

Hotels have two souls: a public one, which presents the dynamic of the arriving and departing zones, brings the idea of encounter, activity and leisure; and an intimate and protected one, designed for the night and the rest. The Hotel Mod05 is the overlap of two buildings, divided without touching each other by a large stripe of glass. The compact but light volume of the rooms floats above the complex and articulated ground floor. The designers made use of the structure of the old building to specify the interior space of the restaurant, which feels open and spacious. With the hard lines of the structure, the space senses a little bit cool. This coolness is offset by the wood textured decoration.

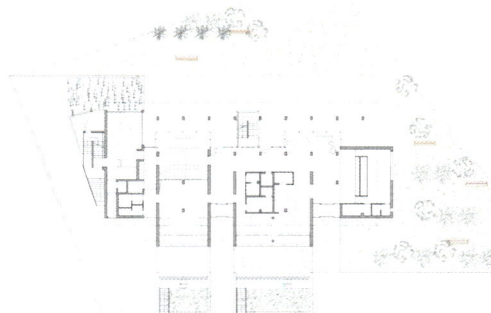

Mod05酒店餐厅

酒店主要包含两部分：公共空间——入口、活动区及娱乐区；私人空间——卧室。Mod05由两栋建筑构成，中间通过大型条状玻璃结构连通。一层空间整洁而流畅，楼上房间排列紧凑而轻盈，如同悬浮在上。设计师巧妙地利用建筑本身的结构安排餐厅内部空间的设计，餐厅带给人的整体感受开敞而通透，配合硬线条的建筑结构有些冷酷的味道，但这种感觉又被木质纹理的装饰良好地缓解了。

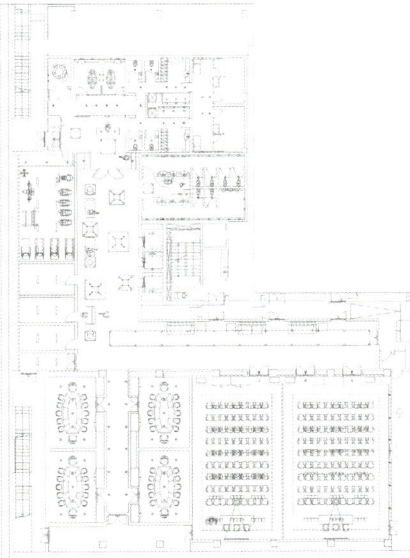

Restaurant in Net Hotel

The Hotel located in the tower from Level -2 to Level 9 represents a 'business focused' hotel (based on the commercial market and business), in the category of 'four stars plus', being an important landmark in the hospitality field not only for Padua city. The restaurant is a world of glass. The glass curtain wall and the glass tableware are all translucent, repetitively reflecting lights to create a gorgeous and romantic space. Various kinds of beverage are available.

内特酒店餐厅

这家酒店在一座高层建筑中占据从地下2层到地上9层的空间，代表了商务酒店（基于商业市场和商务）的典范，4星以上级别。在帕多瓦市乃至更大范围内，这都可以算是一座酒店类的地标建筑。餐厅给人的感觉是玻璃的世界，玻璃幕墙、玻璃餐具，处处晶莹剔透，光线经过层层反射让餐厅显得华丽而浪漫。酒吧有很多种类的酒水供客人选择。

Restaurant

Italy

Lecce

2007

Photo: Marino Mannarini

Scacchetti

Restaurant in Risorgimento Resort

In the Risorgimento Resort, the décor of the areas dedicated to dining and refreshment is elegant and sleek without being minimalist, and fully exploits the spectacular architecture. The large windows, as well as allowing the natural light in, are designed to communicate an openness of the spaces towards the city, allowing them conserve an informal nature to make access desirable also to people not residing in the hotel. Risorgimento Resort is home to three restaurants and an elegant lobby bar, all offering a range of cuisine, from gourmet to bistro to light snacks. There is also a roof garden with panoramic terrace, where you can listen to music, sip a cocktail and gaze out over Lecce's old town.

复兴度假村酒店餐厅

餐厅休闲区的装饰以典雅时尚为主，并充分运用建筑的壮观。大窗的设计不仅使得室内光线充裕，同时增添空间开阔感，让酒店外面的人也能欣赏这里的独特风景。酒店内共有三间餐厅，一个典雅的大堂酒吧，供应各色菜肴。此外，屋顶花园带有露台，坐在这里听听音乐、喝杯茶、欣赏老城的景致，是何等的惬意！

Hi-Café Lounge Bar

The counter is placed in a very important position and is the main and most eye-catching element. It looks like a big, white block, made of Pral (a very smooth and bright material) and shows no joints and cuts. A further innovative element is the bottle holder. It has been designed as a part of the counter top and the inner lighting makes the bottles themselves shine in a suggestive way. The tables are large and at a good distance from each other to guarantee intimacy. A quick service is possible anyway at the high tables with stools and at the counters along the walls.

你好咖啡休闲吧

吧台设计安置在空间最醒目的位置，它看上去就像一个巨大的白色方块，由平滑有光泽的材料制成，看不出任何接缝和切割的痕迹。设计的另一项创新亮点是酒柜。它位于吧台上方，在柜中灯光的照射下，酒瓶发出闪烁的光。宽大的餐桌间隔着很大的距离，保证了顾客的私密性，有助于放松心情。沿着墙边的高脚桌椅区是快餐区。

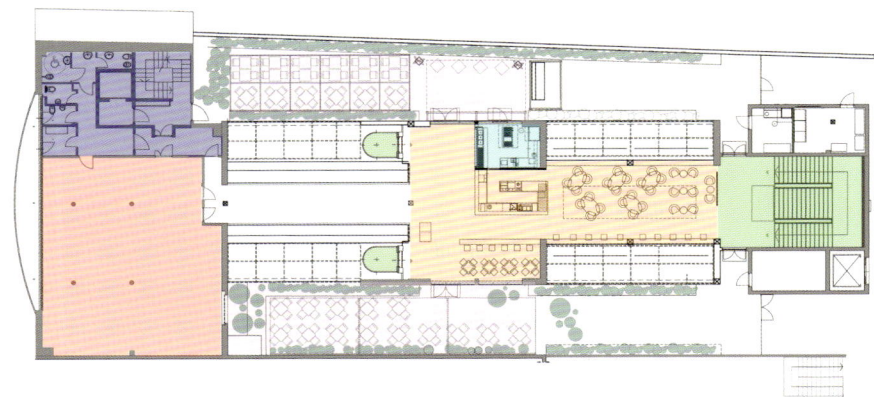

Bar

Italy

Milan

2007

Photo: Matteo Carassale

Arch. Massimo Mussapi

Bar

Italy

Sardegna

2008

Photo: Giorgio Dettori

Architect Pierluigi PIU

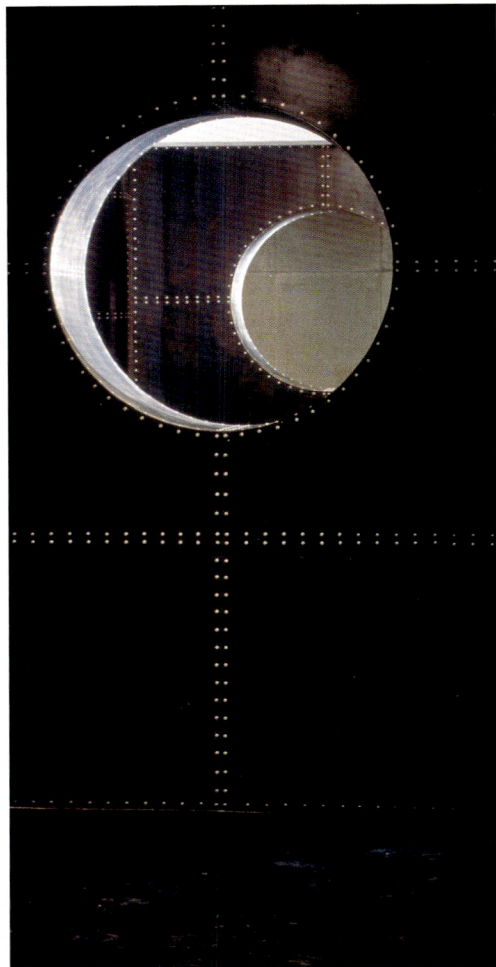

Max Piano-Bar

The whole floor is made of a simple casting of pre-coloured and smoothened cement. All surrounding walls are sound-proof and covered with asbestos-less cement-fibre boards, fixed to a hidden frame with showing bolts. Against such covering, behind the stage, one can see the venue's logotype, made of three-dimensional letters of oxidised brass. The easy-chairs, designed by Xavier Pouchard in 1934, are now re-edited by the French manufacturer 'Fenetre sur Cour', while the coffee tables are made-tomeasure, according to Pierluigi Piu's drawings.

蜜斯陀钢琴酒吧

地板采用原色光滑水泥结构，自然简单。墙壁采用隔音设计，并覆以纤维板。由普沙尔于1934年设计的"安乐椅"经法国制造商"内特雷河畔"改装后，风格独具。精致的咖啡桌根据皮耶路易·吉彪的绘制精心制作，令整个空间情趣盎然。

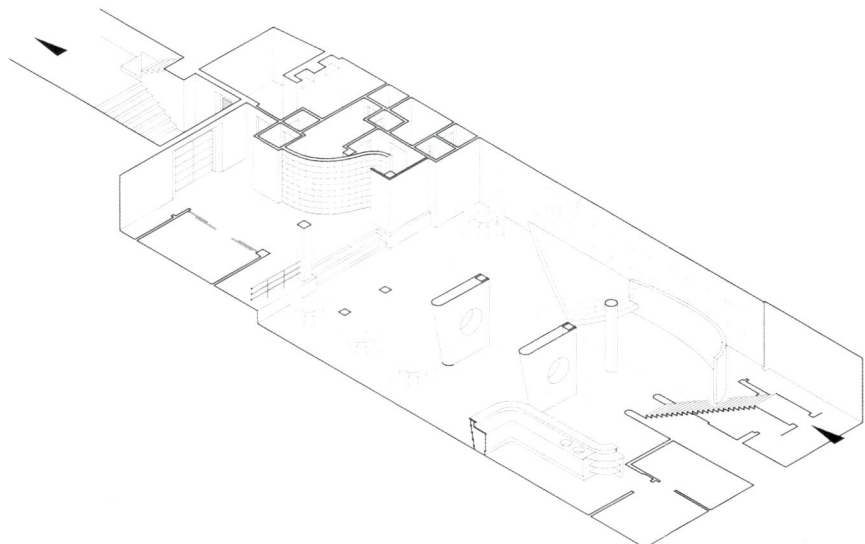

Sosushi Torino

The new take-away restaurant designed for the sushi brand Sosushi had to meet requirements of turning a small old stationery store, deep seated in the historical urban structure of the city, into a trendy place. UAU has been successful in maximising the limited space offering each intersection of materials by using shelves, benches and cubbies. This has helped avoiding nooks and corners to remain unused. Made of clean cut shapes emphasised by matt white finish and furniture details coloured magenta, the space, though overworked, is an harmonious composition balanced through the use of Plexwood.

托里诺如此寿司店

品牌寿司店"如此寿司"的外卖餐厅由一个旧文具店改造而来，地处城市的历史街区，又具有潮流特征。UAU设计公司成功地利用架子、长椅和小隔间将这个有限的空间最大化，避免了角落空间的浪费。白色亚光漆面家具和其间点缀的洋红色形成了和谐的平衡。

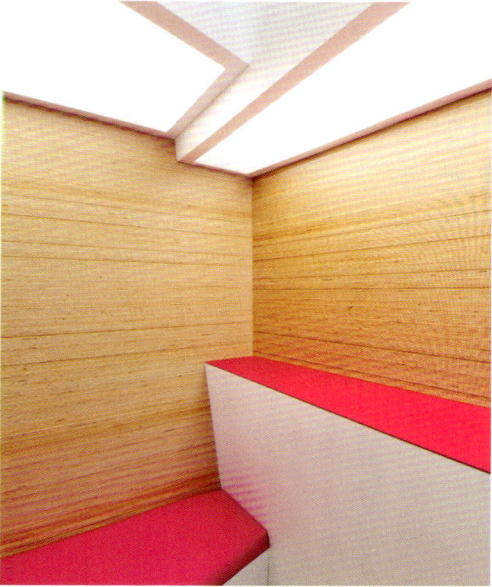

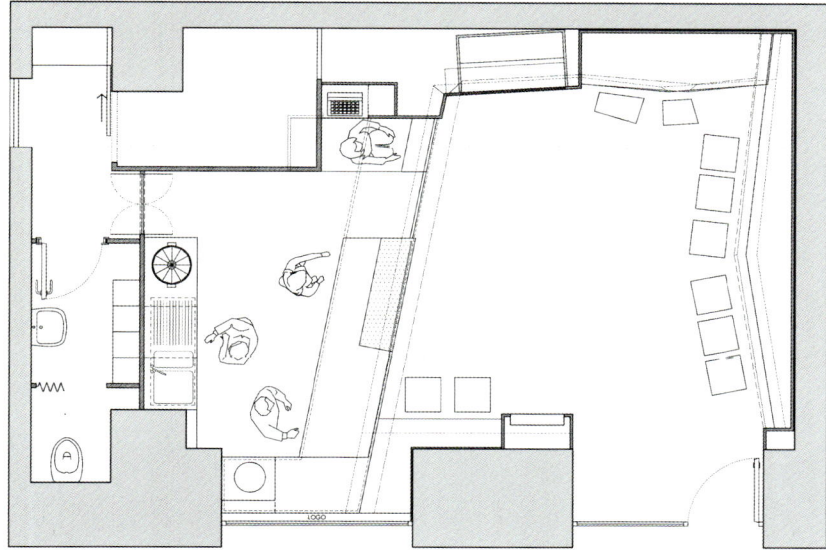

Take-Away Restaurant

Italy

Turin

2009

Photo: Enrico Muraro Photograph

UAU

Restaurant

Italy

Rome

2007

Photo: Dizel&Sate

Dizel&Sate

Bauer Restaurant

The clients took inspiration from the graphic worlds and contemporary art galleries. They wanted to capture the contrasts of gotgatan. This neighbourhood is constantly changing, almost like an art gallery. Designers dizel&sate are renowned for mixing upscale culture and street art. So they were given a free hand. The inspiration behind Bauer's design is the 1920s' Bauhaus style and Berlin's new gallery and bar culture. The animated figures climbing the walls represent different types of pleasure. Here food, drink and music with high ambitions and no pretensions are served.

鲍尔餐厅

这个项目的客户从平面世界和当代画廊里寻求灵感，他们试图展示出gotgatan的对比。这里附近经常改变，就像画廊一样。dizel&sate的设计师以善于融合高层文化和街头艺术而闻名。所以在这个项目中，客户给了设计师足够的空间来自由发挥。鲍尔餐厅的设计灵感是20世纪20年代的包豪斯建筑学派和柏林新美术馆和酒吧文化。挂在墙上的生动的装饰图案代表了不同种类的愉悦。

Santa Marta Restaurant

The project consisted of the restoration of the chapel and its conversion into a restaurant. New restaurant has three floors. Each floor has 110 square metres. The ground floor was separate from the chapel above and used for storage. Now it is the main entrance to the restaurant; there is a bar, the kitchen and the toilets. The first floor (it is the old chapel before) has been transformed into the main dining room. An interior gallery overlooks the dining room. This features the original double height vaulted ceiling. The gallery framework is in steel with wood flooring. To conserve the ancient chapel, the gallery is fixed to the old walls at only four points. The dining room can be seen through two openings which give onto small balconies.

圣玛尔塔餐厅

项目由一座礼拜堂改造而来。新餐厅共分为3层，每层有110平方米。一楼原来是与上方的礼拜堂隔开的仓库，现在则是餐厅的主入口，包括吧台、厨房和洗手间。二楼（原来的礼拜堂）被改造成主就餐区；三楼是室内长廊，以拱顶为特色。长廊的框架由钢铁制成，地面则铺设着木地板。为了保持礼拜堂的结构，长廊与墙壁只通过4个点相连。从通往阳台的窗口可以俯瞰楼下的就餐区。

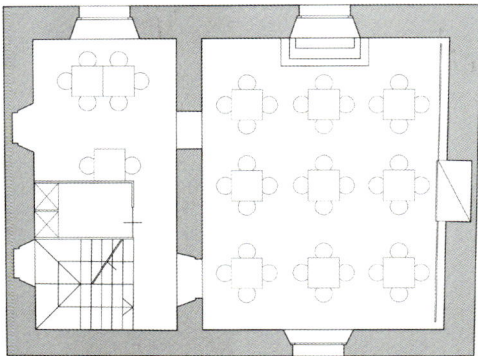

Restaurant

Italy

Torino

2008

Photo: Photophilla.it

Studio Kuadra

Lounge Bar

Italy

Merano

2007

Photo: Jurgen Eheim

Simone Micheli Architectural Hero

'Sketch' Lounge Bar

The Lounge Bar 'Sketch' was created within the Hotel Aurora in Merano. Upon entry the entire space becomes emphatic, almost completely occupied by the enormous bar counter covered in shiny stainless steel and mirrors with circular glazing surrounded by small points of flashing blue LED lights. The opposite wall is completely covered in flexible plasterboard; its sinuosity and fluidity contrasts with the space's decisive furnishings: enormous circular wall lamps covered in mirror and backlit with slim blue neon lights and small coat hangers in shiny stainless steel. The flooring is made of small brushed oak slats with steel lights posed upon it that pulsate, moving the space with a play of shadow and light and later animated with moving images from video projectors that are hidden in the upper veil of the bar counter, projecting and deforming the length of the wall.

素描休闲吧

素描休闲吧位于梅诺拉极光酒店内。整个空间的重点被不锈钢吧台所占据，吧台上的圆形镜面反射着旁边的蓝色LED灯光。对面的墙壁上覆盖着灵活的石膏板，它的弯曲度和流畅性与其他家具形成了对比，如巨大的圆形镜面、蓝色霓虹壁灯和小而闪亮的挂衣钩。地面铺设着小块橡木板，上方的灯光不断闪烁，形成了动感的光影效果，吧台上方隐藏的视频投影机直接投射在对面的墙壁上。

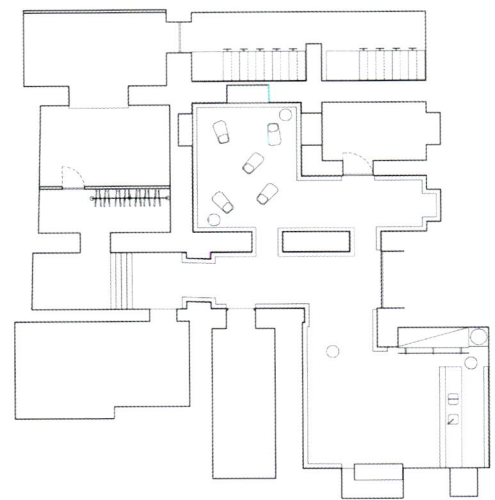

Hamasei Restaurant

Hamasei sushi-restaurant offers to the public a sophisticated environment, warm and pleasant at the same time, where the use of wood and earthly colours is combined to a careful lighting dosage. The light, specifically adjusted on every table like an essential architectonic element, confers an intimate atmosphere to the spaces. The strong character of the project is emphasised by some important elements like the Japanese calligraphic work in the first room. The symbol, printed on backlighted rise paper, creates the background for the tables and of the white sofa running all along the wall, a further accent added to the space. Wooden panels, installed to the walls and ceilings leaving a small distance, which transmit the idea of a decomposable box, express the Japanese concept of modularity, applied both in architecture and philosophy.

哈玛塞餐厅

哈玛塞寿司餐厅为顾客提供精致、温馨而愉悦的就餐环境，使用木材和大地色系的照明。特别调整至每张餐桌的灯光就像基本的建筑元素一样，为空间增添了一种亲密的氛围。项目的重点体现在第一个房间里的日本书法作品。背光照明的纸面为餐桌和彩色的沙发打造了背景，为空间增添了情调。墙上和天花板上的木嵌板之间留有一段距离，让空间看起来像一个可分解的盒子，将日式理念的模块性引用到建筑之中。

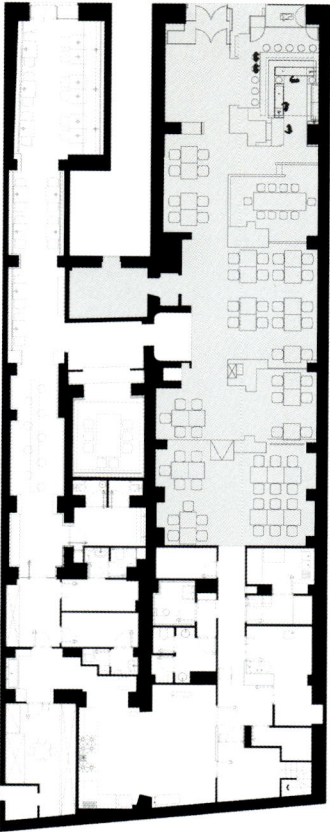

Restaurant

Italy

Roma

2007

Photo: Luigi Filetici

Alvisi Kirimoto + Partners

Café

Italy

Vicenza

2008

Photo: Alessandro Frinzi

Rosolen Sandro

Bricco Café Vicenza

The project involves the renewal of the bar inside the central train station in Vicenza and it's part a wider renewal programme of all Italian train stations bars. The main design concept was to underline where food & beverages are sold and prepared, giving a clear direction to the customers no matter from which entrance they come in. The counter is covered by a red false ceiling that continues on the back wall, creating a sort of open box that contains the different goods on sale. The right quantity and quality of lights over the counter stress that this is the centre of the design concept. The place is very bright to give a lively atmosphere for the travellers who are waiting for their trains. At the same time the materials, like dark wood for the counter, oak for floorings and chairs and the colours like red, dark grey and ivory, connected with the spot light above every table, create a friendly and comfortable atmosphere.

维琴察布里克咖啡吧

项目涉及维琴察中央火车站内餐吧的翻修工程，也是意大利火车站餐吧改造工程的一部分。主要设计理念是强调事物和饮品的贩售和制作，为顾客提供鲜明的方向。吧台上方有一块红色的假吊顶，吊顶一直延伸到后墙，营造出开放式盒子的效果。恰到好处的灯光设计强调了吧台式设计的重点。咖啡吧明亮的环境为等车的旅客带来了活跃的氛围。吧台的黑木、地板和座椅的橡木，红色、深灰、象牙色等色彩，与餐桌上的聚光灯一起，打造了一个友好而舒适的环境。

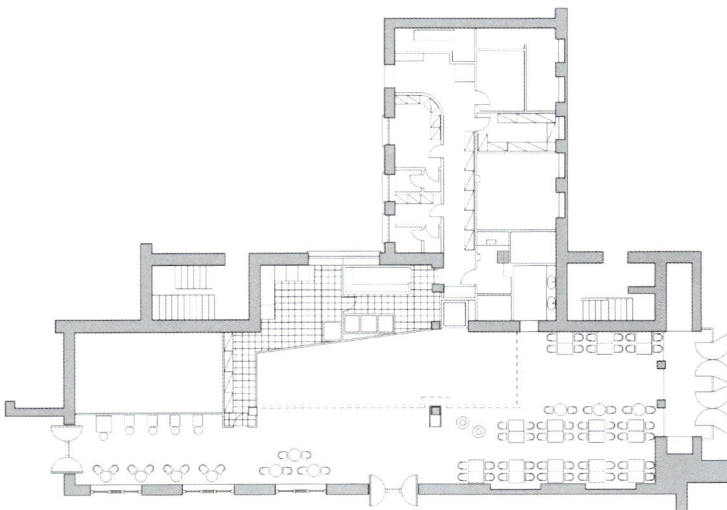

Café

Italy

Monfalcone

2009

Photo: Marco Covi, Trieste (copyright Dimitri Waltritsch)

waltritsch a+u, Arch. Dimitri Waltritsch Trieste, Italy

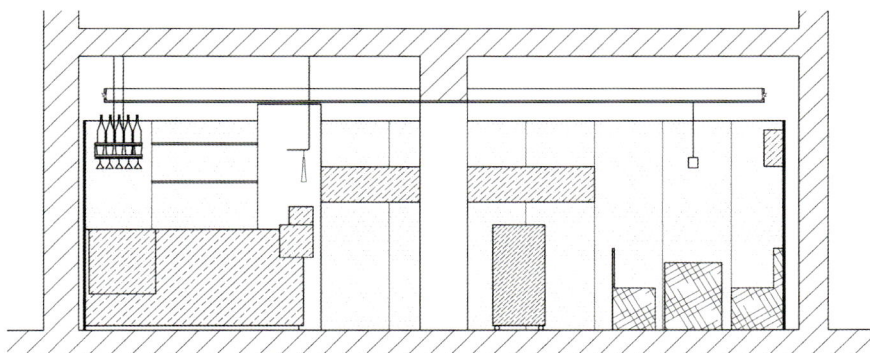

Qubik Bar Monfalcone

Qubik Monfalcone is one of the concept bars of a new brand of selected blends named Qubik Café. This Qubik bar is located inside cinema multiplex, in a context similar to a commercial centre or an airport, without windows or outdoor spaces. The silk-screened coloured glass envelope provides a sober and elegant atmosphere to welcome the passer-by. A long black cantilevered Corian counter acts as catalyst into the atrium of the multiplex, while a bench and a few chairs provide a more intimate alcove. In between there is the Kubo, a monolith-counter on wheels. Particular care was devoted to the presentation of the goods of the brand, following the Quality Boutique Café concept: an illuminated lantern, like a sugar cube terminates the suspended bar counter, while different colour glass shelves are marking the glass cladded walls, and a shopper-rack over the cash desk is ready for those willing to buy the brand products.

魁比克吧蒙法尔科特店

魁比克吧蒙法尔科特店是魁比克品牌的概念餐吧之一。项目位于一座综合性电影院内，周边环境类似于购物中心和机场，没有窗户和户外空间。丝印彩色玻璃墙有一种严肃而优雅的感觉。黑色悬臂大理石吧台为空间增添了多元化特征，长椅和座椅营造出一个私密的小空间。二者之间是一个带轮子的独体大吧台。设计注重魁比克品牌产品的展示：方糖形状的玻璃盒吊在吧台之上，不同颜色的玻璃货架是墙壁的一部分，收银台上方还有一个购物者行李架。

Photo: Coopsette, Pierpaolo Cenacchi, Marina Chiesa, Mario Lamber

ARC, Paolo Lamber, collaborator Pierpaolo Cenacchi

Restaurant in I-Point Hotel

I-Point Hotel has been designed for business people, very dynamic, and always on the move, where people normally spend a very short time, normally a couple of nights per stay. The restaurant was designed extremely warm and exquisite. The colours of orange and red were adopted here too, as well as the wooden decoration. The relatively small space would not feel crowded but quite warm. The tables and the chairs are themselves part of the inteiror decoration for the restaurant. Lighting design is well integrated to enhance the beauty of the decorations.

爱点酒店餐厅

酒店专为那些短期居住的商业人士打造，动感十足而又富于变化。橙、红两色使得房间内活力十足，木质装饰则营造完全不同的温馨恬淡之感。餐厅的设计异常精致温暖，贯穿整个酒店的橙、红两色和木质装饰依然被运用在餐厅中。不大的空间不但不显得拥挤反而更添温馨。桌椅本身也是室内的装饰，同时设计师还巧妙地设计了室内照明，增加室内装饰的美感。

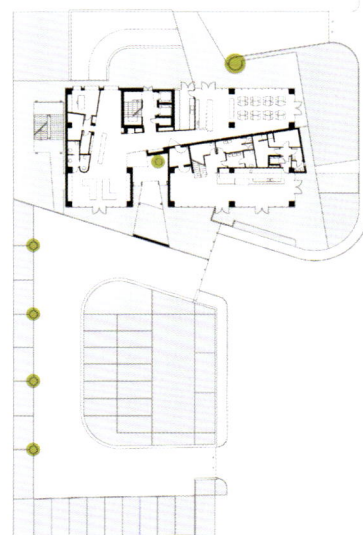

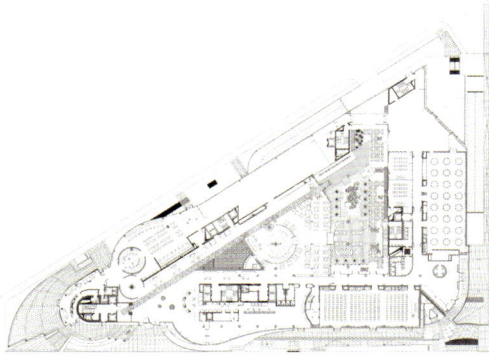

Restaurant in T Hotel

The project was developed around three natural elements, water, stone and light, which were interpreted so as to convey an 'archaic sense' of wellbeing. In particular the use of the stone on floors and walls produces a light relief design which vibrates with the light. In particular, the interiors have been inspired by the neighbouring Teatro Lirico and are reminiscent of theatrical and musical themes, through their lighting effects and the use of colour, materials and finishes. The interiors of the rooms feature particular usage of hues of blue, green, red and orange.

T酒店餐厅

设计师围绕水、石、光三种自然元素着手，向客人们传达古老的感觉。室内设计受到相邻剧院的影响，通过光线、色彩及材料的恰当运用，到处洋溢着戏剧和音乐的主题。值得一提的是，地面和墙壁上的石材营造了浅浮雕的设计效果，随着光线摆动。

Photo: Francesco Bittichesu

Studio Marco Piva

Restaurant in UNA Hotel

For the new hotel project, situated opposite Bologna railway station, designers wanted to develop a particular concept that would enable them to 'construct' a place dedicated to business tourism, located in the throbbing hub of one of the most active cities in Italy, still a pillar of thought and of Western history. Restaurant is a good place to bring out culture. Here the designers decorated the wall with texts. The texts are mirrored on the tables and the chairs, as a unique feature of the restaurant. A common characteristic of the selected texts is that they narrate the adventures, the impressions and also the exploits of great travellers, whether they made a real voyage, like Marco Polo, an interior journey, like Rama or an epic journey like Ulysses.

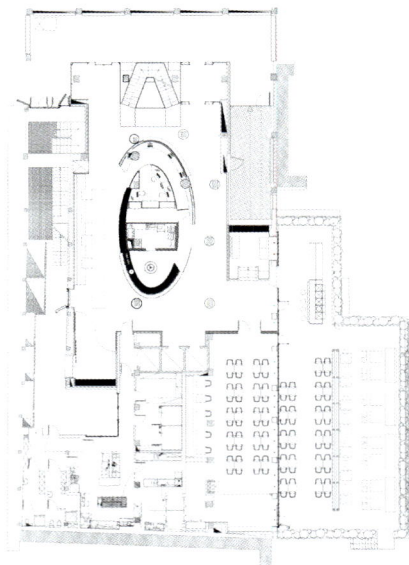

UNA酒店餐厅

酒店选址在博洛尼亚——意大利最繁华的城市中心，与火车站对街相立。Marco Piva工作室旨在构思一个独特的理念，打造一个商务旅游型酒店。餐厅是展示文化的好地方，设计师们将文字作为装饰用在了餐厅的墙壁上，镜面的桌椅上映着文字装饰，成为特色。

Restaurant in UNA Hotel Napoli

A 19th- century historic building, facing on Garibaldi Square in the city centre, has been restored and transformed in a four-star hotel for a prominent Italian hotel chain. The original building, a seven-storey tuff-stone structure, very long and narrow, has suggested the design of a vertical hall crossed by staircases and elevators dangling in space. A comfortable restaurant was created for UNA Hotel. The screens on the wall and on the columns offer more entertainment to the dining guests. The interior decoration here is simple, but you could also find something fresh and novel. Particularly, the colour of the chairs is peculiar, showing the ingenuity of the designers.

UNA那不勒斯酒店餐厅

UNA酒店隶属于意大利知名酒店连锁集团,由一幢始建于19世纪的建筑改建而来,共包括90间客房、一间小会议室以及带有屋顶花园的餐厅。设计师为UNA设计了非常舒适的餐厅。墙面和柱子上的电视非常体贴,让客人可以享受更为闲适的用餐时间。室内装饰是简单的风格,同时设计师也用一些小的变化带来新奇的感受。座椅的颜色很特别,同样也是设计师别出心裁的设计。

Restaurant

Italy

Naples

2006

Photo: Scacchetti

Scacchetti

Restaurant & Bar

Italy

Piedmont

2010

Photo: Daniele Domenicali

Andrea Langhi

The Guinness

A bar and restaurant located in an industrial building, it preserves the size and the original structures visible. The design is inspired by traditional Irish pubs, but in different proportions, much larger, and with some design solutions that combine traditional materials and modern lines. Some elements, like the columns at the entrance, were included in the caskets of crystal to make them look important objects from the museum, illuminated from below. Other elements divide the space into different areas, each with a distinctive feature. To highlight these areas the designers used different kinds of ceiling. The designers used a large painting of Caravaggio to emphasise a central sofa, and a large ceiling panel decorated to highlight an area of tables.

吉尼斯酒吧

酒吧餐厅位于一座工业建筑中，保留了原有的建筑结构和外观。餐厅设计受传统爱尔兰酒吧影响，但是更大，而且在设计中结合了传统材料和现代线条。一些元素（如入口的门柱）被设在水晶盒子里，并且给予了内置照明，看起来像是博物馆里的展品。其他划分区域的元素都具有独特的特征。为了强调这些区域，设计师使用了不同的天花板。设计师用一幅巨型卡拉瓦乔的油画突出了中心沙发，而用一个巨大的吊顶镶板标志了餐桌区域。

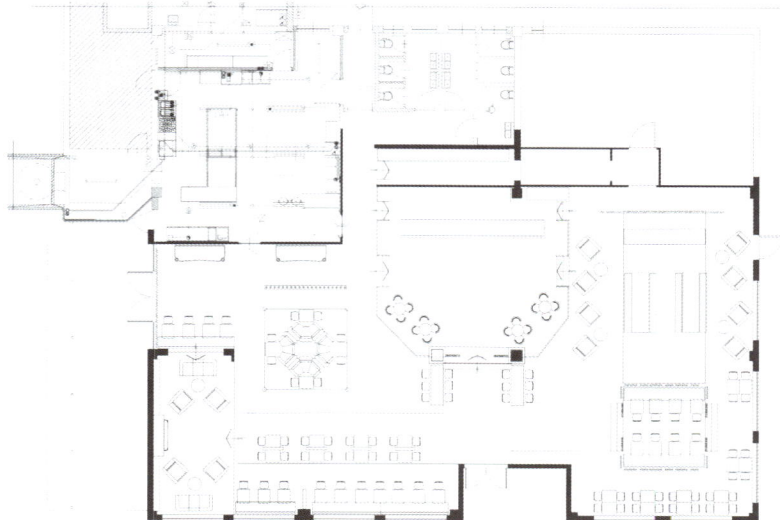

Bar in The Gray

Designed with Milan's most exclusive fashionista set in mind, the simply named Gray treads a careful line between elegance and opulence. Unashamedly elitist and purely residential, only the most established names were granted access to the Gray's private club-like atmosphere, housed in what were once residential buildings. Here the murmur of barely spoken conversations is interspersed with the musical clink of glasses from the bar, in a scene cloaked in style and the scent of eminence. The tables, chairs and other furnishings in the interior got a modern air with their simple and fluid lines. The lighting enhanced the ambience of the space. The peculiar detail designs revealed the distinction of the bar.

格雷酒店酒吧

酒店虽名字简单，设计中却洋溢着高雅时尚的气息，如同私人俱乐部般的氛围只为那些社会名流敞开大门。酒吧内，风格时尚，散发着高雅的气韵，酒杯相碰的清脆声音谱成和谐的乐曲，打破了原有的沉静。桌椅的陈设以及室内装饰，以流畅简洁的线条散发着时尚的气息，灯光设计渲染了室内的氛围，细节处出奇的设计宣告着这里的与众不同。

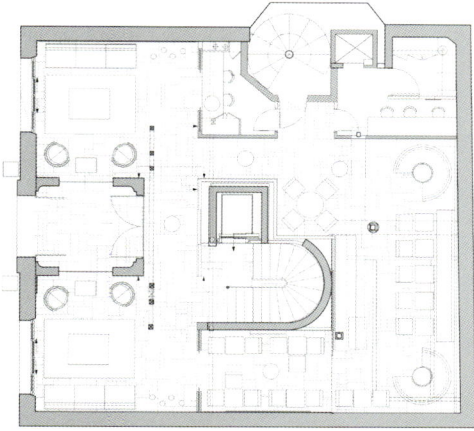

Bar

Italy

Milan

2008

Photo: Guido Ciompi

Guido Ciompi

Restaurant

Ukraine

Kiev

2007

Photo: Andrey Avdeenko

Drozdov&Partners
Oleg Drozdov, Alexander Zhuk

Amici Mi

Downtown Kiev has seen the opening of a new restaurant, which has found its new content in the catering context of the city. The overall aesthetics of the restaurant's interior is subordinated to the priority of food and visitors over the interior, which is perceived as a kind of laconic neutral passepartout for the repast. The project features luxury clothed in the gown of democracy, refinement disguised in sophisticated simplicity – all this being distinctive features of modernism. In fact, there were no purely architectural tasks involved here: the shell structure was set by the layout of the building (the restaurant occupying the ground floor). The authors' intention was to neutralise the existing pylons avoiding their involvement in spatial zoning. The ceiling is also perceived as a neutral element. The overall idea behind that was to interpret the walls, the ceiling and the floor as framing surfaces with the emphasis on what is happening inside.

阿米西米餐厅

阿米西米餐厅为基辅市中心增添了新活力。餐厅的室内设计以顾客和食物为重，为餐饮活动提供了简洁中立的背景。项目奢华、精致而简洁，具有现代主义设计典型风格。设计师力求淡化原有架桥在空间划分中的效果。天花板的设计也很中立。整个设计背后的理念是将墙壁、天花板和地板作为框架表面，而重点是其中所发生的事情。

Yaske Sushi Bar

This total transparency – extending to the smallest detail in the interior including plastic stools and the almost invisible needle-like legs of the tables – turns into something completely different with the coming of evening, as the internal space of the sushi bar is flooded with the bright light of the lamps. Then the glass walls turn into screens which reflect every thing and person inside the bar. In contrast with indistinct masses of the surrounding buildings, the bright lantern of the bar seems to live in the darkness of the night. The area is divided into two functional areas.

雅士客寿司吧

餐厅是完全透明的，连塑料凳子和细得几乎看不见的桌腿都是透明的。到了晚上，餐厅的灯亮起时，这里就变成了一道奇特的风景，玻璃幕墙变成镜子，映出餐厅内的景象；周围的建筑和餐厅内明亮的灯光把黑夜点亮。餐厅分为两个功能区，其中一个区域，包括厕所、厨房和包间；另一个是透明开放的用餐区。

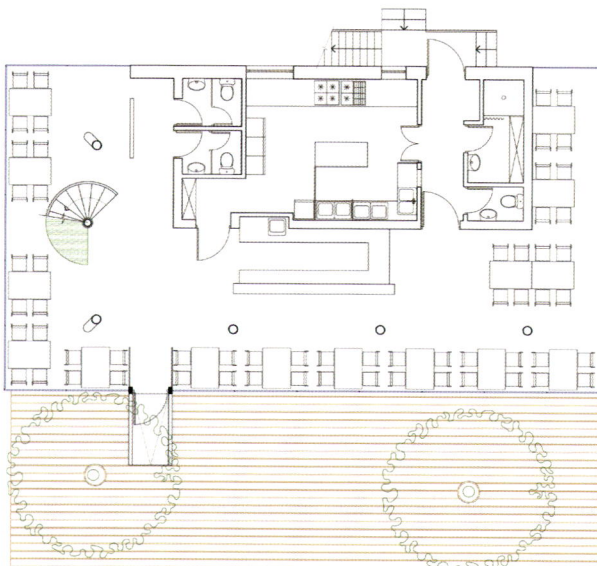

Bar

Ukraine

Kharkov

2006

Photo: Andrey Avdeenko

Drozdov & Partners Ltd.

Restaurant & Bar

Czech Republic

Prague

2009

Photo: RDD

RDD

480

Restaurant & Bar in The Augustine Hotel

Tom's Bar, named after the monastery's patron saint, St. Thomas, is located in a barrel-vaulted double height hall that provides both inspiration and constraint. The intention was to create a cocktail bar that embraced the building's ecclesiastical origins yet sustained elegance appealing to an international clientele. For the Monastery Restaurant, RDD was charged with the conceptual planning, spatial layout and interior design of a double height extension to accommodate 100 all-day and fine dining covers. The structure's orientation was a challenge, given the awkward courtyard space into which the restaurant was to be built, as well as the surrounding building heights and three large, protected trees.

奥古斯丁酒店餐厅酒吧

汤姆酒吧（根据圣人托马斯的名字命名）位于带有桶形穹顶的双层高度大厅内，既为其他空间设计提供灵感，也为其设置了障碍。最初目标是打造一个鸡尾酒吧，保留建筑原有的教会特色的同时营造典雅的氛围，借以吸引来自世界各地的客人。修道院餐厅的设计中，RDD负责整体概念规划、空间格局打造和室内装饰。餐厅的朝向是最大的挑战，既要考虑庭院空间的形状，又要符合周围建筑的高度。最终，设计师运用捷克的传统元素，玻璃制品、锈迹斑斑的黄铜以及铁器和当地手工工艺打造了一个现代化的餐厅。

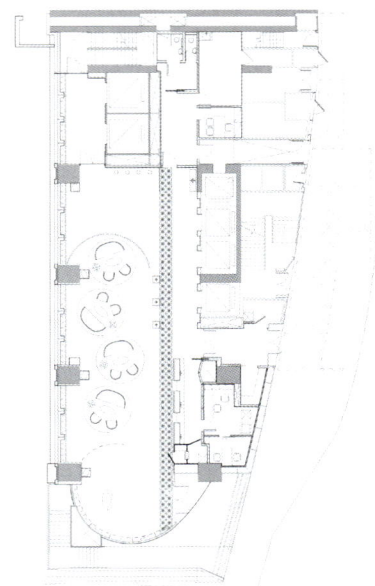

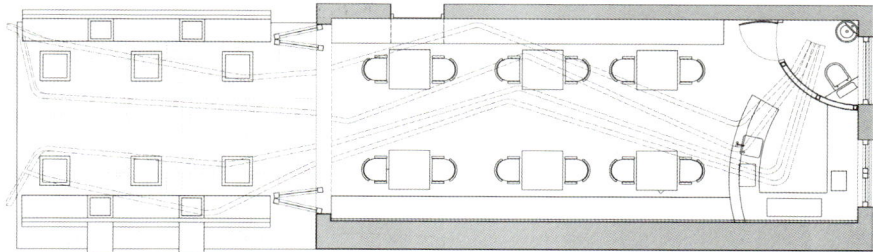

Nature and Society

The interior concept of 'Priroda i dru.tvo', represents an interference between Nature and Society. Regarding this fact, the place is longwise divided into two irregular parts, the Nature on the right side and the Society on the left side. The contrast between these two groups is reflected through various elements, in colour, furniture and in the effect of 'upside-down'. The Nature on the right side is defined through light colours and furniture, unlike the Society, which is represented through darker furniture and colour. The effect 'upside-down' marks the entire interior design and represents today's current state that involves Nature and Society.

自然与社会

餐厅的室内设计理念展现了"自然"与"社会"的冲突，整个空间被分割成不规则的两部分——"社会"在左，"自然"在右。设计师通过色彩、家具等元素的运用，突显两部分空间的对比——右侧空间采用鲜艳的色彩和亮丽的家具，而左侧则恰恰相反。对比的设计手法贯穿整个室内空间，映射出了当今时代的发展状态。

Photo: Munever Salihovic

Normal arhitektura, Emir Salkic, Muhamed Serdarevic, Armin Mesic

Restaurant

Bosnia and Herzegovina

Sarajevo

2009

Photo: Dzenat Drekovic

In/Out Arh Studio

Pasta Zen

The day/night live show cooking concept of Pasta Zen restaurant is aimed at today's business people, who need quick, good quality, healthy food in a welcoming atmosphere where they can meet friends, family and business associates. This restaurant in the city centre soon became known for its warmth, commitment to good food and contemporary setting. The entry space suggests the idea of adaptation, with the atmosphere changing from daytime to night-time as light levels vary. Adaptation is also reflected in the use of materials such as oil wood and rough stone walls, the somewhat brazen simplicity of which enables customers quickly to feel at home in the place, as though it were part of them. The architects also used interior details, greenery and various kitchen utensils to give the place a homely feel.

禅意意面餐厅

禅意意面餐厅为当今的商业人士提供快速、高品质、健康的食品，在这个热情洋溢的环境中，他们可以与亲友和商业伙伴进行会面。这家位于市中心的餐厅以其温馨的环境、美味的菜肴和现代化的装饰背景而闻名。入口空间显示出渐变的理念，随着光线的变化，整体氛围也变得不同。渐变还体现在材料的运用上，如油木和糙石墙面，餐厅让顾客有一种家的归属感。建筑师还运用室内细部、绿色植物和各式各样的餐具打造出家的感觉。

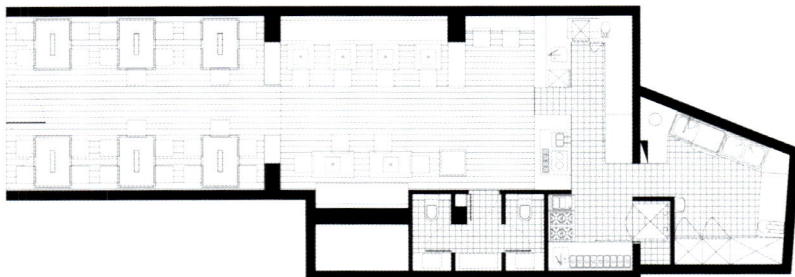

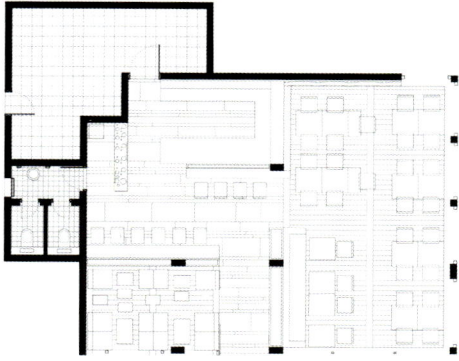

Old Friends

Old Friends is a restaurant and lounge bar that extends over an area of 100 square metres. The project is the renovation of existing bar and intervention on the façade of the existing object. The interior consists of three zones: a lounge bar, an area for fast food and restaurant-wine bar. The designers' desire was to include these areas and visually connect them, and yet they keep their own integrity. The connection between them is nature, so as a line of space limitations of individual zones occur with elements of birch trees and artificial grass. Birch trees make the curtain between spaces, such as algae build curtain of the external façade and make a connection between the outer street façade and interior.

老朋友餐厅

老朋友是餐厅和休闲吧的结合体，总面积为100平方米。项目主要对室内和外墙进行了翻新。室内分为三个区域：休闲吧、快餐区和餐厅酒吧。设计师试图将这些区域在视觉上连接起来，又保持它们各自的个性。它们之间通过自然相连，桦树和人造草坪将各区域隔绝开来。桦树形成了空间之间的帘幕，将外面街道和室内连接起来。

Restaurant & Bar

Bosnia and Herzegovina

Sarajevo

2007

Photo: In/Out Arh Studio

In/Out Arh Studio with Lamila Simisic, architect

Photo: Andrei Margulescu and the magazine Arhitectura

Parasite Studio

Café Mode

Based on the relation between contexts and skin the designers tried to create a contrast between the existent space and the plastic image of the project's interventions. The interior design, like a piece of temporary clothing, is reversible, made out of light panels mounted on the existing structure, which can be easily taken off according to current fashion trends. The designers created a series of key elements of the project: the bar counter, actually a support for the main text meant to welcome the public and the low tables with a red insertion that houses the personalised ashtray.

风尚咖啡厅

项目与原有建筑在外观上形成鲜明对比，营造出强烈的视觉冲击，十分引人注目。室内设计灵活，结构可根据实际情况随意调节，特殊的光板结构令室内历久弥新。吧台设计巧妙，热情活泼的设计时刻期待顾客的光临，而落地桌上盛放烟灰盒的红色地带更是别出心裁，时刻吸引人们的靠近。

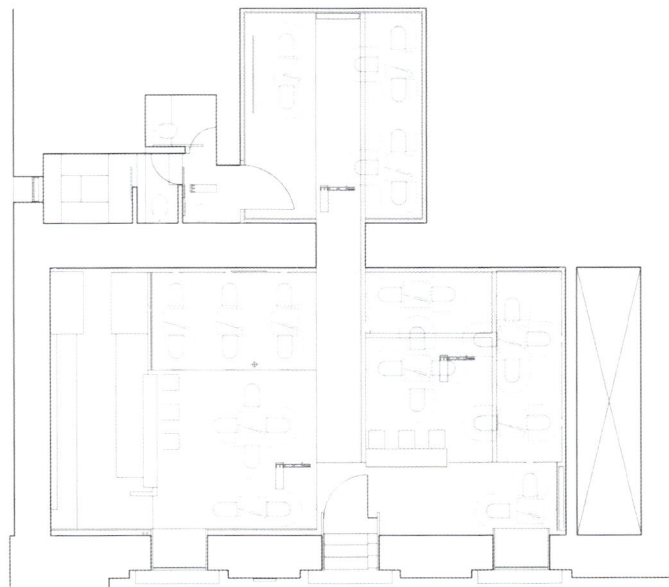

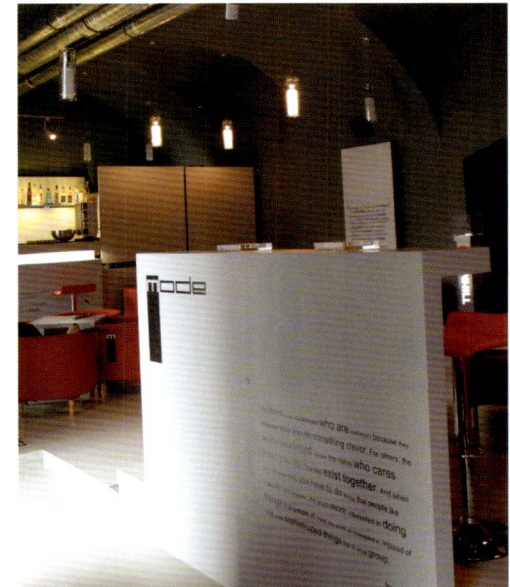

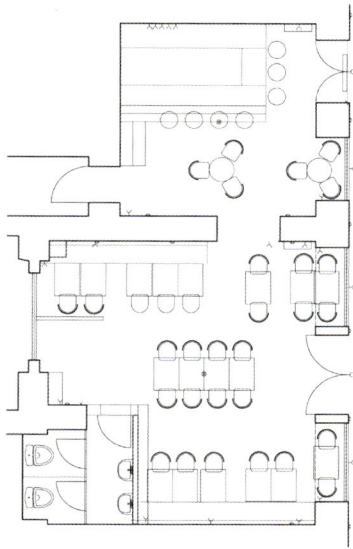

La Bonne Bouche

La Bonne Bouche – Barr a vins et champagnes – situated in the old city centre of Bucharest is a newly opened bar that offers every week a new menu of fine foods with appropriate wine selection. Among the classic French bistro features (banquette, cast iron tables, carreaux ciment), the designers introduced some contemporary elements like the tom dixon wall lamps, the navy chair and the taraxacum. The metallic window between kitchen and restaurant is a reclaimed piece from a 1920's factory and it fitted the arch perfectly. The designers left visible a few original features of the building such as the metal pillars by the entrance, brick arches and 'Paris stone' pillar.

珍品酒吧

珍品酒吧位于布加勒斯特旧城区中心，每周都有新菜单，提供美食和相应的酒品。除了经典法式酒馆特征（如卡座、铸铁餐桌、方砖水泥等）之外，设计师还引入了一些现代元素，如壁灯、海军椅和蒲公英。厨房和餐厅之间的金属窗从一座工厂回收得来，与餐厅的拱门出奇地相适。设计师保留了建筑一些原有的特征，如门口的金属柱子、铜拱门和巴黎石柱。

Bar

Romania

Bucharest

2010

Photo: Corvin Cristian/Liviu Bradean

Corvin Cristian

BB Club

The Croatian Island of Hvar is often regarded as the Monte Carlo of the Adriatic. It is very popular with the rich and famous attracting deluxe yachts that fill the Hvar harbour. The hotel is entered from an extensive terrace positioned on the waters edge. The historic fabric is juxtaposed with trendy, colourful furniture and fixtures intended to appeal to a young, or young at heart, clientele. In addition to the forty-five rooms and nine suites, Jestico + Whiles has designed a new hotel bar with a backlit, green, etched glass counter and a teak veneered top. A fusion restaurant is located on the other side of reception with further seating on the enlarged terrace.

BB 俱乐部

该建筑是中世纪时期的风格，酒店原有的框架，包括瓦楞柱和石墙常年裸露。除了45间客房和9间套房外，贾斯蒂戈+威尔斯建筑事务所还为酒店的新酒吧设计了背光照明的绿色蚀刻玻璃立面和柚木台面的吧台。接待台的另一端是一个一体化的餐厅，它和一个扩大的平台相连，平台也是餐厅的室外就餐区。

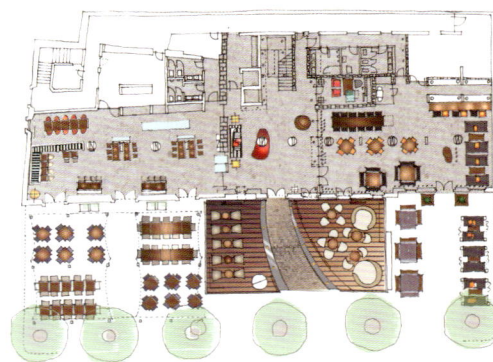

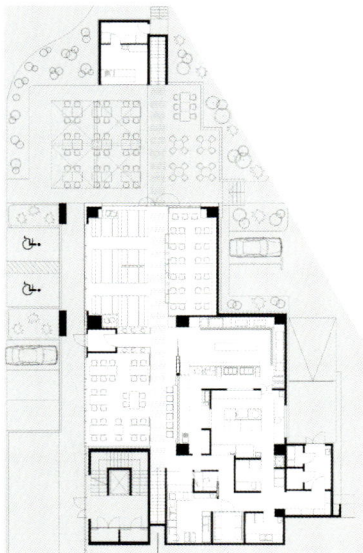

Restaurant

Slovenia

Sencur

2009

Photo: Miran Kambic

Protim Ržišnik Perc

Kantina Cubis

Cubis canteen restaurant is located in the newly constructed facility Business Centre Cubis. With an area of 620 square metres, seating 140 and placed on the ground floor of the Cubis. The restaurant was designed as a buffet and a la carte restaurant and contains various areas: bar/reception with drinks, self-service buffet, a private dining area, large group seating areas, terrace bar and kitchen. The main entrance is located on the north side of the building and makes the main façade even more recognisable. When you enter the canteen restaurant you can easily recognise that even entrance is designed with passion through selected warm colours and natural materials on floor and walls. Very important element in restaurant is play with light and shadows.

丘比斯餐厅

丘比斯餐厅位于新建的丘比斯商业中心一楼，总面积620平方米，拥有140个坐席。餐厅既有自助餐区又有点菜区，分为多个部分：酒吧、自助餐厅、私人就餐区、团体就餐区、露台酒吧和厨房。餐厅的主入口位于建筑北侧，其装饰采用了热情的暖色色调和自然材质。餐厅设计的光影效果也是十分重要的元素。

Restaurant

Slovenia

Maribor

2006

Photo: Miran Kambi

AKSL Arhitekti

Restaurant Rozmarin

The whole space is divided in five levels, but still strongly connected with communications inside glass façade. Cafeteria is situated in the ground floor and it communicates with pedestrian street in front of her. Above cafeteria is a small gallery which has a nice view over Pedestrian Street. It has a more cosy and relaxed atmosphere, wooden floor, and is designed with low seating elements. One level lower, the toilettes for customers and a smoking room are placed. The design is simple and minimalist and it builds on simplicity, hidden details and light. Six metres under street level a wine bar is situated. The aim was to create similar feeling as in wine cellars, but not to literally. Lounge area with low seating elements is situated near staircases in the open volume of glass façade. Modern chandeliers break seriousness and introduce a bit of kitsch.

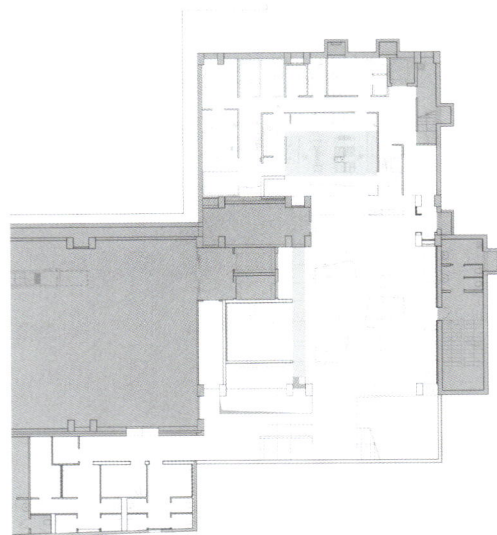

罗兹马林餐厅

整个餐厅空间分为5层，它们之间通过玻璃外墙紧密相连。一层是自助餐厅，与外面的人行道相连。自助餐厅上方是可以俯瞰街景的休闲长廊，具有舒适、放松的氛围，摆放着低矮的椅子。地下一层是客用洗手间和吸烟室。其细部和灯光的设计都极为简单。地下二层的酒吧有一种酒窖的氛围。休息区位于楼梯旁边，靠着玻璃外墙，现代吊灯打破了严肃的气氛，有一种通俗感。

Valvas'or Restaurant

The restaurant is designed as a modern, sober space, within an old, arched tavern, accessible from the ground floor. The idea was to create a space where old and new harmoniously unite to create the restaurant's unique image. To preserve the charm and character of the old space, maintain a feeling of spaciousness, add a metropolitan touch and create a gourmet atmosphere in which to enjoy delicious food – such was the ambition of architects. The elongated restaurant was built on three levels. The ground floor level with the view of the street is the widest. Here the designers find the entrance to the restaurant, following which, on the left-hand side, stands a set of tables placed before a brown leather box-element, which serves as the sitting area. To the right-hand side, beneath a pair of gold metal chandeliers, are two round tables.

瓦尔瓦索尔餐厅

餐厅被设计成一个现代而庄重的空间，里面还设有一个旧式拱顶小酒馆。在这里，新旧元素完美和谐地融合在一起，打造了餐厅独特的形象。建筑师试图保留旧空间的魅力和特点，保持空间感，增加都市感和打造美味气氛。餐厅共分为3层，一层是餐厅的入口，可以看到街道的风景。入口的左侧是棕色皮革座椅组成的休息区，右侧是一对金色吊灯和两张圆桌。

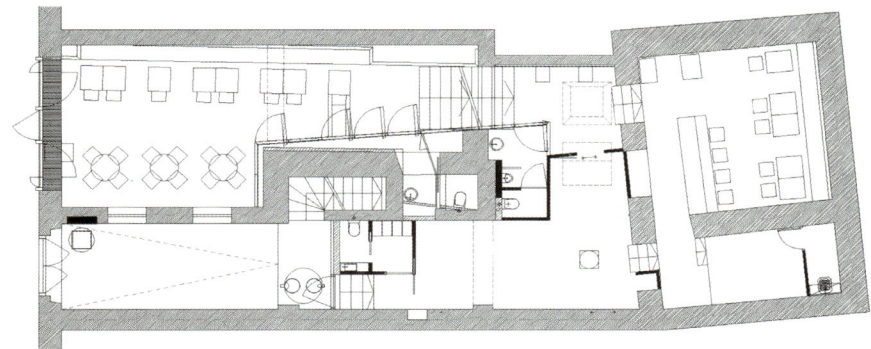

Mixed-Restaurant

Slovenia

Laško

2007

Photo: Bogdan Zupan and Tomaz Gregoric

Borut Rebolj, Studio Rebeka d.o.o.

Lasko Wellness Park

Areas with different functions were designed carefully, including the restaurant, the bar, the café and the tea room, each serving for different needs of guests. Each area carries a specific story embedded in the use of the materials and the colour codes. The café is elegant with neutral colours; the restaurant is warm with yellow, while in the tea room, the adoption of rose pink and wall painting breaks the tradition and adds a bit of modern sense to the space. The delicate wooden floor brings guests a home feeling.

拉斯科公园疗养酒店餐厅

设计师用心设计了餐厅、酒吧、咖啡厅、茶室等不同功能区域，为不同的顾客提供不同的服务。其中，每个区域的设计都运用不同的材料和色彩，讲述着自己的故事。咖啡厅有素色的高贵，餐厅有黄色的温暖，茶室则打破传统，用玫红和墙画增加现代感。精致的木地板给顾客家一般的感受。

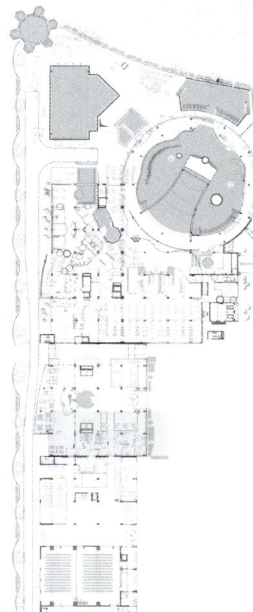

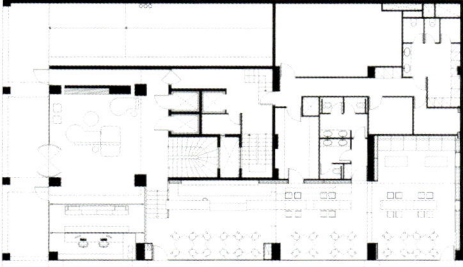

Restaurant & Bar in Fresh Hotel

The Fresh boldly mixes rich natural materials, like oak and walnut wood, with bright pinks and oranges, making it a clean-lined sanctuary from the flurry of downtown Athens just outside its doors. The nine-storey building, with a pool and bar sporting a view of the Acropolis on the top floor, welcomes its guests into a lobby extending over two floors that gives them a taste of the funkiness to come upstairs. A generous fireplace that is surrounded by an imposing black wall stands opposite the reception created by an attractive pink glass box.

鲜明酒店餐厅酒吧

鲜明酒店把丰富的自然资源像橡木和胡桃木与亮粉色和橙色大胆地结合到一起，使它成为一个远离门外喧嚣都市的庇护所。这是一座九层的建筑，顶层配有游泳池和酒吧，在那里可以欣赏到雅典卫城秀美的风光。

Restaurant & Bar

Greece

Athens

2005

Photo: Ch. Louizidis, K. Glinou, Paterakis

Zeppos - Georgiadis & Associates

Restaurant

Greece

Larissa

2009

Photo: Mihajlo Savic

Christina Zerva Architects

Cartel Club Restaurant

The architect's purpose was to create a sense of a non-dimensional space. Self-illuminating objects gave the dramatic lighting that created the impressive atmosphere and the sense of endlessness. Raw concrete and acrylic glass along with the white interior provides an exciting contrast. Dedicated to ecological design, sustainability, clean air, clean water, salads without pesticides, apples with worms, world without violence and bloodshed. Objects hanging from the roof with chains remind us of the consumerist society we live in. Recycled and second hand materials have been used for the building. For the insulation is used cellulose rather than the common building insulation materials that may contain toxics. Also photovoltaic solar panels provide sustainable electricity for any use.

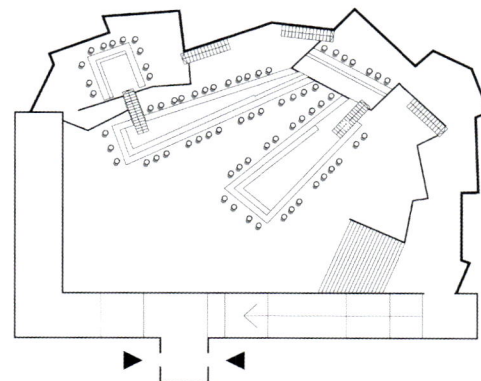

卡特尔俱乐部餐厅

建筑师的目的是打造一个无维度的空间，发光物体让灯光如梦似幻，打造出无限感。未加工的混凝土和有机玻璃与白色的室内空间形成了鲜明对比。餐厅崇尚生态设计、可持续发展、清洁的空气和水、无农药沙拉。天花板上垂下来的链条提醒我们这是一个消费主义社会。建筑采用了二手材料和回收材料。隔热材料采用了纤维素，而不是有毒材料。太阳能电池板为餐厅提供可持续电能。

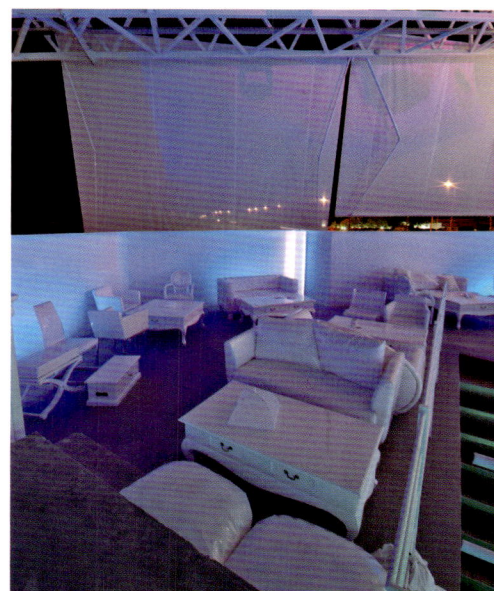

Restaurant

Greece

Athens

2009

Photo: George Fakaros

Kokkinou Kourkoulas Architects & Associates

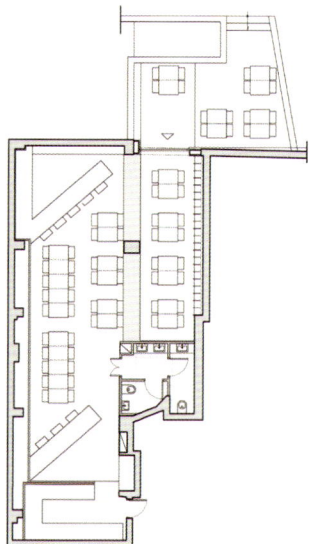

Scala Vinoteca

Scala Vinoteca is a small restaurant, 100 square metres, situated on a pedestrian stepped street in a residential part of the centre of Athens. The main construction materials were steel sheets and wood. They are used as floor and ceiling, tables and chairs, lighting fixtures and bar. These two materials constructed the space of the restaurant in their 'raw' state; no paint and no colour, emphasising their materiality. A mirror, placed opposite the entrance, creates the illusion of a back yard, giving a sense of expanding space. The front door, constructed with steel and glass, when open, slides and disappears within the façade. The whole atmosphere of the restaurant is based on the dialogue between wood and iron – inside and outside – direct and indirect light.

斯卡拉•维诺提卡餐厅

斯卡拉•维诺提卡是一间100平方米的小餐厅，坐落在雅典中心住宅区的一条人行道上。餐厅的主要建筑材料是钢板和木材，它们被广泛运用在地板、天花板、桌椅、灯具和吧台设计上。为了强调质感，这两种材料的应用完全没有涂漆或上色。正对餐厅入口的镜子让餐厅看起来多了一个后院，在视觉上扩大了空间。钢铁和玻璃组成的前门在打开时会消失在墙壁里。餐厅的整体氛围以木材和钢铁、内部与外部、直接采光与间接采光为基础。

Restaurant

Australia

Sydney

2009

Photo: Sharrin Rees

Koichi Takada Architects

Cave Restaurant

The designers aimed to change the way people eat and chat in the restaurant. The acoustic quality of restaurants contributes to the comfort and enjoyment of a dining experience. The designers have experimented with noise levels in relation to the comfort of dining and the ambience a cave-like environment can create. The timber profiles generate a sound studio atmosphere, and a pleasant 'noise' of dining conversation, offering a more intimate experience as well as a visually interesting and complex surrounding. The series of acoustic curvatures were tested and developed with computer modelling and each 'timber grain' profile has been translated and cut from computer-generated 3D data, using Computer Numerical Control (CNC) technology.

洞穴餐厅

设计师试图改变人们在餐厅就餐和聊天的方式。餐厅的隔音效果直接影响就餐的舒适度，设计师就此专门进行了噪声测试，证明洞穴环境更加舒适。木轮廓营造了一个录音棚般的效果，在就餐时打造了悦耳的声音，也提供了一种私密的空间体验和独特的视觉效果。这一系列的声学曲面经过了测试，由电脑建模而成，采用了计算机数控工艺。

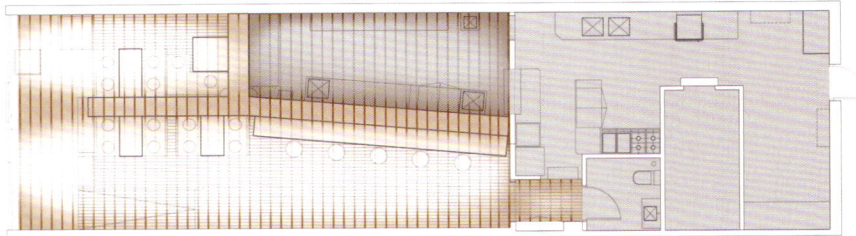

Roslyn Street Bar & Restaurant

The softly modulated interior of the restaurant accommodates the shifting geometry of the site, uses of the room and its services. The palette of timber, off white concrete, graded textiles and delicate white furniture gives the room a calm and slight remove from the intensity of its urban setting. The programmes include four levels plus basement service area. A deep street awning on the north side shades the full height glazing of the restaurant and a roof garden provides insulation and contributes to the greening (and bird life) of Kings Cross.

罗斯林街酒吧&餐厅

餐厅的室内设计适应了场地，保留了原有的房间和服务。不同色彩的木材、灰白混凝土、渐变的纺织品和精致的白色家具让房间看起来十分宁静，远离了喧嚣的城市背景。项目分为4层，外加一个地下服务区。北侧街面上的遮阳篷能够为屋顶花园提供阴凉。屋顶花园既能隔热又能为街道带来绿化。

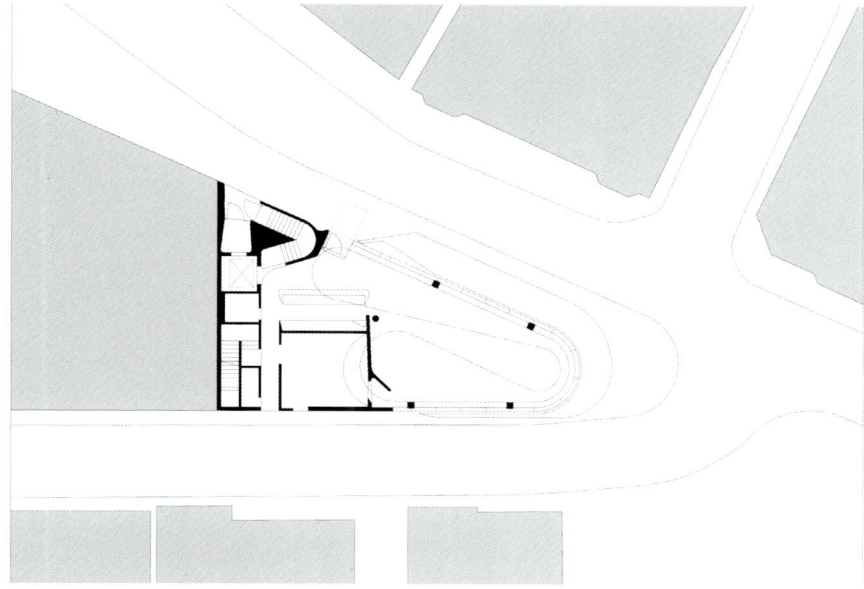

Restaurant & Bar

Australia

Sydney

2009

Photo: Anthony Browell, Peter Bennetts

Durbach Block Architects

Restaurant

Australia

Brisbane

2009

Photo: Scott Burrows

Arkhefield

Urbane Restaurant

The original Urbane Restaurant is an icon of Brisbane dining with numerous awards and a substantial reputation. The redeveloped Urbane extends this philosophy of providing the highest quality food and service over four distinct dining and bar experiences. The entire site of the original Urbane has been extended and redeveloped to incorporate a new Urbane, Euro, Sub-Urbane and The Laneway. Sub-Urbane is a linked cellar dining area, intimately positioned in the basement of the building, flanked by 100 years old stone walls and a well stocked cellar. The Euro is a bistro providing great quality food in tandem with a humming bar experience. The Laneway is a bar perched within the centre of the site, providing a hideaway bar to while away the hours with bar food of a calibre that isn't available in the city.

都市餐厅

原来的都市餐厅是布里斯班餐饮业的标杆，获得过无数奖项，极具声望。新改造的都市餐厅延续了高品质的餐饮服务，提供4种独特的餐厅和酒吧体验：新都市餐厅、欧元酒馆、城郊餐厅和巷道酒吧。城郊餐厅位于建筑的地下室，两侧的石墙拥有百年历史。欧元酒馆提供经典美食和酒品。巷道酒吧设在场地的中心，提供无以伦比的酒吧服务。

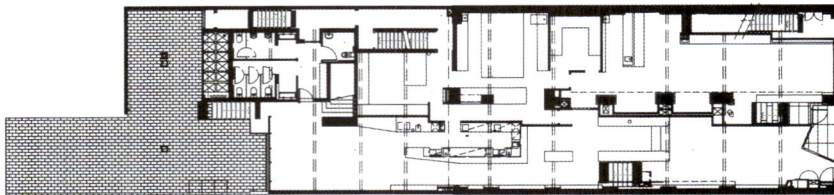

Restaurant & Bar

Australia

Sydney

2009

Photo: Katherine Lu

Facet Studio

Phamish

Phamish is a Vietnamese restaurant located at Darlinghurst, Sydney. The designers were approached by the owners to create a new image for Phamish using elements of 'chrysanthemum' and 'gold'. From the time and space limitations, the designers focused on the 'partition' as a system which softly defines the spaces and is widely utilised in Asia. If it is 'partition' only, then it is possible to prefabricate in the factory prior to construction on site. Soon the designers started hand sketching the patterns of 'chrysanthemum flowers', and by digitising the hand drawn patterns, it became possible to laser cut the partition panels one by one with precision. The designers then coated the panels with brass metal, aged the brass with greenish patina, and finally moved them in on site.

法密施餐厅

法密施是一家越南餐厅，位于悉尼的达令赫斯特港。餐厅主人委托设计师运用"菊花"和"金色"为法密施打造一个新形象。由于时间和空间的限制，设计师采用了广泛应用于亚洲的"屏风"。屏风可以在现场施工之前于工厂定做。设计师手绘了菊花图案，通过激光将其精确地切割到屏风上。屏风上的铜板进行了做旧处理，带有铜绿斑点。

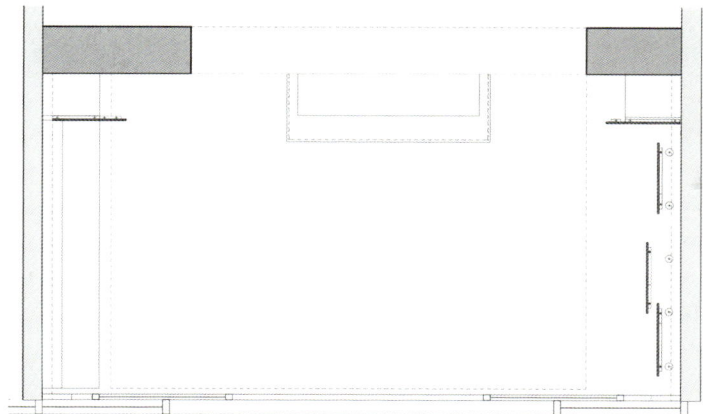

Bar

Australia

Melbourne

2010

Photo: Sonia Mangiapane

Architects EAT

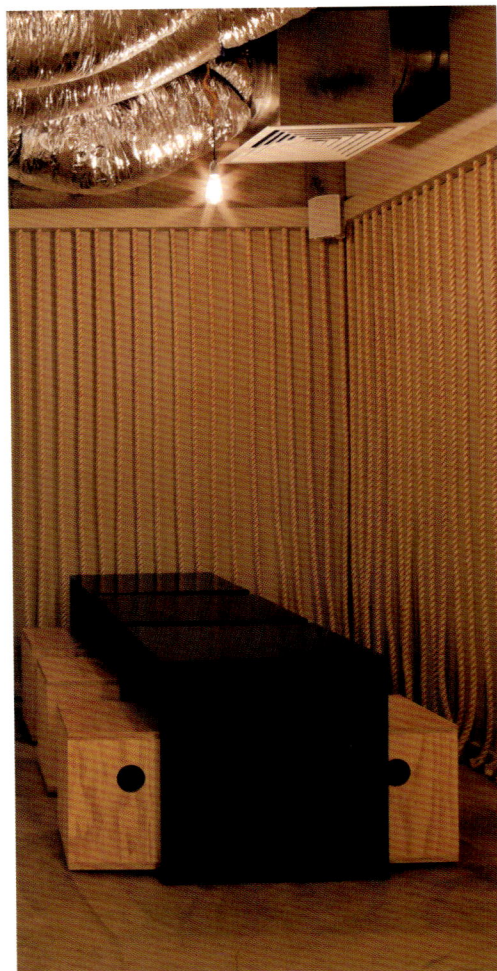

Egg Sake Bistro

Housed in the basement of the University of Melbourne, Egg Sake Bistro is a casual place for staff and students to have a drink, some lunch, or grab some food to go. Specialising in traditional Japanese food and sake, Egg Bistro is another venture by the owners of Melbourne's Maedaya Sake & Grill in Richmond. Whilst being very different concepts, the owners wanted to maintain a link between the two identities. This was achieved through a similar palette of raw materials: timber, concrete and manila rope. These were used to reflect natural elements such as vegetation and earth – an intrinsic ideology of traditional Japanese culture where people seek a connection with nature. The rope idea originated from the classic design of sake bottles which are traditionally secured with ropes and also echoes the bindings of sake barrels.

鸡蛋清酒酒馆

鸡蛋清酒酒馆位于墨尔本大学地下室，是教职工和学生就餐的好去处。酒馆以传统日式料理和清酒为特色，是米达雅清酒烧烤店的姊妹店。尽管二者经营理念不同，店主还是要求二者之间保持联系。设计师运用相似的天然材料：木材、混凝土和马尼拉粗绳实现了这一要求。粗绳的概念源于清酒瓶的经典设计，原来的清酒瓶和酒桶都是用粗绳捆绑保护的。

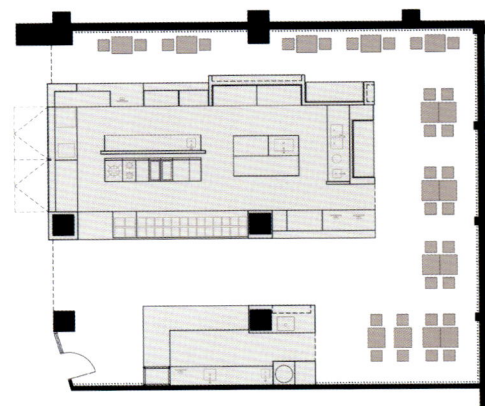

Bar

Australia

Perth

2007

Photo: Doug Blight

Marshall Kusinski Design Consultants

The Champagne Lounge at Must Wine Bar

In a small space above a well-known Mt Lawley bar and bistro is what might be described as a private club. But you can't buy a membership and there's no joining fee. In fact the only prerequisite is that you love to drink Champagne, fine wines, Cognacs or other rare and divine elixirs. The Champagne Lounge will lavish its guests with highly personal service. Zach Nelson – Manager of the Champagne Lounge will greet you at the locked door and show you to your seat. A collection of chaise style long lounges, Louis French vintage inspired single armchairs and parlour chairs in a selection of decadent and detailed fabrics add variety and richness to the space. Clusters of lounges and chairs are teamed with glossy reflective crystal like cubes – used to rest a drink on and individual table lamps to add golden yellow light and warmth.

葡萄汁酒吧香槟休息室

项目在著名的劳里山酒吧上方，具有私人俱乐部性质。进入俱乐部无需购买会员卡或付费，只需热爱香槟、红酒、白兰地和其他美酒。香槟休息室为顾客提供个性化的服务，经理会在门口引领你走向座位。休息室的风格深受传统法式风格影响，大行颓废美学和繁复的织物，为空间增添了多样性和复杂感。沙发和座椅旁配有水晶面矮桌，用于放置饮品和台灯。

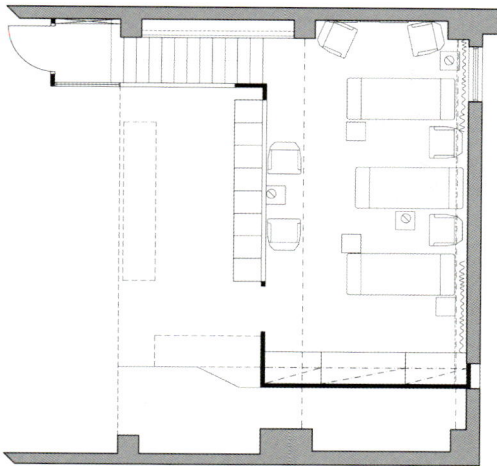

Restaurant

Australia

Soutbank

2006

Photo: Earl Carter

Batessmart

500

Rockpool Bar & Grill

The designer's brief was to create a great produce experience, drawing on a fusion of the Rockpool ethos and a classic North American steakhouse. Guests enter the restaurant through the Produce Hall, where food items are stored and prepared, and which provides glimpses of the chef's vestibule and kitchen area. They are greeted in the cellar bar, where light meals and drinks are served, or in the adjacent casual dining area. The spectacle of arriving in the formal dining room – with its sweeping views over the city skyline – is heralded by passing through a specially designed rope curtain. At the heart of the space, and viewed from all areas, is the substantial display kitchen defined by a monumental beaten copper canopy. Throughout the restaurant there is a strong use of robust materials, in particular the extensive use of oiled or waxed solid timbers.

石池酒吧烧烤店

设计师力求打造一个融合石池餐厅精神和经典北美牛排餐厅特色的美食基地。顾客通过准备大厅（这里摆放着准备好的食材，还可以看到主厨的前厅和厨房）进入餐厅，穿过提供小食和酒品的酒吧或是休闲餐区，拨开特别设计的绳索门帘，最后到达正餐厅，从那里可以俯瞰城市全景。

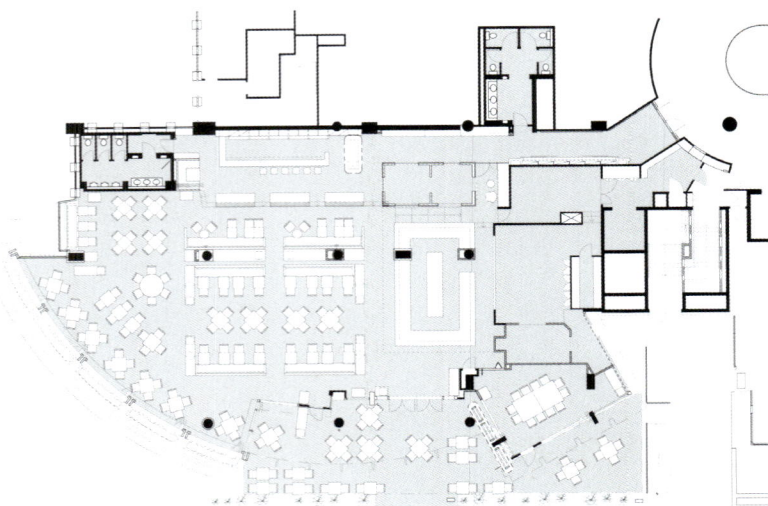

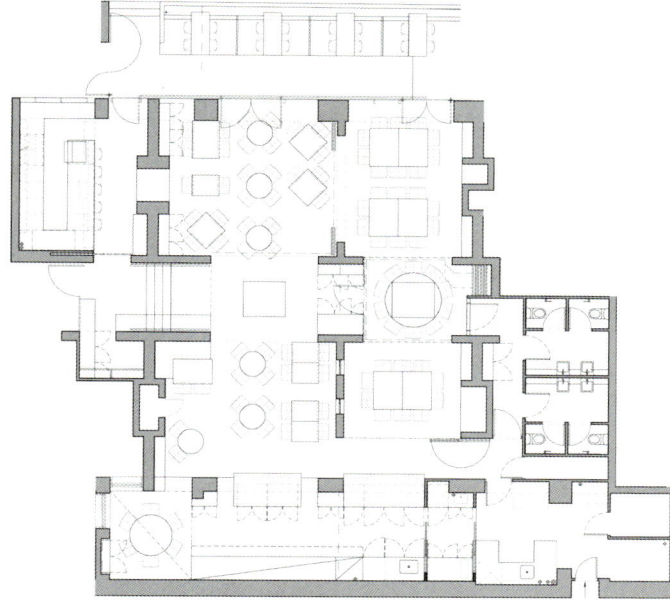

Vue de Monde

The design for Vue de Monde, a contemporary Melbourne restaurant highly regarded for its fine French dining experience, sat to juxtapose the functionality of a hardworking restaurant with the luxury of sampling fine food and wine. Highly patterned stone bench tops and timber floors provide texture to the stark white walls while the large globe light fittings create illuminated highlighters dotted through the space. In contrast, the gleaming stainless kitchen equipment is evident providing a working insight into a functioning kitchen during service. Ceiling mounted mirrors above the preparation benches which divide the kitchen and dining spaces intensify the experience as one can watch head chef and proprietor Shannon Bennett complete the finishing touches to every meal.

世界观餐厅

世界观餐厅以精致的法国菜而闻名，提供奢华的美食和美酒。图案繁杂的石椅上部和木地板为朴实的白色墙壁增添了质感，球形吊灯为室内洒下了点点闪光。擦亮的不锈钢厨房器具为厨房营造了一种工作氛围。准备台上方的镜子增添了客人就餐的经历，让他们可以全程观看主厨的烹饪过程。

Restaurant

Australia

Melbourne

2005

Photo: Dianna Snape

Callum Fraser, Cassian Lau, Marcus Ieraci, Iva Foschia

Bakery

Australia

Sydney

2006

Photo: Design Clarity

Design Clarity

Shepherd's Bakehouse

Shepherd's Bakehouse has bought a traditional bakery with a contemporary twist to the quiet area of Concord West. Formally an old pub, the prominent corner site provided the perfect spot for a gourmet bakery/café. The designers have contrasted the existing old shopfront tiles, traditional awning and new timber-framed windows with a warm yet sophisticated interior. The use of colour, custom-made pendant lights and European style furniture has provided a contemporary design solution to the original client brief. Plantation timber is specified throughout this project for floor, counter front cladding, internal tabletops and joinery units. Lighting is minimal due to the location of apertures. The shopfront also faces east, so the need for excessive lighting wasn't necessary.

牧羊人面包坊

牧羊人面包坊融合时尚元素于传统的面包店之中，独特的设计风格令其在当地独领风骚。该建筑前身是一个老酒馆。原有的瓷砖店面、传统的遮阳篷和木质窗口，与室内温馨、精致的设计风格形成了鲜明对比。造型独特的吊灯、欧洲风格的家具为室内增添了现代气息。地板、吧台、桌面等皆采用热带木质材料凸显空间的高雅。由于店面向东，室内阳光充足，因此，室内无需大量照明设备，从而节约了成本。

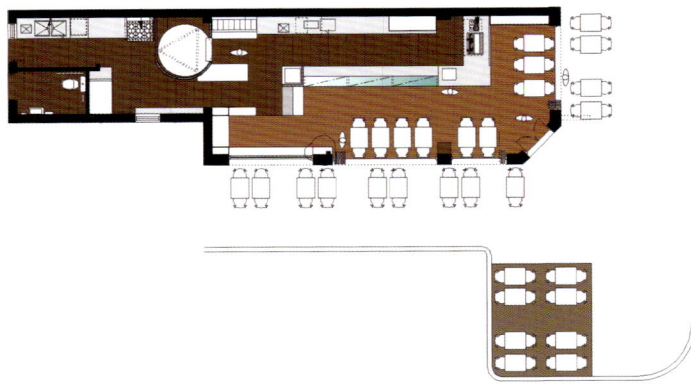

Bakery

Australia

Sydney

2009

Photo: Design Clarity

Design Clarity

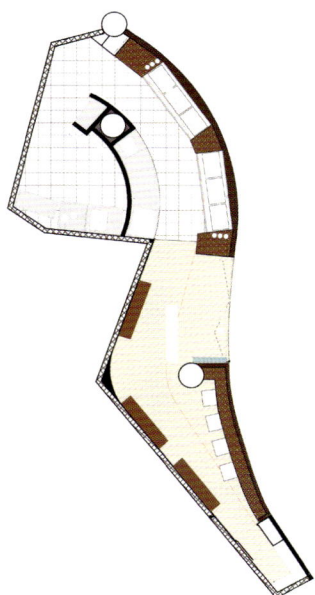

The Cupcake Bakery

A limited palette of materials was selected to complement the lolly pink, blue and chocolate striped wallpaper, wrapping around the curved rear wall as a backdrop to the circular merchandise wall units. The colours were derived from the candy colours of the icing of the cupcakes on display. Beech solid timber floor boards, the inclusion of a Dulux 'Domino' painted dropped bulkhead and 3D edge-lit shopfront disc signage all combine to present a quirky, tasty food environment. Three feature 1,500 millimetre diameter 2-pac finished wall mounted display units have large-scale backdrop 'sex in the city' styled photographs with decorative LED lighting tape to the circumference, allowing for a retail element within the store where the client can display promotional Cupcake Bakery merchandise and complimentary products.

蛋糕面包坊

粉红色、蓝色和巧克力色条纹墙纸与室内糖果色蛋糕交相辉映，妙趣横生。榉木实木地板、多乐士"多米诺"画与3D店铺形象标志一起打造了奇妙、诱人的食品空间。三个1,500毫米直径的壁挂式显示器，装饰性LED照明设备为客户完美呈现出蛋糕面包促销商品和免费产品。

Restaurant

Australia

Melbourne

2007

Photo: Peter Clarke, Latitude

Gray Puksand

Angliss Restaurant, William Angliss Institute of Tafe

This commercially functioning student training restaurant is combined with learning spaces to deliver new outcomes in contemporary interior design. Inherent to this process was the functional layout and design of hospitality spaces combined with educational delivery. The education spaces are located at front of house, leading to restaurant and bar facilities beyond. This provides maximum exposure of education programmes whilst catering for existing clientele and attracting new dining patrons and students. The bar and training areas were fully re-planned and reconfigured from existing outdated facilities, some of which dated back to the time of the original building.

安格力斯餐厅

当具有商业气息的餐厅与教学环境相遇，便带来了全新的现代室内设计方式——服务业的功能规划与设计与教学环境融为了一体。人们通过建筑前部的教学区走近餐厅与其后方酒吧，这让教学项目得到更多的曝光率，也增加了吸引潜在顾客和潜在学生的能力。酒吧和教学区进行了完整的改造，以更适合现在的功能区域划分。

Wild Food

The material palette is simple, fresh and raw. The joinery is a combination of lacquered Hoop Pine and laminate for durability. Bare light bulbs hang over the service counter to create a random chandelier-like installation. Graphics are burned into the eye-catching timber panels emphasising the raw and natural philosophy of Wild Food. Vinyl decals are applied to paint finishes and joinery as a cost effective way of getting maximum graphic effect for minimal cost. Fruit and vegetables are zoned effectively and displayed in an array of wicker baskets and quirky products such as 'Not Another Bloody Bottle of Water' are sold with a sense of humour and fun.

纯天然食品

建筑材料简单、新颖、原生态。漆面南洋杉与层压板的结合，耐久而实用。服务台上方的照明设备随意、简约。强调食品原料和野生自然哲学的图案随处可见，突显出"纯天然食品"所倡导的宗旨。乙烯贴花罩面漆搭配细木工制品为空间营造出十足的艺术气息。水果和蔬菜按照既定的顺序陈列在柳条筐之中，而一些奇异商品则以幽默、玩笑的方式进行销售。

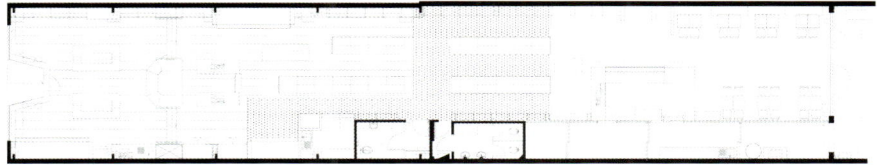

Café

Australia

Sydney

2009

Photo: Design Clarity

Design Clarity

Bar

Australia

Tasmania

2005

Photo: Morris-Nunn + Associates

Morris-Nunn + Associates

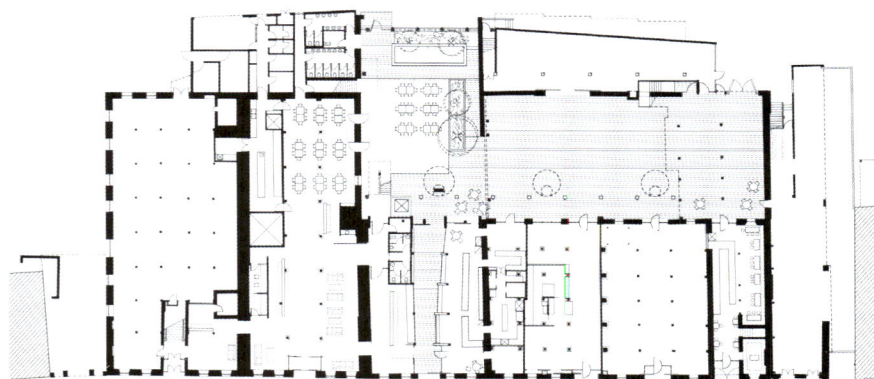

Bar in IXL Development + Henry Jones Art Hotel

An important part of the project was the inclusion of a large toroidal glass roofed atrium on the rear side of the development. There is a bar in the atrium. The atrium becomes a new cultural precinct for the City of Hobart, and, as it is located on the sunny northern side of all the historic warehouse buildings, this dramatic structure is also the means by which the historic structures are naturally heated. Fresh air is warmed by utilising the Greenhouse Effect and in winter trapped and collected within the upper part of the atrium airspace, from which it is sent via huge chutes as warm air, throughout the high thermal mass interiors of the old buildings.

亨利琼斯艺术酒店酒吧

亨利琼斯艺术酒店的设计重点是在酒店的后侧建立一个拥有大型环形玻璃天棚的中庭，酒吧位于中庭之中。中庭坐落于历史悠久的建筑群北端，这一特殊的地理位置为其成为霍巴特市新文化区创造了条件。空间上方的大型斜槽设计将来自原建筑室内的热气输送到中庭中，确保整个空间即使在寒冷冬日也能温暖如春。

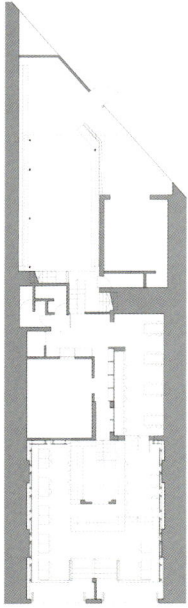

Water Moon

The task is to change the original tired pub with a history over 15 years, to a restaurat which enables the enjoyment of Japanese cuisine alongside traditional sake. It was further requested by the client to communicate to people, even if they are looking on from the street, that Water Moon is a place where one can have both the Japanese food and drink. The designers came up with the idea of having a lightbox in the middle of the restaurant as the 'moon' in the night sky. They spent quite some time researching various materials which would enable them to create this lightbox; it was a fine balance between the lighting effect, the sheer dimension of the lightbox, and cost.

水月餐厅

项目需要将一个拥有15年历史的老酒吧改造成一个提供日本传统美食和清酒的日式餐厅。委托人要求餐厅在外观上吸引顾客，让走在街道上的人们都能够知道水月是一间日式餐厅。设计师在餐厅中心设计了一个发光盒，作为夜空中的月亮。他们花费了大量时间搜寻适合制作发光盒的材料，在灯光效果、发光盒的尺寸和成本之间达成了平衡。

Restaurant

Australia

Sydney

2010

Photo: Katherine Lu

Facet Studio